Game Theory

GAME THEORY

Analysis of Conflict

ROGER B. MYERSON

HARVARD UNIVERSITY PRESS
Cambridge, Massachusetts
London, England

Copyright © 1991 by the President and Fellows of Harvard College
All rights reserved
Printed in the United States of America

First Harvard University Press paperback edition, 1997

Library of Congress Cataloging-in-Publication Data

Myerson, Roger B.
Game theory : analysis of conflict / Roger B. Myerson.
   p.   cm.
  Includes bibliographical references and index.
  ISBN 0-674-34115-5 (cloth)
  ISBN 0-674-34116-3 (pbk.)
  1. Game Theory   I. Title
H61.25.M94   1991
519.3—dc20   90–42901

For Gina, Daniel, and Rebecca
With the hope that a better understanding of conflict
may help create a safer and more peaceful world

Contents

7   Repeated Games   308

8   Bargaining and Cooperation in Two-Person Games   370

9   Coalitions in Cooperative Games   417

Preface

Game theory has a very general scope, encompassing questions that are basic to all of the social sciences. It can offer insights into any economic, political, or social situation that involves individuals who have different goals or preferences. However, there is a fundamental unity and coherent methodology that underlies the large and growing literature on game theory and its applications. My goal in this book is to convey both the generality and the unity of game theory. I have tried to present some of the most important models, solution concepts, and results of game theory, as well as the methodological principles that have guided game theorists to develop these models and solutions.

This book is written as a general introduction to game theory, intended for both classroom use and self-study. It is based on courses that I have taught at Northwestern University, the University of Chicago, and the University of Paris–Dauphine. I have included here, however, somewhat more cooperative game theory than I can actually cover in a first course. I have tried to set an appropriate balance between noncooperative and cooperative game theory, recognizing the fundamental primacy of noncooperative game theory but also the essential and complementary role of the cooperative approach.

The mathematical prerequisite for this book is some prior exposure to elementary calculus, linear algebra, and probability, at the basic undergraduate level. It is not as important to know the theorems that may be covered in such mathematics courses as it is to be familiar with the basic ideas and notation of sets, vectors, functions, and limits. Where more advanced mathematics is used, I have given a short, self-contained explanation of the mathematical ideas.

In every chapter, there are some topics of a more advanced or specialized nature that may be omitted without loss of subsequent comprehension. I have not tried to "star" such sections or paragraphs. Instead, I have provided cross-references to enable a reader to skim or pass over sections that seem less interesting and to return to them if they are needed later in other sections of interest. Page references for the important definitions are indicated in the index.

In this introductory text, I have not been able to cover every major topic in the literature on game theory, and I have not attempted to assemble a comprehensive bibliography. I have tried to exercise my best judgment in deciding which topics to emphasize, which to mention briefly, and which to omit; but any such judgment is necessarily subjective and controversial, especially in a field that has been growing and changing as rapidly as game theory. For other perspectives and more references to the vast literature on game theory, the reader may consult some of the other excellent survey articles and books on game theory, which include Aumann (1987b) and Shubik (1982).

A note of acknowledgment must begin with an expression of my debt to Robert Aumann, John Harsanyi, John Nash, Reinhard Selten, and Lloyd Shapley, whose writings and lectures taught and inspired all of us who have followed them into the field of game theory. I have benefited greatly from long conversations with Ehud Kalai and Robert Weber about game theory and, specifically, about what should be covered in a basic textbook on game theory. Discussions with Bengt Holmstrom, Paul Milgrom, and Mark Satterthwaite have also substantially influenced the development of this book. Myrna Wooders, Robert Marshall, Dov Monderer, Gregory Pollock, Leo Simon, Michael Chwe, Gordon Green, Akihiko Matsui, Scott Page, and Eun Soo Park read parts of the manuscript and gave many valuable comments. In writing the book, I have also benefited from the advice and suggestions of Lawrence Ausubel, Raymond Deneckere, Itzhak Gilboa, Ehud Lehrer, and other colleagues in the Managerial Economics and Decision Sciences department at Northwestern University. The final manuscript was ably edited by Jodi Simpson, and was proofread by Scott Page, Joseph Riney, Ricard Torres, Guangsug Hahn, Jose Luis Ferreira, Ioannis Tournas, Karl Schlag, Keuk-Ryoul Yoo, Gordon Green, and Robert Lapson. This book and related research have been supported by fellowships from the John Simon Guggenheim Memorial Foundation and the Alfred P. Sloan

Foundation, and by grants from the National Science Foundation and the Dispute Resolution Research Center at Northwestern University. Last but most, I must acknowledge the steady encouragement of my wife, my children, and my parents, all of whom expressed a continual faith in a writing project that seemed to take forever.

Evanston, Illinois
December 1990

Game Theory

1

Decision-Theoretic Foundations

1.1 Game Theory, Rationality, and Intelligence

Game theory can be defined as the study of mathematical models of conflict and cooperation between intelligent rational decision-makers. Game theory provides general mathematical techniques for analyzing situations in which two or more individuals make decisions that will influence one another's welfare. As such, game theory offers insights of fundamental importance for scholars in all branches of the social sciences, as well as for practical decision-makers. The situations that game theorists study are not merely recreational activities, as the term "game" might unfortunately suggest. "Conflict analysis" or "interactive decision theory" might be more descriptively accurate names for the subject, but the name "game theory" seems to be here to stay.

Modern game theory may be said to begin with the work of Zermelo (1913), Borel (1921), von Neumann (1928), and the great seminal book of von Neumann and Morgenstern (1944). Much of the early work on game theory was done during World War II at Princeton, in the same intellectual community where many leaders of theoretical physics were also working (see Morgenstern, 1976). Viewed from a broader perspective of intellectual history, this propinquity does not seem coincidental. Much of the appeal and promise of game theory is derived from its position in the mathematical foundations of the social sciences. In this century, great advances in the most fundamental and theoretical branches of the physical sciences have created a nuclear dilemma that threatens the survival of our civilization. People seem to have learned more about how to design physical systems for exploiting radioactive materials than about how to create social systems for moderating human behavior in conflict. Thus, it may be natural to hope that advances in the most fundamental and theoretical branches of the social sciences might be able to provide the understanding that we need to match our great advances in the physical sciences. This hope is one of the motivations that has led many mathematicians and social scientists to work in game theory during the past 50 years. Real proof of the power of game theory has come in recent years from a prolific development of important applications, especially in economics.

Game theorists try to understand conflict and cooperation by studying quantitative models and hypothetical examples. These examples may be unrealistically simple in many respects, but this simplicity may make the fundamental issues of conflict and cooperation easier to see in these examples than in the vastly more complicated situations of real life. Of course, this is the method of analysis in any field of inquiry: to pose one's questions in the context of a simplified model in which many of the less important details of reality are ignored. Thus, even if one is never involved in a situation in which people's positions are as clearly defined as those studied by game theorists, one can still come to understand real competitive situations better by studying these hypothetical examples.

In the language of game theory, a *game* refers to any social situation involving two or more individuals. The individuals involved in a game may be called the *players*. As stated in the definition above, there are two basic assumptions that game theorists generally make about players: they are rational and they are intelligent. Each of these adjectives is used here in a technical sense that requires some explanation.

A decision-maker is *rational* if he makes decisions consistently in pursuit of his own objectives. In game theory, building on the fundamental results of decision theory, we assume that each player's objective is to maximize the expected value of his own payoff, which is measured in some *utility* scale. The idea that a rational decision-maker should make decisions that will maximize his expected utility payoff goes back at least to Bernoulli (1738), but the modern justification of this idea is due to von Neumann and Morgenstern (1947). Using remarkably weak assumptions about how a rational decision-maker should behave, they showed that for any rational decision-maker there must exist some way of assigning utility numbers to the various possible outcomes that he cares about, such that he would always choose the option that maximizes his expected utility. We call this result the *expected-utility maximization theorem*.

It should be emphasized here that the logical axioms that justify the expected-utility maximization theorem are weak consistency assumptions. In derivations of this theorem, the key assumption is generally a *sure-thing* or *substitution* axiom that may be informally paraphrased as follows: "If a decision-maker would prefer option 1 over option 2 when event A occurs, and he would prefer option 1 over option 2 when event A does not occur, then he should prefer option 1 over option 2 even before he learns whether event A will occur or not." Such an assumption, together with a few technical regularity conditions, is sufficient to guarantee that there exists some utility scale such that the decision-maker always prefers the options that give the highest expected utility value.

Consistent maximizing behavior can also be derived from models of evolutionary selection. In a universe where increasing disorder is a physical law, complex organisms (including human beings and, more broadly speaking, social organizations) can persist only if they behave in a way that tends to increase their probability of surviving and reproducing themselves. Thus, an evolutionary-selection argument suggests that individuals may tend to maximize the expected value of some measure of general survival and reproductive fitness or success (see Maynard Smith, 1982).

In general, maximizing expected utility payoff is not necessarily the same as maximizing expected monetary payoff, because utility values are not necessarily measured in dollars and cents. A risk-averse individual may get more incremental utility from an extra dollar when he is poor than he would get from the same dollar were he rich. This observation suggests that, for many decision-makers, utility may be a nonlinear function of monetary worth. For example, one model that is commonly used in decision analysis stipulates that a decision-maker's utility payoff from getting x dollars would be $u(x) = 1 - e^{-cx}$, for some number c that represents his *index of risk aversion* (see Pratt, 1964). More generally, the utility payoff of an individual may depend on many variables besides his own monetary worth (including even the monetary worths of other people for whom he feels some sympathy or antipathy).

When there is uncertainty, expected utilities can be defined and computed only if all relevant uncertain events can be assigned probabilities, which quantitatively measure the likelihood of each event. Ramsey (1926) and Savage (1954) showed that, even where objective probabilities cannot be assigned to some events, a rational decision-maker should be able to assess all the subjective probability numbers that are needed to compute these expected values.

In situations involving two or more decision-makers, however, a special difficulty arises in the assessment of subjective probabilities. For example, suppose that one of the factors that is unknown to some given individual 1 is the action to be chosen by some other individual 2. To assess the probability of each of individual 2's possible choices, individual 1 needs to understand 2's decision-making behavior, so 1 may try to imagine himself in 2's position. In this thought experiment, 1 may realize that 2 is trying to rationally solve a decision problem of her own and that, to do so, she must assess the probabilities of each of 1's possible choices. Indeed, 1 may realize that 2 is probably trying to imagine herself in 1's position, to figure out what 1 will do. So the rational solution to each individual's decision problem depends on the solution to the other individual's problem. Neither problem can be solved without understanding the solution to the other. Thus, when rational decision-makers interact, their decision problems must be analyzed together, like a system of equations. Such analysis is the subject of game theory.

When we analyze a game, as game theorists or social scientists, we say that a player in the game is *intelligent* if he knows everything that we know about the game and he can make any inferences about the situation that we can make. In game theory, we generally assume that players are intelligent in this sense. Thus, if we develop a theory that describes the behavior of intelligent players in some game and we believe that this theory is correct, then we must assume that each player in the game will also understand this theory and its predictions.

For an example of a theory that assumes rationality but not intelligence, consider price theory in economics. In the general equilibrium model of price theory, it is assumed that every individual is a rational utility-maximizing decision-maker, but it is not assumed that individuals understand the whole structure of the economic model that the price theorist is studying. In price-theoretic models, individuals only perceive and respond to some intermediating price signals, and each individual is supposed to believe that he can trade arbitrary amounts at these prices, even though there may not be anyone in the economy actually willing to make such trades with him.

Of course, the assumption that all individuals are perfectly rational and intelligent may never be satisfied in any real-life situation. On the other hand, we should be suspicious of theories and predictions that are not consistent with this assumption. If a theory predicts that some individuals will be systematically fooled or led into making costly mistakes, then this theory will tend to lose its validity when these individuals learn (from experience or from a published version of the theory itself) to better understand the situation. The importance of game theory in the social sciences is largely derived from this fact.

1.2 Basic Concepts of Decision Theory

The logical roots of game theory are in Bayesian decision theory. Indeed, game theory can be viewed as an extension of decision theory (to the case of two or more decision-makers), or as its essential logical fulfillment. Thus, to understand the fundamental ideas of game theory, one should begin by studying decision theory. The rest of this chapter is devoted to an introduction to the basic ideas of Bayesian decision theory, beginning with a general derivation of the expected utility maximization theorem and related results.

At some point, anyone who is interested in the mathematical social sciences should ask the question, Why should I expect that any simple quantitative model can give a reasonable description of people's behavior? The fundamental results of decision theory directly address this question, by showing that any decision-maker who satisfies certain intuitive axioms should always behave so as to maximize the mathematical expected value of some utility function, with respect to some subjective probability distribution. That is, any rational decision-maker's behavior should be describable by a *utility function*, which gives a quantitative characterization of his preferences for outcomes or prizes, and a *subjective probability distribution*, which characterizes his beliefs about all relevant unknown factors. Furthermore, when new information becomes available to such a decision-maker, his subjective probabilities should be revised in accordance with Bayes's formula.

There is a vast literature on axiomatic derivations of the subjective probability, expected-utility maximization, and Bayes's formula, beginning with Ramsey (1926), von Neumann and Morgenstern (1947), and Savage (1954). Other notable derivations of these results have been offered by Herstein and Milnor (1953), Luce and Raiffa (1957), Anscombe and Aumann (1963), and Pratt, Raiffa, and Schlaiffer (1964); for a general overview, see Fishburn (1968). The axioms used here are mainly borrowed from these earlier papers in the literature, and no attempt is made to achieve a logically minimal set of axioms. (In fact, a number of axioms presented in Section 1.3 are clearly redundant.)

Decisions under uncertainty are commonly described by one of two models: a *probability model* or a *state-variable model*. In each case, we speak of the decision-maker as choosing among *lotteries*, but the two models differ in how a lottery is defined. In a probability model, lotteries are probability distributions over a set of prizes. In a state-variable model, lotteries are functions from a set of possible states into a set of prizes. Each of these models is most appropriate for a specific class of applications.

A probability model is appropriate for describing gambles in which the prizes will depend on events that have obvious objective probabilities; we refer to such events as *objective unknowns*. These gambles are the "roulette lotteries" of Anscombe and Aumann (1963) or the "risks" of Knight (1921). For example, gambles that depend on the toss of a fair coin, the spin of a roulette wheel, or the blind draw of a ball out of an urn containing a known population of identically sized but differently colored balls all could be adequately described in a probability model. An important assumption being used here is that two objective unknowns with the same probability are completely equivalent for decision-making purposes. For example, if we describe a lottery by saying that it "offers a prize of $100 or $0, each with probability ½," we are assuming that it does not matter whether the prize is determined by tossing a fair coin or by drawing a ball from an urn that contains 50 white and 50 black balls.

On the other hand, many events do not have obvious probabilities; the result of a future sports event or the future course of the stock market are good examples. We refer to such events as *subjective unknowns*. Gambles that depend on subjective unknowns correspond to the "horse lotteries" of Anscombe and Aumann (1963) or the "uncertainties" of Knight (1921). They are more readily described in a state-variable model, because these models allow us to describe how the prize will be determined by the unpredictable events, without our having to specify any probabilities for these events.

Here we define our lotteries to include both the probability and the state-variable models as special cases. That is, we study lotteries in which the prize may depend on both objective unknowns (which may be directly described by probabilities) and subjective unknowns (which must be described by a state variable). (In the terminology of Fishburn, 1970, we are allowing extraneous probabilities in our model.)

Let us now develop some basic notation. For any finite set Z, we let $\Delta(Z)$ denote the set of probability distributions over the set Z. That is,

$$(1.1) \qquad \Delta(Z) = \{q{:}Z \to \mathbf{R} \mid \sum_{y \in Z} q(y) = 1 \text{ and } q(z) \geq 0, \quad \forall z \in Z\}.$$

(Following common set notation, "|" in set braces may be read as "such that.")

Let X denote the set of possible *prizes* that the decision-maker could ultimately get. Let Ω denote the set of possible *states*, one of which will be the *true state of the world*. To simplify the mathematics, we assume that X and Ω are both finite sets. We define a *lottery* to be any function f that specifies a nonnegative real number $f(x|t)$, for every prize x in X and every state t in Ω, such that $\sum_{x \in X} f(x|t) = 1$ for every t in Ω. Let L denote the set of all such lotteries. That is,

$$L = \{f{:}\Omega \to \Delta(X)\}.$$

For any state t in Ω and any lottery f in L, $f(\cdot|t)$ denotes the probability distribution over X designated by f in state t. That is,

$$f(\cdot|t) = (f(x|t))_{x \in X} \in \Delta(X).$$

Each number $f(x|t)$ here is to be interpreted as the objective conditional probability of getting prize x in lottery f if t is the true state of the world. (Following common probability notation, "|" in parentheses may be interpreted here to mean "given.") For this interpretation to make sense, the state must be defined broadly enough to summarize all subjective unknowns that might influence the prize to be received. Then, once a state has been specified, only objective probabilities will remain, and an objective probability distribution over the possible prizes can be calculated for any well-defined gamble. So our formal definition of a lottery allows us to represent any gamble in which the prize may depend on both objective and subjective unknowns.

A *prize* in our sense could be any commodity bundle or resource allocation. We are assuming that the prizes in X have been defined so that they are mutually exclusive and exhaust the possible consequences of the decision-maker's decisions. Furthermore, we assume that each prize in X represents a complete specification of all aspects that the decision-maker cares about in the situation resulting from his decisions. Thus, the decision-maker should be able to assess a preference ordering over the set of lotteries, given any information that he might have about the state of the world.

The information that the decision-maker might have about the true state of the world can be described by an *event*, which is a nonempty subset of Ω. We let Ξ denote the set of all such events, so that

$$\Xi = \{S \,|\, S \subseteq \Omega \text{ and } S \neq \varnothing\}.$$

For any two lotteries f and g in L and any event S in Ξ, we write $f \succsim_S g$ iff the lottery f would be at least as desirable as g, in the opinion of the decision-maker, if he learned that the true state of the world was in the set S. (Here *iff* means "if and only if.") That is, $f \succsim_S g$ iff the decision-maker would be willing to choose the lottery f when he has to choose between f and g and he knows only that the event S has occurred. Given this relation ($\succsim_S$), we define relations ($\succ_S$) and ($\sim_S$) so that

$$f \sim_S g \text{ iff } f \succsim_S g \text{ and } g \succsim_S f;$$
$$f \succ_S g \text{ iff } f \succsim_S g \text{ and } g \not\succsim_S f.$$

That is, $f \sim_S g$ means that the decision-maker would be indifferent between f and g, if he had to choose between them after learning S; and $f \succ_S g$ means that he would strictly prefer f over g in this situation.

We may write $\succsim$, $\succ$, and $\sim$ for $\succsim_\Omega$, $\succ_\Omega$, and $\sim_\Omega$, respectively. That is, when no conditioning event is mentioned, it should be assumed that we are referring to prior preferences before any states in Ω are ruled out by observations.

Notice the assumption here that the decision-maker would have well-defined preferences over lotteries conditionally on any possible event in Ξ. In some expositions of decision theory, a decision-maker's conditional preferences are derived (using Bayes's formula) from the prior preferences that he would assess before making any observations; but such derivations cannot generate rankings of lotteries conditionally on events that have prior probability 0. In game-theoretic contexts, this omission is not as innocuous as it may seem. Kreps and Wilson (1982) have shown that the characterization of a rational decision-maker's beliefs and preferences after he observes a zero-probability event may be crucial in the analysis of a game.

For any number α such that $0 \leq \alpha \leq 1$, and for any two lotteries f and g in L, $\alpha f + (1 - \alpha)g$ denotes the lottery in L such that

$$(\alpha f + (1 - \alpha)g)(x|t) = \alpha f(x|t) + (1 - \alpha)g(x|t), \quad \forall x \in X, \quad \forall t \in \Omega.$$

To interpret this definition, suppose that a ball is going to be drawn from an urn in which α is the proportion of black balls and $1 - \alpha$ is the proportion of white balls. Suppose that if the ball is black then the decision-maker will get to play lottery f and if the ball is white then the decision-maker will get to play lottery g. Then the decision-maker's ultimate probability of getting prize x if t is the true state is $\alpha f(x|t) + (1 - \alpha)g(x|t)$. Thus, $\alpha f + (1 - \alpha)g$ represents the compound lottery that is built up from f and g by this random lottery-selection process.

For any prize x, we let $[x]$ denote the lottery that always gives prize x for sure. That is, for every state t,

(1.2) $[x](y|t) = 1$ if $y = x$, $[x](y|t) = 0$ if $y \neq x$.

Thus, $\alpha[x] + (1 - \alpha)[y]$ denotes the lottery that gives either prize x or prize y, with probabilities α and $1 - \alpha$, respectively.

1.3 Axioms

Basic properties that a rational decision-maker's preferences may be expected to satisfy can be presented as a list of axioms. Unless otherwise stated, these axioms are to hold for all lotteries e, f, g, and h in L, for all events S and T in Ξ, and for all numbers α and β between 0 and 1.

Axioms 1.1A and 1.1B assert that preferences should always form a complete transitive order over the set of lotteries.

AXIOM 1.1A (COMPLETENESS). $f \succsim_S g$ or $g \succsim_S f$.

AXIOM 1.1B (TRANSITIVITY).   If $f \succsim_S g$ and $g \succsim_S h$ then $f \succsim_S h$.

It is straightforward to check that Axiom 1.1B implies a number of other transitivity results, such as if $f \sim_S g$ and $g \sim_S h$ then $f \sim_S h$; and if $f >_S g$ and $g \succsim_S h$ then $f >_S h$.

Axiom 1.2 asserts that only the possible states are relevant to the decision-maker, so, given an event S, he would be indifferent between two lotteries that differ only in states outside S.

AXIOM 1.2 (RELEVANCE).   If $f(\cdot|t) = g(\cdot|t)$ $\forall t \in S$, then $f \sim_S g$.

Axiom 1.3 asserts that a higher probability of getting a better lottery is always better.

AXIOM 1.3 (MONOTONICITY). *If $f >_S h$ and $0 \le \beta < \alpha \le 1$, then* $\alpha f + (1 - \alpha)h >_S \beta f + (1 - \beta)h$.

Building on Axiom 1.3, Axiom 1.4 asserts that $\gamma f + (1 - \gamma)h$ gets better in a continuous manner as γ increases, so any lottery that is ranked between f and h is just as good as some randomization between f and h.

AXIOM 1.4 (CONTINUITY). *If $f \gtrsim_S g$ and $g \gtrsim_S h$, then there exists some number γ such that $0 \le \gamma \le 1$ and $g \sim_S \gamma f + (1 - \gamma)h$.*

The substitution axioms (also known as independence or sure-thing axioms) are probably the most important in our system, in the sense that they generate strong restrictions on what the decision-maker's preferences must look like even without the other axioms. They should also be very intuitive axioms. They express the idea that, if the decision-maker must choose between two alternatives and if there are two mutually exclusive events, one of which must occur, such that in each event he would prefer the first alternative, then he must prefer the first alternative before he learns which event occurs. (Otherwise, he would be expressing a preference that he would be sure to want to reverse after learning which of these events was true!) In Axioms 1.5A and 1.5B, these events are objective randomizations in a random lottery-selection process, as discussed in the preceding section. In Axioms 1.6A and 1.6B, these events are subjective unknowns, subsets of Ω.

AXIOM 1.5A (OBJECTIVE SUBSTITUTION). *If $e \gtrsim_S f$ and $g \gtrsim_S h$ and $0 \le \alpha \le 1$, then $\alpha e + (1 - \alpha)g \gtrsim_S \alpha f + (1 - \alpha)h$.*

AXIOM 1.5B (STRICT OBJECTIVE SUBSTITUTION). *If $e >_S f$ and $g \gtrsim_S h$ and $0 < \alpha \le 1$, then $\alpha e + (1 - \alpha)g >_S \alpha f + (1 - \alpha)h$.*

AXIOM 1.6A (SUBJECTIVE SUBSTITUTION). *If $f \gtrsim_S g$ and $f \gtrsim_T g$ and $S \cap T = \varnothing$, then $f \gtrsim_{S \cup T} g$.*

AXIOM 1.6B (STRICT SUBJECTIVE SUBSTITUTION). *If $f >_S g$ and $f >_T g$ and $S \cap T = \varnothing$, then $f >_{S \cup T} g$.*

To fully appreciate the importance of the substitution axioms, we may find it helpful to consider the difficulties that arise in decision theory when we try to drop them. For a simple example, suppose an individual would prefer x over y, but he would also prefer $.5[y]+.5[z]$ over $.5[x] + .5[z]$, in violation of substitution. Suppose that w is some other prize that he would consider better than $.5[x] + .5[z]$ and worse than $.5[y] + .5[z]$. That is,

$$x > y \text{ but } .5[y] + .5[z] > [w] > .5[x] + .5[z].$$

Now consider the following situation. The decision-maker must first decide whether to take prize w or not. If he does not take prize w, then a coin will be tossed. If it comes up Heads, then he will get prize z; and if it comes up Tails, then he will get a choice between prizes x and y.

What should this decision-maker do? He has three possible decision-making strategies: (1) take w, (2) refuse w and take x if Tails, (3) refuse w and take y if Tails. If he follows the first strategy, then he gets the lottery $[w]$; if he follows the second, then he gets the lottery $.5[x] + .5[z]$; and if he follows the third, then he gets the lottery $.5[y] + .5[z]$. Because he likes $.5[y] + .5[z]$ best among these lotteries, the third strategy would be best for him, so it may seem that he should refuse w. However, if he refuses w and the coin comes up Tails, then his preferences stipulate that he should choose x instead of y. So if he refuses w, then he will actually end up with z if Heads or x if Tails. But this lottery $.5[x] + .5[z]$ is worse than w. So we get the contradictory conclusion that he should have taken w in the first place.

Thus, if we are to talk about "rational" decision-making without substitution axioms, then we must specify whether rational decision-makers are able to commit themselves to follow strategies that they would subsequently want to change (in which case "rational" behavior would lead to $.5[y] + .5[z]$ in this example). If they cannot make such commitments, then we must also specify whether they can foresee their future inconstancy (in which case the outcome of this example should be $[w]$) or not (in which case the outcome of this example should be $.5[x]+.5[z]$). If none of these assumptions seem reasonable, then to avoid this dilemma we must accept substitution axioms as a part of our definition of rationality.

Axiom 1.7 asserts that the decision-maker is never indifferent between all prizes. This axiom is just a regularity condition, to make sure that there is something of interest that could happen in each state.

AXIOM 1.7 (INTEREST).   *For every state t in* Ω, *there exist prizes y and z in X such that* $[y] >_{\{t\}} [z]$.

Axiom 1.8 is optional in our analysis, in the sense that we can state a version of our main result with or without this axiom. It asserts that the decision-maker has the same preference ordering over objective gambles in all states of the world. If this axiom fails, it is because the same prize might be valued differently in different states.

AXIOM 1.8 (STATE NEUTRALITY).   *For any two states r and t in* Ω, *if* $f(\cdot|r) = f(\cdot|t)$ *and* $g(\cdot|r) = g(\cdot|t)$ *and* $f \gtrsim_{\{r\}} g$, *then* $f \gtrsim_{\{t\}} g$.

1.4 The Expected-Utility Maximization Theorem

A *conditional-probability function* on Ω is any function $p:\Xi \to \Delta(\Omega)$ that specifies nonnegative conditional probabilities $p(t|S)$ for every state t in Ω and every event S, such that

$$p(t|S) = 0 \text{ if } t \notin S, \text{ and } \sum_{r \in S} p(r|S) = 1.$$

Given any such conditional-probability function, we may write

$$p(R|S) = \sum_{r \in R} p(r|S), \quad \forall R \subseteq \Omega, \quad \forall S \in \Xi.$$

A *utility function* can be any function from $X \times \Omega$ into the real numbers **R**. A utility function $u:X \times \Omega \to \mathbf{R}$ is *state independent* iff it does not actually depend on the state, so there exists some function $U:X \to \mathbf{R}$ such that $u(x,t) = U(x)$ for all x and t.

Given any such conditional-probability function p and any utility function u and given any lottery f in L and any event S in Ξ, we let $E_p(u(f)|S)$ denote the expected utility value of the prize determined by f, when $p(\cdot|S)$ is the probability distribution for the true state of the world. That is,

$$E_p(u(f)|S) = \sum_{t \in S} p(t|S) \sum_{x \in X} u(x,t) f(x|t).$$

THEOREM 1.1.   *Axioms 1.1AB, 1.2, 1.3, 1.4, 1.5AB, 1.6AB, and 1.7 are jointly satisfied if and only if there exists a utility function* $u:X \times \Omega \to \mathbf{R}$ *and a conditional-probability function* $p:\Xi \to \Delta(\Omega)$ *such that*

(1.3) $\max\limits_{x \in X} u(x,t) = 1$ and $\min\limits_{x \in X} u(x,t) = 0, \quad \forall t \in \Omega;$

(1.4) $p(R\,|\,T) = p(R\,|\,S)p(S\,|\,T), \quad \forall R, \ \forall S, \ and \ \forall T \ such \ that$

$R \subseteq S \subseteq T \subseteq \Omega \ and \ S \neq \varnothing;$

(1.5) $f \gtrsim_S g$ if and only if $E_p(u(f)\,|\,S) \geq E_p(u(g)\,|\,S),$

$\forall f,g \in L, \quad \forall S \in \Xi.$

Furthermore, given these Axioms 1.1AB–1.7, Axiom 1.8 is also satisfied if and only if conditions (1.3)–(1.5) here can be satisfied with a state-independent utility function.

In this theorem, condition (1.3) is a normalization condition, asserting that we can choose our utility functions to range between 0 and 1 in every state. (Recall that X and Ω are assumed to be finite.) Condition (1.4) is a version of Bayes's formula, which establishes how conditional probabilities assessed in one event must be related to conditional probabilities assessed in another. The most important part of the theorem is condition (1.5), however, which asserts that the decision-maker always prefers lotteries with higher expected utility. By condition (1.5), once we have assessed u and p, we can predict the decision-maker's optimal choice in any decision-making situation. He will choose the lottery with the highest expected utility among those available to him, using his subjective probabilities conditioned on whatever event in Ω he has observed. Notice that, with X and Ω finite, there are only finitely many utility and probability numbers to assess. Thus, the decision-maker's preferences over all of the infinitely many lotteries in L can be completely characterized by finitely many numbers.

To apply this result in practice, we need a procedure for assessing the utilities $u(x,t)$ and the probabilities $p(t\,|\,S)$, for all x, t, and S. As Raiffa (1968) has emphasized, such procedures do exist, and they form the basis of practical decision analysis. To define one such assessment procedure, and to prove Theorem 1.1, we begin by defining some special lotteries, using the assumption that the decision-maker's preferences satisfy Axioms 1.1AB–1.7.

Let a_1 be a lottery that gives the decision-maker one of the best prizes in every state; and let a_0 be a lottery that gives him one of the worst prizes in every state. That is, for every state t, $a_1(y\,|\,t) = 1 = a_0(z\,|\,t)$ for some prizes y and z such that, for every x in X, $y \gtrsim_{\{t\}} x \gtrsim_{\{t\}} z$. Such best and worst prizes can be found in every state because the preference relation $(\succsim_{\{t\}})$ forms a transitive ordering over the finite set X.

For any event S in Ξ, let b_S denote the lottery such that

$$b_S(\cdot|t) = a_1(\cdot|t) \text{ if } t \in S,$$

$$b_S(\cdot|t) = a_0(\cdot|t) \text{ if } t \notin S.$$

That is, b_S is a "bet on S" that gives the best possible prize if S occurs and gives the worst possible prize otherwise.

For any prize x and any state t, let $c_{x,t}$ be the lottery such that

$$c_{x,t}(\cdot|r) = [x](\cdot|r) \text{ if } r = t,$$

$$c_{x,t}(\cdot|r) = a_0(\cdot|r) \text{ if } r \notin t.$$

That is, $c_{x,t}$ is the lottery that always gives the worst prize, except in state t, when it gives prize x.

We can now define a procedure to assess the utilities and probabilities that satisfy the theorem, given preferences that satisfy the axioms. For each x and t, first ask the decision-maker, "For what number β would you be indifferent between $[x]$ and $\beta a_1 + (1 - \beta)a_0$, if you knew that t was the true state of the world?" By the continuity axiom, such a number must exist. Then let $u(x,t)$ equal the number that he specifies, such that

$$[x] \sim_{\{t\}} u(x,t)a_1 + (1 - u(x,t))a_0.$$

For each t and S, ask the decision-maker, "For what number γ would you be indifferent between $b_{\{t\}}$ and $\gamma a_1 + (1 - \gamma)a_0$ if you knew that the true state was in S?" Again, such a number must exist, by the continuity axiom. (The subjective substitution axiom guarantees that $a_1 \succsim_S b_{\{t\}} \succsim_S a_0$.) Then let $p(t|S)$ equal the number that he specifies, such that

$$b_{\{t\}} \sim_S p(t|S)a_1 + (1 - p(t|S))a_0.$$

In the proof of Theorem 1.1, we show that defining u and p in this way does satisfy the conditions of the theorem. Thus, finitely many questions suffice to assess the probabilities and utilities that completely characterize the decision-maker's preferences.

Proof of Theorem 1.1. Let p and u be as constructed above. First, we derive condition (1.5) from the axioms. The relevance axiom and the definition of $u(x,t)$ implies that, for every state r,

$$c_{x,t} \sim_{\{t\}} u(x,t)b_{\{t\}} + (1 - u(x,t))a_0.$$

Then subjective substitution implies that, for every event S,

$$c_{x,t} \sim_S u(x,t)b_{\{t\}} + (1 - u(x,t))a_0.$$

Axioms 1.5A and 1.5B together imply that $f \gtrsim_S g$ if and only if

$$\left(\frac{1}{|\Omega|}\right)f + \left(1 - \frac{1}{|\Omega|}\right)a_0 \gtrsim_S \left(\frac{1}{|\Omega|}\right)g + \left(1 - \frac{1}{|\Omega|}\right)a_0.$$

(Here, $|\Omega|$ denotes the number of states in the set Ω.) Notice that

$$\left(\frac{1}{|\Omega|}\right)f + \left(1 - \frac{1}{|\Omega|}\right)a_0 = \left(\frac{1}{|\Omega|}\right)\sum_{t \in \Omega}\sum_{x \in X} f(x|t)c_{x,t}.$$

But, by repeated application of the objective substitution axiom,

$$\left(\frac{1}{|\Omega|}\right)\sum_{t \in \Omega}\sum_{x \in X} f(x|t)c_{x,t}$$

$$\sim_S \left(\frac{1}{|\Omega|}\right)\sum_{t \in \Omega}\sum_{x \in X} f(x|t)\big(u(x,t)b_{\{t\}} + (1 - u(x,t))a_0\big)$$

$$\sim_S \left(\frac{1}{|\Omega|}\right)\sum_{t \in \Omega}\sum_{x \in X} f(x|t)\big(u(x,t)\big(p(t|S)a_1$$

$$+ (1 - p(t|S))a_0\big) + (1 - u(x,t))a_0\big)$$

$$= \left(\frac{1}{|\Omega|}\right)\sum_{t \in \Omega}\sum_{x \in X} f(x|t)u(x,t)p(t|S)a_1$$

$$+ \left(1 - \sum_{t \in \Omega}\sum_{x \in X} f(x|t)u(x,t)p(t|S)/|\Omega|\right)a_0$$

$$= \big(E_p(u(f)|S)/|\Omega|\big)a_1 + \big(1 - (E_p(u(f)|S)/|\Omega|)\big)a_0.$$

Similarly,

$$(1/|\Omega|)g + (1 - (1/|\Omega|))a_0$$

$$\sim_S \big(E_p(u(g)|S)/|\Omega|\big)a_1 + \big(1 - (E_p(u(g)|S)/|\Omega|)\big)a_0.$$

Thus, by transitivity, $f \gtrsim_S g$ if and only if

$$\left(E_p(u(f)|S)/|\Omega|\right)a_1 + \left(1 - \left(E_p(u(f)|S)/|\Omega|\right)\right)a_0$$
$$\succeq_S \left(E_p(u(g)|S)/|\Omega|\right)a_1 + \left(1 - \left(E_p(u(g)|S)/|\Omega|\right)\right)a_0.$$

But by monotonicity, this final relation holds if and only if

$$E_p(u(f)|S) \geq E_p(u(g)|S),$$

because interest and strict subjective substitution guarantee that $a_1 >_S a_0$. Thus, condition (1.5) is satisfied.

Next, we derive condition (1.4) from the axioms. For any events R and S,

$$\left(\frac{1}{|R|}\right) b_R + \left(1 - \frac{1}{|R|}\right) a_0 = \left(\frac{1}{|R|}\right) \sum_{r \in R} b_{\{r\}}$$

$$\sim_S \left(\frac{1}{|R|}\right) \sum_{r \in R} \left(p(r|S)a_1 + (1 - p(r|S))a_0\right)$$

$$= \left(\frac{1}{|R|}\right) \left(p(R|S)a_1 + (1 - p(R|S))a_0\right) + \left(1 - \frac{1}{|R|}\right) a_0,$$

by objective substitution. ($|R|$ is the number of states in the set R.) Then, using Axioms 1.5A and 1.5B, we get

$$b_R \sim_S p(R|S)a_1 + (1 - p(R|S))a_0.$$

By the relevance axiom, $b_S \sim_S a_1$ and, for any r not in S, $b_{\{r\}} \sim_S a_0$. So the above formula implies (using monotonicity and interest) that $p(r|S) = 0$ if $r \notin S$, and $p(S|S) = 1$. Thus, p is a conditional-probability function, as defined above.

Now, suppose that $R \subseteq S \subseteq T$. Using $b_S \sim_S a_1$ again, we get

$$b_R \sim_S p(R|S)b_S + (1 - p(R|S))a_0.$$

Furthermore, because b_R, b_S, and a_0 all give the same worst prize outside S, relevance also implies

$$b_R \sim_{T \setminus S} p(R|S)b_S + (1 - p(R|S))a_0.$$

(Here $T \setminus S = \{t | t \in T, t \notin S\}$.) So, by subjective and objective substitution,

$$b_R \sim_T p(R|S)b_S + (1 - p(R|S))a_0$$

$$\sim_T p(R|S)\left(p(S|T)a_1 + (1 - p(S|T))a_0\right) + (1 - p(R|S))a_0$$

$$= p(R|S)p(S|T)a_1 + (1 - p(R|S)p(S|T))a_0.$$

But $b_R \sim_T p(R|T)a_1 + (1 - p(R|T))a_0$. Also, $a_1 >_T a_0$, so monotonicity implies that $p(R|T) = p(R|S)p(S|T)$. Thus, Bayes's formula (1.4) follows from the axioms.

If y is the best prize and z is the worst prize in state t, then $[y] \sim_{\{t\}} a_1$ and $[z] \sim_{\{t\}} a_0$, so that $u(y,t) = 1$ and $u(z,t) = 0$ by monotonicity. So the range condition (1.3) is also satisfied by the utility function that we have constructed.

If state neutrality is also given, then the decision-maker will give us the same answer when we assess $u(x,t)$ as when we assess $u(x,r)$ for any other state r (because $[x] \sim_{\{t\}} \beta a_1 + (1 - \beta)a_0$ implies $[x] \sim_{\{r\}} \beta a_1 + (1 - \beta)a_0$, and monotonicity and interest guarantee that his answer is unique). So Axiom 1.8 implies that u is state-independent.

To complete the proof of the theorem, it remains to show that the existence of functions u and p that satisfy conditions (1.3)–(1.5) in the theorem is sufficient to imply all the axioms (using state independence only for Axiom 1.8). If we use the basic mathematical properties of the expected-utility formula, verification of the axioms is straightforward. To illustrate, we show the proof of one axiom, subjective substitution, and leave the rest as an exercise for the reader.

Suppose that $f \gtrsim_S g$ and $f \gtrsim_T g$ and $S \cap T = \varnothing$. By (1.5), $E_p(u(f)|S) \geq E_p(u(g)|S)$ and $E_p(u(f)|T) \geq E_p(u(g)|T)$. But Bayes's formula (1.4) implies that

$$E_p(u(f)|S \cup T) = \sum_{t \in S \cup T} \sum_{x \in X} p(t|S \cup T)f(x|t)u(x,t)$$

$$= \sum_{t \in S} \sum_{x \in X} p(t|S)p(S|S \cup T)f(x|t)u(x,t)$$

$$+ \sum_{t \in T} \sum_{x \in X} p(t|T)p(T|S \cup T)f(x|t)u(x,t)$$

$$= p(S|S \cup T)E_p(u(f)|S) + p(T|S \cup T)E_p(u(f)|S)$$

and

$$E_p(u(g)|S \cup T) = p(S|S \cup T)E_p(u(g)|S) + p(T|S \cup T)E_p(u(g)|S).$$

So $E_p(u(f)|S \cup T) \geq E_p(u(g)|S \cup T)$ and $f \gtrsim_{S \cup T} g$. ∎

1.5 Equivalent Representations

When we drop the range condition (1.3), there can be more than one pair of utility and conditional-probability functions that represent the same decision-maker's preferences, in the sense of condition (1.5). Such equivalent representations are completely indistinguishable in terms of their decision-theoretic properties, so we should be suspicious of any theory of economic behavior that requires distinguishing between such equivalent representations. Thus, it may be theoretically important to be able to recognize such equivalent representations.

Given any subjective event S, when we say that a utility function v and a conditional-probability function q *represent* the preference ordering $\gtrsim_S$, we mean that, for every pair of lotteries f and g, $E_q(v(f)|S) \geq E_q(v(g)|S)$ if and only if $f \gtrsim_S g$.

THEOREM 1.2. *Let S in Ξ be any given subjective event. Suppose that the decision-maker's preferences satisfy Axioms 1.1AB through 1.7, and let u and p be utility and conditional-probability functions satisfying (1.3)–(1.5) in Theorem 1.1. Then v and q represent the preference ordering $\gtrsim_S$ if and only if there exists a positive number A and a function $B{:}S \to \mathbf{R}$ such that*

$$q(t|S)v(x,t) = Ap(t|S)u(x,t) + B(t), \quad \forall t \in S, \quad \forall x \in X.$$

Proof. Suppose first that A and $B(\cdot)$ exist as described in the theorem. Then, for any lottery f,

$$E_q(v(f)|S) = \sum_{t \in S} \sum_{x \in X} f(x|t)q(t|S)v(x,t)$$

$$= \sum_{t \in S} \sum_{x \in X} f(x|t)(Ap(t|S)u(x,t) + B(t))$$

$$= A \sum_{t \in S} \sum_{x \in X} f(x|t)p(t|S)u(x,t) + \sum_{t \in S} B(t) \sum_{x \in X} f(x|t)$$

$$= AE_p(u(f)|S) + \sum_{t \in S} B(t),$$

because $\sum_{x \in X} f(x|t) = 1$. So expected v-utility with respect to q is an increasing linear function of expected u-utility with respect to p, because $A > 0$. Thus, $E_q(v(f)|S) \geq E_q(v(g)|S)$ if and only if $E_p(u(f)|S) \geq E_p(u(g)|S)$, and so v and q together represent the same preference ordering over lotteries as u and p.

Conversely, suppose now that v and q represent the same preference ordering as u and p. Pick any prize x and state t, and let

$$\lambda = \frac{E_q(v(c_{x,t})|S) - E_q(v(a_0)|S)}{E_q(v(a_1)|S) - E_q(v(a_0)|S)}.$$

Then, by the linearity of the expected-value operator,

$$E_q\big(v(\lambda a_1 + (1 - \lambda)a_0)|S\big) = E_q(v(a_0)|S) + \lambda\big(E_q(v(a_1)|S) - E_q(v(a_0)|S)\big)$$
$$= E_q(v(c_{x,t})|S),$$

so $c_{x,t} \sim_S \lambda a_1 + (1 - \lambda)a_0$. In the proof of Theorem 1.1, we constructed u and p so that

$$c_{x,t} \sim_S u(x,t)b_{\{t\}} + (1 - u(x,t))a_0$$
$$\sim_S u(x,t)\big(p(t|S)a_1 + (1 - p(t|S))a_0\big) + (1 - u(x,t))a_0$$
$$\sim_S p(t|S)u(x,t)a_1 + (1 - p(t|S)u(x,t))a_0.$$

The monotonicity axiom guarantees that only one randomization between a_1 and a_0 can be just as good as $c_{x,t}$, so

$$\lambda = p(t|S)u(x,t).$$

But $c_{x,t}$ differs from a_0 only in state t, where it gives prize x instead of the worst prize, so

$$E_q(v(c_{x,t})|S) - E_q(v(a_0)|S) = q(t|S)\left(v(x,t) - \min_{z \in X} v(z,t)\right).$$

Thus, going back to the definition of λ, we get

$$p(t|S)u(x,t) = \frac{q(t|S)(v(x,t) - \min\limits_{z \in X} v(z,t))}{E_q(v(a_1)|S) - E_q(v(a_0)|S)}.$$

Now let

$$A = E_q(v(a_1)|S) - E_q(v(a_0)|S),$$

and let

$$B(t) = q(t|S) \min_{z \in X} v(z,t).$$

Then

$$Ap(t|S)u(x,t) + B(t) = q(t|S)v(x,t).$$

Notice that A is independent of x and t and that $B(t)$ is independent of x. In addition, $A > 0$, because $a_1 >_S a_0$ implies $E_q(v(a_1)|S) > E_q(v(a_0)|S)$. ∎

It is easy to see from Theorem 1.2 that more than one probability distribution can represent the decision-maker's beliefs given some event S. In fact, we can make the probability distribution $q(\cdot|S)$ almost anything and still satisfy the equation in Theorem 1.2, as long as we make reciprocal changes in v, to keep the left-hand side of the equation the same. The way to eliminate this indeterminacy is to assume Axiom 1.8 and require utility functions to be state independent.

THEOREM 1.3. *Let S in Ξ be any given subjective event. Suppose that the decision-maker's preferences satisfy Axioms 1.1AB through 1.8, and let u and p be the state-independent utility function and the conditional-probability function, respectively, that satisfy conditions (1.3)–(1.5) in Theorem 1.1. Let v be a state-independent utility function, let q be a conditional-probability function, and suppose that v and q represent the preference ordering $\gtrsim_S$. Then*

$$q(t|S) = p(t|S), \quad \forall t \in S,$$

and there exist numbers A and C such that $A > 0$ and

$$v(x) = Au(x) + C, \quad \forall x \in X.$$

(For simplicity, we can write $v(x)$ and $u(x)$ here, instead of $v(x,t)$ and $u(x,t)$, because both functions are state independent.)

Proof. Let $A = E_q(v(a_1)|S) - E_q(v(a_0)|S)$, and let $C = \min_{z \in X} v(z)$. Then, from the proof of Theorem 1.2,

$$Ap(t|S)u(x) + q(t|S)C = q(t|S)v(x), \quad \forall x \in X, \quad \forall t \in S.$$

Summing this equation over all t in S, we get $Au(x) + C = v(x)$. Then, substituting this equation back, and letting x be the best prize so $u(x) = 1$, we get

$$Ap(t|S) + q(t|S)C = Aq(t|S) + q(t|S)C.$$

Because $A > 0$, we get $p(t|S) = q(t|S)$. ∎

1.6 Bayesian Conditional-Probability Systems

We define a *Bayesian conditional-probability system* (or simply a *conditional-probability system*) on the finite set Ω to be any conditional-probability function p on Ω that satisfies condition (1.4) (Bayes's formula). That is, if p is a Bayesian conditional-probability system on Ω, then, for every S that is a nonempty subset of Ω, $p(\cdot|S)$ is a probability distribution over Ω such that $p(S|S) = 1$ and

$$p(R|T) = p(R|S)p(S|T), \quad \forall R \subseteq S, \quad \forall T \supseteq S.$$

We let $\Delta^*(\Omega)$ denote the set of all Bayesian conditional-probability systems on Ω.

For any finite set Z, we let $\Delta^0(Z)$ denote the set of all probability distributions on Z that assign positive probability to every element in Z, so

(1.6) $$\Delta^0(Z) = \{q \in \Delta(Z)|q(z) > 0, \quad \forall z \in Z\}.$$

Any probability distribution $\hat{p}$ in $\Delta^0(\Omega)$ generates a conditional-probability system p in $\Delta^*(\Omega)$ by the formula

$$p(t|S) = \frac{\hat{p}(t)}{\sum_{r \in S} \hat{p}(r)} \text{ if } t \in S,$$

$$p(t|S) = 0 \text{ if } t \notin S.$$

The conditional-probability systems that can be generated in this way from distributions in $\Delta^0(\Omega)$ do not include all of $\Delta^*(\Omega)$, but any other Bayesian conditional-probability system in $\Delta^*(\Omega)$ can be expressed as the limit of conditional-probability systems generated in this way. This fact is asserted by the following theorem. For the proof, see Myerson (1986b).

THEOREM 1.4. *The probability function p is a Bayesian conditional-probability system in $\Delta^*(\Omega)$ if and only if there exists a sequence of probability distributions $\{\hat{p}^k\}_{k=1}^{\infty}$ in $\Delta^0(\Omega)$ such that, for every nonempty subset S of Ω and every t in Ω,*

$$p(t|S) = \lim_{k \to \infty} \frac{\hat{p}^k(t)}{\sum_{r \in S} \hat{p}^k(r)} \text{ if } t \in S,$$

$$p(t|S) = 0 \text{ if } t \notin S.$$

1.7 Limitations of the Bayesian Model

We have seen how expected-utility maximization can be derived from axioms that seem intuitively plausible as a characterization of rational preferences. Because of this result, mathematical social scientists have felt confident that mathematical models of human behavior that are based on expected-utility maximization should have a wide applicability and relevance. This book is largely motivated by such confidence.

It is important to try to understand the range of applicability of expected-utility maximization in real decision-making. In considering this question, we must remember that any model of decision-making can be used either descriptively or prescriptively. That is, we may use a model to try to describe and predict what people will do, or we may use a model as a guide to apply to our own (or our clients') decisions. The predictive validity of a model can be tested by experimental or empirical data. The prescriptive validity of a decision model is rather harder to test; one can only ask whether a person who understands the model would feel that he would be making a mistake if he did not make decisions according to the model.

Theorem 1.1, which derives expected-utility maximization from intuitive axioms, is a proof of the prescriptive validity of expected-utility maximization, if any such proof is possible. Although other models of decision-making have been proposed, few have been able to challenge the logical appeal of expected-utility maximization for prescriptive purposes.

There is, of course, a close relationship between the prescriptive and predictive roles of any decision-making model. If a model is prescriptively valid for a decision-maker, then he diverges from the model only when he is making a mistake. People do make mistakes, but they try not to. When a person has had sufficient time to learn about a situation and think clearly about it, we can expect that he will make relatively few mistakes. Thus, we can expect expected-utility maximization to be predictively accurate in many situations.

However, experimental research on decision-making has revealed some systematic violations of expected-utility maximization (see Allais and Hagen, 1979; Kahneman and Tversky, 1979; and Kahneman, Slovic, and Tversky, 1982). This research has led to suggestions of new models of decision-making that may have greater descriptive accuracy (see Kahneman and Tversky, 1979; and Machina, 1982). We discuss here three of the best-known examples in which people often seem to violate expected-utility maximization: one in which utility functions seem inapplicable, one in which subjective probability seems inapplicable, and one in which any economic model seems inapplicable.

Consider first a famous paradox, due to M. Allais (see Allais and Hagen, 1979). Let $X = \{\$12 \text{ million}, \$1 \text{ million}, \$0\}$, and let

$$f_1 = .10[\$12 \text{ million}] + .90[\$0],$$
$$f_2 = .11[\$1 \text{ million}] + .89[\$0],$$
$$f_3 = [\$1 \text{ million}],$$
$$f_4 = .10[\$12 \text{ million}] + .89[\$1 \text{ million}] + .01[\$0].$$

Many people will express the preferences $f_1 > f_2$ and $f_3 > f_4$. (Recall that no subscript on $>$ means that we are conditioning on Ω.) Such people may feel that $12 million is substantially better than $1 million, so the slightly higher probability of winning in f_2 compared with f_1 is not worth the lower prize. On the other hand, they would prefer to take the sure $1 million in f_3, rather than accept a probability .01 of getting nothing in exchange for a probability .10 of increasing the prize to $12 million in f_4.

Such preferences cannot be accounted for by any utility function. To prove this, notice that

$$.5f_1 + .5f_3 = .05[\$12 \text{ million}] + .5[\$1 \text{ million}] + .45[\$0]$$
$$= .5f_2 + .5f_4.$$

Thus, the common preferences $f_1 > f_2$ and $f_3 > f_4$ must violate the strict objective substitution axiom.

Other paradoxes have been generated that challenge the role of subjective probability in decision theory, starting with a classic paper by Ellsberg (1961). For a simple example of this kind, due to Raiffa (1968), let $X = \{-\$100, \$100\}$, let $\Omega = \{A, N\}$, and let

$$b_A(\$100|A) = 1 = b_A(-\$100|N),$$
$$b_N(-\$100|A) = 1 = b_N(\$100|N).$$

That is, b_A is a $100 bet in which the decision-maker wins if A occurs, and b_N is a $100 bet in which the decision-maker wins if N occurs. Suppose that A represents the state in which the American League will win the next All-Star game (in American baseball) and that N represents the state in which the National League will win the next All-Star game. (One of these two leagues must win the All-Star game, because the rules of baseball do not permit ties.)

Many people who feel that they know almost nothing about American baseball express the preferences $.5[\$100] + .5[-\$100] > b_A$ and $.5[\$100] + .5[-\$100] > b_N$. That is, they would strictly prefer to bet $100 on Heads in a fair coin toss than to bet $100 on either league in the All-Star game. Such preferences cannot be accounted for by any subjective probability distribution over Ω. At least one state in Ω must have probability greater than or equal to .5, and the bet on the league that wins in that state must give expected utility that is at least as great as the bet on the fair coin toss. To see it another way, notice that

$$.50b_A + .50b_N = .5[\$100] + .5[-\$100]$$
$$= .50(.5[\$100] + .5[-\$100]) + .50(.5[\$100] + .5[-\$100]),$$

so the common preferences expressed above must violate the strict objective substitution axiom.

To illustrate the difficulty of constructing a model of decision-making that is both predictively accurate and prescriptively appealing, Kahneman and Tversky (1982) have suggested the following example. In Situation A, you are arriving at a theatrical performance, for which you have bought a pair of tickets that cost $40. You suddenly realize that your tickets have fallen out of your pocket and are lost. You must decide whether to buy a second pair of tickets for $40 (there are some similar seats still available) or simply go home. In Situation B, you are arriving at a theatrical performance for which a pair of tickets costs $40. You did not buy tickets in advance, but you put $40 in your pocket when you left home. You suddenly realize that the $40 has fallen out of your pocket and is lost. You must decide whether to buy a pair of tickets for $40 with your charge card (which you still have) or simply go home.

As Kahneman and Tversky (1982) report, most people say that they would simply go home in Situation A but would buy the tickets in Situation B. However, in each of these situations, the final outcomes resulting from the two options are, on the one hand, seeing the performance and being out $80 and, on the other hand, missing the performance and being out $40. Thus, it is impossible to account for such behavior in any economic model that assumes that the levels of monetary wealth and theatrical consumption are all that should matter to the decision-maker in these situations.

Any analytical model must derive its power from simplifying assumptions that enable us to see different situations as analytically equivalent, but such simplifying assumptions are always questionable. A model that correctly predicts the common behavior in this example must draw distinctions between situations on the basis of fine details in the order of events that have no bearing on the final outcome. Such distinctions, however, would probably decrease the normative appeal of the model if it were applied for prescriptive purposes. (What would you think of a consultant who told you that you should make a point of behaving differently in Situations A and B?)

The explanatory power of expected-utility maximization can be extended to explain many of its apparent contradictions by the analysis of *salient perturbations*. A perturbation of a given decision problem is any other decision problem that is very similar to it (in some sense). For any given decision problem, we say that a perturbation is salient if people who actually face the given decision problem are likely to act as if they think that they are in this perturbation. A particular perturbation of a decision problem may be salient when people find the decision problem to be hard to understand and the perturbation is more like the kind of situations that they commonly experience. If we can predict the salient perturbation of an individual's decision problem, then the decision that maximizes his expected utility in this salient perturbation may be a more accurate prediction of his behavior.

For example, let us reconsider the problem of betting on the All-Star game. To get a decision-maker to express his preference ordering ($\succsim_\Omega$) over $\{b_A, b_N, .5[\$100] + .5[-\$100]\}$, we must ask him, for each pair in this set, which bet would he choose if this pair of bet-options were offered to him uninformatively, that is, in a manner that does not give him any new information about the true state in Ω. That is, when we ask him whether he would prefer to bet $100 on the American League or on a fair coin toss, we are assuming that the mere fact of offering this option to him does not change his information about the All-Star game. However, people usually offer to make bets only when they have some special information or beliefs. Thus, when someone who knows little about baseball gets an offer from another person to bet on the American League, it is usually because the other person has information suggesting that the American League is likely to lose. In such situations, an opportunity to bet on one side of the All-Star game should (by Bayes's formula) make someone who knows little about baseball decrease his subjective probability of the event that this side will win, so he may well prefer to bet on a fair coin toss. We can try to offer bets uninformatively in controlled experiments, and we can even tell our experimental subjects that the bets are being offered uninformatively, but this is so unnatural that the experimental subjects may instead respond to the salient perturbation in which we would only offer baseball bets that we expected the subject to lose.

1.8 Domination

Sometimes decision-makers find subjective probabilities difficult to assess. There are fundamental theoretical reasons why this should be particularly true in games. In a game situation, the unknown environment or "state of the world" that confronts a decision-maker may include the outcome of decisions that are to be made by other people. Thus, to assess his subjective probability distribution over this state space, the decision-maker must think about everything that he knows about other people's decision-making processes. To the extent that these other people are concerned about his own decisions, his beliefs about their behavior may be based at least in part on his beliefs about what they believe that he will do himself. So assessing subjective probabilities about others' behavior may require some understanding of the predicted outcome of his own decision-making process, part of which is his probability assessment itself. The resolution of this seeming paradox is the subject of game theory, to be developed in the subsequent chapters of this book.

Sometimes, however, it is possible to say that some decision-options could not possibly be optimal for a decision-maker, no matter what his beliefs may be. In this section, before turning from decision theory to game theory, we develop some basic results to show when such probability-independent statements can be made.

Consider a decision-maker who has a state-dependent utility function $u:X \times \Omega \to \mathbf{R}$ and can choose any x in X. That is, let us reinterpret X as the set of decision-options available to the decision-maker. If his subjective probability of each state t in Ω were $p(t)$ (that is, $p(t) = p(t|\Omega)$, $\forall t \in \Omega$), then the decision-maker would choose some particular y in X only if

(1.7) $\quad \sum_{t \in \Omega} p(t)u(y,t) \geq \sum_{t \in \Omega} p(t)u(x,t), \quad \forall x \in X.$

Convexity is an important property of many sets that arise in mathematical economics. A set of vectors is *convex* iff, for any two vectors p and q and any number λ between 0 and 1, if p is in the set and q is in the set then the vector $\lambda p + (1 - \lambda)q$ must also be in the set. Geometrically, convexity means that, for any two points in the set, the whole line segment between them must also be contained in the set.

THEOREM 1.5. *Given $u:X \times \Omega \to \mathbf{R}$ and given y in X, the set of all p in $\Delta(\Omega)$ such that y is optimal is convex.*

Proof. Suppose that y would be optimal for the decision-maker with beliefs p and q. Let λ be any number between 0 and 1, and let $r = \lambda p + (1 - \lambda)q$. Then for any x in X

$$\sum_{t \in \Omega} r(t)u(y,t) = \lambda \sum_{t \in \Omega} p(t)u(y,t) + (1 - \lambda) \sum_{t \in \Omega} q(t)u(y,t)$$

$$\geq \lambda \sum_{t \in \Omega} p(t)u(x,t) + (1 - \lambda) \sum_{t \in \Omega} q(t)u(x,t)$$

$$= \sum_{t \in \Omega} r(t)u(x,t).$$

So y is optimal for beliefs r. ∎

For example, suppose $X = \{\alpha,\beta,\gamma\}$, $\Omega = \{\theta_1,\theta_2\}$, and the utility function u is as shown in Table 1.1. With only two states, $p(\theta_1) = 1 - p(\theta_2)$. The decision α is optimal for the decision-maker iff the following two inequalities are both satisfied:

$$8p(\theta_1) + 1(1 - p(\theta_1)) \geq 5p(\theta_1) + 3(1 - p(\theta_1))$$
$$8p(\theta_1) + 1(1 - p(\theta_1)) \geq 4p(\theta_1) + 7(1 - p(\theta_1)).$$

Table 1.1   Expected utility payoffs for states θ_1 and θ_2

Decision	θ_1	θ_2
α	8	1
β	5	3
γ	4	7

The first of these inequalities asserts that the expected utility payoff from α must be at least as much as from β, and the second asserts that the expected utility payoff from α must be at least as much as from γ. By straightforward algebra, these inequalities imply that α is optimal when $p(\theta_1) \geq 0.6$.

Similarly, decision γ would be optimal iff

$$4p(\theta_1) + 7(1 - p(\theta_1)) \geq 8p(\theta_1) + 1(1 - p(\theta_1))$$
$$4p(\theta_1) + 7(1 - p(\theta_1)) \geq 5p(\theta_1) + 3(1 - p(\theta_1)),$$

and these two inequalities are both satisfied when $p(\theta_1) \leq 0.6$. Decision β would be optimal iff

$$5p(\theta_1) + 3(1 - p(\theta_1)) \geq 8p(\theta_1) + 1(1 - p(\theta_1))$$
$$5p(\theta_1) + 3(1 - p(\theta_1)) \geq 4p(\theta_1) + 7(1 - p(\theta_1)),$$

but there is no value of $p(\theta_1)$ that satisfies both of these inequalities. Thus, each decision is optimal over some convex interval of probabilities, except that the interval where β is optimal is the empty set.

Thus, even without knowing p, we can conclude that β cannot possibly be the optimal decision for the decision-maker. Such a decision-option that could never be optimal, for any set of beliefs, is said to be *strongly dominated*.

Recognizing such dominated options may be helpful in the analysis of decision problems. Notice that α would be best if the decision-maker were sure that the state was θ_1, and γ would be best if the decision-maker were sure that the state was θ_2, so it is easy to check that neither α nor γ is dominated in this sense. Option β is a kind of intermediate decision, in that it is neither best nor worst in either column of the payoff table. However, such intermediate decision-options are not necessarily dominated. For example, if the utility payoffs from decision β were changed to 6 in both states, then β would be the optimal decision whenever $5/7 \geq p(\theta_1) \geq 1/3$. On the other hand, if the payoffs from decision β were changed to 3 in both states, then it would be obvious that β could never be optimal, because choosing γ would always be better than choosing β.

There is another way to see that β is dominated, for the original payoff table shown above. Suppose that the decision-maker considered the following randomized strategy for determining his decision: toss a coin, and choose α if it comes out Heads, and choose γ if it comes out

Tails. We may denote this strategy by $.5[\alpha] + .5[\gamma]$, because it gives a probability of .5 to α and γ each. If the true state were θ_1, then this randomized strategy would give the decision-maker an expected utility payoff of $.5 \times 8 + .5 \times 4 = 6$, which is better than the payoff of 5 that he would get from β. (Recall that, because these payoffs are utilities, higher expected values are always more preferred by the decision-maker.) If the true state were θ_2, then this randomized strategy would give an expected payoff of $.5 \times 1 + .5 \times 7 = 4$, which is better than the payoff of 3 that he would get from β. So no matter what the state may be, the expected payoff from $.5[\alpha] + .5[\gamma]$ is strictly higher than the payoff from β. Thus, we may argue that the decision-maker would be irrational to choose β because, whatever his beliefs about the state might be, he would get a higher expected payoff from the randomized strategy $.5[\alpha] + .5[\gamma]$ than from choosing β. We may say that β is strongly dominated by the randomized strategy $.5[\alpha] + .5[\gamma]$.

In general, a *randomized strategy* is any probability distribution over the set of decision options X. We may denote such a randomized strategy in general by $\sigma = (\sigma(x))_{x \in X}$, where $\sigma(x)$ represents the probability of choosing x. Given the utility function $u:X \times \Omega \to \mathbf{R}$, we may say that a decision option y in X is *strongly dominated* by a randomized strategy σ in $\Delta(X)$ such that

$$(1.8) \qquad \sum_{x \in X} \sigma(x)u(x,t) > u(y,t), \quad \forall t \in \Omega.$$

That is, y is strongly dominated by σ if, no matter what the state might be, σ would always be strictly better than y under the expected-utility criterion.

We have now used the term "strongly dominated" in two different senses. The following theorem asserts that they are equivalent.

THEOREM 1.6. *Given $u:X \times \Omega \to \mathbf{R}$, where X and Ω are nonempty finite sets, and given any y in X, there exists a randomized strategy σ in $\Delta(X)$ such that y is strongly dominated by σ, in the sense of condition (1.8), if and only if there does not exist any probability distribution p in $\Delta(\Omega)$ such that y is optimal in the sense of condition (1.7).*

The proof is deferred to Section 1.9.

Theorem 1.6 gives us our first application of the important concept of a randomized strategy. Notice, however, that this result itself does not assert that a rational decision-maker should necessarily use a randomized strategy. It only asserts that, if we want to show that there are no beliefs about the state in Ω that would justify using a particular option y in X, we should try to find a randomized strategy that would be better than y in every state. Such a dominating randomized strategy would not necessarily be the best strategy for the decision-maker; it would only be clearly better than y.

A decision option y in X is *weakly dominated* by a randomized strategy σ in $\Delta(X)$ iff

$$\sum_{x \in X} \sigma(x)u(x,t) \geq u(y,t), \quad \forall t \in \Omega,$$

and there exists at least one state s in Ω such that

$$\sum_{x \in X} \sigma(x)u(x,s) > u(y,s).$$

That is, y is weakly dominated by σ if using σ would never be worse than y in any state and σ would be strictly better than y in at least one possible state. For example, suppose $X = \{\alpha,\beta\}$, $\Omega = \{\theta_1,\theta_2\}$, and $u(\cdot,\cdot)$ is as shown in Table 1.2. In this case β is weakly dominated by α (that is, by $[\alpha]$, the randomized strategy that puts probability one on choosing α). Notice that β would be optimal (in a tie with α) if the decision-maker believed that θ_1 was the true state with probability one. However, if he assigned any positive probability of θ_2, then he would not choose β. This observation is generalized by the following analogue of Theorem 1.6.

THEOREM 1.7. *Given $u:X \times \Omega \to \mathbf{R}$, where X and Ω are nonempty finite sets, and given any y in X, there exists a randomized strategy σ in $\Delta(X)$ such that y is weakly dominated by σ if and only if there does not exist any probability*

Table 1.2 Expected utility payoffs for states θ_1 and θ_2

Decision	θ_1	θ_2
α	5	3
β	5	1

distribution p in $\Delta^0(\Omega)$ such that y is optimal in the sense of condition (1.7). (Recall that $\Delta^0(\Omega)$ is the set of probability distributions on Ω that assign strictly positive probability to every state in Ω.)

1.9 Proofs of the Domination Theorems

Theorems 1.6 and 1.7 are proved here using the duality theorem of linear programming. A full derivation of this result can be found in any text on linear programming (see Chvatal, 1983; or Luenberger, 1984). A statement of this theorem is presented here in Section 3.7, following the discussion of two-person zero-sum games. Readers who are unfamiliar with the duality theorem of linear programming should defer reading the proofs of Theorems 1.6 and 1.7 until after reading Section 3.8.

Proof of Theorem 1.6. Consider the following two linear programming problems. In the first problem, the variables are δ and $(p(t))_{t \in \Omega}$:

minimize δ subject to

$$p(s) \geq 0, \quad \forall s \in \Omega,$$

$$\sum_{t \in \Omega} p(t) \geq 1,$$

$$-\sum_{t \in \Omega} p(t) \geq -1,$$

$$\delta + \sum_{t \in \Omega} p(t)(u(y,t) - u(x,t)) \geq 0, \quad \forall x \in X.$$

In the second problem, the variables are $(\eta(t))_{t \in \Omega}$, $(\varepsilon_1, \varepsilon_2)$, and $(\sigma(x))_{x \in X}$:

maximize $\varepsilon_1 - \varepsilon_2$ subject to $\eta \in \mathbf{R}_+^\Omega$, $\varepsilon \in \mathbf{R}_+^2$, $\sigma \in \mathbf{R}_+^X$,

$$\sum_{x \in X} \sigma(x) = 1,$$

$$\eta(t) + \varepsilon_1 - \varepsilon_2 + \sum_{x \in X} \sigma(x)(u(y,t) - u(x,t)) = 0, \quad \forall t \in \Omega.$$

(Here $\mathbf{R}_+$ denotes the set of nonnegative real numbers, so $\mathbf{R}_+^\Omega$ is the set of vectors with nonnegative components indexed on Ω.) There exists some p such that y is optimal if and only if there is a solution to the first problem that has an optimal value (of δ) that is less than or equal to 0.

On the other hand, there exists some randomized strategy σ that strongly dominates y if and only if the second problem has an optimal value (of $\varepsilon_1 - \varepsilon_2$) that is strictly greater than 0. The second problem is the dual of the first (see Section 3.8), and the constraints of both problems can be satisfied. So the optimal values of both problems must be equal, by the duality theorem of linear programming. Thus, y is strongly dominated by some randomized strategy (and both problems have strictly positive value) if and only if there is no probability distribution in $\Delta(\Omega)$ that makes y optimal.   ∎

Proof of Theorem 1.7.   Consider the following two linear programming problems. In the first problem, the variables are δ and $(p(t))_{t \in \Omega}$:

minimize δ subject to

$$p(s) + \delta \geq 0, \quad \forall s \in \Omega,$$

$$- \sum_{t \in \Omega} p(t) \geq -1,$$

$$\sum_{t \in \Omega} p(t)(u(y,t) - u(x,t)) \geq 0, \quad \forall x \in X.$$

In the second problem, the variables are $(\eta(t))_{t \in \Omega}$, ε, and $(\sigma(x))_{x \in X}$:

maximize $-\varepsilon$ subject to $\eta \in \mathbf{R}_+^{\Omega}$, $\varepsilon \geq 0$, $\sigma \in \mathbf{R}_+^X$,

$$\sum_{t \in \Omega} \eta(t) = 1,$$

$$\eta(s) + -\varepsilon + \sum_{x \in X} \sigma(x)(u(y,t) - u(x,t)) = 0, \quad \forall s \in \Omega.$$

There exists some p in $\Delta^0(\Omega)$ such that y is optimal if and only if there is a solution to the first problem that has an optimal value (of δ) that is strictly less than 0. On the other hand, there exists some randomized strategy σ that weakly dominates y if and only if the second problem has an optimal value (of $-\varepsilon$) that is greater than or equal to 0. (The vector σ that solves the second problem may be a positive scalar multiple of the randomized strategy that weakly dominates y.) The second problem is the dual of the first, and the constraints of both problems can be satisfied. So the optimal values of both problems must be equal, by the duality theorem of linear programming. Thus, y is weakly dominated by some randomized strategy (and both problems have nonnegative value) if and only if there is no probability distribution in $\Delta^0(\Omega)$ that makes y optimal. ∎

Exercises

Exercise 1.1.   Suppose that the set of prizes X is a finite subset of **R**, the set of real numbers, and a prize x denotes an award of x dollars. A decision-maker says that, if he knew that the true state of the world was in some set T, then he would weakly prefer a lottery f over another lottery g (that is, $f \gtrsim_T g$) iff

$$\min_{s \in T} \sum_{x \in X} xf(x|s) \geq \min_{s \in T} \sum_{x \in X} xg(x|s).$$

(That is, he prefers the lottery that gives the higher expected payoff in the worst possible state.) Which of our axioms (if any) does this preference relation violate?

Exercise 1.2.   Alter Exercise 1.1 by supposing instead that $f \gtrsim_T g$ iff

$$\sum_{s \in T} \min \{x | f(x|s) > 0\} \geq \sum_{s \in T} \min \{x | g(x|s) > 0\}.$$

Which of our axioms (if any) does this preference relation violate?

Exercise 1.3.   Show that Axiom 1.1B implies:

if $f \sim_s g$ and $g \sim_s h$ then $f \sim_s h$; and

if $f >_s g$ and $g \gtrsim_s h$ then $f >_s h$.

Exercise 1.4.   Show that Axioms 1.1A and 1.5B together imply Axiom 1.1B and Axiom 1.3.

Exercise 1.5.   A decision-maker expresses the following preference ordering for monetary lotteries

$$[\$600] > [\$400] > 0.90[\$600] + 0.10[\$0]$$
$$> 0.20[\$600] + 0.80[\$0]$$
$$> 0.25[\$400] + 0.75[\$0] > [\$0].$$

Are these preferences consistent with any state-independent utility for money? If so, show a utility function that applies. If not, show an axiom that this preference ordering violates.

Exercise 1.6. Consider a decision-maker whose subjective probability distribution over the set of possible states Ω is $p = (p(s))_{s \in \Omega}$. We ask him to tell us his subjective probability distribution, but he can lie and report any distribution in $\Delta(\Omega)$ that he wants. To guide his reporting decision, we plan to give him some reward $Y(q, \tilde{s})$ that will be a function of the probability distribution q that he reports and the true state of nature $\tilde{s}$ that will be subsequently observed.

a. Suppose that his utility for our reward is $u(Y(q, s), s) = q(s)$, for every q in $\Delta(\Omega)$ and every s in Ω. Will his report q be his true subjective probability distribution p? If not, what will he report?

b. Suppose that his utility for our reward is $u(Y(q, s), s) = \log_e(q(s))$, for every q and s. Will his report q be his true subjective probability distribution p? If not, what will he report?

Exercise 1.7. Suppose that utility payoffs depend on decisions and states as shown in Table 1.3. Let $(p(\theta_1), p(\theta_2))$ denote the decision-maker's subjective probability distribution over $\Omega = \{\theta_1, \theta_2\}$.

a. Suppose first that $B = 35$. For what range of values of $p(\theta_1)$ is α optimal? For what range is β optimal? For what range is γ optimal? Is any decision strongly dominated? If so, by what randomized strategies?

b. Suppose now that $B = 20$. For what range of values of $p(\theta_1)$ is α optimal? For what range is β optimal? For what range is γ optimal? Is any decision strongly dominated? If so, by what randomized strategies?

c. For what range of values for the parameter B is the decision β strongly dominated?

Table 1.3   Expected utility payoffs for states θ_1 and θ_2

Decision	θ_1	θ_2
α	15	90
β	B	75
γ	55	40

Exercise 1.8. Suppose that a function $W:\Delta(\Omega) \to \mathbf{R}$ satisfies

$$W(p) = \max_{x \in X} \sum_{t \in \Omega} p(t)u(x,t), \quad \forall p \in \Delta(\Omega).$$

Show that W is a convex function, that is,

$$W(\lambda p + (1 - \lambda)q) \le \lambda W(p) + (1 - \lambda)W(q)$$

for any p and q in $\Delta(\Omega)$ and any λ in such that $0 \le \lambda \le 1$.

Exercise 1.9. In this exercise, we consider some useful conditions that are sufficient to guarantee that observing a higher signal would not lead a decision-maker to choose a lower optimal decision.

Suppose that X and Ω are nonempty finite sets, $X \subseteq \mathbf{R}$, $\Omega = \Omega_1 \times \Omega_2$, $\Omega_1 \subseteq \mathbf{R}$, and $\Omega_2 \subseteq \mathbf{R}$. A decision-maker has utility and probability functions $u:X \times \Omega \to \mathbf{R}$ and $p:\Omega \to \mathbf{R}$ that satisfy the following properties, for each x and y in X, each s_1 and t_1 in Ω_1, and each s_2 and t_2 in Ω_2:

> if $x > y$, $s_1 \ge t_1$, $s_2 \ge t_2$, and $(s_1,s_2) \ne (t_1,t_2)$,
>
> then $u(x,s_1,s_2) - u(y,s_1,s_2) > u(x,t_1,t_2) - u(y,t_1,t_2)$;
>
> if $s_1 > t_1$ and $s_2 > t_2$, then $p(s_1,s_2)p(t_1,t_2) \ge p(s_1,t_2)p(t_1,s_2)$;
>
> and $p(s_1,s_2) > 0$.

The condition on u asserts that the net benefit of an increase in X increases as the components in Ω increase (that is, u has *increasing differences*). By Bayes's formula, if the decision-maker observed that the second component of the true state of the world was s_2, then he would assign conditional probability

$$p(s_1|s_2) = \frac{p(s_1,s_2)}{\displaystyle\sum_{r_1 \in \Omega_1} p(r_1,s_2)}$$

to the event that the unknown first component of the true state was s_1.

a. Show that if $s_1 > t_1$ and $s_2 > t_2$ then $p(s_1|s_2)/p(t_1|s_2) \ge p(s_1|t_2)/p(t_1|t_2)$. (This is called the *monotone likelihood ratio property*. See also Milgrom, 1981; and Milgrom and Weber, 1982.)

b. Suppose that y would be optimal in X for the decision-maker if he observed that the second component of the true state was s_2, and x

would be optimal in X for the decision-maker if he observed that the second component of the true state was t_2. That is,

$$\sum_{r_1 \in \Omega_1} p(r_1 | s_2) u(y, r_1, s_2) = \max_{z \in X} \sum_{r_1 \in \Omega_1} p(r_1 | s_2) u(z, r_1, s_2),$$

$$\sum_{r_1 \in \Omega_1} p(r_1 | t_2) u(x, r_1, t_2) = \max_{z \in X} \sum_{r_1 \in \Omega_1} p(r_1 | t_2) u(z, r_1, t_2).$$

Show that if $s_2 > t_2$, then $y \geq x$.

2

Basic Models

2.1 Games in Extensive Form

The analysis of any game or conflict situation must begin with the specification of a model that describes the game. Thus, the general form or structure of the models that we use to describe games must be carefully considered. A model structure that is too simple may force us to ignore vital aspects of the real games that we want to study. A model structure that is too complicated may hinder our analysis by obscuring the fundamental issues. To avoid these two extremes, several different general forms are used for representing games, the most important of which are the *extensive* form and the *strategic* (or *normal*) form. The extensive form is the most richly structured way to describe game situations. The definition of the extensive form that is now standard in most of the literature on game theory is due to Kuhn (1953), who modified the earlier definition used by von Neumann and Morgenstern (1944) (see also Kreps and Wilson, 1982, for an alternative way to define the extensive form). The strategic form and its generalization, the *Bayesian* form, are conceptually simpler forms that are more convenient for purposes of general analysis but are generally viewed as being derived from the extensive form.

To introduce the extensive form, let us consider a simple card game that is played by two people, whom we call "player 1" and "player 2." (Throughout this book, we follow the convention that odd-numbered players are male and even-numbered players are female. When players are referred to by variables and gender is uncertain, generic male pronouns are used.)

At the beginning of this game, players 1 and 2 each put a dollar in the pot. Next, player 1 draws a card from a shuffled deck in which half the cards are red (diamonds and hearts) and half are black (clubs and spades). Player 1 looks at his card privately and decides whether to raise or fold. If player 1 folds then he shows the card to player 2 and the game ends; in this case, player 1 takes the money in the pot if the card is red, but player 2 takes the money in the pot if the card is black. If player 1 raises then he adds another dollar to the pot and player 2 must decide whether to meet or pass. If player 2 passes, then the game ends and player 1 takes the money in the pot. If player 2 meets, then she also must add another dollar to the pot, and then player 1 shows the card to player 2 and the game ends; in this case, again, player 1 takes the money in the pot if the card is red, and player 2 takes the money in the pot if the card is black.

Figure 2.1 is a tree diagram that shows the possible events that could occur in this game. The tree consists of a set of *branches* (or line segments), each of which connects two points, which are called *nodes*. The leftmost node in the tree is the *root* of the tree and represents the beginning of the game. There are six nodes in the tree that are not followed to the right by any further branches; these nodes are called *terminal nodes* and represent the possible ways that the game could end. Each possible sequence of events that could occur in the game is represented by a path of branches from the root to one of these terminal nodes. When the game is actually played, the path that represents the actual sequence of events that will occur is called the *path of play*. The goal of game-theoretic analysis is to try to predict the path of play.

At each terminal node, Figure 2.1 shows a pair of numbers, which represent the payoffs that players 1 and 2 would get if the path of play

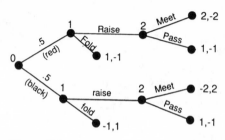

Figure 2.1

ended at this node. For example, in one possible sequence of events, player 1 might get a red card, player 1 might raise, and player 2 might meet, in which case the payoffs would be +2 for player 1 and −2 for player 2. This sequence of events is represented in Figure 2.1 by the path from the root to the terminal node at the top right, where the payoff vector $(2,-2)$ is shown. In another possible sequence of events, player 1 might get a black card and then fold; this sequence is represented in Figure 2.1 by the path from the root to the terminal node at the bottom near the middle of the figure, where the payoffs are −1 for player 1 and +1 for player 2.

At each nonterminal node that is followed to the right by more than one branch, the branches represent alternative events, of which at most one can occur. The determination of which of these alternative events would occur is controlled either by a player or by chance. If the event is determined by chance, then we give the node a label "0" (zero). That is, a nonterminal node with label "0" is a *chance node*, where the next branch in the path of play would be determined by some random mechanism, according to probabilities that are shown on the branches that follow the chance node. In Figure 2.1, the root has label "0" because the color of the card that player 1 draws is determined by chance. (Player 1 cannot look at his card until after he draws it.) Each of the two branches following the root has probability .5, because half of the cards in the deck are red and half are black. A nonterminal node with a label other than zero is a *decision node*, where the next branch in the path of play would be determined by the player named by the label. After drawing his card, player 1 decides whether to raise or fold, so the two nodes that immediately follow the root are controlled by player 1 and have the label "1."

Figure 2.1 is *not* an adequate representation of our simple card game, however. Nowhere in Figure 2.1 have we indicated the crucial fact that player 1 knows the color of the card but player 2 does not know the color. If you only looked at Figure 2.1, you might expect that player 2 would pass if 1 raised with a red card (because she would then prefer payoff −1 to −2), but that player 2 would meet if 1 raised with a black card (because she would then prefer payoff 2 to −1). However, player 2's expected behavior must be the same at these two nodes, because she does not know the color of 1's card when she chooses between meeting and passing. On the other hand, player 1 could plan to raise with a red card but to fold with a black card, because he can distinguish the two nodes that he controls. To indicate each player's information when he or she moves in this game, we need to augment the tree diagram as shown in Figure 2.2.

In Figure 2.2, each decision node has two labels, separated by a decimal point. To the left of the decimal point, we write the *player label*, which indicates the name of the player who controls the node. To the right of the decimal point, we write an *information label*, which indicates the *information state* of the player when he or she moves at this node. So the label "1.a" indicates a node where player 1 moves in information state "a," and the label "2.0" indicates a node where player 2 moves in information state "0." The assignment of letters and numbers to name the various information states may be quite arbitrary. In Figure 2.2, 1's information state "a" is the state of having a red card, 1's information state "b" is the state of having a black card, and 2's information state "0" is the state of knowing that 1 has raised. The only significance of the information labels is to indicate sets of nodes that cannot be distinguished by the player who controls them. Thus, because player 1's nodes have different information labels but player 2's nodes have the same information labels, the reader of Figure 2.2 knows that player 1 can distinguish his two nodes when he moves, but player 2 cannot distinguish her two nodes when she moves. To emphasize the sets of nodes that cannot be distinguished, we may also draw a dashed curve around sets of nodes that have the same player and information labels.

Figure 2.2 is a complete description of our simple card game as a game in extensive form. Notice that "red" and "black" labels on the branches following the chance node in Figure 2.1 are omitted in Figure

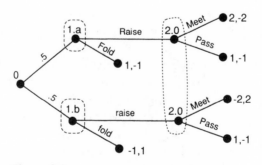

Figure 2.2

2.2, but the labels for the other branches are retained. It is easy to see that the actual color of the card that puts player 1 in a winning position should not matter to the analysis of the game, so we do not need labels on branches that represent chance events. However, *move labels* on branches that represent decisions by players are an essential part of the description of the game.

A person cannot make a meaningful choice without knowing what the options are among which he or she must choose. Thus, when player 2 makes a decision without knowing which node she is playing at, she cannot choose any specific branch. What she must actually choose is a *move*, "Meet" or "Pass," and the next branch in the path of play will be the branch at the current node that has this move label. To guarantee that a player always knows the options available to him or her at any point in the game, the sets of move labels following two nodes must be the same if the two nodes are controlled by the same player in the same information state. For example, the set of move labels on the branches following 2's upper node must be the same as the set of move labels on the branches following 2's lower node. (If we added a third branch labeled "Punt" after her lower node but we added no such branch after her upper node, it would mean that she could punt if and only if player 1 had a black card. But how could she exercise this option when, not knowing the color of player 1's card, she could not know whether the option existed? To have a meaningful game, we would have to include a "Try to punt" branch at both of 2's decision nodes, as long as they have the same information label.) On the other hand, the move labels on the branches following the "1.a" node do not have to match the move labels on the branches following the "1.b" node, because player 1 can distinguish these two nodes. If the move labels did not match, player 1 would still know what his options were, because he would know at which node he was playing.

Because moves are the actual object of choice for a player who has two or more nodes with the same information label, the way that move labels are assigned to branches can be very important. The interest in this game derives from the fact that Pass is better for 2 at her upper node but Meet is better for 2 at her lower node, and 2 does not know which node is actually in the path of play when she decides whether to Meet or Pass. On the other hand, if we switched the move labels at player 2's upper node, leaving everything else the same as in Figure 2.2, then the resulting game would be completely different. The modified tree would represent a very uninteresting game in which player 2 should clearly choose Meet, because Meet would give her a payoff of -1 against a red card and (as before) 2 against a black card, whereas Pass would give her a payoff of -2 against a red card and -1 against a black card.

To give a rigorous general definition of a game in the extensive form, we need to make our terminology more precise. In the language of mathematical graph theory, a *graph* is a finite set of points or nodes, together with a set of branches, each of which connects a pair of nodes. Set-theoretically, a branch may be identified with the set consisting of the two nodes that it connects. Then a *path* is a set of branches of the form

$$\{\{x_1,x_2\},\{x_2,x_3\}, \ldots ,\{x_{m-1},x_m\}\} = \{\{x_k,x_{k+1}\}|\ k = 1, \ldots ,m - 1\},$$

where $m \geq 2$ and each x_k is a different node in the graph. We may say that such a path *connects* the nodes x_1 and x_m. A *tree* is a graph in which each pair of nodes is connected by exactly one path of branches in the graph. A *rooted tree* is a tree in which one special node is designated as the *root* of the tree. In this book, the root of a tree is always shown leftmost in the diagram. When we speak of *the path* to a given node, we mean the unique path that connects this node and the root. An *alternative* at a node in a rooted tree is any branch that connects it with another node and is not in the path to this node. A node or branch x *follows* another node or branch y iff y is in the path to x. A node x *immediately follows* a node y iff x follows y and there is an alternative at y that connects x and y. A *terminal node* in a rooted tree is a node with no alternatives following it.

For any positive integer n, an *n-person extensive-form game* Γ^e is a rooted tree together with functions that assign labels to every node and branch, satisfying the following five conditions.

1. Each nonterminal node has a *player label* that is in the set $\{0,1,2, \ldots ,n\}$. Nodes that are assigned a player label 0 are called *chance nodes*. The set $\{1,2, \ldots ,n\}$ represents the set of *players* in the game, and, for each i in this set, the nodes with the player-label i are *decision nodes* that are controlled by player i.

2. Every alternative at a chance node has a label that specifies its probability. At each chance node, these *chance probabilities* of the alternatives are nonnegative numbers that sum to 1.

3. Every node that is controlled by a player has a second label that specifies the *information state* that the player would have if the path of play reached this node. When the path of play reaches a node controlled by a player, he knows only the information state of the current node. That is, two nodes that belong to the same player should have the same information state if and only if the player would be unable to distinguish between the situations represented by these nodes when either occurs in the play of the game. In our notation, the player label and the information label at any node are separated by a decimal point, with the player label to the left and the information label to the right, so "*i.k*" would indicate a node where player i moves with information state k.

4. Each alternative at a node that is controlled by a player has a *move label*. Furthermore, for any two nodes x and y that have the same player label and the same information label, and for any alternative at node x, there must be exactly one alternative at node y that has the same move label.

5. Each terminal node has a label that specifies a vector of n numbers $(u_1, \ldots, u_n) = (u_i)_{i \in \{1, \ldots, n\}}$. For each player i, the number u_i is interpreted as the *payoff* to player i, measured in some utility scale, when this node is the outcome of the game.

We will generally assume that extensive-form games satisfy an additional condition known as *perfect recall*. This condition asserts that whenever a player moves, he remembers all the information that he knew earlier in the game, including all of his own past moves. This assumption of perfect recall can be expressed formally as follows.

6. For any player i, for any nodes x, y, and z that are controlled by i, and for any alternative b at x, if y and z have the same information state and if y follows x and b, then there exists some node w and some alternative c at w such that z follows w and c, w is controlled by player i, w has the same information state as x, and c has the same move label as b. (It may be that $w = x$ and $c = b$, of course.) That is, if i's decision nodes y and z are indistinguishable to him, then for any past decision node and move that i recalls at y, there must be an indistinguishable decision node and move that he recalls at z. (For an example of a game without perfect recall, suppose that the node labels "2.0" were changed to "1.0" in Figure 2.2. The resulting game would represent a rather bizarre situation in which player 1 controls all the decision nodes, but when he decides whether to meet or pass he cannot recall the infor-

mation that he knew earlier when he decided to raise. See also Figure 4.3 in Chapter 4.)

If no two nodes have the same information state, then we say that the game has *perfect information*. That is, in a game with perfect information, whenever a player moves, he knows the past moves of all other players and chance, as well as his own past moves. The game in Figure 2.4 has perfect information, but the games in Figures 2.2 and 2.3 do not have perfect information.

A *strategy* for a player in an extensive-form game is any rule for determining a move at every possible information state in the game. Mathematically, a strategy is a function that maps information states into moves. For each player i, let S_i denote the set of possible information states for i in the game. For each information state s in S_i, let D_s denote the set of moves that would be available to player i when he moved at a node with information state s. Then the set of strategies for player i in the extensive-form game is $\times_{s \in S_i} D_s$.

In our simple card game, player 1 has four possible strategies. We denote the set of strategies for player 1 in this game as

$$\{Rr, Rf, Fr, Ff\},$$

where we write first in upper case the initial letter of the designated move at the 1.a node (with a red card), and second in lower case the initial letter of the designated move at the 1.b node (with a black card). For example, Rf denotes the strategy "Raise if the card is red, but fold if the card is black," and Rr denotes the strategy "Raise no matter what the color of the card may be." Notice that a strategy for player 1 is a complete rule that specifies a move for player 1 in all possible contingencies, even though only one contingency will actually arise. Player 2 has only two possible strategies, which may be denoted "M" (for "Meet if 1 raises") and "P," because player 2 has only one possible information state.

For two additional examples, consider Figures 2.3 and 2.4. Figure 2.3 shows a game in which player 2 must choose between L and R without observing 1's move. Against either of 2's possible moves, L or R, player 1 would be better off choosing T, so player 1 should choose T in the game represented by Figure 2.3. When player 1 chooses T, player 2 can get payoff 2 from choosing L.

Figure 2.4 differs from 2.3 only in that player 2's two nodes have different information labels, so 2 observes 1's actual choice before she

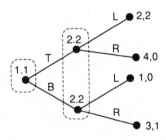

Figure 2.3

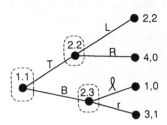

Figure 2.4

chooses between L and R (or ℓ and r) in this game. Thus, player 1 has an opportunity to influence 2's choice in this game. It would be better for player 2 to choose L if she observed T (because 2 > 0), and it would be better for player 2 to choose r if she observed B (because 1 > 0). So player 1 should expect to get payoff 2 from choosing T and payoff 3 from choosing B. Thus, player 1 should choose B in the game represented by Figure 2.4. After player 1 chooses B, player 2 should choose r and get payoff 1.

Notice that the set of strategies for player 2 is only {L,R} in Figure 2.3, whereas her set of strategies in Figure 2.4 is {Lℓ, Lr, Rℓ, Rr} (writing the move at the 2.2 node first and the move at the 2.3 node second). Thus, a change in the informational structure of the game that increases the set of strategies for player 2 may change the optimal move for player 1, and so may actually decrease player 2's expected payoff.

If we watched the game in Figure 2.4 played once, we would observe player 2's move (L or R or ℓ or r), but we would not be able to observe 2's strategy, because we would not see what she would have done at her other information state. For example, if player 1 did B then player 2's observable response would be the same (r) under both the strategy Rr and the strategy Lr. To explain why player 1 should choose B in Figure 2.4, however, the key is to recognize that player 2 should rationally follow the strategy Lr ("L if T, r if B"), and that player 1 should intelligently expect this. If player 2 were expected to follow the strategy Rr ("R if T, r if B"), then player 1 would be better off choosing T to get payoff 4.

2.2 Strategic Form and the Normal Representation

A simpler way to represent a game is to use the strategic form. To define a game in strategic form, we need only to specify the set of players in the game, the set of options available to each player, and the way that players' payoffs depend on the options that they choose.

Formally, a *strategic-form* game is any Γ of the form

(2.1) $\Gamma = (N, (C_i)_{i \in N}, (u_i)_{i \in N})$,

where N is a nonempty set, and, for each i in N, C_i is a nonempty set and u_i is a function from $\times_{j \in N} C_j$ into the set of real numbers **R**. Here, N is the set of players in the game Γ. For each player i, C_i is the set of *strategies* (or *pure strategies*) available to player i. When the strategic-form game Γ is played, each player i must choose one of the strategies in the set C_i. A *strategy profile* is a combination of strategies that the players in N might choose. We let C denote the set of all possible strategy profiles, so that

$$C = \underset{j \in N}{\times} C_j.$$

For any strategy profile $c = (c_j)_{j \in N}$ in C, the number $u_i(c)$ represents the expected utility payoff that player i would get in this game if c were the combination of strategies implemented by the players. When we study a strategic-form game, we usually assume that the players all choose their strategies simultaneously, so there is no element of time in the analysis of strategic-form games.

A strategic-form game is *finite* if the set of players N and all the strategy sets C_i are finite. In developing the basic ideas of game theory, we generally assume finiteness here, unless otherwise specified.

An extensive-form game is a dynamic model, in the sense that it includes a full description of the sequence in which moves and events may occur over time in an actual play of the game. On the other hand, a strategic-form game is a static model, in the sense that it ignores all questions of timing and treats players as if they choose their strategies simultaneously. Obviously, eliminating the time dimension from our models can be a substantial conceptual simplification, if questions of timing are not essential to our analysis. To accomplish such a simplification, von Neumann and Morgenstern suggested a procedure for constructing a game in strategic form, given any extensive-form game Γ^e.

To illustrate this procedure, consider again the simple card game, shown in Figure 2.2. Now suppose that players 1 and 2 know that they are going to play this game tomorrow, but today each player is planning his or her moves in advance. Player 1 does not know today what color his card will be, but he can plan now what he would do with a red card, and what he would do with a black card. That is, as we have seen, the set of possible strategies for player 1 in this extensive-form game is $C_1 = \{Rr, Rf, Fr, Ff\}$, where the first letter designates his move if his card is red (at the node labeled 1.a) and the second letter designates his move if his card is black (at the node labeled 1.b). Player 2 does not know today whether player 1 will raise or fold, but she can plan today whether to meet or pass if 1 raises. So the set of strategies that player 2 can choose among today is $C_2 = \{M,P\}$, where M denotes the strategy "meet if 1 raises," and P denotes the strategy "pass if 1 raises."

Even if we knew the strategy that each player plans to use, we still could not predict the actual outcome of the game, because we do not know whether the card will be red or black. For example, if player 1 chooses the strategy Rf (raise if the card is red, fold if the card is black) and player 2 chooses the strategy M, then player 1's final payoff will be either $+2$, if the card is red (because 1 will then raise, 2 will meet, and 1 will win), or -1, if the card is black (because 1 will then fold and lose). However, we can compute the expected payoff to each player when these strategies are used in the game, because we know that red and black cards each have probability ½. So when player 1 plans to use the strategy Rf and player 2 plans to use the strategy M, the expected payoff to player 1 is

$$u_1(Rf, M) = 2 \times \tfrac{1}{2} + -1 \times \tfrac{1}{2} = 0.5.$$

Similarly, player 2's expected payoff from the strategy profile (Rf, M) is

$$u_2(Rf, M) = -2 \times \tfrac{1}{2} + 1 \times \tfrac{1}{2} = -0.5.$$

We can similarly compute the expected payoffs to each player from each pair of strategies. The resulting expected payoffs $(u_1(c), u_2(c))$ depend on the combination of strategies $c = (c_1, c_2)$ in $C_1 \times C_2$ according to Table 2.1.

The strategic-form game shown in Table 2.1 is called the *normal representation* of our simple card game. It describes how, at the beginning of the game, the expected utility payoffs to each player would depend on their strategic plans. From decision theory (Theorem 1.1 in Chapter 1), we know that a rational player should plan his strategy to maximize his expected utility payoff. Thus, Table 2.1 represents the decision-theoretic situation that confronts the players as they plan their strategies at or before the beginning of the game.

In general, given any extensive-form game Γ^e as defined in Section 2.1, a representation of Γ^e in strategic form can be constructed as follows. Let N, the set of players in the strategic-form game, be the same as the set of players $\{1, 2, \ldots, n\}$ in the given extensive-form game Γ^e. For any player i in N, let the set of strategies C_i for player i in the strategic-form game Γ be the same as the set of strategies for player i in the extensive-form game Γ^e, as defined at the end of Section 2.1. That is, any strategy c_i in C_i is a function that specifies a move $c_i(r)$ for every information state r that player i could have during the game.

For any strategy profile c in C and any node x in the tree of Γ^e, we define $P(x|c)$ to be the probability that the path of play will go through node x, when the path of play starts at the root of Γ^e and, at any decision node in the path, the next alternative to be included in the path is determined by the relevant player's strategy in c, and, at any chance node in the path, the next alternative to be included in the path is determined by the probability distribution given in Γ^e. (This definition can be formalized mathematically by induction as follows. If x is the

Table 2.1 The simple card game in strategic form, the normal representation

C_1	C_2	
	M	P
Rr	0,0	1,−1
Rf	0.5,−0.5	0,0
Fr	−0.5,0.5	1,−1
Ff	0,0	0,0

root of Γ^e, then $P(x|c) = 1$. If x immediately follows a chance node y, and q is the chance probability of the branch from y to x, then $P(x|c) = qP(y|c)$. If x immediately follows a decision node y that belongs to player i in the information state r, then $P(x|c) = P(y|c)$ if $c_i(r)$ is the move label on the alternative from y to x, and $P(x|c) = 0$ if $c_i(r)$ is not the move label on the alternative from y to x.) At any terminal node x, let $w_i(x)$ be the utility payoff to player i at node x in the game Γ^e. Let Ω^* denote the set of all terminal nodes of the game Γ^e. Then, for any strategy profile c in C, and any i in N, let $u_i(c)$ be

$$u_i(c) = \sum_{x \in \Omega^*} P(x|c)w_i(x).$$

That is, $u_i(c)$ is the expected utility payoff to player i in Γ^e when all the players implement the strategies designated for them by c. When $\Gamma = (N, (C_i)_{i \in N}, (u_i)_{i \in N})$ is derived from Γ^e in this way, Γ is called the *normal representation* of Γ^e.

We have seen that Table 2.1 is the normal representation of the extensive-form game in Figure 2.2. It may also be helpful to consider and compare the normal representations of the extensive-form games in Figures 2.3 and 2.4. In both games, the set of strategies for player 1 is $C_1 = \{T,B\}$, but the sets of strategies for player 2 are very different in these two games. In Figure 2.3, player 2 has only one possible information state, so $C_2 = \{L,R\}$. The resulting utility payoffs (u_1,u_2) in normal representation of the game in Figure 2.3 are shown in Table 2.2.

In Figure 2.4, where player 2 has two possible information states, her set of strategies is $C_2 = \{L\ell, Lr, R\ell, Rr\}$. The utility payoffs in the normal representation of Figure 2.4 are shown in Table 2.3. Thus, the seemingly minor change in information-state labels that distinguishes Figure

Table 2.2   A game in strategic form

	C_2	
C_1	L	R
T	2,2	4,0
B	1,0	3,1

Table 2.3   A game in strategic form

C_1	C_2			
	Lℓ	Lr	Rℓ	Rr
T	2,2	2,2	4,0	4,0
B	1,0	3,1	1,0	3,1

2.4 from Figure 2.3 leads to a major change in the normal representation.

Von Neumann and Morgenstern argued that, in a very general sense, the normal representation should be all that we need to study in the analysis of any game. The essence of this argument is as follows. We game theorists are trying to predict what rational players should do at every possible stage in a given game. Knowing the structure of the game, we should be able to do our analysis and compute our predictions before the game actually begins. But if the players in the game are intelligent, then each player should be able to do the same computations that we can and determine his rational plan of action before the game begins. Thus, there should be no loss of generality in assuming that all players formulate their strategic plans simultaneously at the beginning of the game; so the actual play of the game is then just a mechanistic process of implementing these strategies and determining the outcome according to the rules of the game. That is, we can assume that all players make all substantive decisions simultaneously at the beginning of the game, because the substantive decision of each player is supposed to be the selection of a complete plan or strategy that specifies what moves he would make under any possible circumstance, at any stage of the game. Such a situation, in which players make all their strategic decisions simultaneously and independently, is exactly described by the normal representation of the game.

This argument for the sufficiency of the normal representation is one of the most important ideas in game theory, although the limitations of this argument have been reexamined in the more recent literature (see Section 2.6 and Chapters 4 and 5).

As a corollary of this argument, the assumption that players choose their strategies independently, in a strategic-form game, can be defended as being without loss of generality. Any opportunities for communication that might cause players' decisions to not be independent can, in principle, be explicitly represented by moves in the extensive-form game. (Saying something is just a special kind of move.) If all communication opportunities are included as explicit branches in the extensive-form game, then strategy choices "at the beginning of the game" must be made without prior communication. In the normal representation of such an extensive-form game, each strategy for each player includes a specification of what he should say in the communication process and how he should respond to the messages that he may receive; and these are the strategies that players are supposed to choose independently. (When possibilities for communication are very rich and complex, however, it may be more convenient in practice to omit such opportunities for communication from the structure of the game and to account for them in our solution concept instead. This alternative analytical approach is studied in detail in Chapter 6.)

Another implication of the argument for the sufficiency of the normal representation arises when a theorist tries to defend a proposed general solution concept for strategic-form games by arguing that players would converge to this solution in a game that has been repeated many times. Such an argument would ignore the fact that a process of "repeating a game many times" may be viewed as a single game, in which each repetition is one stage. This overall game can be described by an extensive-form game, and by its normal representation in strategic form. If a proposed general solution concept for strategic-form games is valid then, we may argue, it should be applied in this situation to this overall game, not to the repeated stages separately. (For more on repeated games, see Chapter 7.)

2.3 Equivalence of Strategic-Form Games

From the perspective of decision theory, utility numbers have meaning only as representations of individuals' preferences. Thus, if we change the utility functions in a game model in such a way that the underlying preferences represented by these functions is unchanged, then the new game model must be considered equivalent to the old game model. It is important to recognize when two game models are equivalent, because our solution concepts must not make different predictions for two equivalent game models that could really represent the same situation. That is, the need to generate the same solutions for equivalent games may be a useful criterion for identifying unsatisfactory solution theories.

From Theorem 1.3 in Chapter 1, we know that two utility functions $u(\cdot)$ and $\hat{u}(\cdot)$ are equivalent, representing identical preferences for the decision-maker, iff they differ by a linear transformation of the form $\hat{u}(\cdot) = Au(\cdot) + B$, for some constants $A > 0$ and B. Thus, we say that two games in strategic form, $\Gamma = (N, (C_i)_{i \in N}, (u_i)_{i \in N})$ and $\hat{\Gamma} = (N, (C_i)_{i \in N}, (\hat{u}_i)_{i \in N})$, are *fully equivalent* iff, for every player i, there exist numbers A_i and B_i such that $A_i > 0$ and

$$\hat{u}_i(c) = A_i u_i(c) + B_i, \quad \forall c \in C.$$

That is, two strategic-form games with the same player set and the same strategy sets are fully equivalent iff each player's utility function in one game is decision-theoretically equivalent to his utility function in the other game.

The most general way to describe (predicted or prescribed) behavior of the players in a strategic-form game is by a probability distribution over the set of strategy profiles $C = \times_{i \in N} C_i$. Recall (from equation 1.1 in Chapter 1) that, for any finite set Z, we let $\Delta(Z)$ denote the set of all probability distributions over the set Z. So $\Delta(C)$ denotes the set of all probability distributions over the set of strategy profiles for the players in Γ or $\hat{\Gamma}$ as above. In the game Γ, with utility function u_i, player i would prefer that the players behave according to a probability distribution $\mu = (\mu(c))_{c \in C}$ in $\Delta(C)$ (choosing each profile of strategies c with probability $\mu(c)$) rather than according to some other probability distribution λ iff μ gives him a higher expected utility payoff than λ, that is

$$\sum_{c \in C} \mu(c) u_i(c) \geq \sum_{c \in C} \lambda(c) u_i(c).$$

In game $\hat{\Gamma}$, similarly, i would prefer μ over λ iff

$$\sum_{c \in C} \mu(c) \hat{u}_i(c) \geq \sum_{c \in C} \lambda(c) \hat{u}_i(c).$$

It follows from Theorem 1.3 that Γ and $\hat{\Gamma}$ are fully equivalent iff, for each player i in N, for each μ in $\Delta(C)$, and for each λ in $\Delta(C)$, player i would prefer μ over λ in the game Γ if and only if he would prefer μ over λ in the game $\hat{\Gamma}$.

Other definitions of equivalence have been proposed for strategic-form games. One weaker definition of equivalence (under which more pairs of games are equivalent) is called *best-response equivalence*. Best-response equivalence is based on the narrower view that a player's utility function serves only to characterize how he would choose his strategy once his beliefs about the other players' behavior are specified. To formally define best-response equivalence, we must first develop additional notation.

For any player i, let C_{-i} denote the set of all possible combinations of strategies for the players other than i; that is,

$$C_{-i} = \underset{j \in N-i}{\times} C_j.$$

(Here, $N-i$ denotes the set of all players other than i.) Given any $e_{-i} = (e_j)_{j \in N-i}$ in C_{-i} and any d_i in C_i, we let (e_{-i}, d_i) denote the strategy profile in C such that the i-component is d_i and all other components are as in e_{-i}.

For any set Z and any function $f : Z \to \mathbf{R}$, $\mathrm{argmax}_{y \in Z} f(y)$ denotes the set of points in Z that maximize the function f, so

$$\underset{y \in Z}{\mathrm{argmax}}\, f(y) = \{ y \in Z \,|\, f(y) = \underset{z \in Z}{\max}\, f(z) \}.$$

If player i believed that some distribution η in $\Delta(C_{-i})$ predicted the behavior of the other players in the game, so each strategy combination e_{-i} in C_{-i} would have probability $\eta(e_{-i})$ of being chosen by the other players, then player i would want to choose his own strategy in C_i to maximize his own expected utility payoff. So in game Γ, player i's set of *best responses* to η would be

$$\underset{d_i \in C_i}{\mathrm{argmax}} \sum_{e_{-i} \in C_{-i}} \eta(e_{-i}) u_i(e_{-i}, d_i).$$

In game $\hat{\Gamma}$, similarly, i's best responses would be defined by replacing the utility function u_i by $\hat{u}_i$. The games Γ and $\hat{\Gamma}$ are *best-response equivalent* iff these best-response sets always coincide, for every player and every possible probability distribution over the others' strategies; that is,

$$\underset{d_i \in C_i}{\mathrm{argmax}} \sum_{e_{-i} \in C_{-i}} \eta(e_{-i}) u_i(e_{-i}, d_i)$$

$$= \underset{d_i \in C_i}{\mathrm{argmax}} \sum_{e_{-i} \in C_{-i}} \eta(e_{-i}) \hat{u}_i(e_{-i}, d_i), \quad \forall i \in N, \quad \forall \eta \in \Delta(C_{-i}).$$

For example, consider the two games shown in Tables 2.4 and 2.5.

These two games are best-response equivalent because, in each game, each player i's best response would be to choose y_i if he thought that the other player would choose the y-strategy with probability $\frac{1}{8}$ or more. However, the two games are obviously not fully equivalent. (For ex-

Table 2.4   A game in strategic form

C_1	C_2	
	x_2	y_2
x_1	9,9	0,8
y_1	8,0	7,7

Table 2.5   A game in strategic form

C_1	C_2	
	x_2	y_2
x_1	1,1	0,0
y_1	0,0	7,7

ample, each player prefers (x_1,x_2) over (y_1,y_2) in the first game but not in the second game.) The distinction between these two equivalence concepts depends on whether we admit that one player's preferences over the possible strategies of another player may be meaningful and relevant to our analysis of the game.

2.4 Reduced Normal Representations

There are some strategic-form games that we can simplify by eliminating redundant strategies. For example, consider the extensive-form game in Figure 2.5. Its normal representation is shown in Table 2.6.

It is not a coincidence that the payoffs in the top three rows of Table 2.6 are identical. In the given extensive-form game, if player 1 chose a_1 at his 1.1 node, then the path of play could never get to his 1.3 nodes, so the expected payoffs cannot depend on what move (among x_1, y_i, and z_1) he would plan to choose if the path of play did reach a 1.3 node. Thus, the distinction between player 1's strategies a_1x_1, a_1y_1, and a_1z_1 may seem unnecessary.

In general, given any strategic-form game $\Gamma = (N, (C_i)_{i \in N}, (u_i)_{i \in N})$, for any player i and any two strategies d_i and e_i in C_i, we may say that d_i and e_i are *payoff equivalent* iff

$$u_j(c_{-i},d_i) = u_j(c_{-i},e_i), \quad \forall c_{-i} \in C_{-i}, \quad \forall j \in N.$$

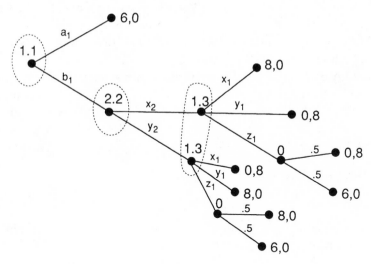

Figure 2.5

Table 2.6   A game in strategic form, the normal representation of Figure 2.5

C_1	C_2	
	x_2	y_2
a_1x_1	6,0	6,0
a_1y_1	6,0	6,0
a_1z_1	6,0	6,0
b_1x_1	8,0	0,8
b_1y_1	0,8	8,0
b_1z_1	3,4	7,0

That is, two of i's strategies d_i and e_i are payoff equivalent iff, no matter what the players other than i may do, no player would ever care whether i used strategy d_i or e_i. In Table 2.6, a_1x_1, a_1y_1, and a_1z_1 are payoff equivalent to one another.

When there are payoff-equivalent strategies, we can simplify the normal representation by merging together payoff-equivalent strategies and replacing each such set of payoff-equivalent strategies by a single strategy. The result of simplifying the normal representation is called the *purely reduced normal representation*.

For example, we may merge and replace a_1x_1, a_1y_1, and a_1z_1 in Table 2.6 by a single strategy $a_1\cdot$, which may be interpreted as the plan "do a_1 at 1.1, move unspecified at 1.3." Then the purely reduced normal representation is as shown in Table 2.7.

Although no two strategies are payoff equivalent in Table 2.7, there is a sense in which the strategy b_1z_1 is redundant. To see why, suppose that player 1 considered choosing between the strategies $a_1\cdot$ and b_1y_1 on the basis of a coin toss. The resulting randomized strategy, which can be denoted $.5[a_1\cdot] + .5[b_1y_1]$, would give expected-payoff allocations

$$.5(6,0) + .5(0,8) = (3,4) \text{ against } x_2,$$
$$.5(6,0) + .5(8,0) = (7,0) \text{ against } y_2.$$

Thus, b_1z_1 is redundant, because it is payoff equivalent to a randomized strategy that player 1 could implement using strategies other than b_1z_1.

In general, given any strategic-form game $\Gamma = (N, (C_i)_{i \in N}, (u_i)_{i \in N})$, a *randomized strategy* for player i is any probability distribution over the set of C_i. Thus, $\Delta(C_i)$ denotes the set of all randomized strategies for player i. To emphasize the distinction from these randomized strategies, the strategies in the set C_i may also be called *pure strategies*. For any pure strategy c_i in C_i and any randomized strategy σ_i in $\Delta(C_i)$, $\sigma_i(c_i)$ denotes the probability that i would choose c_i if he were implementing the randomized strategy σ_i in $\Delta(C_i)$.

A strategy d_i in C_i is *randomly redundant* iff there is a probability distribution σ_i in $\Delta(C_i)$ such that $\sigma_i(d_i) = 0$ and

$$u_j(c_{-i},d_i) = \sum_{e_i \in C_i} \sigma_i(e_i)u_j(c_{-i},e_i), \quad \forall c_{-i} \in C_{-i}, \quad \forall j \in N.$$

Table 2.7 A game in strategic form, the purely reduced normal representation of Figure 2.5

	C_2	
C_1	x_2	y_2
$a_1\cdot$	6,0	6,0
b_1x_1	8,0	0,8
b_1y_1	0,8	8,0
b_1z_1	3,4	7,0

That is, d_i is randomly redundant iff there is some way for player i to randomly choose among his other strategies such that, no matter what combination of actions might be chosen by the other players, every player would get the same expected payoff when i uses d_i as when i randomizes in this way.

The *fully reduced normal representation* is derived from the purely reduced normal representation by eliminating all strategies that are randomly redundant in the purely reduced normal representation. The fully reduced normal representation of the extensive-form game in Figure 2.5 is shown in Table 2.8. Unless specified otherwise, the *reduced normal representation* of an extensive-form game may be taken to mean the fully reduced normal representation.

2.5 Elimination of Dominated Strategies

Concepts of domination, defined for one-person decision problems in Section 1.8, can also be applied to strategic-form games. Given any strategic-form game $\Gamma = (N, (C_i)_{i \in N}, (u_i)_{i \in N})$, any player i in N, and any strategy d_i in C_i, d_i is *strongly dominated* for player i iff there exists some randomized strategy σ_i in $\Delta(C_i)$ such that

$$u_i(c_{-i}, d_i) < \sum_{e_i \in C_i} \sigma_i(e_i) u_i(c_{-i}, e_i), \quad \forall c_{-i} \in C_{-i}.$$

For example, in the strategic form of our simple card game (Table 2.1), the strategy Ff is strongly dominated for player 1 by the randomized strategy $.5[Rr] + .5[Rf]$.

By Theorem 1.6, d_i is strongly dominated for player i if and only if d_i can never be a best response for i, no matter what he may believe about the other players' strategies. This fact may suggest that eliminat-

Table 2.8   A game in strategic form, the fully reduced normal representation of Figure 2.5

C_1	C_2	
	x_2	y_2
$a_1 \cdot$	6,0	6,0
$b_1 x_1$	8,0	0,8
$b_1 y_1$	0,8	8,0

ing a strongly dominated strategy for any player i should not affect the analysis of the game, because player i would never use this strategy, and this fact should be evident to all the other players if they are intelligent.

After one or more strongly dominated strategies have been eliminated from a game, other strategies that were not strongly dominated in the original game may become strongly dominated in the game that remains. For example, consider the game in Table 2.9. In this game, z_2 is strongly dominated for player 2 by $.5[x_2] + .5[y_2]$ (the randomized strategy that gives probability 0.5 to x_2 and y_2 each). None of the other strategies for either player are strongly dominated in this game, because each is a best response to at least one conjecture about how the other player may behave. (Strategy a_1 is best for 1 against x_2, b_1 is best for 1 against z_2, x_2 is best for 2 against a_1, and y_2 is best for 2 against b_1.) However, in the game that remains after eliminating z_2, b_1 is strongly dominated by a_1 for player 1 (because $0 < 2$ and $1 < 3$). Furthermore, when z_2 and b_1 are both eliminated, we are left with a game in which y_2 is strongly dominated by x_2 for player 2 (because $0 < 3$). Then eliminating y_2 leaves only one strategy for each player in the game: a_1 for player 1 and x_2 for player 2. Thus, iterative elimination of strongly dominated strategies leads to a unique prediction as to what the players should do in this game.

This process of iterative elimination of strongly dominated strategies may be formalized for a general strategic-form game $\Gamma = (N, (C_i)_{i \in N}, (u_i)_{i \in N})$ as follows. For any player i, let $C_i^{(1)}$ denote the set of all strategies in C_i that are not strongly dominated for i. Then let $\Gamma^{(1)}$ be the strategic-form game

$$\Gamma^{(1)} = \left(N, (C_i^{(1)})_{i \in N}, (u_i)_{i \in N} \right).$$

(In the game $\Gamma^{(1)}$, each u_i function is, of course, actually the restriction of the original utility function to the new smaller domain $\times_{j \in N} C_j^{(1)}$.)

Table 2.9 A game in strategic form

	C_2		
C_1	x_2	y_2	z_2
a_1	2,3	3,0	0,1
b_1	0,0	1,6	4,2

Then, by induction, for every positive integer k, we can define the strategic-form game $\Gamma^{(k)}$ to be

$$\Gamma^{(k)} = \left(N, (C_i^{(k)})_{i \in N}, (u_i)_{i \in N}\right),$$

where, for each player i, $C_i^{(k)}$ is the set of all strategies in $C_i^{(k-1)}$ that are not strongly dominated for i in the game $\Gamma^{(k-1)}$ (and u_i is reinterpreted as the restriction of the original utility function for i to the smaller domain $\times_{j \in N} C_j^{(k)}$). Clearly, for each i, $C_i \supseteq C_i^{(1)} \supseteq C_i^{(2)} \supseteq \ldots$, and it can be shown that all of these sets are nonempty. Thus, since we started with a finite game Γ, there must exist some number K such that

$$C_i^{(K)} = C_i^{(K+1)} = C_i^{(K+2)} = \ldots, \quad \forall i \in N.$$

Given this number K, we let $\Gamma^{(\infty)} = \Gamma^{(K)}$ and $C_i^{(\infty)} = C_i^{(K)}$ for every i in N. The strategies in $C_i^{(\infty)}$ are *iteratively undominated*, in the strong sense, for player i. The game $\Gamma^{(\infty)}$ may be called the *residual game* generated from Γ by iterative strong domination.

Because all players are supposed to be rational decision-makers, in the given strategic-form game Γ we may conclude that no player could possibly use any strongly dominated strategy. That is, each player i must be expected to choose a strategy in $C_i^{(1)}$. Because all players are assumed to be intelligent, they should know as we do that no player i will use a strategy outside of $C_i^{(1)}$. Thus, each player i must choose a strategy in $C_i^{(2)}$, because $C_i^{(2)}$ is the set of all strategies that are best responses to probability distributions over $\times_{j \in N-i} C_j^{(1)}$. But then, because each player i is intelligent, he must also know that every player j will use a strategy in $C_j^{(2)}$, and so i must use a strategy in $C_i^{(3)}$, the set of his best responses to probability distributions over $\times_{j \in N-i} C_j^{(2)}$. Repeatedly using the assumptions of rationality and intelligence in this way, this argument can be extended to show that each player i must use a strategy in $C_i^{(\infty)}$.

For each player i, every strategy in $C_i^{(\infty)}$ is a best response to some probability distribution over $\times_{j \in N-i} C_j^{(\infty)}$. In fact, these sets $(C_i^{(\infty)})_{i \in N}$ are the largest subsets of $(C_i)_{i \in N}$ respectively for which this condition holds.

Given any strategic-form game $\Gamma = (N, (C_i)_{i \in N}, (u_i)_{i \in N})$, any player i in N, and any strategy d_i in C_i, d_i is *weakly dominated* for player i iff there exists some randomized strategy σ_i in $\Delta(C_i)$ such that

$$u_i(c_{-i}, d_i) \leq \sum_{e_i \in C_i} \sigma_i(e_i) u_i(c_{-i}, e_i), \quad \forall c_{-i} \in C_{-i},$$

and, for at least one strategy combination $\hat{c}_{-i}$ in C_{-i},

$$u_i(\hat{c}_{-i}, d_i) < \sum_{e_i \in C_i} \sigma_i(e_i) u_i(\hat{c}_{-i}, e_i).$$

It is harder to argue that eliminating a weakly dominated strategy should not affect the analysis of the game, because weakly dominated strategies could be best responses for a player, if he feels confident that some strategies of other players have probability 0. (Recall Theorem 1.7.) Furthermore, there are technical difficulties with iterative elimination of weakly dominated strategies that do not arise with strong domination. Consider, for example, the game shown in Table 2.10 (due to Kohlberg and Mertens, 1986). If we first eliminate the strongly dominated strategy z_1, then we are left with a game in which y_2 is weakly dominated. On the other hand, if we first eliminate the strongly dominated strategy y_1 from Table 2.10, then we are left with a game in which x_2 is weakly dominated. If we begin by eliminating both y_1 and z_1 from Table 2.10, then neither of player 2's strategies would be weakly dominated in the game that remains (nor would they be payoff equivalent, in the sense of Section 2.4, because player 1's payoffs from x_2 and y_2 are different). Thus, which of player 2's strategies would be eliminated by a process of iterative elimination of weakly dominated strategies depends on the order in which we eliminate player 1's dominated strategies.

This order-dependence problem does not arise if we only eliminate strongly dominated strategies. That is, if we keep eliminating strongly dominated strategies until we have a residual game in which no strongly dominated strategies can be found, then the residual game will be the same no matter what order of elimination is used. Eliminating strategies for other players can never cause a strongly dominated strategy for player i to cease being strongly dominated, but it can cause a weakly

Table 2.10 A game in strategic form

C_1	C_2	
	x_2	y_2
x_1	3,2	2,2
y_1	1,1	0,0
z_1	0,0	1,1

dominated strategy to cease being weakly dominated (see Gilboa, Kalai, and Zemel, 1989).

Nevertheless, weak domination and iterative elimination of weakly dominated strategies are useful concepts for analysis of games. In our simple card game (Table 2.1), the fact that Fr and Ff are both weakly dominated for player 1 is a formal expression of our natural intuition that player 1 should not fold when he has a winning card. In Table 2.3, iterative weak domination can first eliminate 2's strategies Lℓ, Rℓ, and Rr (all weakly dominated by Lr) and then eliminate 1's strategy T in the game that remains. On the other hand, in Table 2.2, iterative (strong) domination can first eliminate 1's strategy B and then eliminate 2's strategy R. So iterative elimination of dominated strategies leads to the conclusions discussed at the end of Section 2.2: that we should expect to observe the moves T and L in the game shown in Figure 2.3, but we should expect to observe the moves B and r in the game shown in Figure 2.4.

2.6 Multiagent Representations

The normal representation of von Neumann and Morgenstern effectively defines a mapping from games in extensive form to games in strategic form. Another such mapping was proposed by Selten (1975); he called it the *agent-normal* form. We use here a slight modification of Selten's terminology and call this mapping the *multiagent representation*. The idea behind Selten's multiagent representation is that it should not matter if a given player in Γ^e were represented by a different agent in each of his possible information states, provided that these agents all share the same preferences and information of the original player.

Let Γ^e be any given game in extensive form, and let N denote the set of players in Γ^e. For any i in N, we let S_i denote the set of information states for player i that occur at the various nodes belonging to i in the game. Without loss of generality (relabeling if necessary), we may assume that these S_i sets are disjoint, so $S_i \cap S_j = \emptyset$ if $i \neq j$.

The set of players in the multiagent representation of this extensive-form game is $S^* = \cup_{i \in N} S_i$. That is, in the multiagent representation of Γ^e, there is one player for every possible information state of every player in Γ^e itself. We may refer to the players in the multiagent representation as *temporary agents*. A temporary agent r representing player i is responsible for choosing the move that i would make when the path of play reaches a node that is controlled by i with the information state r. We may imagine that all of the temporary agents plan their moves simultaneously at the beginning of the game, although some agents' plans may never be implemented (if the path of play does not go through the nodes with the information states for which they are responsible).

For any player i in N and information state r in S_i, we let D_r be the set of move labels on the alternatives at the nodes that are controlled by player i in the information state r. This set D_r is the set of strategies available to the temporary agent r in the multiagent representation of Γ^e. The utility functions v_r for the temporary agents in the multiagent representation are defined to coincide with the utility functions u_i of the corresponding players in the normal representation. That is, for any player i in N and any r in S_i, we define $v_r: \times_{s \in S*} D_s \to \mathbf{R}$ so that, for any $(d_s)_{s \in S*}$ in $\times_{s \in S*} D_s$, if $(c_j)_{j \in N}$ is the strategy profile for the normal representation such that $c_j(t) = d_t$ for every j in N and every t in S_j, then $v_r((d_s)_{s \in S*}) = u_i((c_j)_{j \in N})$.

Together these structures $(S^*, (D_r)_{r \in S*}, (v_r)_{r \in S*})$ define the multiagent representation of Γ^e. Like the normal representation, the multiagent representation is a game in strategic form.

For example, consider the game in Figure 2.6. In the normal representation, the set of players is $N = \{1,2\}$, the strategy sets are $C_1 = \{a_1 w_1, a_1 x_1, b_1 w_1, b_1 x_1\}$ and $C_2 = \{y_2, z_2\}$, and the payoffs (u_1, u_2) are shown in Table 2.11. In the multiagent representation, on the other hand, the set of players is $S^* = \{1,2,3\}$, of whom agents 1 and 2 represent different information states of the original player 1, and agent 3 represents the original player 2. The strategy sets in the multiagent representation are $D_1 = \{a_1, b_1\}$, $D_2 = \{w_1, x_1\}$, $D_3 = \{y_2, z_2\}$. The payoffs (v_1, v_2, v_3) in the

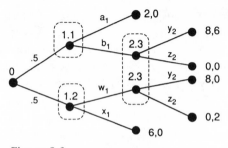

Figure 2.6

Table 2.11   A game in strategic form, the normal representation of Figure 2.6

	C_2	
C_1	y_2	z_2
a_1w_1	5,0	1,1
a_1x_1	4,0	4,0
b_1w_1	8,3	0,1
b_1x_1	7,3	3,0

Table 2.12   A game in strategic form, the multiagent representation of Figure 2.6

	y_2		z_2	
	w_1	x_1	w_1	x_1
a_1	5,5,0	4,4,0	1,1,1	4,4,0
b_1	8,8,3	7,7,3	0,0,1	3,3,0

multiagent representation are shown in Table 2.12. Of course, the first two payoffs are the same in each cell of Table 2.12, because temporary agents 1 and 2 both represent the original player 1, acting at different information states in the given extensive-form game.

To appreciate some of the technical significance of the multiagent representation, notice that the strategy a_1w_1 is strongly dominated (by b_1x_1) for player 1 in the normal representation. Furthermore, iterative elimination of weakly dominated strategies in the normal representation would lead to the conclusion that player 2 should use the strategy y_2 (because z_2 becomes weakly dominated once the strategy a_1w_1 is eliminated), and so player 1 should use strategy b_1w_1. However, no strategies are dominated (weakly or strongly) in the multiagent representation. (For each temporary agent, each of his two strategies is a unique best response to some combination of strategies by the other two agents.) Thus, a domination argument that may seem rather convincing when we only consider the normal representation becomes more questionable when we consider the multiagent representation.

2.7   Common Knowledge

Suppose that after player 1 has drawn a black card in our simple card game he asks a consultant to help him decide whether to raise or fold.

Given the information that the card is black, the consultant knows that the payoffs actually would be $(-1,1)$ if 1 folds, $(-2,2)$ if 1 raises and 2 meets, and $(1,-1)$ if 1 raises and 2 passes. Thus, he might be tempted to model this situation by the game shown in Figure 2.7.

However, Figure 2.7 would be seriously inadequate as a representation of this card game. Looking only at Figure 2.7, the consultant might reason as follows: "Player 2 should obviously be expected to meet, because meeting gives her a payoff of 2 whereas passing gives her a payoff of -1. Thus, it is better for player 1 to pass and get -1, rather than raise and get -2." The error in this reasoning is that player 2 does not know that the payoffs are those shown in Figure 2.7. Her ignorance about the color of 1's card is crucial for understanding why she might pass when 1 raises, and this ignorance is shown in Figure 2.2 but not in Figure 2.7. Thus, even if the consultant actually knows the color of the card drawn by player 1, the chance node with both possible outcomes of the draw must be included in his model of the game, as shown in Figure 2.2, because player 2's behavior may be influenced by her uncertainty about the color of the card.

Following Aumann (1976), we say that a fact is *common knowledge* among the players if every player knows it, every player knows that every player knows it, and so on; so every statement of the form "(every player knows that)^k every player knows it" is true, for $k = 0,1,2, \ldots$. A player's *private information* is any information that he has that is not common knowledge among all the players in the game. In the simple card game, after player 1 has drawn his card, it is common knowledge that the path of play must have reached one of the two nodes controlled by player 1 in Figure 2.2; the actual node that has been reached is player 1's private information.

In general, whatever model of a game we may choose to study, the methods of game theory compel us to assume that this model must be

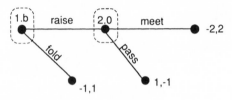

Figure 2.7

common knowledge among the players. To understand why, recall our proposed definition of game theory (from Section 1.1) as the study of mathematical models of conflict and cooperation between intelligent rational decision-makers. The intelligence assumption means that whatever we may know or understand about the game must be known or understood by the players of the game, because they are as intelligent as we are. Thus, the intelligence assumption implies that, whatever model of the game we may study, we must assume that the players know this model, too. Furthermore, because we know that the players all know the model, the intelligent players must also know that they all know the model. Having established this fact, we also recognize that the intelligent players also know that they all know that they all know the model, and so on.

Thus, to do analysis as a game theorist, the consultant to player 1 who knows that the card is actually black must disregard this information, because it is not common knowledge among the players. To represent this game in extensive form, he must use the game as originally shown in Figure 2.2, because Figure 2.2 shows the maximum information about the game that is common knowledge. (In general, a larger game tree like Figure 2.2 expresses more uncertainty, and hence less information, about what may happen in the game than a smaller tree like Figure 2.7.) Under these circumstances, the root of the game in Figure 2.2 may be called a *historical chance node* because, at the time when the game model is formulated and analyzed, the outcome of this node has already occurred and is known to some (but not all) players. The historical chance node is needed to represent the uncertainty of player 2.

In general, a historical node is needed whenever we represent in extensive form a situation in which the players already have different private information. Because the extensive form begins with a single node, this node (the root) must represent a situation at some time in the past before the players learned their private information, so everything that any player then knew about the game was common knowledge. All relevant private information that the players may have now must be accounted for in the extensive-form game by nodes and branches representing the past events that the players may have observed.

To illustrate the importance of common knowledge, I cite a well-known fable. This story concerns a village of 100 married couples, who were all perfect logicians but had somewhat peculiar social customs. Every evening the men of the village would have a meeting, in a great circle around a campfire, and each would talk about his wife. If when the meeting began a man had any reason to hope that his wife had always been faithful to him, then he would praise her virtue to all of the assembled men. On the other hand, if at any time before the current meeting he had ever gotten proof that his wife had been unfaithful, then he would moan and wail and invoke the terrible curse of the (male) gods on her. Furthermore, if a wife was ever unfaithful, then she and her lover would immediately inform all of the other men in the village except her husband. All of these traditions were common knowledge among the people of this village.

In fact, every wife had been unfaithful to her husband. Thus, every husband knew of every infidelity except for that of his own wife, whom he praised every evening.

This situation endured for many years, until a traveling holy man visited the village. After sitting through a session around the campfire and hearing every man praise his wife, the holy man stood up in the center of the circle of husbands and said in a loud voice, "A wife in this village has been unfaithful." For ninety-nine evenings thereafter, the husbands continued to meet and praise their wives, but on the hundredth evening they all moaned and wailed and invoked the terrible curse.

To understand what happened in this fable, notice first that, if there had been only one unfaithful wife, her husband would have moaned and wailed on the first evening after the holy man's visit, because (knowing of no other infidelities and knowing that he would have known of them if they existed) he would have known immediately that the unfaithful wife was his. Furthermore, one can show by induction that, for any integer k between 1 and 100, if there were exactly k unfaithful wives, then all husbands would praise their wives for $k-1$ evenings after the holy man's visit and then, on the kth evening, the k husbands of unfaithful wives would moan and wail. Thus, on the hundredth evening, after 99 more evenings of praise, every husband knew that there must be 100 unfaithful wives, including his own.

Now let us ask, What did this holy man tell the husbands that they did not already know? Every husband already knew of 99 unfaithful wives, so that was not news to anyone. But the holy man's statement also made it common knowledge among the men that there was an unfaithful wife, since it was common knowledge that he announced this to all the men. Before the holy man's announcement, every statement of the form "(every husband knows that)^k there is an unfaithful wife" was true for $k \leq 99$, but it was not true for $k = 100$. For example, if we number the husbands from 1 to 100, 1 knew that 2 knew that 3 knew that . . . that 99 knew that 100's wife was unfaithful; but 1 did not know that 2 knew that 3 knew that . . . that 99 knew that 100 knew that 1's wife was unfaithful. Thus, the lesson to be drawn from this fable is that the consequences that follow if a fact is common knowledge can be very different from the consequences that would follow if (for example) it were merely known that everyone knew that everyone knew it.

2.8 Bayesian Games

A game with *incomplete information* is a game in which, at the first point in time when the players can begin to plan their moves in the game, some players already have private information about the game that other players do not know. For example, our simple card game would become a game with incomplete information if player 1 drew his card and looked at it before he learned the actual rules of the game in which the card was to be used. Games with incomplete information often arise in practical situations. We often want to study situations in which the various players currently have different private information that they have known for a long time, and it may be unnatural to define the beginning of the game to be some point in the distant past before they learned their private information. Furthermore, some parts of a player's private information may be so basic to his identity that it is not even meaningful to talk about him planning his actions before learning this information (e.g., information about the player's gender, native language, and level of risk aversion). Thus, we should admit the possibility that a player may have some private information already at the first point in time when he begins to plan his moves in the game. The initial private information that a player has at this point in time is called the *type* of the player.

Games with incomplete information can be modeled in extensive form by using a historical chance node to describe the random determination of the players' types. As noted earlier, however, defining the beginning of the game to be some time in the past, before players learned their types, may be interpretively awkward. Furthermore, there may be serious concerns that the argument in Section 2.2 for the sufficiency of the normal representation in strategic form should not apply to games with incomplete information. The key to that argument was the assumption that each player should be able to plan his moves at or before the point in time represented by the root of the extensive-form game. But for a game with incomplete information, that assumption would be simply false (because the root represents a point in time before players learn their types, and a player's type is the private information that he has when he begins to think about the game). Thus, Harsanyi (1967–68) argued that a generalization of the strategic form, called the Bayesian form, is needed to represent games with incomplete information. The Bayesian form gives us a way of representing games that is almost as simple as the strategic form (at least in comparison to the extensive form), but which does not require us to pretend that players choose their strategies before learning any private information.

To define a *Bayesian game* (or a *game in Bayesian form*), we must specify a set of players N and, for each player i in N, we must specify a set of possible *actions* C_i, a set of possible *types* T_i, a probability function p_i, and a utility function u_i. We let

$$C = \underset{i \in N}{\times} C_i, \quad T = \underset{i \in N}{\times} T_i.$$

That is, C is the set of all possible profiles or combinations of actions that the players may use in the game, and T is the set of all possible profiles or combinations of types that the players may have in the game. For each player i, we let T_{-i} denote the set of all possible combinations of types for the players other than i, that is,

$$T_{-i} = \underset{j \in N-i}{\times} T_j.$$

The probability function p_i in the Bayesian game must be a function from T_i into $\Delta(T_{-i})$, the set of probability distributions over T_{-i}. That is, for any possible type t_i in T_i, the probability function must specify a probability distribution $p_i(\cdot | t_i)$ over the set T_{-i}, representing what player i would believe about the other players' types if his own type were t_i. Thus, for any t_{-i} in T_{-i}, $p_i(t_{-i} | t_i)$ denotes the subjective probability that i would assign to the event that t_{-i} is the actual profile of types for the other players, if his own type were t_i.

For any player i in N, the utility function u_i in the Bayesian game must be a function from $C \times T$ into the real numbers **R**. That is, for any profile of actions and types (c,t) in $C \times T$, the function u_i must specify a number $u_i(c,t)$ that represents the payoff that player i would get, in some von Neumann-Morgenstern utility scale, if the players' actual types were all as in t and the players all chose their actions as specified in c.

These structures together define a Bayesian game Γ^b, so we may write

(2.2) $\Gamma^b = (N, (C_i)_{i \in N}, (T_i)_{i \in N}, (p_i)_{i \in N}, (u_i)_{i \in N})$.

Γ^b is *finite* iff the sets N and C_i and T_i (for every i) are all finite. When we study such a Bayesian game Γ^b, we assume that each player i knows the entire structure of the game (as defined above) and his own actual type in T_i and this fact is common knowledge among all the players in N.

To avoid confusion, we refer to the object of choice in a Bayesian game as an "action" rather than a "strategy." Relative to some underlying extensive-form game, an action for a player in a Bayesian game may represent a plan that specifies a move for every contingency that the player would consider possible after he has learned his type. On the other hand, a strategy would normally be thought of as a complete plan covering all contingencies that the player would consider possible, before he learns his type. Thus, a *strategy* for player i in the Bayesian game Γ^b is defined to be a function from his set of types T_i into his set of actions C_i.

As noted earlier, our simple card game becomes a game with incomplete information if we assume that player 1 already knows the color of the card when the game begins. The representation of this game in Bayesian form then has $N = \{1,2\}$, $T_1 = \{1.a,1.b\}$, $T_2 = \{2\}$, $C_1 = \{R,F\}$, and $C_2 = \{M,P\}$. The probability functions are

$$p_2(1.a|2) = .5 = p_2(1.b|2),$$

because player 2 thinks that red and black cards are equally likely and

$$p_1(2|1.a) = 1 = p_1(2|1.b),$$

because player 1 has no uncertainty about 2's type. The utility functions $(u_1(c,t), u_2(c,t))$ depend on $(c,t) = (c_1,c_2,t_1)$, as shown in Table 2.13.

For another example of a Bayesian game, consider the following bargaining game, where player 1 is the seller of some object and player

Table 2.13 Expected payoffs for all types and action profiles

$t_1 = 1.a$

	M	P
R	2,−2	1,−1
F	1,−1	1,−1

$t_1 = 1.b$

	M	P
R	−2,2	1,−1
F	−1,1	−1,1

2 is the only potential buyer. Each player knows what the object is worth to himself but thinks that its value (in dollars) to the other player may be any integer from 1 to 100, each with probability $1/100$. In the bargaining game, each player will simultaneously name a bid (in dollars) between 0 and 100 for trading the object. If the buyer's bid is greater than or equal to the seller's bid, then they will trade the object at a price equal to the average of their bids; otherwise no trade will occur. Let us assume that the players are risk neutral, so that we can identify utility payoffs with monetary profits from trading. Then this game may be formulated as a Bayesian game as follows. The set of players is $N = \{1,2\}$. For each player i, the set of his possible types is $T_i = \{1,2,\ldots,100\}$ (here we identify a player's type with his value for the object), and the set of his possible actions (bids) is $C_i = \{0,1,2,\ldots,100\}$. The probability functions are

$$p_i(t_{-i}|t_i) = 1/100, \quad \forall i \in N, \quad \forall t = (t_{-i},t_i) \in T.$$

The utility payoffs, for any c in C and any t in T, are

$$u_1(c,t) = (c_1 + c_2)/2 - t_1 \text{ if } c_2 \geq c_1,$$
$$u_2(c,t) = t_2 - (c_1 + c_2)/2 \text{ if } c_2 \geq c_1,$$
$$u_1(c,t) = 0 = u_2(c,t) \text{ if } c_2 < c_1.$$

Although it is easier to develop general notation for finite games, it is often easier to analyze examples with infinite type sets than those with large finite type sets. The only notational complication is that, in the infinite case, the probability distributions $p_i(\cdot|t_i)$ must be defined on all measurable subsets of T_{-i}, instead of just individual elements of T_{-i}. (So if R_{-i} is a subset of T_{-i}, then we let $p_i(R_{-i}|t_i)$ denote the subjective probability that i would assign to the event that the profile of others' types is in R_{-i}, if his own type were t_i.) For example, consider a modified version of the above buyer–seller game in which each player's type set expanded to include all real numbers between 0 and 100, such that $T_1 = T_2 = [0,100]$. For each player i and each t_i in T_i, let $p_i(\cdot|t_i)$ be the uniform distribution over $[0,100]$. Then for any two numbers x and y such that $0 \leq x \leq y \leq 100$, the probability that any type t_i of player i would assign to the event that the other player's type is between x and y would be

$$p_i([x,y]|t_i) = (y - x)/100.$$

(Here $[x,y]$ denotes the closed interval from x to y in the real number line.) This game has been studied by Chatterjee and Samuelson (1983).

We say that beliefs $(p_i)_{i \in N}$ in a Bayesian game are *consistent* iff there is some common prior distribution over the set of type profiles t such that each players' beliefs given his type are just the conditional probability distribution that can be computed from the prior distribution by Bayes's formula. That is (in the finite case), beliefs are consistent iff there exists some probability distribution P in $\Delta(T)$ such that,

$$p_i(t_{-i}|t_i) = \frac{P(t)}{\sum_{s_{-i} \in T_{-i}} P(s_{-i},t_i)}, \quad \forall t \in T, \quad \forall i \in N.$$

(Here (s_{-i},t_i) denotes the profile of types in T such that the i-component is t_i and all other components are as in s_{-i}. Whenever t, t_{-i}, and t_i appear together in the same formula, t is the profile of types in which the i-component is t_i and all other components are as in t_{-i}, so $t = (t_{-i},t_i)$.) For example, beliefs in our simple card game are consistent with the prior distribution

$$P(1.a, 2) = P(1.b, 2) = .5.$$

Beliefs in the finite buyer–seller game described earlier are consistent with the the prior

$$P(t) = 1/10000, \quad \forall t \in T = \{1,2, \ldots ,100\} \times \{1,2, \ldots ,100\}.$$

In the infinite version, beliefs are consistent with a uniform prior on $[0,100]^2$.

Most of the Bayesian games that have been studied in applied game theory, and all of the formal examples that appear in this book, have beliefs that are consistent with a common prior in this sense. (Indeed, the definitions in Section 2.1 implicitly imposed a consistency requirement on games in extensive form, although it would be easy to define a generalized extensive form in which a different subjective probability distribution for each player may be specified for the set of alternatives following each chance node.) One reason for this tendency to use consistent models is that consistency simplifies the definition of the model. The common prior on T determines all the probability functions, and it is simpler to specify one probability distribution than many probability functions that depend on types. Furthermore, inconsistency often seems like a strikingly unnatural feature of a model. In a consistent model, differences in beliefs among players can be explained by differences in information, whereas inconsistent beliefs involve differences of opinion that cannot be derived from any differences in observations and must be simply assumed a priori. Nevertheless, it is possible to imagine games with inconsistent beliefs. For example, in a sports match, if it is common knowledge among the coaches of two teams that each believes that his own team has a $\frac{2}{3}$ probability of winning its next game against the other team, then the coaches' beliefs cannot be consistent with a common prior. In a consistent model, it can happen that each coach believes that his team has a $\frac{2}{3}$ probability of winning, but this difference of beliefs cannot be common knowledge among the coaches (see Aumann, 1976).

Bayesian games contain both probability and utility functions, so equivalence for Bayesian games is derived from Theorem 1.2 in Chapter 1. Thus, we may say that two Bayesian games $(N, (C_i)_{i \in N}, (T_i)_{i \in N}, (p_i)_{i \in N}, (u_i)_{i \in N})$ and $(N, (C_i)_{i \in N}, (T_i)_{i \in N}, (q_i)_{i \in N}, (w_i)_{i \in N})$ are (fully) equivalent iff, for every i in N, there exist functions $A_i : T_i \to \mathbf{R}$ and $B_i : T \to \mathbf{R}$ such that, for every t_i in T_i, $A_i(t_i) > 0$ and

$$q(t_{-i} | t_i) w(c, t) = A(t_i) p(t_{-i} | t_i) u_i(c, t) + B_i(t),$$
$$\forall c \in C, \quad \forall t_{-i} \in T_{-i}.$$

That is, the Bayesian games are equivalent iff, for every possible type of every player, the two games impute probability and utility functions that are decision-theoretically equivalent in the sense of Theorem 1.2. (Notice that the multiplicative constant $A_i(t_i)$ depends on i's type alone, whereas the additive constant $B_i(t)$ can depend on the types of all players.)

Using this equivalence criterion, we find that any Bayesian game with finite type sets is equivalent to a Bayesian game with consistent beliefs. Given any game Γ^b as defined above, we can construct such an equivalent Bayesian game by letting

$$(2.3) \qquad q_i(t_{-i}|t_i) = 1/|T_{-i}| \text{ and } w_i(c,t) = |T_{-i}| \ p_i(t_{-i}|t_i)u_i(c,t)$$

for every i in N, t in T, and c in C. (Here, for any finite set X, $|X|$ denotes the number of elements in the set X.) Notice, in fact, that the types are independent and uniformly distributed in the consistent prior of the game $(N, (C_i)_{i \in N}, (T_i)_{i \in N}, (q_i)_{i \in N}, (w_i)_{i \in N})$. Thus, consistency of beliefs and independence of types cannot be a problematic assumption in our analysis as long as we consider finite Bayesian games with general utility functions.

Consistency and independence of types can be important, however, if we want to restrict our attention to utility functions with special structure. For example, we might want to assume that each player's utility payoff depends only on his own type (this is called the *private values assumption*), or we might want to assume that monetary wealth enters into each player's utility function in a simple additively separable way (as when we make an assumption of *transferable utility*). The utility functions w_i constructed in formula (2.3) may fail to satisfy these conditions, even when the u_i functions do satisfy them. Thus, consistency may be an important and useful assumption to make about Γ^b in the context of such additional assumptions about the utility functions. Furthermore, a construction using formula (2.3) is not possible when type sets are infinite, so consistency in infinite Bayesian games may also be a nontrivial assumption.

Harsanyi (1967–68), following a suggestion by R. Selten, discussed a way to represent any Bayesian game Γ^b (as defined in equation (2.2)) by a game in strategic form, which we call the *type-agent representation*. (Harsanyi called this the *Selten game* or the *posterior-lottery model*.) In the type-agent representation, there is one player or agent for every possible type of every player in the given Bayesian game. By relabeling types if necessary, we may assume without loss of generality that the sets T_i are disjoint, so that $T_i \cap T_j = \varnothing$ if $i \neq j$. Then, given the Bayesian game Γ^b as above, the set of players in the type-agent representation is

$$T^* = \bigcup_{i \in N} T_i.$$

For any i in N and any t_i in T_i, the set of strategies available to agent t_i in the type-agent representation is $D_{t_i} = C_i$. The idea is that the agent for t_i in the type-agent representation is responsible for selecting the action that player i will use in Γ^b if t_i is i's actual type. In the type-agent representation, the utility payoff to any agent t_i in T_i is defined to be the conditionally expected utility payoff to player i in Γ^b given that t_i is i's actual type. Formally, for any player i in N and any type t_i in T_i, the utility function $v_{t_i}: \times_{s \in T^*} D_s \to \mathbf{R}$ in the type-agent representation is defined so that, for any $d = (d(s))_{s \in T^*}$ in $\times_{s \in T^*} D_s$,

$$v_{t_i}(d) = \sum_{t_{-i} \in T_{-i}} p_i(t_{-i} | t_i) u_i\big((d(t_j))_{j \in N}, (t_j)_{j \in N}\big).$$

(Here, for any j in $N - i$, whenever t_j and t_{-i} appear in the same formula, t_j is the j-component of t_{-i}.) With these definitions, the type-agent representation $(T^*, (D_r)_{r \in T^*}, (v_r)_{r \in T^*})$ is indeed a game in strategic form and may be viewed as a representation of the given Bayesian game.

2.9 Modeling Games with Incomplete Information

The models discussed in this chapter give us a general framework for analyzing conflicts that arise in any economic, political, and social situation. However, there are fundamental reasons to be concerned about the possibility of accurately describing realistic situations exactly by simple game models. In particular, practical modeling difficulties often arise when players' beliefs are characterized by subjective probabilities, so the question of what one player might believe about another player's subjective probabilities becomes problematic.

For example, suppose that we change our simple card game into a "trivia quiz" game, in which the outcome depends on whether player 1 knows the correct answer to some randomly selected question (e.g., "In whose honor was America named?" or "Who wrote *Theory of Games and Economic Behavior?*") rather than on the color of a card. That is, the sequence of play is (1) both players put $1 in the pot, (2) a question is drawn at random from a large stack and is announced to both players, (3) player 1 can raise $1 or fold, (4) if player 1 raises then player 2 must meet (adding $1 more of her own) or pass, and then (5) player 1 attempts to answer the question. Player 1 wins the money if he answers correctly or if he raises and 2 passes. Player 2 wins the money if she does not pass and player 1 answers incorrectly.

In this game, after the question has been announced, there is basic uncertainty about whether player 1 knows the answer to the announced question. Bayesian decision theory tells us that player 2 must be able to describe her beliefs about this basic uncertainty by some number Q, between 0 and 1, that represents her subjective probability of the event that player 1 knows the answer. In the simple card game, the probability of a red card at the beginning of the game was objectively ½, so it was reasonable to assume that this number was common knowledge. In this trivia quiz, however, it is reasonable to suppose that player 1 may have some uncertainty about 2's subjective probability Q. Again, Bayesian decision theory tells us that player 1 should be able to describe his beliefs about Q by some subjective probability distribution over the interval from 0 to 1. Furthermore, if we do not assume that 1's beliefs about Q are common knowledge, then player 2 must be able to describe her beliefs about 1's beliefs about Q by some probability distribution over the set of all probability distributions over the interval from 0 to 1. If in turn these beliefs are not common knowledge, then player 2's type, which includes a specification of everything that she knows that is not common knowledge, must include a specification of a distribution over the set of distributions over the interval from 0 to 1, which is a point in a very complicated infinite-dimensional vector space!

It might be hoped that these beliefs about beliefs about beliefs might be irrelevant for the fundamental problem of predicting or explaining the players' behavior, but (as the fable in Section 2.7 illustrated) we cannot count on such irrelevance. When player 1 does not know the answer, he would want to raise (as a bluff) if he thought that the probability of player 2 meeting was less than ⅓. It is reasonable to expect that, other things being equal, player 2 should be more likely to meet a raise if Q is lower. So the more probability that player 1 puts on low values of Q, the less likely he is to raise if he does not know the answer. So if 2 thinks that 1 puts high probability on low values of Q, then 2 may take a raise by 1 as strong evidence that 1 does know the answer, which should in turn decrease her willingness to meet a raise. (Notice that Q is 2's prior probability that 1 knows the answer, assessed before she learns whether 1 will raise. Her beliefs about 1 after he raises depend by Bayes's formula both on Q and on the probability that 1 would raise if he did not know the answer.) Thus, 2's beliefs about 1's beliefs about 2's beliefs about whether 1 knows the answer may indeed be an important factor in determining whether 2 would meet a raise in this game!

In general, we have the paradoxical result that, the less common knowledge is, the larger the sets of possible types must be, because a player's type is a summary of everything that he knows that is not common knowledge. But these sets of possible types, as a part of the structure of the Bayesian game, are supposed to be common knowledge among players. Thus, to describe a situation in which many individuals have substantial uncertainty about one another's information and beliefs, we may have to develop a very complicated Bayesian-game model with large type sets and assume that this model is common knowledge among the players.

This result begs the question, Is it possible to construct a situation for which there are no sets of types large enough to contain all the private information that players are supposed to have, so that no Bayesian game could represent this situation? This question was considered by Mertens and Zamir (1985), who built on the seminal work of Harsanyi (1967–68). Mertens and Zamir showed, under some technical assumptions, that no such counterexample to the generality of the Bayesian game model can be constructed, because a *universal belief space* can be constructed that is always big enough to serve as the set of types for each player. Unfortunately, this universal belief space is an enormously complicated mathematical object. So practical analysis requires that there be enough common knowledge to let us use some smaller and more tractable sets as the sets of types in our Bayesian game. Thus, although constructing an accurate model for any given situation may be extremely difficult, we can at least be confident that no one will ever be able to prove that some specific conflict situation cannot be described by any sufficiently complicated Bayesian game.

To understand this result, suppose that there is some specific conflict situation that we want to represent by a game in Bayesian form. Suppose that we can interview the various players or individuals who are involved in this situation, and that they will answer our questions honestly, as long as we make sure that we ask meaningful questions. (Let us assume that they are willing to be honest because they accept us as disinterested scientific observers who will not leak any information back to anyone else.)

A player's *type* in this game must account for all the relevant private information (information that is not common knowledge) that he now has about the game. We define an *action* for a player in this game to be any plan that specifies a move for the player in every possible future contingency, given his current type. So we can suppose that all players simultaneously choose actions now (when each player knows his own type) and they then make no further decisions in the game.

There are several basic issues in a conflict situation about which players might have different information: How many players are actually in the game? What moves or actions are feasible for each player? How will the outcome depend on the actions chosen by the players? And what are the players' preferences over the set of possible outcomes? Harsanyi (1967–68) argued that all of these issues can be modeled in a unified way. Uncertainty about whether a player is "in the game" can be converted into uncertainty about his set of feasible actions, by allowing him only one feasible action ("nonparticipation") when he is supposed to be "out of the game." Uncertainty about whether a particular action is feasible for player i can in turn be converted into uncertainty about how outcomes depend on actions, by saying that player i will get some very bad outcomes if he uses an action that is supposed to be infeasible. Alternatively, whenever an action c_i is supposed to be infeasible for player i, we can simply identify some feasible other action d_i and suppose that the outcome from using c_i in our game model is the same as for d_i (so c_i in our game model is reinterpreted as "Do c_i if you can, otherwise do d_i"). Uncertainty about how outcomes depend on actions and uncertainty about preferences over outcomes can be unified by modeling each player's utility as a function directly from the set of profiles of players' actions to the set of possible utility payoffs (representing the composition of the function from actions to outcomes with the function from outcomes to utility payoffs). Thus, we can model all the basic uncertainty in the game as uncertainty about how utility payoffs depend on profiles of actions. This uncertainty can be represented formally by introducing an unknown parameter $\tilde{\theta}$ into the utility functions.

So let N denote the set of all players whom anyone might consider to be involved in this situation. For each i in N, let C_i denote the set of all possible actions that anyone might believe to be feasible for player i. Let Θ denote the set of possible values for the parameter $\tilde{\theta}$. We call Θ the domain of basic uncertainty in the game. Writing $C = \times_{j \in N} C_j$, we can describe the dependence of each player i's utility payoff on the players' actions and the unknown parameter $\tilde{\theta}$ by a function $w_i{:}C \times \Theta \to \mathbf{R}$.

In some situations, the domain of basic uncertainty may be straightforward to define. In our simple card game, $\tilde{\theta}$ could be identified with the color of the randomly selected card, so we could let $\Theta = \{$Red, Black$\}$. In the trivia quiz, the basic uncertainty would be about whether player 1 knows the answer, so we could let $\Theta = \{$Knows answer, Does not know answer$\}$. In a game of bargaining between a seller and buyer of some object, the basic uncertainty could be about the value of the object to each player, so Θ could be a subset of $\mathbf{R}^2$, where the first component of any vector in Θ represents the value (say, in dollars) of the object to the seller and the second component represents the value of the object to the buyer.

In general, it is always possible to identify Θ with a subset of $\mathbf{R}^{C \times N}$, because the only role of $\tilde{\theta}$ in our analysis is to identify how each player's expected utility should depend on C. If we make this identification, then we can simply define each w_i function so that $w_i(c,\theta) = \theta_{i,c}$ for every i in N, c in C, and θ in Θ. Furthermore, if we assume that C is finite, then there is no loss of generality in assuming that players have bounded utility functions, so Θ is a closed and bounded (compact) subset of a finite-dimensional vector space.

To continue the construction of Mertens and Zamir's universal belief space, we need some more sophisticated mathematics (topology and measure theory as in Royden, 1968, and Kolmogorov and Fomin, 1970; see also Section 3.13 in Chapter 3). Readers with less mathematics are encouraged to skim or omit this construction (through formula 2.4 below), as nothing later in the book will depend on it.

Given any metric space Z, let $\Delta(Z)$ denote the set of all probability distributions on Z that are defined on the set of Borel-measurable subsets of Z. We give Z the weak topology, which is defined so that, for every bounded continuous function $f{:}Z \to \mathbf{R}$, the integral $\int_{x \in Z} f(x)p(dx)$ is a continuous function of p in $\Delta(Z)$. If Z is compact, then $\Delta(Z)$ is also compact and metrizable (with the Prohorov metric). Billingsley (1968) gives a full development of this result.

The structures $(N, (C_i)_{i \in N}, \Theta, (w_i)_{i \in N})$ are not sufficient to describe the situation in question, because they do not tell us what the players' beliefs or information about the unknown $\tilde{\theta}$ might be. Bayesian decision theory tells us that each player must have a subjective probability distribution over Θ, which we shall call his *first-order beliefs*. Let

$$B_i^1 = \Delta(\Theta),$$

so that B_i^1 contains all possible first-order beliefs of player i.

In a game, a player's optimal decision will generally depend on what he expects the other players to do. And what he expects the other players to do will depend on what he thinks they believe. Thus, we must now ask, What does player i think are the other players' first-order beliefs? Furthermore, to describe whether player i believes that other players' beliefs are accurate or inaccurate, we must ask, What does player i believe about the relationship between other players' first-order beliefs and the true value of $\tilde{\theta}$? Bayesian decision theory implies that each player i's beliefs about these questions must be describable by some subjective probability distribution over $\Theta \times (\times_{j \in N-i} B_j^1)$, which we may call his *second-order beliefs*. Let

$$B_i^2 = \Delta \left(\Theta \times \left(\underset{j \in N-i}{\times} B_j^1 \right) \right),$$

so B_i^2 contains all possible second-order beliefs of player i.

Defined in this way, i's second-order beliefs implicitly determine his first-order beliefs, which are just the marginal distribution over Θ. That is, for any second-order beliefs β_i^2 in B_i^2, the first-order beliefs that correspond to β_i^2 may be denoted $\phi_i^2(\beta_i^2)$, where

$$(\phi_i^2(\beta_i^2))(\Psi) = \beta_i^2 \left(\Psi \times \left(\underset{j \in N-i}{\times} B_j^1 \right) \right), \quad \forall \Psi \subseteq \Theta.$$

Now we can inductively define, for each k in $\{3,4,5,\ldots\}$, a player's *k-order beliefs* to be his beliefs about the other players' $(k-1)$-order beliefs and about their relationship to the basic unknown $\tilde{\theta}$. We inductively define

$$B_i^k = \Delta \left(\Theta \times \left(\underset{j \in N-i}{\times} B_j^{k-1} \right) \right)$$

for each player i and each positive integer k. Then player i's k-order beliefs can be described by a point in B_i^k. As before, player i's k-order beliefs determine his $(k-1)$-order beliefs (and so, by induction, his beliefs of all lower orders), by the function $\phi_i^{k-1}:B_i^k \to B_i^{k-1}$ that is inductively defined so that, for any β_i^k in B_i^k and any Ψ that is a Borel-measurable subset of $\Theta \times (\times_{j \in N-i} B_j^{k-2})$,

$$(\phi_i^{k-1}(\beta_i^k))(\Psi) = \beta_i^k(\{(\theta,(\beta_j^{k-1})_{j \in N-i}) \,|\, (\theta,(\phi_j^{k-2}(\beta_j^{k-1}))_{j \in N-i}) \in \Psi\}).$$

(We use here the assumption that it is common knowledge that every player's beliefs satisfy the laws of probability.)

Then the *universal belief space* for player i is

$$B_i^\infty = \{\beta_i = (\beta_i^1, \beta_i^2, \ldots) \in \underset{k=1}{\overset{\infty}{\times}} B_i^k \,|\, \beta_i^{k-1} = \phi_i^{k-1}(\beta_i^k), \quad \forall k \geq 2\}.$$

Under the assumptions that the domain of basic uncertainty Θ is a compact metric space and the set of players N is finite, B_i^∞ is also a compact set (with the product topology).

Any β_i in B_i^∞ induces a probability distribution over $\Theta \times (\times_{j \in N-i} B_j^\infty)$, and we let $q_i(\cdot | \beta_i)$ denote this distribution. If Ψ is any closed subset of $\Theta \times (\times_{j \in N-i} B_j^\infty)$, then

$$q_i(\Psi | \beta_i) = \lim_{k \to \infty} \beta_i^k(\{(\theta, (\beta_j^{k-1})_{j \in N-i}) \,|\, (\theta, (\beta_j)_{j \in N-i}) \in \Psi\}).$$

In fact, Mertens and Zamir have shown that $q_i(\cdot | \cdot)$ is a homeomorphism between B_i^∞ and $\Delta(\Theta \times (\times_{j \in N-i} B_j^\infty))$. That is, player i's universal belief space includes all possible (Borel-measurable) beliefs about the basic uncertainty in Θ and the other players' beliefs.

Notice now that the random variable $\tilde{\theta}$ cannot directly influence any player i's behavior in the game, except to the extent that he may have information about $\tilde{\theta}$ that is reflected in his actual beliefs, which we denote $\tilde{\beta}_i$. So we can integrate the basic uncertainty variable $\tilde{\theta}$ out of the probability and utility functions without losing any structures relevant to predicting the player's behavior. For any β_i in B_i^∞, let $p_i(\cdot | \beta_i)$ be the marginal probability distribution of $q_i(\cdot | \beta_i)$ on $\times_{j \in N-i} B_j^\infty$. For any profile of beliefs $\beta = (\beta_j)_{j \in N}$ in $\times_{j \in N} B_j^\infty$, let $\bar{q}_i(\cdot | \beta)$ denote the conditional distribution on Θ that player i would derive (by Bayes's formula) from the prior distribution $q_i(\cdot | \beta_i)$ on $\Theta \times (\times_{j \in N-i} B_j^\infty)$ if i learned that every other player's actual beliefs were as specified in the profile β. For any profile of actions c in C and any profile $\beta = (\beta_j)_{j \in N}$ in $\times_{j \in N} B_j^\infty$, let $u_i(c, \beta)$ denote the conditionally expected value of $w_i(c, \tilde{\theta})$, using the distribution $\bar{q}_i(\cdot | \beta)$ on Θ. That is,

$$p_i(\Psi | \beta_i) = q_i(\Theta \times \Psi | \beta_i), \quad \forall \Psi \subseteq \underset{j \in N-i}{\times} B_j^\infty,$$

$$u_i(c, \beta) = \int_{\theta \in \Theta} w_i(c, \theta) \bar{q}_i(d\theta | \beta).$$

Thus, at last we get the universal Bayesian game,

(2.4) $(N, (C_i)_{i \in N}, (B_i^\infty)_{i \in N}, (p_i)_{i \in N}, (u_i)_{i \in N})$.

By construction, each type set B_i^∞ is large enough to include all possible states of i's beliefs about the basic unknown parameter $\tilde{\theta}$ in Θ and about all other players' information or beliefs that are relevant to this basic unknown parameter.

As noted earlier, these universal belief spaces are so large that, in practice, we have little hope of analyzing the situation unless some other Bayesian-game model with smaller type spaces can be found that also describes the conflict situation. These results about the existence of universal belief spaces and universal Bayesian games constitute a theoretical proof that there should exist a Bayesian-game model that describes all the relevant structure of information and incentives in any given situation with incomplete information, so we can be confident that there is nothing intrinsically restrictive about the structure of the Bayesian game. We can be sure that no one will ever be able to take a real-life conflict situation and prove that it would be impossible to describe by any Bayesian-game model.

We say that $\times_{i \in N} R_i$ is a *belief-closed* subset of $\times_{i \in N} B_i^\infty$ iff, for every player i, $R_i \subseteq B_i^\infty$ and, for every β_i in R_i, the probability distribution $p_i(\cdot | \beta_i)$ assigns probability one to the set $\times_{j \in N-i} R_j$. If the profile of players' types is in the belief-closed subset $\times_{i \in N} R_i$, then this fact can be made common knowledge without affecting any player's beliefs. Thus, belief-closed subsets correspond to events that are effectively common knowledge when they occur.

To get a Bayesian game that is tractable for analysis, we must assume that there is enough common knowledge about the structure of the game so that each player's private information can be represented by a variable that ranges over a tractably small set of possible types. In effect, we must assume that players' types can be represented by points in some small belief-closed subset of the players' universal belief spaces.

However, for practical modeling, it is really not very helpful to think about types as points in universal belief space. The way that tractable models are usually constructed in practice is to assume that each player's private information and beliefs depend only on some random variable that he observes and that has a suitably small range, so this random variable may then be identified with his type. If we also assume that these random variables and the basic unknown $\tilde{\theta}$ are drawn from a joint prior distribution that is common knowledge, then Bayes's formula will enable us to derive each player's beliefs of all orders, as a function of his type.

For example, to get a tractable model of the trivia quiz, we might suppose that each player either is sure that he knows the correct answer or is sure that he does not know and has no chance of even guessing the answer correctly. We might then assume that each player's private information and beliefs depend only on whether he knows the answer or not. Notice that, if player 2 knows the answer, then she might naturally think it relatively more likely that many other people also know the answer, and so she might be relatively more pessimistic about winning the bet with player 1.

So we might let $T_1 = \{a_1, b_1\}$ and $T_2 = \{a_2, b_2\}$, where, for each i, a_i is the type that knows the answer and b_i is the type that does not know the answer. To generate a consistent prior distribution over these random variables, we might suppose that the actual proportion of people who know the answer (in the population that includes players 1 and 2) is a random variable π drawn from a uniform distribution on the interval from 0 to 1. From this assumption, we can derive the following prior distribution

$$P(a_1, a_2) = \frac{1}{3}, \quad P(a_1, b_2) = \frac{1}{6}, \quad P(b_1, a_2) = \frac{1}{6}, \quad P(b_1, b_2) = \frac{1}{3}.$$

(For example, the probability that both know the answer is $\int_0^1 \pi^2 \, d\pi = \frac{1}{3}$.) In effect, each player can make Bayesian inferences about this unknown proportion on the basis of his own type, as a statistical sample of size 1.

In this model, the basic unknown $\tilde{\theta}$ that affects payoffs is just player 1's type itself, so player 1's first-order beliefs always put probability 1 on the true value of $\tilde{\theta}$. Player 2's first-order beliefs put probability either $\frac{2}{3}$ or $\frac{1}{3}$ on the event that $\tilde{\theta} = a_1$, depending on whether player 2 knows the answer or not. If 1's type is a_1, then his second-order beliefs put probability $\frac{2}{3}$ on the event that 2's subjective probability of a_1 is $\frac{2}{3}$ (because her type is a_2), and put probability $\frac{1}{3}$ on the event that 2's subjective probability of a_1 is $\frac{1}{3}$ (because her type is b_2). On the other hand, if 1's type is b_1, then his second-order beliefs put probability $\frac{1}{3}$ on the event that 2's subjective probability of a_1 is $\frac{2}{3}$, and put probability $\frac{2}{3}$ on the event that 2's subjective probability of a_1 is $\frac{1}{3}$. These and all other beliefs of all orders can be computed by Bayes's formula from the prior distribution P. In effect, assuming that the players' beliefs depend on random variables with small ranges is equivalent to assuming that, for every k, the set of possible k-order beliefs for any player i is in a small subset of $\mathbf{B}_i^k$.

However, there is no need to actually compute all these higher order beliefs for the various types in T_1 and T_2. We only need to compute the probability functions $p_1 : T_1 \to \Delta(T_2)$ and $p_2 : T_2 \to \Delta(T_1)$ that are part of the structure of the Bayesian game. For this example, these probabilities are

$$p_1(a_2 | a_1) = \tfrac{2}{3}, \quad p_1(b_2 | a_1) = \tfrac{1}{3},$$
$$p_1(a_2 | b_1) = \tfrac{1}{3}, \quad p_1(b_2 | b_1) = \tfrac{2}{3},$$
$$p_2(a_1 | a_2) = \tfrac{2}{3}, \quad p_2(b_1 | a_2) = \tfrac{1}{3},$$
$$p_2(a_1 | b_2) = \tfrac{1}{3}, \quad p_2(b_1 | b_2) = \tfrac{2}{3}.$$

Of course, we must be prepared to question the assumption that this particular model is an accurate representation of the trivia quiz, when it is actually played by two specific people. For example, it might be considered important to expand the T_i sets to include types that are not sure about the answer but can make good guesses that have a positive probability of being correct. However, this kind of sensitivity analysis is needed in any area of applied mathematical modeling. Real-life situations are almost always more complicated than any mathematical model that we can work with, so there is a trade-off between analytical tractability and modeling accuracy. As in any analytical approach to real-life problems, the best that we can hope for is to have a class of models sufficiently rich and flexible that, if anyone objects that our model has neglected some important aspect of the situation we are trying to analyze, we can generate a more complicated extension of our model that takes this aspect into account.

Exercises

Exercise 2.1. The new widget production process that firm 1 is developing is equally likely to have high or low costs. Firm 1 will learn whether the production process has high costs or low costs at the beginning of next quarter. Then firm 1 can choose whether to build a new plant or not. Firm 2 will not be able to observe the costs of firm 1's new process, but firm 2 will be able to observe whether firm 1 builds a new plant or not. Firm 2 will subsequently decide whether to enter the widget market against firm 1 or not. Firm 2 will make $2 million (in present discounted value of long-run profits) from entering the widget market if firm 1's process has high costs, but firm 2 will lose $4 million from entering the widget market if firm 1's process has low costs. Lower costs in the new process will increase firm 1's profits by $4 million. Building a new plant would add $2 million more to firm 1's profits if the new process has low costs (because conversion to the new process would be much easier in a new plant), but building a new plant would subtract $4 million from firm 1's profits if the new process has high costs. In any event (whether the new process has high or low costs, whether firm 1 builds a new plant or not), firm 2's entry into the widget market would lower firm 1's profits by $6 million. (Both firms are risk neutral, so we may identify utility with monetary profit.)

a. Describe this game in extensive form.

b. Construct the normal representation of this game in strategic form.

c. Analyze the normal representation by iterative elimination of weakly dominated strategies.

d. Show the multiagent representation of the extensive-form game. Does any agent have a dominated strategy in the multiagent representation?

e. Suppose now that the game starts at the point in time just after firm 1 learns whether its costs are high or low (but before firm 1 decides whether to build the new plant). Write down a Bayesian game with incomplete information to describe this game.

f. Write down the type-agent representation of the Bayesian game from (e). What is the set of players in this strategic-form game?

Exercise 2.2. Recall the simple card game introduced in Section 2.1. Out of 52 cards in the deck, there are 20 cards that are "ten or higher" (the tens, jacks, queens, kings, and aces). Suppose that the rules were changed so that player 1 wins the money if he has a card that is ten or higher or (as before) if player 2 passes following a raise. As before, player 1 sees the card before he decides whether to raise, but now the majority of the cards are favorable to player 2.

a. Model this game in extensive form and construct the normal representation of this game.

b. If you were going to play this game, would you prefer to take the role of player 1 or player 2? (Just express your intuition. For a formal analysis, see Exercise 4.7—in Chapter 4.)

Exercise 2.3. Suppose that the simple card game from Section 2.1 is changed by allowing player 1, after he looks at his card, to either raise $1.00, raise $0.75, or pass. If player 1 raises, then player 2 knows the amount that player 1 has added and must choose either to meet the raise by putting in the same additional amount or to pass. As before, player 1 wins if he has a red card or if player 2 passes after a raise.

    a. Model this game in extensive form.

    b. Show the normal representation of this game.

    c. Identify all strategies that can be iteratively eliminated by weak domination.

Exercise 2.4. Construct the normal representation of the game shown in Figure 2.8.

Exercise 2.5. Construct the normal representation of the game shown in Figure 4.10 (in Chapter 4).

Exercise 2.6. Consider the strategic-form game presented in Table 2.14. Analyze this game by iterative elimination of weakly dominated strategies.

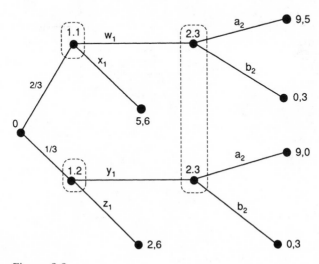

Figure 2.8

Exercise 2.7. Show that, if two Bayesian games are equivalent, then their type-agent representations are fully equivalent.

Exercise 2.8. (A voting game from Moulin, 1983.) Players 1, 2, and 3 are voting in a committee to choose among three options, called α, β, and γ. First, each player submits a secret vote for one of the three options. Then the votes are opened. If any option gets two votes, then it is the outcome of the game. Otherwise, if there is a (necessarily three-way) tie, then player 1 (who is the chairman of the committee) will choose the outcome. The players' payoffs depend on the outcome and are shown in Table 2.15. Analyze this game by iterative elimination of weakly dominated strategies.

Exercise 2.9. A nonnegative integer $\tilde{X}$ is randomly determined according to a geometric distribution with parameter 0.9. That is, for any nonnegative integer k, and the probability that $\tilde{X}$ equals k is $P(\tilde{X} = k) = 0.9^k \times 0.1$. $\tilde{Y}_1$ and $\tilde{Y}_2$ are random variables that depend on $\tilde{X}$ so that

$$(\tilde{Y}_1, \tilde{Y}_2) = (\tilde{X}, \tilde{X} + 1) \text{ with probability } \tfrac{1}{2},$$
$$= (\tilde{X} + 1, \tilde{X}) \text{ with probability } \tfrac{1}{2}.$$

Table 2.14 A game in strategic form

		C_2	
C_1	x_2	y_2	z_2
x_1	2,4	9,4	1,3
y_1	3,9	8,1	2,9
z_1	5,5	6,6	7,6

Table 2.15 Payoffs for all outcomes in a voting game

		Player	
Option	1	2	3
α	8	0	4
β	4	8	0
γ	0	4	8

Player 1 observes only $\tilde{Y}_1$, and player 2 observes only $\tilde{Y}_2$. That is, one of the players observes $\tilde{X}$ and the other player observes $\tilde{X} + 1$, and each player has an equal chance of being the one to observe $\tilde{X}$.

a. Show that, if a player i observes that Y_i is not 0, then he thinks that the probability that he has observed $\tilde{X}$ itself is less than $\frac{1}{2}$. That is, for any k in $\{1,2,3, \ldots\}$, show that $P(\tilde{X} = \tilde{Y}_i | \tilde{Y}_i = k) < \frac{1}{2}$.

b. The player i who sees the higher number $\tilde{Y}_i$ will win \$1 from the other player. Consider the statement

> Both players know that . . . both players know that each player thinks that his probability of winning the dollar is strictly greater than $\frac{1}{2}$.

If $\tilde{X}$ actually equals 5, but it is common knowledge among the players that each player i has only observed his own $\tilde{Y}_i$, then what would be the maximum number of times that the phrase "both players know that" can be repeated in this sentence such that this sentence will be true?

c. If $\tilde{Y}_1$ actually equals 5, but it is common knowledge among the players that each player i has only observed his own $\tilde{Y}_i$, then what is the smallest event (the event with smallest prior probability) that is common knowledge among the players?

3

Equilibria of Strategic-Form Games

3.1 Domination and Rationalizability

Of the three different forms for mathematical games that were introduced in the preceding chapter, the simplest conceptually is the strategic form. Furthermore, we have seen that any game in extensive form or Bayesian form can also be represented in strategic form (via the normal, multiagent, or type-agent representation). Thus, it is natural for us to begin studying solution concepts for games in strategic form.

We can denote any strategic-form game Γ as

$$\Gamma = (N, (C_i)_{i \in N}, (u_i)_{i \in N}),$$

where N is the set of players, C_i is the set of strategies for player i, and $u_i : C \to \mathbf{R}$ is the utility payoff function for player i. Here C denotes the set of all possible combinations or *profiles* of strategies that may be chosen by the various players, when each player i chooses one of his strategies in C_i; that is,

$$C = \underset{i \in N}{\times} C_i.$$

Unless otherwise specified, we shall generally assume that Γ is a finite game.

The simplest kind of solution concept is one that specifies the set of strategies that each player might reasonably be expected to use, without making any attempt to assess the probability of the various strategies. That is, such a solution would specify, for each player i, a set D_i that is a nonempty subset of C_i and is to be interpreted as the set of strategies that player i might actually choose.

Recall our basic assumption that each player is rational, in the sense that his goal is to maximize his expected utility payoff, and is intelligent, in the sense that he understands the game at least as well as we do. So if every player j is expected to choose a strategy in D_j, then every player i should understand this fact and use some strategy that is a best response to some probability distribution over D_{-i}, where

$$D_{-i} = \underset{j \in N-i}{\times} D_j.$$

Let $G_i(D_{-i})$ denote the set of all strategies that are such best responses; that is, $d_i \in G_i(D_{-i})$ iff there exists some η in $\Delta(D_{-i})$ such that

$$d_i \in \underset{c_i \in C_i}{\operatorname{argmax}} \sum_{d_{-i} \in D_{-i}} \eta(d_{-i}) u_i(d_{-i}, c_i),$$

(Recall that, for any set Z, $\Delta(Z)$ denotes the set of probability distributions over Z, and (d_{-i}, c_i) denotes the strategy profile in which the i-component is c_i and all other components are as in $d_{-i} = (d_j)_{j \in N-i}$.) So if player i knows that no player j would ever use a strategy outside of D_j, then player i would never use a strategy outside of $G(D_{-i})$. Thus, our solution concept should satisfy

(3.1) $D_i \subseteq G_i(D_{-i}), \quad \forall i \in N.$

As in Section 2.5, let $C_i^{(\infty)}$ denote the set of all of player i's strategies that remain after iterative elimination of strongly dominated strategies. It can be shown, as a corollary of Theorem 1.6, that

$$C_i^{(\infty)} = G_i \left(\underset{j \in N-i}{\times} C_j^{(\infty)} \right),$$

because each iteratively undominated strategy is a best response to some probability distribution over the set of iteratively undominated strategies (or else there would be more dominated strategies to eliminate). Furthermore, condition (3.1) implies that each set D_i must be a subset of $C_i^{(\infty)}$. (The proof is to show, by induction in k, that $D_i \subseteq C_i^{(k)}$ for every player i and every positive integer k.) So, when all players are rational and intelligent, no player can be expected to use any strategy that is iteratively eliminated by strong domination. Thus, our first and weakest solution concept predicts only that the outcome of the game should be some profile of iteratively undominated strategies in $\times_{i \in N} C_i^{(\infty)}$.

We have seen examples like Table 2.2 and Table 2.9 where iterative strong domination leads to us to predict a unique strategy profile. In general, however, iterative strong domination is an extremely weak solution concept. For the simple card game (Table 2.1), iterative strong domination only tells us that player 1 should not use the strategy Ff (always fold). It cannot even rule out the strategy Fr, because player 1 would be willing to fold with a winning card if he thought that player 2 would surely pass.

One way to eliminate more strategies is to do iterative elimination of weakly dominated strategies. As we saw in Section 2.5, iterative elimination of weakly dominated strategies enables us to eliminate Fr in Table 2.1 and to eliminate all strategies except B and Lr in Table 2.3. However, the order in which weakly dominated strategies are eliminated may matter, in general. To avoid this order problem, Samuelson (1989) suggested that we might instead look for the largest sets $(D_i)_{i \in N}$ such that, for each player i, D_i is the set of all strategies in C_i that would not be weakly dominated for player i if he knew that the other players would only use strategy combinations in $\times_{j \in N - i} D_j$. However, Samuelson showed that there exist games for which no such sets exist. One such game is shown in Table 3.1. (If our sets exclude y_1, for this example, then they must exclude y_2, but then y_1 cannot be excluded. If our sets include y_1, then they must include y_2, but then y_1 must be excluded.)

In Section 2.2, we argued that there should be no loss of generality in assuming that all players choose their strategies independently, because any possibilities for communication and coordination could be built into the definition of a strategy. Thus, we can suppose that the players' strategies are independent random variables. So, if our solution concept tells us that each player j will choose his strategy in the set D_j, and if player i is intelligent, then he should choose a strategy that is a best response for him when the other players randomize independently

Table 3.1   A game in strategic form

C_1	C_2	
	x_2	y_2
x_1	1,1	1,0
y_1	1,0	0,1

over their D_j sets. Let $H_i(D_{-i})$ denote the set of such best responses. That is, d_i is in $H_i(D_{-i})$ iff there exist probability distributions $(\sigma_j)_{j \in N-i}$ such that

$$\sigma_j \in \Delta(D_j), \quad \forall j \in N - i,$$

and

$$d_i \in \underset{c_i \in C_i}{\operatorname{argmax}} \sum_{d_{-i} \in D_{-i}} \left(\prod_{j \in N-i} \sigma_j(d_j) \right) u_i(d_{-i}, c_i).$$

So if all players are intelligent, in the sense that they understand the game and our solution concept $(D_j)_{j \in N}$ as well as we do, then our solution concept should satisfy

(3.2) $D_i \subseteq H_i(D_{-i}), \quad \forall i \in N.$

Bernheim (1984) and Pearce (1984) have shown that there exist sets $(D_j^*)_{j \in N}$ that satisfy (3.2) such that, for any other $(D_j)_{j \in N}$ that satisfies (3.2),

$$D_i \subseteq D_i^*, \quad \forall i \in N.$$

The strategies in D_i^* are called *rationalizable strategies*. Any rationalizable strategy is a best response to some combination of independent randomizations over rationalizable strategies by the other players; that is,

$$D_i^* = H_i(D_{-i}^*), \quad \forall i \in N.$$

It is straightforward to check that $D_i^* \subseteq C_i^{(\infty)}$ in general and that $D_i^* = C_i^{(\infty)}$ for two-player games.

3.2 Nash Equilibrium

Given any strategic-form game $\Gamma = (N, (C_i)_{i \in N}, (u_i)_{i \in N})$, a *randomized strategy* for any player i is a probability distribution over C_i. We let $\Delta(C_i)$ denote the set of all possible randomized strategies for player i. To emphasize the distinction from randomized strategies, the strategies in C_i can be called *pure strategies*.

A *randomized-strategy profile* is any vector that specifies one randomized strategy for each player; so the set of all randomized-strategy profiles is $\times_{i \in N} \Delta(C_i)$. That is, σ is a randomized-strategy profile in $\times_{i \in N} \Delta(C_i)$ iff, for each player i and each pure strategy c_i in C_i, σ specifies a nonnegative real number $\sigma_i(c_i)$, representing the probability that player i would choose c_i, such that

$$\sum_{d_i \in C_i} \sigma_i(d_i) = 1, \quad \forall i \in N.$$

We can write $\sigma = (\sigma_i)_{i \in N}$, where $\sigma_i = (\sigma_i(c_i))_{c_i \in C_i}$ for each i. If the players choose their pure strategies independently, according to the randomized-strategy profile σ, then the probability that they will choose the pure-strategy profile $c = (c_i)_{i \in N}$ is $\prod_{i \in N} \sigma_i(c_i)$, the multiplicative product of the individual-strategy probabilities.

Bayesian decision theory tells us that, although we may be uncertain about what profile of pure strategies will be chosen by the players when Γ is played, there should exist some probability distribution over the set of pure-strategy profiles $C = \times_{i \in N} C_i$ that quantitatively expresses our beliefs about the players' strategy choices. Furthermore, following the argument in Section 2.2, we can assume that the players choose their strategies independently. Thus, our beliefs about the game should correspond to some randomized-strategy profile σ in $\times_{i \in N} \Delta(C_i)$.

For any randomized-strategy profile σ, let $u_i(\sigma)$ denote the expected payoff that player i would get when the players independently choose their pure strategies according to σ. That is,

$$u_i(\sigma) = \sum_{c \in C} \left(\prod_{j \in N} \sigma_j(c_j) \right) u_i(c), \quad \forall i \in N.$$

For any τ_i in $\Delta(C_i)$, we let (σ_{-i}, τ_i) denote the randomized-strategy profile in which the i-component is τ_i and all other components are as in σ. Thus,

$$u_i(\sigma_{-i}, \tau_i) = \sum_{c \in C} \left(\prod_{j \in N-i} \sigma_j(c_j) \right) \tau_i(c_i) u_i(c).$$

Using notation introduced in Section 1.2 (Equation 1.2), we let $[d_i]$ denote the randomized strategy in $\Delta(C_i)$ that puts probability 1 on the pure strategy c_i. Thus, using standard linear algebra notation, we may write

$$\sigma_i = \sum_{c_i \in C_i} \sigma_i(c_i)[c_i].$$

If player i used the pure strategy d_i, while all other players behaved independently according to the randomized-strategy profile σ, then player i's expected payoff would be

$$u_i(\sigma_{-i},[d_i]) = \sum_{c_{-i} \in C_{-i}} \left(\prod_{j \in N-i} \sigma_j(c_j) \right) u_i(c_{-i},d_i),$$

where $C_{-i} = \times_{j \in N-i} C_j$.

Now suppose that our beliefs are well founded and that all players are intelligent enough to share these beliefs. Then each player i would want to choose the pure strategies that maximize his expected payoff, and there should be zero probability of his choosing any strategy that does not achieve this maximum. That is,

(3.3)   if $\sigma_i(c_i) > 0$, then $c_i \in \underset{d_i \in C_i}{\mathrm{argmax}}\ u_i(\sigma_{-i},[d_i])$.

A randomized-strategy profile σ is a *(Nash) equilibrium* of Γ (or a *strategic equilibrium*) iff it satisfies this condition (3.3) for every player i and every c_i in C_i (see Nash, 1951).

Thus, a randomized-strategy profile is a Nash equilibrium iff no player could increase his expected payoff by unilaterally deviating from the prediction of the randomized-strategy profile. That is, σ is a Nash equilibrium of Γ iff

(3.4) $u_i(\sigma) \geq u_i(\sigma_{-i},\tau_i), \quad \forall i \in N, \quad \forall \tau_i \in \Delta(C_i)$.

The fact that condition (3.3) (for all i and c_i) is equivalent to condition (3.4) is a consequence of the following useful lemma.

LEMMA 3.1. *For any σ in $\times_{j \in N} \Delta(C_j)$ and any player i in N,*

$$\max_{d_i \in C_i} u_i(\sigma_{-i},[d_i]) = \max_{\tau_i \in \Delta(C_i)} u_i(\sigma_{-i},\tau_i).$$

Furthermore, $\rho_i \in \mathrm{argmax}_{\tau_i \in \Delta(C_i)}\ u_i(\sigma_{-i},\tau_i)$ if and only if $\rho_i(c_i) = 0$ for every c_i such that $c_i \notin \mathrm{argmax}_{d_i \in C_i}\ u_i(\sigma_{-i},[d_i])$.

Proof. Notice that, for any τ_i in $\Delta(C_i)$,

$$u_i(\sigma_{-i},\tau_i) = \sum_{d_i \in C_i} \tau_i(d_i) u_i(\sigma_{-i},[d_i]).$$

Thus, $u_i(\sigma_{-i}, \tau_i)$ is a weighted average of the terms $u_i(\sigma_{-i}, [d_i])$, where the weights $\tau_i(d_i)$ are nonnegative and sum to 1. Such a weighted average cannot be greater than the maximum of the terms being averaged and is strictly less than this maximum whenever any nonmaximal term gets positive weight. ∎

So the highest expected utility that player i can get against any combination of other players' randomized strategies does not depend on whether i himself uses randomized strategies or only pure strategies. Furthermore, the optimal randomized strategies for i are just those that assign positive probability only to his optimal pure strategies.

We can say that a pure-strategy profile c in C is an equilibrium (or, more precisely, an *equilibrium in pure strategies*) iff

$$u_i(c) \geq u_i(c_{-i}, d_i), \quad \forall i \in N, \quad \forall d_i \in C_i.$$

(Here (c_{-i}, d_i) denotes the pure-strategy profile in C, such that the i-component is d_i and all other components are as in $c = (c_j)_{j \in N}$.) Lemma 3.1 implies that c is a Nash equilibrium in pure strategies iff the randomized-strategy profile $([c_i])_{i \in N}$ (which puts probability 1 on c) is a Nash equilibrium in $\times_{i \in N} \Delta(C_i)$.

It may be helpful to compare Nash equilibrium with the strategy-elimination concepts discussed in Section 3.1. Whereas rationalizability and iterative dominance only specify which strategies are supposed to be reasonable possibilities, a Nash equilibrium specifies a numerical probability for each strategy. When we thus increase the information that a solution specifies, we implicitly strengthen the restrictions that it must satisfy if it is common knowledge among rational players. So, for many games, the set of Nash equilibria may be significantly smaller than the sets of rationalizable or iteratively undominated strategy profiles. For example, consider the game in Table 3.2.

Table 3.2   A game in strategic form

C_1	C_2		
	x_2	y_2	z_2
x_1	3,0	0,2	0,3
y_1	2,0	1,1	2,0
z_1	0,3	0,2	3,0

In this game, no strategies are dominated (weakly or strongly), and all strategies are rationalizable. In fact, every strategy in this game is a best response to one of the other player's strategies. There are some who might argue that, therefore, any strategy in this game could be rationally used by a player. For example, player 1 might choose x_1 because he expects 2 to choose x_2. To explain to himself why 2 should choose x_2, 1 might suppose that 2 believes that 1 is planning to use z_1. Thus, 1's choice of x_1 can be explained by a theory that describes the strategies that the players will choose, their beliefs about each other's choices, their beliefs about each other's beliefs about each other's choices, and so on. If this theory is right, however, then at least some of the players' beliefs must be wrong. Certainly, if player 1 plans to choose x_1 because he believes that 2 believes that he will choose z_1, either player 1 or player 2 must be wrong. When, as game theorists, we try to describe what rational players should actually do in this game, we cannot subscribe to this theory (that player 1 should choose x_1 because 2 will believe that 1 will choose z_1), because such a theory would violate our assumption that player 2 is intelligent enough to understand everything about the game that we do.

If player 1 believes that 2 will choose y_2, however, then 1 should choose y_1, and 1 can explain to himself why 2 would choose y_2 by assuming that she understands that he will choose y_1. Thus, we can suggest that player 1 should choose y_1 and player 2 should choose y_2 in this game, without any need to suppose that either player is making an error or lacks our intelligence. In fact, $([y_1],[y_2])$ is the unique equilibrium of this game. Thus, the theory that player 1 will choose y_1 for sure and player 2 will choose y_2 for sure is the only theory about the players' behavior that specifies probabilities for all pure-strategy profiles and that could be understood by all players without invalidating itself.

We can now state the general existence theorem of Nash (1951). The proof is presented in Section 3.12.

THEOREM 3.1. *Given any finite game Γ in strategic form, there exists at least one equilibrium in $\times_{i \in N} \Delta(C_i)$.*

Randomized strategies are needed to prove this existence theorem, because many games have no equilibria in pure strategies. For example, recall the simple card game from Chapter 2. Its normal representation is shown in Table 3.3.

Table 3.3   The simple card game in strategic form

C_1	C_2	
	M	P
Rr	0,0	1,−1
Rf	0.5,−0.5	0,0
Fr	−0.5,0.5	1,−1
Ff	0,0	0,0

It is straightforward to verify that there are no equilibria in pure strategies for this game. (Just check all eight possibilities.) Thus, there must be an equilibrium that uses properly randomized strategies. To find this equilibrium, notice first that the pure strategy Ff is strongly dominated and Fr is weakly dominated in this game. It can be shown that any equilibrium of the residual game generated by iterative elimination of weakly or strongly dominated strategies is also an equilibrium of the original game. Thus, we can expect to find an equilibrium that involves randomization between Rr and Rf and between M and P.

So let $q[Rr] + (1 − q)[Rf]$ and $s[M] + (1 − s)[P]$ denote the equilibrium strategies for players 1 and 2, where q and s are some numbers between 0 and 1. That is, let q denote the probability that player 1 would raise even with a losing card, and let s denote the probability that player 2 would meet if 1 raised. Player 1 would be willing to randomize between Rr and Rf only if Rr and Rf give him the same expected utility against $s[M] + (1 − s)[P]$, so

$$0s + 1(1 − s) = 0.5s + 0(1 − s),$$

which implies that $s = \frac{2}{3}$. Thus, player 2's strategy in the equilibrium must be $\frac{2}{3}[M] + \frac{1}{3}[P]$ to make player 1 willing to randomize between [Rr] and [Rf]. Similarly, to make player 2 willing to randomize between M and P, M and P must give her the same expected utility against $q[Rr] + (1 − q)[Rf]$, so

$$0q + −0.5(1 − q) = −1q + 0(1 − q),$$

which implies that $q = \frac{1}{3}$. Thus, the equilibrium must be

$$(\tfrac{1}{3}[Rr] + \tfrac{2}{3}[Rf], \tfrac{2}{3}[M] + \tfrac{1}{3}[P]).$$

That is, player 1 raises for sure when he has a winning (red) card, 1 raises with probability $\frac{1}{3}$ when he has a losing (black) card, and player

2 meets with probability ⅔ when she sees 1 raise in this equilibrium. In fact, this randomized-strategy profile is the unique equilibrium of the simple card game.

The expected utility payoffs from the unique equilibrium are ⅓ for player 1 and −⅓ for player 2. This fact suggests that (when payoffs are in dollars) a risk-neutral person who can refuse to play this game should be willing to take the unfavorable role of player 2 in this game for a preplay bribe of $0.34, but not for any amount of money less than $0.33 (assuming that the bribe is offered before the color of 1's card is determined).

Two general observations about Nash equilibria are now in order. A game may have equilibria that are *inefficient*, and a game may have *multiple equilibria*.

An outcome of a game is (*weakly*) *Pareto efficient* iff there is no other outcome that would make all players better off. For an example of equilibria that are not efficient, consider the game in Table 3.4, known as the *Prisoners' Dilemma*. The story behind this game (taken from Luce and Raiffa, 1957) may be described as follows. The two players are accused of conspiring in two crimes, one minor crime for which their guilt can be proved without any confession, and one major crime for which they can be convicted only if at least one confesses. The prosecutor promises that, if exactly one confesses, the confessor will go free now but the other will go to jail for 6 years. If both confess, then they both go to jail for 5 years. If neither confesses then they will both go to jail for only 1 year. So each player i has two possible strategies: to confess (f_i) or to not confess (g_i). The payoffs, measured in the number of years of freedom that the player will enjoy over the next 6 years, are as shown in Table 3.4.

In this game, $([f_1],[f_2])$ is the unique Nash equilibrium. (In fact, f_1 and f_2 are the only strategies that are not strongly dominated.) However, the outcome resulting from (f_1,f_2) is the only outcome of the game that

Table 3.4 Prisoners' Dilemma game in strategic form

	C_2	
C_1	g_2	f_2
g_1	5,5	0,6
f_1	6,0	1,1

is not Pareto efficient. Thus, if Nash equilibria can be interpreted as describing how rational players should play a game, then rational individuals should expect to all do relatively badly in this game. This example has been very influential as a simple illustration of how people's rational pursuit of their individual best interests can lead to outcomes that are bad for all of them.

For an example of a game with multiple equilibria, consider the game shown in Table 3.5, known as the *Battle of the Sexes*. The story behind this game (again, taken from Luce and Raiffa, 1957) is that players 1 and 2 are husband and wife, respectively, and that they are deciding where to go on Saturday afternoon. Each f_i strategy represents going to the football match, whereas s_i represents going to the shopping center. Neither spouse would derive any pleasure from being without the other, but the husband would prefer meeting at the football match, whereas the wife would prefer meeting at the shopping center.

There are three equilibria of this game: $([f_1],[f_2])$, which gives the payoffs allocation (3,1) to players 1 and 2, respectively; $([s_1],[s_2])$, which gives the payoff allocation (1,3); and

$$(.75[f_1] + .25[s_1], .25[f_2] + .75[s_2]),$$

which gives each player an expected payoff of 0.75. In the first equilibrium, the players both go to the football match, which is player 1's favorite outcome. In the second equilibrium, the players both go to the shopping center, which is player 2's favorite outcome. So each player prefers a different equilibrium.

In the third equilibrium, the players behave in a random and uncoordinated manner, which neither player can unilaterally improve on. Each player is uncertain about where the other player will go, and gets the same expected payoff ($.75 \times 1 = .25 \times 3$) from going either way. The third equilibrium is worse for both players than either of the other two equilibria, so it is also an inefficient equilibrium.

Table 3.5   Battle of the Sexes game in strategic form

C_1	C_2	
	f_2	s_2
f_1	3,1	0,0
s_1	0,0	1,3

3.3 Computing Nash Equilibria

In a Nash equilibrium, if two different pure strategies of player i both have positive probability, then they must both give him the same expected payoff in the equilibrium, because otherwise he would never use the strategy that gave him a lower expected payoff. That is, in equilibrium, a player must be indifferent between any of his strategies that have positive probability in his randomized strategy.

Thus, the set of strategies that have positive probability is an important qualitative aspect of an equilibrium. In general, the *support* of a randomized-strategy profile σ in $\times_{i \in N} \Delta(C_i)$ is the set of all pure-strategy profiles in C that would have positive probability if the players chose their strategies according to σ. That is, the support of σ is the set

$$\underset{i \in N}{\times} \{c_i \in C_i \,|\, \sigma_i(c_i) > 0\}.$$

For example, consider again the Battle of the Sexes game. The support of the $([f_1],[f_2])$ equilibrium is obviously $\{f_1\} \times \{f_2\}$. In this equilibrium, each player i strictly prefers using the f_i strategy to the s_i strategy, given that the other player is expected to go to the football match. Similarly, the support of the $([s_1],[s_2])$ equilibrium is $\{s_1\} \times \{s_2\}$; and in this equilibrium each player i strictly prefers s_i to f_i, given that the other player is expected to go shopping.

As we have seen, this game also has a third equilibrium with support $\{f_1,s_1\} \times \{f_2,s_2\}$. In an equilibrium with this support, each player i must be indifferent between f_i and s_i, given the anticipated behavior of the other player. Notice that player 2's expected payoff against player 1's randomized strategy σ_1 depends on her pure strategy choice as follows:

$$u_2(\sigma_1,[f_2]) = 1\sigma_1(f_1) + 0\sigma_1(s_1) \text{ if } 2 \text{ chooses } f_2,$$

$$u_2(\sigma_1,[s_2]) = 0\sigma_1(f_1) + 3\sigma_1(s_1) \text{ if } 2 \text{ chooses } s_1.$$

So for player 2 to be willing to choose either f_2 or s_2, each with positive probability, these two expected payoffs must be equal. That is, we must have $\sigma_1(f_1) = 3\sigma_1(s_1)$. This equation, together with the probability equation $\sigma_1(f_1) + \sigma_1(s_1) = 1$, implies that $\sigma_1(f_1) = .75$ and $\sigma_1(s_1) = .25$. Notice that we have just derived player 1's randomized strategy from the condition that player 2 should be willing to randomize between her pure strategies in the support. Similarly, to make player 1 willing to choose either f_1 or s_1, each with positive probability, we must have

$$3\sigma_2(f_2) + 0\sigma_2(s_2) = 0\sigma_2(f_2) + 1\sigma_2(s_2),$$

which implies that $\sigma_2(f_2) = .25$ and $\sigma_2(s_2) = .75$. Thus, the equilibrium with support $\{f_1,s_1\} \times \{f_2,s_2\}$ must be

$$(\sigma_1,\sigma_2) = (.75[f_1] + .25[s_1], .25[f_2] + .75[s_2]).$$

We now describe a general procedure for finding equilibria of any finite strategic-form game $\Gamma = (N, (C_i)_{i\in N}, (u_i)_{i\in N})$. Although there are infinitely many randomized-strategy profiles, there are only finitely many subsets of C that can be supports of equilibria. So we can search for equilibria of Γ by sequentially considering various guesses as to what the support may be and looking for equilibria with each guessed support.

For each player i, let D_i be some nonempty subset of player i's strategy set C_i, which will represent our current guess as to which strategies of player i have positive probability in equilibrium. If there is an equilibrium σ with support $X_{i\in N} D_i$, then there must exist numbers $(\omega_i)_{i\in N}$ such that the following equations are satisfied:

(3.5) $$\sum_{c_{-i}\in C_{-i}} \left(\prod_{j\in N-i} \sigma_j(c_j) \right) u_i(c_{-i},d_i) = \omega_i, \quad \forall i \in N, \quad \forall d_i \in D_i,$$

(3.6) $$\sigma_i(e_i) = 0, \quad \forall i \in N, \quad \forall e_i \in C_i \backslash D_i,$$

(3.7) $$\sum_{c_i\in D_i} \sigma_i(c_i) = 1, \quad \forall i \in N.$$

Condition (3.5) asserts that each player must get the same payoff, denoted by ω_i, from choosing any of his pure strategies that have positive probability under σ_i. Conditions (3.6) and (3.7) follow from the assumption that σ is a randomized-strategy profile with support $X_{i\in N} D_i$. Condition (3.6) asserts that i's pure strategies outside of D_i get zero probability, and condition (3.7) asserts that the probabilities of pure strategies in D_i sum to 1. Conditions (3.5)–(3.7) together imply that ω_i is player i's expected payoff under σ, because

$$u_i(\sigma) = \sum_{d_i\in C_i} \sigma_i(d_i)u_i(\sigma_{-i},[d_i]) = \omega_i.$$

These conditions (3.5)–(3.7) give us $\sum_{i\in N} (|C_i| + 1)$ equations in the same number of unknowns (the probabilities $\sigma_i(c_i)$ and the payoffs ω_i).

(Here $|C_i|$ denotes the number of pure strategies in the set C_i.) Thus, we can hope to be able to solve these equations. For two-player games, these equations are all linear in σ and ω; but (3.5) becomes nonlinear in σ when there are more than two players, so the task of solving these equations may be quite difficult.

Let us assume, however, that we can find all solutions to conditions (3.5)–(3.7), given the guessed support $\times_{i \in N} D_i$. These solutions do not necessarily give us equilibria of Γ, because there are three difficulties that may arise. First, no solutions may exist. Second, a solution may fail to be a randomized-strategy profile, if some of the $\sigma_i(c_i)$ numbers are negative. So we must require

(3.8) $\sigma_i(d_i) \geq 0, \quad \forall i \in N, \quad \forall d_i \in D_i.$

Third, a solution that satisfies (3.5)–(3.8) may fail to be an equilibrium if some player i has some other pure strategy outside of D_i that would be better for him against σ_{-i} than any strategy in D_i. So we must also require

(3.9) $\omega_i \geq \sum_{c_{-i} \in C_{-i}} \left(\prod_{j \in N-i} \sigma_j(c_j) \right) u_i(c_{-i}, e_i), \quad \forall i \in N, \quad \forall e_i \in C_i \backslash D_i.$

If we find a solution (σ, ω) to equations (3.5)–(3.7) that also satisfies the inequalities (3.8) and (3.9), then σ is a Nash equilibrium of Γ, and ω_i is the expected payoff to player i in this equilibrium.

On the other hand, if there is no solution that satisfies (3.5)–(3.9), then there is no equilibrium with support $\times_{i \in N} D_i$. To find an equilibrium, we must guess other supports and repeat this procedure. Nash's existence theorem guarantees that there will be at least one support $\times_{i \in N} D_i$ for which conditions (3.5)–(3.9) can be satisfied.

To illustrate this procedure, consider the game in Table 3.6. We can begin by looking for nonrandomized equilibria, that is, equilibria where

Table 3.6   A game in strategic form

C_1	C_2		
	L	M	R
T	7,2	2,7	3,6
B	2,7	7,2	4,5

the support includes just one strategy for each player. If player 1 were expected to choose T for sure, then player 2 would want to choose M, but B would be player 1's best response to M. Thus, there is no equilibrium in which player 1 chooses T for sure. If player 1 were expected to choose B for sure, then player 2 would want to choose L, but T would be player 1's best response to L. Thus, there is no equilibrium in which player 1 chooses B for sure. So there are no nonrandomized equilibria of this game. Any equilibrium of this game must include both of player 1's strategies T and B in the support.

Similarly, there is no equilibrium for which the support includes only one strategy for player 2, because player 1 has a unique best response to each of player 2's three strategies (T to L, B to M or R), and we have seen that player 1 must be willing to randomize between T and B in any equilibrium. Thus, in any equilibrium of this game, player 1 must give positive probability to both T and B, and player 2 must give positive probability to at least two of her three strategies. There are four different supports of this kind for us to try.

As a first guess, let us try the support $\{T,B\} \times \{L,M,R\}$. That is, let us guess that both of player 1's strategies and all three of player 2's strategies will get positive probability in the randomized-strategy profile σ. To make player 1 willing to choose randomly between T and B, he must get the same expected payoff from both strategies, so we need

$$\omega_1 = 7\sigma_2(L) + 2\sigma_2(M) + 3\sigma_2(R) = 2\sigma_2(L) + 7\sigma_2(M) + 4\sigma_2(R).$$

To make player 2 willing to choose randomly among L and M and R, she must get the same expected payoff from all three strategies, so we need

$$\omega_2 = 2\sigma_1(T) + 7\sigma_2(B) = 7\sigma_1(T) + 2\sigma_1(B) = 6\sigma_1(T) + 5\sigma_1(B).$$

In addition, σ must satisfy the two probability equations

$$\sigma_1(T) + \sigma_1(B) = 1, \quad \sigma_2(L) + \sigma_2(M) + \sigma_2(R) = 1.$$

So our seven unknowns $(\sigma_1(T), \sigma_1(B), \sigma_2(L), \sigma_2(M), \sigma_2(R), \omega_1, \omega_2)$ must satisfy a system of seven equations. Unfortunately, given the probability equations, $2\sigma_1(T) + 7\sigma_2(B) = 7\sigma_1(T) + 2\sigma_1(B)$ implies that $\sigma_1(T) = \sigma_1(B) = \frac{1}{2}$, whereas $7\sigma_1(T) + 2\sigma_1(B) = 6\sigma_1(T) + 5\sigma_1(B)$ implies that $\sigma_1(T) = 3\sigma_1(B)$. Thus, there is no solution to this system of equations, and so there is no equilibrium of this game with support $\{T,B\} \times \{L,M,R\}$.

As a second guess, let us try the support $\{T,B\} \times \{M,R\}$. That is, let us assume that $\sigma_2(L) = 0$, but every other strategy may be chosen with positive probability. With this support, the probability equations are

$$\sigma_1(T) + \sigma_1(B) = 1, \quad \sigma_2(M) + \sigma_2(R) = 1.$$

To make player 1 indifferent between T and B, so that he should be willing to choose randomly between T and B, we need

$$\omega_1 = 2\sigma_2(M) + 3\sigma_2(R) = 7\sigma_2(M) + 4\sigma_2(R).$$

To make player 2 indifferent between M and R, we need

$$\omega_2 = 7\sigma_1(T) + 2\sigma_1(B) = 6\sigma_1(T) + 5\sigma_1(B).$$

(Eliminating L from the support has eliminated one unknown, by setting $\sigma_2(L)$ equal to 0, but has also eliminated one indifference equation, because it is no longer necessary that player 2 be indifferent between L and her other strategies.) These equations have a unique solution in σ, and it is

$$\sigma_2(M) = -\tfrac{1}{4}, \quad \sigma_2(R) = \tfrac{5}{4}, \quad \sigma_1(T) = \tfrac{3}{4}, \quad \sigma_1(B) = \tfrac{1}{4}.$$

But player 2 cannot have a negative probability of choosing M. Thus, there is no equilibrium with support $\{T,B\} \times \{M,R\}$.

As a third guess, let us try the support $\{T,B\} \times \{L,M\}$. That is, let us assume that $\sigma_2(R) = 0$, but every other strategy may be chosen with positive probability. With this support, the probability equations are

$$\sigma_1(T) + \sigma_1(B) = 1, \quad \sigma_2(L) + \sigma_2(M) = 1.$$

To make player 1 indifferent between T and B, so that he should be willing to choose randomly between T and B, we need

$$\omega_1 = 7\sigma_2(L) + 2\sigma_2(M) = 2\sigma_2(L) + 7\sigma_2(M).$$

To make player 2 indifferent between L and M, we need

$$\omega_2 = 2\sigma_1(T) + 7\sigma_1(B) = 7\sigma_1(T) + 2\sigma_1(B).$$

These equations have a unique solution, and it is

$$\sigma_2(L) = .5 = \sigma_2(M), \quad \sigma_1(T) = .5 = \sigma_1(B), \quad \omega_1 = \omega_2 = 4.5.$$

Now we have a solution, and it gives us no negative probabilities; so things look good so far. But is this solution an equilibrium? There is one more thing to check. We have made sure that player 2 is indifferent between L and M in this solution, but we have not yet checked whether player 2 actually prefers these two strategies to the strategy R that she is giving zero probability. When player 1 chooses T with probability .5, the expected payoff to player 2 from her strategy R would be $6 \times .5 + 5 \times .5 = 5.5$, but the expected payoff to player 2 from her strategy L or M would be $\omega_2 = 4.5$. So player 2 would not really be willing to choose randomly between L and M when player 1 is expected to choose T with probability .5, because she would prefer to choose R instead. Thus, the solution is not an equilibrium, and so there is no equilibrium with support $\{T,B\} \times \{L,M\}$.

As our fourth (and last possible) guess, let us try the support $\{T,B\} \times \{L,R\}$. That is, let us assume that $\sigma_2(M) = 0$, but every other strategy may be chosen with positive probability. With this support, the probability equations are

$$\sigma_1(T) + \sigma_1(B) = 1, \quad \sigma_2(L) + \sigma_2(R) = 1.$$

To make player 1 indifferent between T and B, we need

$$\omega_1 = 7\sigma_2(L) + 3\sigma_2(R) = 2\sigma_2(L) + 4\sigma_2(R).$$

To make player 2 indifferent between L and R, we need

$$\omega_2 = 2\sigma_1(T) + 7\sigma_1(B) = 6\sigma_1(T) + 5\sigma_1(B).$$

These equations have a unique solution, and it is

$$\sigma_2(L) = \tfrac{1}{6}, \quad \sigma_2(R) = \tfrac{5}{6}, \quad \sigma_1(T) = \tfrac{1}{3}, \quad \sigma_1(B) = \tfrac{2}{3},$$
$$\omega_1 = 3\tfrac{2}{3}, \omega_2 = 5\tfrac{1}{3}.$$

This solution does not give us any negative probabilities. Furthermore, when player 1 behaves according to this solution, the expected payoff to player 2 from choosing M would be

$$u_2(\sigma_1,[M]) = 7 \times \tfrac{1}{3} + 2 \times \tfrac{2}{3} = \tfrac{11}{3} \le \omega_2 = \tfrac{16}{3}.$$

Thus, player 2 would indeed be willing to choose randomly between L and R, and she would not choose M. So the randomized-strategy profile

$$(\tfrac{1}{3}[T] + \tfrac{2}{3}[B], \tfrac{1}{6}[L] + \tfrac{5}{6}[R])$$

is the unique equilibrium of this game. In this equilibrium, the expected payoff to player 1 is $3\tfrac{2}{3}$, and the expected payoff to player 2 is $5\tfrac{1}{3}$.

For games with many players and large strategy sets, the algorithm presented here may become unworkable, as the number of possible supports becomes intractably large and the indifference equations (3.5) become nonlinear in σ. For such games, more sophisticated algorithms, such as those of Scarf (1973) and Wilson (1971), may be needed to find equilibria.

3.4 Significance of Nash Equilibria

Nash's (1951) concept of equilibrium is probably the most important solution concept in game theory. The general argument for the importance of this concept may be summarized as follows. Suppose that we are acting either as theorists, trying to predict the players' behavior in a given game, or as social planners, trying to prescribe the players' behavior. If we specify which (possibly randomized) strategies should be used by the players and if the players understand this specification also (recall that they know everything that we know about the game), then we must either specify an equilibrium or impute irrational behavior to some players. If we do not specify an equilibrium, then some player could gain by changing his strategy to something other than what we have specified for him. Thus, a nonequilibrium specification would be a self-denying prophecy if all the players believed it.

This argument uses the assumption that the players in a strategic-form game choose their strategies independently, so one player's change of strategy cannot cause a change by any other players. As we argued in Section 2.2, this independence assumption is without loss of generality. If there are rounds of communication between the players, then the set of strategies for each player can be redefined to include all plans for what to say in these rounds of communication and what moves to choose as a function of the messages received. That is, in principle, any opportunities for communication can be written into the extensive form of the game and thus can be subsumed in the definition of pure strategies in the normal representation. Thus, there is no loss of generality in assuming that the players have no opportunities to communicate before choosing their strategies in the game. (However, other analytical approaches that do not use this assumption may also be valuable; see Chapter 6.)

Aumann and Maschler (1972) reexamined the argument for Nash equilibrium as a solution concept for games like the example in Table 3.7. The unique equilibrium of this game is

$$(.75[T] + .25[B], .5[L] + .5[R]).$$

It has been suggested that player 1 might prefer to choose T and player 2 might prefer to choose L (each with probability 1), because these strategies are optimal responses to the equilibrium strategies and they guarantee each player his expected equilibrium payoff of 0. But if such behavior were correctly anticipated, then player 1 would be irrational not to choose B in this game, because it is his unique best response to L. Thus, a theory that predicts the actions T and L in this game would destroy its own validity, because ([T],[L]) is not a Nash equilibrium.

Notice that we have not directly argued that intelligent rational players must use equilibrium strategies in a game. When asked why players in a game should behave as in some Nash equilibrium, my favorite response is to ask "Why not?" and to let the challenger specify what he thinks the players should do. If this specification is not a Nash equilibrium, then (as above) we can show that it would destroy its own validity if the players believed it to be an accurate description of each other's behavior.

Notice, however, that this argument only implies that any outcome which is not an equilibrium would necessarily be unreasonable as a description of how decision-makers should behave; it does not imply that any particular equilibrium must be a reasonable prediction in any particular situation. In effect, the concept of Nash equilibrium imposes a constraint on social planners and theorists, in that they cannot prescribe or predict nonequilibrium behavior.

To put this argument in a more general framework, it may be helpful to introduce here some terminology for the evaluation of solution con-

Table 3.7   A game in strategic form

	C_2	
C_1	L	R
T	0,0	0,−1
B	1,0	−1,3

cepts. In general, a *solution concept* is any rule for specifying predictions as to how players might be expected to behave in any given game. However, when we represent a real conflict situation by a mathematical model (in any of the three forms introduced in Chapter 2), we must suppress or omit many "non–game-theoretic" details of the actual situation, which we may call *environmental variables*. For example, in the strategic form, there is no indication of any player's height, weight, socioeconomic status, or nationality. We cannot a priori rule out the possibility that the outcome of the game might depend in some way on such environmental variables. So we should allow that a solution concept may specify more than one prediction for any given game, where the selection among these predictions in a specific real situation may depend on the environment.

Thus, a solution concept can be viewed as a mapping that determines, for every mathematical game Γ in some domain, a set $\phi(\Gamma)$ that is a subset of some range R of mathematical descriptions as to how the players might behave. For example, the domain can be the set of all games in strategic form, and the range R can be the set of all randomized strategy profiles. The goal of game-theoretic analysis is to generate a solution concept that has the following two properties.

1. For any game Γ, for each prediction π in the solution set $\phi(\Gamma)$, there exist environments where π would be an accurate prediction of how rational intelligent players would behave in the game Γ.

2. For any game Γ, for any π in the range R that is *not* in the solution set $\phi(\Gamma)$, there is no environment where π would be an accurate prediction of how rational intelligent players would behave in the game Γ.

Any solution concept that satisfies these two properties can be called an *exact solution*. That is, an exact solution concept should include all predictions that might be considered reasonable in at least some situations represented by the given mathematical game and should exclude all predictions that could never be considered reasonable.

Unfortunately, finding an exact solution concept may be very difficult, so we should also be interested in solution concepts that satisfy one of these two properties. Let us say that a *lower solution* is any solution concept that satisfies property (1) above and an *upper solution* is any solution concept that satisfies property (2) above. That is, a lower solution excludes all unreasonable predictions but may also exclude some reasonable predictions. An upper solution includes all reasonable predictions but may also include some unreasonable predictions. In these terms, it may be best to think of Nash equilibrium as an upper solution rather than as an exact solution, because being a Nash equilibrium is only a necessary condition for a theory to be a good prediction of the behavior of intelligent rational players.

3.5 The Focal-Point Effect

As the Battle of the Sexes game in Table 3.5 shows, a game may have more than one equilibrium. Indeed, it is easy to construct games in which the set of equilibria is quite large. In Chapters 4 and 5, we show that some natural refinements of the Nash equilibrium concept may eliminate some equilibria for some games, but even these refinements still allow a wide multiplicity of equilibria to occur in many games.

When a game has multiple equilibria, the constraint imposed by the concept of Nash equilibrium on social planners and theorists becomes weaker. We know that any one of these equilibria, if it were expected by all players, could become a self-fulfilling prophecy. For example, in the Battle of the Sexes game, suppose that (for some reason) players 1 and 2 expected each other to implement the (f_1, f_2) equilibrium. Then each player would expect to maximize his own payoff by fulfilling this expectation. Player 2 would prefer the (s_1, s_2) equilibrium, but choosing s_2 instead of f_2 would reduce her payoff from 1 to 0, given that she expects player 1 to choose f_1.

Thus, to understand games with multiple equilibria, we must ask what might cause the players in a game to expect each other to implement some specific equilibrium. This question was considered in detail by Schelling (1960). Schelling argued that, in a game with multiple equilibria, anything that tends to focus the players' attention on one equilibrium may make them all expect it and hence fulfill it, like a self-fulfilling prophecy. Schelling called this phenomenon the *focal-point effect*. That is, a *focal equilibrium* is an equilibrium that has some property that conspicuously distinguishes it from all the other equilibria. According to the focal-point effect, if there is one focal equilibrium in a game, then we should expect to observe that equilibrium.

For example, suppose that players 1 and 2 in the Battle of the Sexes game are a husband and wife who live in a society where women have traditionally deferred to their husbands in such situations. Then, even if this couple feels no compulsion to conform to this tradition, this tradition makes the (f_1, f_2) equilibrium more focal and hence more likely to be implemented. Because of this sexist tradition, the wife will expect that her husband will presume that he should choose f_1, so she will reluctantly choose f_2; and the husband will expect his wife to choose f_2, so f_1 is better than s_1 for him. (Remember that we are assuming here that the two players make their choices independently and simultaneously.)

In some situations, even seemingly trivial aspects of the way that a game is presented could determine the focal equilibrium that the players implement. Suppose, for example, that there is no strong cultural bias toward any equilibrium in the Battle of the Sexes game, but the players learned the structure of the game from Table 3.8 instead of from Table 3.5. Although the asterisks and boldface in Table 3.8 have no role in the mathematical definition of this game, they tend to focus the players' attention on the (s_1, s_2) equilibrium, so rational players in this game would be likely to play (s_1, s_2) in this case.

Similarly, (s_1, s_2) may become the focal equilibrium if it is perceived to be the status quo, as it may if the players in the Battle of the Sexes are already at the shopping center, even though costs of going to the football match would be negligible.

Another way that the players could become focused on one equilibrium is by some process of preplay communication. For example, if the day before the Battle of the Sexes game, player 1 told player 2, "Let's meet at the football match," we might well expect them to play (f_1, f_2).

Of course, such preplay communication could also be modeled as a part of the game (as we remarked earlier). To see what we get when we

Table 3.8 Battle of the Sexes game in strategic form, with emphasis

C_1	C_2	
	f_2	s_2
f_1	3,1	0,0
s_1	0,0	* **1,3** *

do so, suppose (for simplicity) that the only possible "preplay" communication is that player 1 can say either "Let's meet at the football match" or "Let's meet at the shopping center" to player 2 on Friday, and then each player must simultaneously decide whether to go to football or to the shopping center on Saturday. Player 1's statement on Friday is not assumed to bind his Saturday decision; he can go shopping on Saturday even if he suggested football on Friday. We can model the interaction between the players during both Friday and Saturday as an extensive-form game. In the reduced normal representation of this game, player 1's strategy set can be written

$$C_1 = \{Ff_1, Fs_1, Sf_1, Ss_1\}.$$

Here the capitalized first letter denotes the option that he suggests on Friday, and the second letter denotes the move that he then makes on Saturday after making this suggestion. (In the reduced normal representation, we suppress the irrelevant specification of what move player 1 would choose on Saturday if he remembered making the statement that his strategy does not specify for Friday.) Player 2's strategy set may be written

$$C_2 = \{f_2 f_2, f_2 s_2, s_2 f_2, s_2 s_2\},$$

where we list first the decision that she would make if 1 suggested football and second the decision that she would make if 1 suggested shopping. The reduced normal representation of this game is shown in Table 3.9. For each equilibrium of the original Battle of the Sexes game, there are equilibria of this game that give the same allocation of expected payoffs. For example, $(Sf_1, s_2 f_2)$ is an equilibrium of Table 3.9 and gives payoffs $(3,1)$. In this equilibrium, player 1 says, "Let's meet at the shopping center," and then goes to the football match, and player

Table 3.9 Battle of the Sexes game with communication, in strategic form

	C_2			
C_1	$f_2 f_2$	$f_2 s_2$	$s_2 f_2$	$s_2 s_2$
Ff_1	3,1	3,1	0,0	0,0
Fs_1	0,0	0,0	1,3	1,3
Sf_1	3,1	0,0	3,1	0,0
Ss_1	0,0	1,3	0,0	1,3

2 goes to the football match but would have gone shopping if player 1 had suggested football. Another equilibrium is $(Fs_1, s_2 s_2)$, which gives payoffs (1,3). In this equilibrium, player 1 suggests football but actually goes shopping, and player 2 would go shopping no matter what player 1 suggests. However, the obvious focal equilibrium of this game is $(Ff_1, f_2 s_2)$, in which player 1 says, "Let's meet at the football match," and then goes to the football match, and player 2 would try to meet player 1 wherever he suggested. This last equilibrium is the unique equilibrium in which player 1's statement on Friday is interpreted according to its literal meaning. So the players' shared understanding of language, which is part of their cultural heritage, may make focal the equilibrium of Table 3.9 in which player 1 can effectively select among the (f_1, f_2) and (s_1, s_2) equilibria of Table 3.5. (See Farrell, 1988, and Myerson, 1989, for a general formal development of this idea.)

In general, when such preplay communication possibilities exist, we say that an individual is a *focal arbitrator* if he can determine the focal equilibrium in a game by publicly suggesting to the players that they should all implement this equilibrium. Even though this suggestion may have no binding force, if each player believes that every other player will accept the arbitrator's suggestion, then each player will find it best to do as the arbitrator suggests, provided the arbitrator's suggestion is an equilibrium. Thus, a game with a large set of equilibria is a game in which an arbitrator or social planner can substantially influence players' behavior.

The focal equilibrium of a game can also be determined by intrinsic properties of the utility payoffs. For example, consider the *Divide the Dollars* game, in which there are two players who can each make a demand for some amount of money between \$0 and \$100. If their demands sum to \$100 or less, then each gets his demand, otherwise both get \$0. That is, in this game the pure strategy sets are

$$C_1 = C_2 = \{x \in \mathbf{R} \mid 0 \le x \le 100\} = [0, 100],$$

and the payoff functions are

$$u_i(c_1, c_2) = 0 \text{ if } c_1 + c_2 > 100,$$
$$= c_i \text{ if } c_1 + c_2 \le 100.$$

For any number x between 0 and 100, the pure strategy pair $(x, 100 - x)$ is an equilibrium in which the players will get to divide the

$100 for sure. There is also an equilibrium, (100,100), in which each player is sure to get a payoff of 0. Furthermore, there are many randomized equilibria in which the probability of dividing the available $100 may be quite small; for example, $((1/99)[1] + (98/99)[99], (1/99)[1] + (98/99)[99])$ is an equilibrium in which each player gets an expected payoff of 1. In this game, an impartial arbitrator would probably suggest that the players should implement the (50,50) equilibrium, because it is Pareto efficient and equitable for the two players. But this fact may make the (50,50) equilibrium focal even when such an impartial arbitrator is not actually present. That is, because both players know that (50,50) is the efficient and equitable equilibrium that an impartial arbitrator would be most likely to suggest, (50,50) has a strong intrinsic focal property even when no arbitrator is present. Thus, welfare properties of *equity* and *efficiency* may determine the focal equilibrium in a game.

A focal equilibrium might also be determined in some games by properties of the strategies themselves. For example, the game in Table 3.10 has three equilibria, but only one of these is in pure (unrandomized) strategies. It can be argued that the players in the game might therefore be more likely to focus on this pure-strategy equilibrium (y_1, y_2) than on the other two equilibria that involve some randomization. In repeated games (see Chapter 7), simplicity or stationarity of the strategies in an equilibrium may make that equilibrium more focal, other things being equal.

Notice that the focal-point effect cannot lead intelligent rational players to implement a strategy profile that is not an equilibrium. For example, suppose that a focal arbitrator for the players in Table 3.11 recommended that the players play (y_1, x_2). He might even reinforce his recommendation by arguments about the efficiency and equity of the

Table 3.10   A game in strategic form

C_1	C_2		
	x_2	y_2	z_2
x_1	3,0	0,0	0,3
y_1	0,0	1,1	0,0
z_1	0,3	0,0	3,0

Table 3.11   A game in strategic form

	C_2	
C_1	x_2	y_2
x_1	5,1	0,0
y_1	4,4	1,5

(4,4) payoff allocation that these strategies generate. But this recommendation could not be a self-fulfilling prophecy. If player 1 believed that player 2 would do x_2 as the arbitrator recommends, then player 1 would prefer to do x_1 instead of y_1. With intelligent rational players, a focal factor can be effective only if it points to an equilibrium.

When different factors tend to focus attention on different equilibria, the question of which equilibrium would be focused on and played by real individuals in a given situation can be answered only with reference to the psychology of human perception and the cultural background of the players. From a game-theoretic perspective, *cultural norms* can be defined to be the rules that a society uses to determine focal equilibria in game situations. There may be some situations where people of a given culture might look to equity principles to determine a focal equilibrium (e.g., in bargaining between people of equal and independent status), other situations where a specific individual might be understood to be an effective focal arbitrator (e.g., a supervisor in a job conflict), and yet other situations where traditional modes of behavior determine equilibria that no one's suggestions or arguments can overturn. There may be a culture (or local subcultures) where a tradition of doing what the wife wants on weekends is so strong that even in the game with preplay communication, where the husband gets to tell the wife on Friday what he thinks they should do, the focal equilibrium that is played may be (Fs_1, s_2s_2), where the husband suggests that they should meet at the football match but his suggestion is ignored. (Such cultural norms can be studied by experimental games; for example, see Roth and Schoumaker, 1983; Roth, 1985; Cooper, DeJong, Forsythe, and Ross, 1989, 1990; and van Huyck, Battalio, and Beil, 1990.)

Thus, the focal-point effect defines both an essential limit on the ability of mathematical game theory to predict people's behavior in real conflict situations and an important agenda for research in social psychology and cultural anthropology. It should not be surprising that game theory cannot provide a complete theory of human behavior without complementary theories from other disciplines. On the other hand, the focal-point effect of game theory can offer a useful perspective on the role of culture and perception. In particular, game theory can predict that cultural norms may be less important in situations where the set of equilibria is small.

3.6 The Decision-Analytic Approach to Games

In game-theoretic analysis, we try to understand the behavior of all of the players in a game, assuming that they are all rational and intelligent individuals. Raiffa (1982) and Kadane and Larkey (1982) have advocated an alternative decision-analytic approach to the study of games, when our task is to advise some particular player i (the *client*) as to what strategy he should use in a given game. Player i's optimal strategy should maximize his expected payoff with respect to his subjective probability distribution over the possible strategies of the other players. The *decision-analytic approach* to player i's decision problem is first to assess some subjective probability distribution to summarize i's beliefs about what strategies will be used by the other players and then to select a strategy for i that maximizes his expected payoff with respect to these beliefs. That is, the decision-analytic approach to i's problem is to try to predict the behavior of the players other than i first and then to solve i's decision problem last. In contrast, the usual game-theoretic approach is to analyze and solve the decision problems of all players together, like a system of simultaneous equations in several unknowns.

A fundamental difficulty may make the decision-analytic approach impossible to implement, however. To assess his subjective probability distribution over the other players' strategies, player i may feel that he should try to imagine himself in their situations. When he does so, he may realize that the other players cannot determine their optimal strategies until they have assessed their subjective probability distributions over i's possible strategies. Thus, player i may realize that he cannot predict his opponents' behavior until he understands what an intelligent person would expect him rationally to do, which is, of course, the problem that he started with. This difficulty would force i to abandon the decision-analytic approach and instead to undertake a game-theoretic approach, in which he tries to solve all players' decision problems simultaneously.

Even if the decision-analytic approach is feasible for player i, there still is a role for Nash equilibrium analysis in his thinking. After following the decision-analytic approach, player i would be well advised at least to make note of whether his predicted strategies for the other players, together with the strategy that he has chosen for himself, form an equilibrium of the game. If they do not form an equilibrium, then there must be some other player who is not making an optimal strategy choice, according to these predictions and plans. Player i should ask himself why this person might make this mistake. For example, is there something that i knows that this other player does not? Or is this a common mistake that naive decision-makers often make? If player i can find such an explanation, then he may feel confident using his selected strategy. On the other hand, if he cannot find any such explanation, then he should probably think further about how the other players might react if they understood his current plans, and he should consider revising his predictions and plans accordingly.

For an example in which the decision-analytic approach may seem much more promising than equilibrium analysis, Raiffa has suggested the following *Dollar Auction* game. There are two risk-neutral players, each of whom must choose a bid that can be any real number between 0 and 1. The high bidder pays the amount of his bid and then wins \$1. (In case of a tie, each has a probability 0.5 of winning and buying the dollar for his bid.) Thus, the strategy sets are $C_1 = C_2 = \{x \in \mathbf{R} \mid 0 \le x \le 1\}$, and the utility functions are

$$u_i(c_1, c_2) = 0 \qquad \text{if } i \notin \underset{j \in \{1,2\}}{\operatorname{argmax}} \, c_j,$$

$$= 1 - c_i \qquad \text{if } \{i\} = \underset{j \in \{1,2\}}{\operatorname{argmax}} \, c_j,$$

$$= (1 - c_i)/2 \quad \text{if } c_1 = c_2.$$

In the unique equilibrium of this game, both players bid 1 for sure, so each player gets a net payoff of 0. Notice that bidding 1 is actually a weakly dominated strategy for each player, because it is the only bid that cannot give him a positive payoff (and no bid ever gives a negative payoff). Thus, it is hard to see how these equilibrium strategies can be recommended to players in this game. If we were advising player 1 how to play this game, surely we would do better trying to follow the decision-analytic approach. As long as he assigns some positive probability to the event that player 2 will bid lower than \$1, player 1 will find that his optimal bid is less than \$1.

Properties of equilibria that seem unreasonable often are rather unstable, in the sense that these properties may disappear if the game is perturbed slightly. Indeed, the unique equilibrium of the Dollar Auction would change significantly with some small perturbations of the game. Suppose, for example, that we changed the rules of the game by stipulating that, for each player j, there is an independent probability .1 that j's bid will be determined, not by a rational intelligent decision-maker, but by a naive agent who chooses the bid from a uniform distribution over the interval from 0 to 1. Let us then interpret $\hat{u}_i(c_1,c_2)$ as the conditionally expected payoff that i would get in this perturbed game given that player i's bid is not being determined by such a naive agent and given that, for each player j, c_j is the bid that player j would make if his bid were not being otherwise determined by such a naive agent. Then the utility functions in this perturbed game are

$$\hat{u}_i(c_1,c_2) = .1c_i(1 - c_i) \qquad \text{if } i \notin \operatorname*{argmax}_{j \in \{1,2\}} c_j,$$

$$= .1c_i(1 - c_i) + .9(1 - c_i) \quad \text{if } \{i\} = \operatorname*{argmax}_{j \in \{1,2\}} c_j,$$

$$= .1c_i(1 - c_i) + .9(1 - c_i)/2 \text{ if } c_1 = c_2.$$

There is an equilibrium of this perturbed game in which each player randomly chooses a bid from the interval between 0.5 and 0.975 in such a way that, for any number x in this interval, the cumulative probability of his bidding below x is $(.025 + .1x^2 - .1x)/(.9(1 - x))$. The median bid for a player under this distribution is 0.954; so the bids tend to be high, but they are lower than 1.

Of course, this equilibrium depended on the particular way that we perturbed the original Dollar Auction game. One way to get around this difficulty is to combine some element of the decision-analytic approach with the game-theoretic approach. The difficulty with the decision-analytic approach arises when player 1 recognizes that player 2 is sophisticated enough to be thinking intelligently about 1's own decision problem. So, as consultants for player 1 in a real Dollar Auction game, we could ask him first, if he were told that player 2 was an especially "naive" decision-maker, then what conditional probability distribution would he assess to describe his beliefs about the bid that such a naive decision-maker would submit; and we could ask him next to assess his subjective probability of the event that player 2 is in fact so naive. That is, after admitting that the assumption that player 2 is rational and intelligent may be inaccurate, we could ask player 1 both to assess the probability that this assumption is inaccurate and to describe how he would expect 2 to behave if it were inaccurate. Assuming that player 2, when she is not naive, would have similar perceptions of player 1, we could then use these answers to construct a perturbed version of the Dollar Auction similar to that of the preceding paragraph, but using the naive bid distribution and probability of naivety that he assessed. It might then be reasonable to recommend to player 1 that he implement his equilibrium strategy of this perturbed game.

3.7 Evolution, Resistance, and Risk Dominance

Axelrod (1984) tried to identify good strategies for games by a kind of biological evolutionary criterion (see also Maynard Smith, 1982). We can describe Axelrod's procedure as follows. First, for each player i in the game, choose a list or set L_i of randomized strategies that may seem promising or interesting for some reason, so

$$L_i \subseteq \Delta(C_i).$$

Then, for each player i and each strategy σ_i in L_i, suppose that there is a given number of "i-animals" that are instinctively programmed to behave according to this σ_i strategy whenever they play the game. We call this the *first-generation population*. Now, we can create a sequence of generations by induction as follows. Each i-animal must play the game many times; and each time, he is in the role of player i while the role of every other player j in the game is filled by a j-animal drawn independently and at random from the population of j-animals in the current generation. So, letting $q_j^k(\sigma_j)$ denote the proportion of j-animals in the generation k who are programmed to use σ_j, we find that the expected payoff to an i-animal programmed to use σ_i is $\bar{u}_i^k(\sigma_i) = u_i(\bar{\sigma}_{-i}^k, \sigma_i)$, where $\bar{\sigma}^k = (\bar{\sigma}_j^k)_{j \in N}$ and

$$\bar{\sigma}_j^k(c_j) = \sum_{\sigma_j \in L_j} q_j^k(\sigma_j) \sigma_j(c_j), \quad \forall j \in N, \quad \forall c_j \in C_j.$$

Then each animal in generation k will have a number of children in the next generation $k+1$ that is proportional to this expected payoff. That is,

$$q_i^{k+1}(\sigma_i) = \frac{q_i^k(\sigma_i)\bar{u}_i(\sigma_i)}{\sum_{\tau_i \in L_i} q_i^k(\tau_i)\bar{u}_i(\tau_i)} .$$

(To avoid difficulties, suppose that all payoffs are positive numbers.) So once $q_i^1(\sigma_i)$ has been specified for each i and each σ_i in L_i, we can then compute the evolution of this imaginary ecological system and see which strategies will tend to dominate the population in the long run, after many generations. (Axelrod actually only looked at symmetric two-person games, but the above is a natural generalization of his procedure to arbitrary strategic-form games.)

The potential significance of such analysis comes from the fact that people actually are the result of a long evolutionary process, and the payoffs in many games of interest might be supposed to have some relationship to the outcomes that would make for reproductive fitness of our ancestors during this process. On the other hand, the outcome of any such evolutionary simulation may be very dependent on the assumptions about the distribution of strategies used in the first generation. In Axelrod's work, even strategies that did quite poorly could be crucial to determining which strategy reproduced best. To the extent that the outcome may depend on the assumed presence in the first generation of animals that play dominated or otherwise foolish strategies, one may argue that the outcome of this evolutionary simulation is not relevant to the players' decision problems, when it is common knowledge that all players are intelligent and rational.

Furthermore, for any Nash equilibrium σ, if almost all of the first-generation i-animals are programmed to use the equilibrium strategy σ_i, for all i, then no animal programmed to play any other strategy can do significantly better. Given any game Γ in strategic form and any equilibria σ and τ in $\times_{i \in N} \Delta(C_i)$, we may say that the *resistance* of σ against τ is the largest number λ such that $0 \leq \lambda \leq 1$ and

$$u_i\big((\lambda\tau_j + (1 - \lambda)\sigma_j)_{j \in N-i}, \sigma_i\big)$$
$$\geq u_i\big((\lambda\tau_j + (1 - \lambda)\sigma_j)_{j \in N-i}, \tau_i\big), \quad \forall i \in N.$$

That is, the resistance of an equilibrium σ against another equilibrium τ is the maximal fraction of τ_i-programmed animals that could be put into a population of σ_i-programmed animals, for all i, such that the τ_i-programmed animals would have no reproductive advantage over the σ_i-programmed animals for any i, according to Axelrod's evolutionary story. Thus defined, the resistance can provide a crude measure of the relative evolutionary stability of one equilibrium in comparison with another. (The crudeness of the resistance index can become evident in games with many players, where the near indifference of one minor player, whose strategies may be of no consequence to any other players, can lower an equilibrium's "resistance," when it is measured this way.)

Harsanyi and Selten (1988) have sought to develop a procedure to identify, for every finite game in strategic form, a unique equilibrium to be considered as the solution to the game. The Harsanyi-Selten solution can be thought of as the limit of an evolutionary process that has some similarities to Axelrod's. However, Harsanyi and Selten offer a sophisticated criterion for defining the initial distributions of strategies in a way that avoids arbitrariness or dependence on irrelevant aspects of the game. Also, the evolutionary process that Harsanyi and Selten use, called the *tracing procedure*, is better interpreted as describing the evolution of tentative plans considered over time by a fixed set of rational intelligent players rather than the reproductive evolution of an ecological system of instinctual animals. The reader is referred to Harsanyi and Selten (1988) for the precise definition of their solution concept.

In the development of their solution concept, Harsanyi and Selten define a *risk dominance* relation between equilibria. They develop this concept axiomatically for the special case of pure-strategy equilibria of two-player games in which each player has two pure strategies. In this case, one equilibrium c risk dominates another equilibrium d iff the resistance of c against d is greater than the resistance of d against c. (In the more general case, Harsanyi and Selten's definition of risk dominance cannot be simply stated in terms of resistances and is not axiomatically derived.)

Consider the example shown in Table 3.12, due to Aumann. Among the two pure-strategy equilibria, the resistance of (y_1,y_2) against (x_1,x_2) is $7/8$, whereas the resistance of (x_1,x_2) against (y_1,y_2) is only $1/8$, so (y_1,y_2) risk-dominates (x_1,x_2). In effect, the range of randomized strategies for which y_1 and y_2 are optimal is wider than that for x_1 and x_2. Furthermore, the randomized equilibrium of this game has a resistance of 0 against either pure-strategy equilibrium.

Just as welfare properties of equity and efficiency might determine the equilibrium that players focus on, so also resistance numbers and risk dominance relations could also determine the focal equilibrium in a game. The game in Table 3.12 is particularly interesting because the property of Pareto-efficiency could tend to make the (x_1,x_2) equilibrium focal, whereas the property of risk dominance could tend to make the (y_1,y_2) equilibrium focal. A tradition of being concerned with one property or the other could determine which equilibrium is focal in any particular situation.

Harsanyi and Selten offer an interesting theoretical argument for risk dominance. They suggest that, for ideally rational players, the determination of the focal equilibrium should not depend on transformations of the game that leave the best-response correspondences unchanged. As shown in Section 2.3, the game in Table 3.12 is best-response equivalent (but not fully equivalent) to the game in Table 2.5, where (y_1,y_2) is both Pareto efficient and risk dominant.

A tendency toward Pareto-efficiency in games can be derived from other biological evolutionary models where matching in the population is characterized by a kind of *viscosity* (in the sense of Hamilton, 1964, and Pollock, 1989). For example, let us modify the evolutionary story that was described earlier by supposing that the animals who play the game live in herds and that each animal lives in the same herd as its ancestors and is much more likely to play against a member of its own herd than against a member of another herd. Suppose that these animals are playing the game in Table 3.12. Within any given herd, the animals who play y_i would tend to multiply faster, unless the initial proportion of animals in the herd who play x_i is larger than $7/8$. So we may anticipate that, after many generations, almost all of the animals within any given herd would be playing the same equilibrium strategy, and the majority of herds (including all those where the proportions of

Table 3.12   A game in strategic form

C_1	C_2	
	x_2	y_2
x_1	9,9	0,8
y_1	8,0	7,7

animals in the initial population who play x_i and y_i are approximately equal) would converge to the risk-dominant equilibrium (y_1, y_2). In the long run, however, the herds that do converge to the Pareto-efficient equilibrium (x_1, x_2) would tend to increase in size faster than the herds that converge to (y_1, y_2). Thus, in the long run, viscosity in the matching process may tend to create a population where most individuals play the efficient equilibrium (x_1, x_2).

In general, biological games differ from economic games in that the choice of strategy is at the genetic level, rather than at the level of individual cognitive choice. Thus, a strategy in a biological game is chosen (through mutation) by a genotype or subspecies. This distinction is important because members of one subspecies can meet and play games against each other. Even when a subspecies is an infinitesimal fraction of the overall species population, members of the subspecies may meet each other with a positive frequency, because kin groups do not disperse uniformly over the entire range of the species.

So suppose that, when two individuals of some given species meet, they play a game $\Gamma = (\{1,2\}, C_1, C_2, u_1, u_2)$ that is *symmetric*, in the sense that

$$C_1 = C_2 \text{ and } u_1(\alpha, \beta) = u_2(\beta, \alpha), \quad \forall \alpha \in C_1, \quad \forall \beta \in C_1.$$

(Here, an individual's payoff represents some contribution to the individual's reproductive fitness.) Let us represent the relative frequency of interactions within small subspecies by some *viscosity parameter* δ, such that $0 < \delta < 1$. Then we may say that σ is a δ-*viscous equilibrium* of Γ iff σ is a randomized-strategy profile that is symmetric, in the sense that $\sigma = (\sigma_1, \sigma_1)$ where $\sigma_1 \in \Delta(C_1)$, and, for each pure strategy c_1 in C_1,

$$\text{if } \sigma_1(c_1) > 0 \text{ then } c_1 \in \underset{e_1 \in C_1}{\text{argmax}} \, u_1([e_1], (1 - \delta)\sigma_1 + \delta[e_1]).$$

That is, the only pure strategies that are used by a positive fraction of the population σ_1 are those that would be optimal for a subspecies if, whenever an individual in the subspecies plays this game, the probability is δ that its opponent is drawn from the same subspecies, and otherwise its opponent is drawn at random from the overall species population. For example, when $\delta > \frac{7}{9}$, the symmetric game with payoffs as in Table 3.12 has a unique δ-viscous equilibrium that yields the Pareto-efficient outcome (9,9).

As the viscosity parameter δ goes to zero, δ-viscous equilibria must approach Nash equilibria. For some symmetric two-player games, however, there exist symmetric Nash equilibria that cannot be approached by any such sequence of δ-viscous equilibria. (See Myerson, Pollock, and Swinkels, 1991.)

The concept of an evolutionary stable strategy is an important solution concept in biological game theory. (See Maynard Smith, 1982.) For a symmetric two-player game Γ as described above, a randomized strategy σ_1 is called an *evolutionary stable strategy* iff (σ_1, σ_1) is a symmetric Nash equilibrium of the game and, for each randomized strategy τ_1, if $\tau_1 \neq \sigma_1$ then there exists some δ between 0 and 1 such that

$$u_1(\tau_1, (1 - \delta)\sigma_1 + \delta\tau_1) < u_1(\sigma_1, (1 - \delta)\sigma_1 + \delta\tau_1).$$

That is, the equilibrium strategy σ_1 should be strictly better than any alternative strategy τ_1 when there is a small positive probability of meeting an individual who is using the alternative strategy τ_1.

Some symmetric two-player games have no evolutionary stable strategies. For example, consider a symmetric two-player game where $C_1 = C_2 = \{x, y, z\}$,

$$u_1(x,x) = u_1(y,y) = u_1(z,z) = 1,$$
$$u_1(x,y) = u_1(y,z) = u_1(z,x) = 3,$$
$$u_1(y,x) = u_1(z,y) = u_1(x,z) = -3,$$

and $u_2(c_1, c_2) = u_1(c_2, c_1)$ for every c_1 and c_2 in C_1. This game has a unique Nash equilibrium (σ_1, σ_1) such that $\sigma_1 = (\frac{1}{3})[x] + (\frac{1}{3})[y] + (\frac{1}{3})[z]$, but the definition of evolutionary stability would be violated for this equilibrium by letting $\tau_1 = [x]$, for example. It has been argued that this game may represent an inherently unstable situation. Notice, however, that the Nash equilibrium of this game is a δ-viscous equilibrium for any positive δ.

3.8 Two-Person Zero-Sum Games

Much of the early work in game theory was on two-person zero-sum games. A *two-person zero-sum game* in strategic form is any Γ of the form $\Gamma = (\{1,2\}, C_1, C_2, u_1, u_2)$ such that

$$u_2(c_1, c_2) = -u_1(c_1, c_2), \quad \forall c_1 \in C_1, \quad \forall c_2 \in C_2.$$

Two-person zero-sum games describe situations in which two individuals are in pure opposition to each other, where one's gain is always the other's loss. In these games, because $u_2 = -u_1$, we could equivalently say that player 2's objective is to minimize the expected payoff to player 1. The card game shown in Table 3.3 is an example of a two-person zero-sum game. The following theorem summarizes some of the important mathematical properties of such games.

THEOREM 3.2. (σ_1,σ_2) *is an equilibrium of a finite two-person zero-sum game* $(\{1,2\}, C_1, C_2, u_1, -u_1)$ *if and only if*

$$\sigma_1 \in \operatorname*{argmax}_{\tau_1 \in \Delta(C_1)} \ \min_{\tau_2 \in \Delta(C_2)} \ u_1(\tau_1,\tau_2)$$

and

$$\sigma_2 \in \operatorname*{argmin}_{\tau_2 \in \Delta(C_2)} \ \max_{\tau_1 \in \Delta(C_1)} \ u_1(\tau_1,\tau_2).$$

Furthermore, if (σ_1,σ_2) *is an equilibrium of this game, then*

$$u_1(\sigma_1,\sigma_2) = \max_{\tau_1 \in \Delta(C_1)} \ \min_{\tau_2 \in \Delta(C_2)} \ u_1(\tau_1,\tau_2) = \min_{\tau_2 \in \Delta(C_2)} \ \max_{\tau_1 \in \Delta(C_1)} \ u_1(\tau_1,\tau_2).$$

Proof. Suppose first that (σ_1,σ_2) is an equilibrium. Then

$$u_1(\sigma_1,\sigma_2) = \max_{\tau_1 \in \Delta(C_1)} \ u_1(\tau_1,\sigma_2) \geq \max_{\tau_1 \in \Delta(C_1)} \ \min_{\tau_2 \in \Delta(C_2)} \ u_1(\tau_1,\tau_2)$$

and

$$u_1(\sigma_1,\sigma_2) = \min_{\tau_2 \in \Delta(C_2)} \ u_1(\sigma_1,\tau_2) \leq \min_{\tau_2 \in \Delta(C_2)} \ \max_{\tau_1 \in \Delta(C_1)} \ u_1(\tau_1,\tau_2).$$

(In each of the two lines above, the equality follows from the definition of an equilibrium and the inequality follows from the fact that, if $f(x) \geq g(x)$ for every x, then max $f \geq$ max g and min $g \leq$ min f.) But

$$\max_{\tau_1 \in \Delta(C_1)} \ \min_{\tau_2 \in \Delta(C_2)} \ u_1(\tau_1,\tau_2) \geq \min_{\tau_2 \in \Delta(C_2)} \ u_1(\sigma_1,\tau_2)$$

and

$$\min_{\tau_2 \in \Delta(C_2)} \ \max_{\tau_1 \in \Delta(C_1)} \ u_1(\tau_1,\tau_2) \leq \max_{\tau_1 \in \Delta(C_1)} \ u_1(\tau_1,\sigma_2).$$

Thus, all of these expressions are equal, a result that implies the three equalities in the last line of the theorem. The two inclusions in the theorem follow from the fact that

$$\min_{\tau_2 \in \Delta(C_2)} u_1(\sigma_1, \tau_2) = \max_{\tau_1 \in \Delta(C_1)} \min_{\tau_2 \in \Delta(C_2)} u_1(\tau_1, \tau_2)$$

and

$$\max_{\tau_1 \in \Delta(C_1)} u_1(\tau_1, \sigma_2) = \min_{\tau_2 \in \Delta(C_2)} \max_{\tau_1 \in \Delta(C_1)} u_1(\tau_1, \tau_2).$$

Now suppose, conversely, that σ_1 and σ_2 satisfy the two inclusions in the theorem. Because we know that there exists an equilibrium of the game, by Theorem 3.1, the last equality in the theorem (the equality between max–min and min–max) still holds. Thus,

$$u_1(\sigma_1, \sigma_2) \geq \min_{\tau_2 \in \Delta(C_2)} u_1(\sigma_1, \tau_2) = \max_{\tau_1 \in \Delta(C_1)} u_1(\tau_1, \sigma_2) \geq u_1(\sigma_1, \sigma_2);$$

so all of these expressions are equal, and (σ_1, σ_2) is an equilibrium of the two-person zero-sum game. ∎

Thus, all equilibria of a two-person zero-sum game give the same expected payoff to player 1. That is, although there may be more than one equilibrium for such a game, both players are indifferent between all the equilibria.

To appreciate the significance of the equality between the max–min and the min–max stated in Theorem 3.2, notice that the proof of this result relied on the existence of an equilibrium. Without randomized strategies, the existence of an equilibrium cannot be guaranteed and the equality may fail. For example, when we allow only pure strategies in the card game (Table 3.3), we get

$$\max_{c_1 \in C_1} \left(\min_{c_2 \in C_2} u_1(c_1, c_2) \right) = \max \{0, 0, -0.5, 0\} = 0,$$

$$\min_{c_2 \in C_2} \left(\max_{c_1 \in C_1} u_1(c_1, c_2) \right) = \min \{0.5, 1\} = 0.5 \neq 0.$$

Throughout this chapter, we have assumed that the players in a strategic-form game choose their strategies simultaneously and independently, so a change of plans by one player cannot affect the strategy chosen by the other player. Suppose for a moment, however, that player 1 is afraid that, whatever randomized strategy he might choose, 2 would be able to discover that he had chosen this and would respond in the way that is worst for 1. Under this assumption, if player 1 chose the randomized strategy τ_1 in $\Delta(C_1)$, then his expected payoff would be $\min_{\tau_2 \in \Delta(C_2)} u_1(\tau_1,\tau_2)$. Theorem 3.2 asserts that player 1's equilibrium strategy in a two-person zero-sum game also maximizes 1's expected payoff under this pessimistic assumption. Because it maximizes his minimum expected payoff, an equilibrium strategy for player 1 in a two-person zero-sum game may be called a *maximin* strategy for player 1. Similarly, any equilibrium strategy for player 2 in a two-person zero-sum game is a *minimax* strategy against player 1.

There is a close connection between the theory of two-person zero-sum games and the concept of duality in optimization theory. To see why, consider the optimization problem

(3.10) $$\underset{x \in \mathbf{R}^n}{\text{minimize}} \; f(x), \quad \text{subject to } g_k(x) \geq 0, \quad \forall k \in \{1,2,\ldots,m\},$$

where $f(\cdot), g_1(\cdot), \ldots, g_m(\cdot)$ are functions from $\mathbf{R}^n$ to $\mathbf{R}$. This problem is equivalent to the problem

(3.11) $$\underset{x \in \mathbf{R}^n}{\text{minimize}} \left(\underset{y \in \mathbf{R}_+^m}{\text{maximum}} \; f(x) - \sum_{k=1}^m y_k g_k(x) \right).$$

(Here $\mathbf{R}_+^m$ is the set of all vectors $y = (y_1,\ldots,y_m)$ such that $y_1 \geq 0, \ldots, y_m \geq 0$.) To see that these two problems are equivalent, notice that the objective function being minimized in (3.11) is equal to $f(x)$ if x satisfies the constraints of (3.10) (where $y = (0,\ldots,0)$ achieves the maximum), and is equal to $+\infty$ if any of the constraints are violated. By reversing the order of maximization and minimization, we can construct the *dual* of this optimization problem, which is

(3.12) $$\underset{y \in \mathbf{R}_+^m}{\text{maximize}} \left(\underset{x \in \mathbf{R}^n}{\text{minimum}} \; f(x) - \sum_{k=1}^m y_k g_k(x) \right).$$

Classic results in optimization theory show that, with some additional assumptions, the optimal values of an optimization problem and its dual are equal (see Geoffrion, 1971). These results can be understood as generalizations of the equality between max–min and min–max in Theorem 3.2. The most important special case of duality theory is the case of linear programming problems.

Linear programming problems are optimization problems in which the objective function is linear and all constraint functions are affine (that is, linear plus a constant). That is, to make (3.10) a linear programming problem, we let

$$f(x) = \sum_{l=1}^{n} c_l x_l, \quad \text{and}$$

$$g_k(x) = \sum_{l=1}^{n} a_{kl} x_l - b_k, \quad \forall k \in \{1, \ldots, m\},$$

where, the numbers a_{kl}, b_k, and c_l are given parameters. Then the optimization problem (3.10) becomes

(3.13) $\displaystyle \minimize_{x \in \mathbf{R}^n} \sum_{l=1}^{n} c_l x_l$

subject to $\displaystyle \sum_{l=1}^{n} a_{kl} x_l \geq b_k, \quad \forall k \in \{1, \ldots, m\}$.

The dual problem (3.12) can then be written

(3.14) $\displaystyle \maximize_{y \in \mathbf{R}^m_+} \left(\minimize_{x \in \mathbf{R}^n} \sum_{l=1}^{n} \left(c_l - \sum_{k=1}^{m} y_k a_{kl} \right) x_l + \sum_{k=1}^{m} y_k b_k \right),$

which is equivalent to the problem

(3.15) $\displaystyle \maximize_{y \in \mathbf{R}^m_+} \sum_{k=1}^{m} y_k b_k$

subject to $\displaystyle \sum_{k=1}^{m} y_k a_{kl} = c_l, \quad \forall l \in \{1, \ldots, n\}$.

(The objective function being maximized in (3.14) would equal $-\infty$ if any constraints in (3.15) were violated, because the components of x in (3.14) can be positive or negative.) Notice that (3.15) has one decision variable (y_k) for each constraint in (3.13), and (3.15) has one constraint for each decision variable (x_l) in (3.13). In (3.13) the constraints are inequalities and the variables can be any real numbers, whereas in (3.15) the constraints are equalities and the variables are required to be non-negative.

The proof of the following version of the duality theorem can be found in any book on linear programming (see, for example, Chvatal, 1983).

DUALITY THEOREM OF LINEAR PROGRAMMING. *Suppose that there exists at least one vector $\hat{x}$ in $\mathbf{R}^n$ such that*

$$\sum_{l=1}^{n} a_{kl}\hat{x}_l \geq b_k, \quad \forall k \in \{1, \ldots, m\},$$

and there exists at least one nonnegative vector $\hat{y}$ in $\mathbf{R}_+^m$ such that

$$\sum_{k=1}^{m} \hat{y}_k a_{kl} = c_l, \quad \forall l \in \{1, \ldots, n\}.$$

Then the optimization problems (3.13) and (3.15) both have optimal solutions, and at these optimal solutions the values of the objective functions $\sum_{l=1}^{n} c_l x_l$ and $\sum_{k=1}^{m} y_k b_k$ are equal.

3.9 Bayesian Equilibria

For a Bayesian game with incomplete information, Harsanyi (1967–68) defined a *Bayesian equilibrium* to be any Nash equilibrium of the type-agent representation in strategic form (as defined in Section 2.8). That is, a Bayesian equilibrium specifies an action or randomized strategy for each type of each player, such that each type of each player would be maximizing his own expected utility when he knows his own given type but does not know the other players' types. Notice that, in a Bayesian equilibrium, a player's action can depend on his own type but not on the types of the other players. (By definition, a player's type is supposed to subsume all of his private information at the beginning of the game, when he chooses his action.) We must specify what every possible type of every player would do, not just the actual types, because otherwise we could not define the expected utility payoff for a player who does not know the other players' actual types.

To formally state the definition of a Bayesian equilibrium, let Γ^b be a Bayesian game as defined in Section 2.8, so

$$\Gamma^b = (N, (C_i)_{i \in N}, (T_i)_{i \in N}, (p_i)_{i \in N}, (u_i)_{i \in N}).$$

A *randomized-strategy profile* for the Bayesian game Γ^b is any σ in the set $\times_{i \in N} \times_{t_i \in T_i} \Delta(C_i)$, that is, any σ such that

$$\sigma = \left((\sigma_i(c_i|t_i))_{c_i \in C_i} \right)_{t_i \in T_i, i \in N},$$

$$\sigma_i(c_i|t_i) \geq 0, \quad \forall c_i \in C_i, \quad \forall t_i \in T_i, \quad \forall i \in N, \quad \text{and}$$

$$\sum_{c_i \in C_i} \sigma_i(c_i|t_i) = 1, \quad \forall t_i \in T_i, \quad \forall i \in N.$$

In such a randomized-strategy profile σ, the number $\sigma_i(c_i|t_i)$ represents the conditional probability that player i would use action c_i if his type were t_i. In the randomized-strategy profile σ, the randomized strategy for type t_i of player i is

$$\sigma_i(\cdot|t_i) = (\sigma_i(c_i|t_i))_{c_i \in C_i}.$$

A Bayesian equilibrium of the game Γ^b is any randomized-strategy profile σ such that, for every player i in N and every type t_i in T_i,

$$\sigma_i(\cdot|t_i) \in \underset{\tau_i \in \Delta(C_i)}{\operatorname{argmax}} \sum_{t_{-i} \in T_{-i}} p_i(t_{-i}|t_i) \sum_{c \in C} \left(\prod_{j \in N-i} \sigma_j(c_j|t_j) \right) \tau_i(c_i) u_i(c,t).$$

For a simple, two-player example, suppose that $C_1 = \{x_1, y_1\}$, $C_2 = \{x_2, y_2\}$, $T_1 = \{1.0\}$ (so player 1 has only one possible type and no private information), $T_2 = \{2.1, 2.2\}$, $p_1(2.1|1.0) = .6$, $p_1(2.2|1.0) = .4$, and the utility payoffs (u_1, u_2) depend on the actions and player 2's type as in Table 3.13.

In this game, y_2 is a strongly dominated action for type 2.1, and x_2 is a strongly dominated action for type 2.2, so 2.1 must choose x_2 and 2.2 must choose y_2 in a Bayesian equilibrium. Player 1 wants to get either (x_1, x_2) or (y_1, y_2) to be the outcome of the game, and he thinks that 2.1 is more likely than 2.2. Thus, the unique Bayesian equilibrium of this game is

$$\sigma_1(\cdot|1.0) = [x_1], \quad \sigma_2(\cdot|2.1) = [x_2], \quad \sigma_2(\cdot|2.2) = [y_2].$$

This example illustrates the danger of analyzing each matrix separately, as if it were a game with complete information. If it were common

Table 3.13   Expected payoffs for all types and action profiles in a Bayesian game

$t_2 = 2.1$				$t_2 = 2.2$		
		C_2				C_2
C_1	x_2	y_2		C_1	x_2	y_2
x_1	1,2	0,1		x_1	1,3	0,4
y_1	0,4	1,3		y_1	0,1	1,2

knowledge that player 2's type was 2.1, then the players would be in the matrix on the left in Table 3.13, in which the unique equilibrium is (x_1, x_2). If it were common knowledge that player 2's type was 2.2, then the players would be in the matrix on the right in Table 3.13, in which the unique equilibrium is (y_1, y_2). Thus, if we only looked at the full-information Nash equilibria of these two matrices, then we might make the prediction, "the outcome of the game will be (x_1, x_2) if 2's type is 2.1 and will be (y_1, y_2) if 2's type is 2.2."

This prediction would be absurd, however, for the actual Bayesian game in which player 1 does not initially know player 2's type. Notice first that this prediction ascribes two different actions to player 1, depending on 2's type (x_1 if 2.1, and y_1 if 2.2). So player 1 could not behave as predicted unless he received some information from player 2. That is, this prediction would be impossible to fulfill unless some kind of communication between the players is added to the structure of the game. Now notice that player 2 prefers (y_1, y_2) over (x_1, x_2) if her type is 2.1, and she prefers (x_1, x_2) over (y_1, y_2) if her type is 2.2. Thus, even if communication between the players were allowed, player 2 would not be willing to communicate the information that is necessary to fulfill this prediction, because it would always give her the outcome that she prefers less. She would prefer to manipulate her communications to get the outcomes (y_1, y_2) if 2.1 and (x_1, x_2) if 2.2. Games with communication are discussed in detail in Chapter 6.

3.10 Purification of Randomized Strategies in Equilibria

Equilibria in randomized strategies sometimes seem difficult to interpret. Consider again the example in Table 3.7. The unique equilibrium of this game is $(.75[T] + .25[B], .5[L] + .5[R])$. But the necessity for player 1 to randomly choose among T and B with probabilities .75 and .25, respectively, might not seem to coincide with any compulsion that people experience in real life. Of course, if player 1 thinks that player 2 is equally likely to choose either L or R, then player 1 is willing to randomize in any way between T and B. But what could make player 1 actually want to use the exact probabilities .75 and .25 in his randomization?

Harsanyi (1973) showed that Nash equilibria that involve randomized strategies can be interpreted as limits of Bayesian equilibria in which each player is (almost) always choosing his uniquely optimal action (see also Milgrom and Weber, 1985). Harsanyi's idea is to modify the game slightly so that each player has some minor private information about his own payoffs. For example, suppose that the game in Table 3.7 were modified slightly, to a game with incomplete information (Table 3.14). Here ε is some given number such that $0 < \varepsilon < 1$, and $\tilde{\alpha}$ and $\tilde{\beta}$ are independent and identically distributed random variables, each drawn from a uniform distribution over the interval from 0 to 1. When the game is played, player 1 knows the value of $\tilde{\alpha}$ but not $\tilde{\beta}$, and player 2 knows the value of $\tilde{\beta}$ but not $\tilde{\alpha}$. If ε is 0, then Table 3.14 becomes the same as Table 3.7, so let us think of ε as some very small positive number (say, .0001). Then $\tilde{\alpha}$ and $\tilde{\beta}$ represent minor factors that have a small influence on the players' payoffs when T or L is chosen.

Given ε, there is an essentially unique Bayesian equilibrium for this game. Player 1 chooses T if he observes $\tilde{\alpha}$ greater than $(2 + \varepsilon)/(8 + \varepsilon^2)$, and he chooses B otherwise. Player 2 chooses L if she observes $\tilde{\beta}$ greater than $(4 - \varepsilon)/(8 + \varepsilon^2)$, and she chooses R otherwise. That is, the Bayesian equilibrium σ satisfies

$$\sigma_1(\cdot|\tilde{\alpha}) = [T] \text{ if } \tilde{\alpha} > (2 + \varepsilon)/(8 + \varepsilon^2),$$
$$\sigma_1(\cdot|\tilde{\alpha}) = [B] \text{ if } \tilde{\alpha} < (2 + \varepsilon)/(8 + \varepsilon^2),$$

and

$$\sigma_2(\cdot|\tilde{\beta}) = [L] \text{ if } \tilde{\beta} > (4 - \varepsilon)/(8 + \varepsilon^2),$$
$$\sigma_2(\cdot|\tilde{\beta}) = [R] \text{ if } \tilde{\beta} < (4 - \varepsilon)/(8 + \varepsilon^2).$$

$(\sigma_1(\cdot|\tilde{\alpha})$ and $\sigma_2(\cdot|\tilde{\beta})$ can be chosen arbitrarily in the zero-probability events that $\tilde{\alpha} = (2 + \varepsilon)/(8 + \varepsilon^2)$ or $\tilde{\beta} = (4 - \varepsilon)/(8 + \varepsilon^2)$.) When player 2 uses the equilibrium strategy σ_2 in this game, player 1 would be indifferent between T and B only if $\tilde{\alpha} = (2 + \varepsilon)/(8 + \varepsilon^2)$; otherwise his expected utility is uniquely maximized by the action designated for him by $\sigma_1(\cdot|\tilde{\alpha})$. Similarly, when player 1 uses the equilibrium strategy σ_1 in

Table 3.14 Payoffs for all strategy profiles

| | C_2 | |
C_1	L	R
T	$\varepsilon\tilde{\alpha}, \varepsilon\tilde{\beta}$	$\varepsilon\tilde{\alpha}, -1$
B	$1, \varepsilon\tilde{\beta}$	$-1, 3$

this game, player 2 would be indifferent between L and R only if $\tilde{\beta} = (4 - \varepsilon)/(8 + \varepsilon^2)$; otherwise her expected utility is uniquely maximized by the action designated for her by $\sigma_2(\cdot\,|\,\tilde{\beta})$. Thus, each player's expected behavior makes the other player almost indifferent between his two actions; so the minor factor that he observes independently can determine the unique optimal action for him. Notice that, as ε goes to 0, this equilibrium converges to the unique equilibrium of the game in Table 3.7, in which player 1 chooses T with probability .75, and player 2 chooses L with probability .50.

To see how the Bayesian equilibrium for Table 3.14 is computed, notice first that T becomes better for player 1 as $\tilde{\alpha}$ increases, and L becomes better for player 2 as $\tilde{\beta}$ increases. Thus, there should exist some numbers p and q such that player 1 chooses T if $\tilde{\alpha}$ is greater than p and chooses B if $\tilde{\alpha}$ is less than p, and player 2 chooses L if $\tilde{\beta}$ is greater than q and chooses R if $\tilde{\beta}$ is less than q. Then, from 1's perspective, the probability that 2 will choose L is $1 - q$. To make player 1 indifferent between T and B at the critical value of $\tilde{\alpha} = p$, we need

$$\varepsilon p = (1)(1 - q) + (-1)q.$$

Similarly, to make player 2 indifferent between L and R at the critical value of $\tilde{\beta} = q$, we need

$$\varepsilon q = (-1)(1 - p) + (3)p.$$

The solution to these two equations is $p = (2 + \varepsilon)/(8 + \varepsilon^2)$ and $q = (4 - \varepsilon)/(8 + \varepsilon^2)$.

In general, when we study an equilibrium that involves randomized strategies, we can interpret each player's randomization as depending on minor factors that have been omitted from the description of the game. When a game has no equilibrium in pure strategies, we should expect that a player's optimal strategy may be determined by some minor factors that he observes independently of the other players.

It may be interesting to compare this conclusion with the focal-point effect. In games with multiple equilibria, factors that have little or no impact on payoffs but are conspicuously observed by all the players may have a significant impact on the outcome of the game by determining the focal equilibrium. Thus, the theory of randomized equilibria identifies situations where minor private information may be decisive, and the focal-point effect identifies situations where minor public information may be decisive.

3.11 Auctions

The concept of Bayesian equilibrium is important in the analysis of auctions (see, for example, Milgrom, 1985, 1987; Milgrom and Weber, 1982; McAfee and McMillan, 1987; and Wilson, 1987). We consider here some simple introductory examples.

Consider first an example where the players are n bidders in an auction for a single indivisible object. Each player knows privately how much the object would be worth to him, if he won the auction. Suppose that there is some positive number M and some increasing differentiable function $F(\cdot)$ such that each player considers the values of the object to the other $n - 1$ players to be independent random variables drawn from the interval $[0,M]$ with cumulative distribution F. (So $F(v)$ is the probability that any given player has a value for the object that is less than v.) In the auction, each player i simultaneously submits a sealed bid b_i, which must be a nonnegative number. The object is delivered to the player whose bid is highest, and this player must pay the amount of his bid. No other player pays anything. Thus, letting $b = (b_1, \ldots, b_n)$ denote the profile of bids and letting $v = (v_1, \ldots, v_n)$ denote the profile of values of the object to the n players, the expected payoff to player i is

$$u_i(b,v) = v_i - b_i \text{ if } \{i\} = \operatorname*{argmax}_{j \in \{1, \ldots, n\}} b_j,$$

$$= 0 \qquad \text{if } i \notin \operatorname*{argmax}_{j \in \{1, \ldots, n\}} b_j.$$

This is a Bayesian game where each player's type is his value for the object. We now show how to find a Bayesian equilibrium in which every player chooses his bid according to some function β that is differentiable and increasing.

In the equilibrium, player i expects the other players to be choosing bids between 0 and $\beta(M)$, so player i would never want to submit a bid higher than $\beta(M)$. Suppose that, when i's value is actually v_i, he submits a bid equal to $\beta(w_i)$. Another player j will submit a bid that is less than $\beta(w_i)$ iff j's value $\bar{v}_j$ satisfies $\beta(\bar{v}_j) < \beta(w_i)$, which happens iff $\bar{v}_j < w_i$, because β is increasing. So the probability that the bid $\beta(w_i)$ will win the auction is $F(w_i)^{n-1}$, because the types of the other $n - 1$ players are independently distributed according to F. Thus, the expected payoff to player i from bidding $\beta(w_i)$ with value v_i would be

$$(v_i - \beta(w_i))F(w_i)^{n-1}.$$

However, by the definition of an equilibrium, the optimal bid for i with value v_i should be $\beta(v_i)$. So the derivative of this expected payoff with respect to w_i should equal 0 when w_i equals v_i. That is,

$$0 = (v_i - \beta(v_i))F'(v_i)(n - 1)F(v_i)^{n-2} - \beta'(v_i)F(v_i)^{n-1}.$$

This differential equation implies that, for any x in $[0,M]$

$$\beta(x)F(x)^{n-1} = \int_0^x y(n - 1)F(y)^{n-2}F'(y)dy.$$

This and other related results have been derived by Vickrey (1961) and Ortega-Reichert (1968). In the case where the types are uniformly distributed, so that $F(y) = y/M$ for any y in $[0,M]$, this formula implies

$$\beta(v_i) = (1 - 1/n)v_i, \quad \forall v_i \in [0,M].$$

Notice that the margin of profit that each player allows himself in his bid decreases as the number of bidders increases.

This game is an example of an auction with *independent private values*, in the sense that each bidder knows privately the actual value of the object to himself and considers the others' values to be independent random variables. In many auctions, however, the value of the object would be the same to all bidders, although the bidders may have different estimates of this value, because they may have different private information about its quality. Such auctions are called *common value* auctions. For example, auctions for oil drilling rights on a given tract of land may be common value auctions, where the common value is the value of the oil under the land, about which the bidders may have different information and different estimates.

Consider now a simple example of a two-bidder auction for a single object with an unknown common value. This object has a monetary value that depends on three independent random variables, $\tilde{x}_0$, $\tilde{x}_1$, and $\tilde{x}_2$, each of which is drawn from a uniform distribution on the interval from 0 to 1. The bidder who gets the object will ultimately derive benefits from owning it that are worth $A_0\tilde{x}_0 + A_1\tilde{x}_1 + A_2\tilde{x}_2$, where A_0, A_1, and A_2 are given nonnegative constants that are commonly known by the bidders. At the time of the auction, player 1 has observed $\tilde{x}_0$ and $\tilde{x}_1$, but does not know $\tilde{x}_2$; and player 2 has observed $\tilde{x}_0$ and $\tilde{x}_2$, but does not know $\tilde{x}_1$. So player 1's type is $(\tilde{x}_0,\tilde{x}_1)$, and 2's type is $(\tilde{x}_0,\tilde{x}_2)$.

In the auction, the players simultaneously submit sealed bids c_1 and c_2 that are nonnegative numbers. The high bidder wins the object and pays the amount of his bid, and the low bidder pays nothing. In case of a tie, each player has a .5 probability of getting the object at the price of his bid. We assume that both players have risk-neutral utility for money. Thus, the utility payoff function for player i is (let j denote the player other than i)

$$
\begin{aligned}
u_i(c_1, c_2, (\tilde{x}_0, \tilde{x}_1), (\tilde{x}_0, \tilde{x}_2)) &= A_0 \tilde{x}_0 + A_1 \tilde{x}_1 + A_2 \tilde{x}_2 - c_i && \text{if } c_i > c_j, \\
&= (A_0 \tilde{x}_0 + A_1 \tilde{x}_1 + A_2 \tilde{x}_2 - c_i)/2 && \text{if } c_i = c_j, \\
&= 0 && \text{if } c_i < c_j.
\end{aligned}
$$

There is a unique linear Bayesian equilibrium of this game, in which player 1's bid is

$$A_0 \tilde{x}_0 + 0.5(A_1 + A_2)\tilde{x}_1$$

and player 2's bid is

$$A_0 \tilde{x}_0 + 0.5(A_1 + A_2)\tilde{x}_2.$$

To prove that this strategy profile is an equilibrium, suppose that player 1 expects 2 to use this strategy, but 1 is considering some other bid b, given the values $\tilde{x}_0 = x_0$ and $\tilde{x}_1 = x_1$ that he has observed. Such a bid would win the object for 1 if

$$b > A_0 x_0 + 0.5(A_1 + A_2)\tilde{x}_2,$$

that is, if

$$2(b - A_0 \tilde{x}_0)/(A_1 + A_2) > \tilde{x}_2.$$

So, given 1's information, a bid of b will win him the object with probability $Y(b)$, where

$$Y(b) = 2(b - A_0 x_0)/(A_1 + A_2),$$

when this number $Y(b)$ is between 0 and 1. Player 1 should not use any bid b that makes $Y(b)$ greater than 1, because such a bid would be larger than necessary to win the object for sure; and he should not use any bid that makes $Y(b)$ less than 0, because such a bid would be sure to lose. Then, given 1's type $(\tilde{x}_0, \tilde{x}_1)$, the conditionally expected payoff for player 1, when 1 submits a bid of b and 2 implements her equilibrium strategy, is

$$\int_0^{Y(b)} (A_0x_0 + A_1x_1 + A_2y_2 - b)dy_2$$

$$= Y(b)(A_0x_0 + A_1x_1 + A_2Y(b)/2 - b).$$

To interpret this formula, notice that $A_0x_0 + A_1x_1 + A_2Y(b)/2$ would be the conditionally expected value of the object, given both that player 1's type is (x_0,x_1) and that player 1 could win the object by bidding b. Substituting the definition of $Y(b)$ into this formula and choosing b to maximize this expected payoff for 1, we get

$$b = A_0x_0 + 0.5(A_1 + A_2)x_1.$$

Substituting this value of b into the definition of $Y(b)$, we get $Y(b) = x_1$; so this bid satisfies our requirement that $0 \le Y(b) \le 1$, for every x_0 and x_1 between 0 and 1.

A similar argument shows that player 2's optimal bid is as specified by her equilibrium strategy, given that player 1 is expected to follow his equilibrium strategy.

Suppose, for example, that $A_0 = A_1 = A_2 = 100$, and $\tilde{x}_0 = 0$ and $\tilde{x}_1 = 0.01$. Then 1's equilibrium strategy tells him to submit a bid of 1. This bid may seem surprisingly low. After all, the expected value of $\tilde{x}_2$ is 0.5, so the expected monetary value of the object in the auction given 1's current information is $100 \times 0 + 100 \times 0.01 + 100 \times 0.5 = 51$. Why is player 1 submitting a bid that is less than 2% of the expected value of the object to him? To understand why player 1 must bid so conservatively, notice what would happen if he submitted a bid of 50 instead. It might seem that this bid would give player 1 a positive expected profit, because his bid is lower than the expected value of the object to him and is high enough to have a substantial probability of winning. But according to our formulas above, 1's expected utility payoff from this bid is -12 (because $Y(50) = .5$). Bidder 1 would expect such serious losses from a bid of 50 because, if 1's bid of 50 were to win the auction, then 2's equilibrium bid of $100\tilde{x}_2$ would be not greater than 50; so $\tilde{x}_2$ would have to be between 0 and 0.50. Thus, the conditionally expected value of the object, given 1's information at the time of the auction and the additional information that a bid of 50 would win the auction, is $100 \times 0.01 + 100 \times 0.5/2 = 26$. So a bid of 50 would give player 1 a probability .5 of buying an object for 50 that would have an expected value of 26 when he gets to buy it at this price, so that 1's expected profit is indeed $0.5(26 - 50) = -12 < 0$.

Thus, when computing the expected profit from a particular bid in an auction, it is important that the bidder estimate the value of the object by its conditionally expected value given his current information *and the additional information that could be inferred if this bid won the auction.* This conditionally expected value is often significantly less than the expected value of the object given the bidder's information at the time that he submits the bid. This fact is called the *winner's curse.*

Now consider the case in which $A_0 = A_1 = \varepsilon$ and $A_2 = 100 - \varepsilon$, where ε is some very small positive number. In this case, player 1's equilibrium strategy is to bid $\varepsilon \tilde{x}_0 + 50\tilde{x}_1$, and player 2's equilibrium strategy is to bid $\varepsilon \tilde{x}_0 + 50\tilde{x}_2$. Thus, even though $\tilde{x}_0$ and $\tilde{x}_1$ have the same small effect on the value of the object, $\tilde{x}_1$ has a much greater effect on player 1's optimal bid, because $\tilde{x}_1$ is information that he knows privately, independently of anything that player 2 knows, whereas $\tilde{x}_0$ is public information. As ε goes to 0, this game converges to one in which player 2 knows the value of the object, and player 1 only knows that this value was drawn from a uniform distribution over the interval from 0 to 100. The limit of the Bayesian equilibria (in the sense of Milgrom and Weber, 1985), as ε goes to 0, is a randomized equilibrium in which the informed player 2 bids one-half of the value of the object and player 1 submits a bid drawn from a uniform distribution over the interval from 0 to 50 and based on minor factors that are independent of anything 2 can observe at the time of the auction.

3.12 Proof of Existence of Equilibrium

The Kakutani (1941) fixed-point theorem is a useful mathematical tool for proving existence of many solution concepts in game theory including Nash equilibrium. To state the Kakutani fixed-point theorem, we must first develop some terminology. (For a more thorough introduction to these mathematical concepts, see Debreu, 1959, pp. 1–27.)

For any finite set M, $\mathbf{R}^M$ is the set of all vectors of the form $x = (x_m)_{m \in M}$ such that, for each m in M, the m-component x_m is a real number in $\mathbf{R}$. When it is more convenient, we can also interpret $\mathbf{R}^M$ equivalently as the set of all functions from M into the real numbers $\mathbf{R}$, in which case the m-component of x in $\mathbf{R}^M$ can be written $x(m)$ instead of x_m. $\mathbf{R}^M$ is a finite-dimensional vector space.

Let S be a subset of the finite-dimensional vector space $\mathbf{R}^M$. S is *convex* iff, for every two vectors x and y and every number λ such that $0 \leq \lambda$

≤ 1, if $x \in S$ and $y \in S$, then $\lambda x + (1 - \lambda)y \in S$. (Here $x = (x_m)_{m \in M}$, $y = (y_m)_{m \in M}$, and $\lambda x + (1 - \lambda)y = (\lambda x_m + (1 - \lambda)y_m)_{m \in M}$.) S is *closed* iff, for every convergent sequence of vectors $(x(j))_{j=1}^{\infty}$, if $x(j) \in S$ for every j, then $\lim_{j \to \infty} x(j) \in S$. (A subset of $\mathbf{R}^M$ is *open* iff its complement is closed.) S is *bounded* iff there exists some positive number K such that, for every vector x in S, $\Sigma_{m \in M} |x_m| \leq K$.

A point-to-set *correspondence* $G{:}X \longrightarrow\!\!\!\!\rightarrow Y$ is any mapping that specifies, for every point x in X, a set $G(x)$ that is a subset of Y. Suppose that X and Y are any metric spaces, so the concepts of *convergence* and *limit* have been defined for sequences of points in X and in Y. A correspondence $G{:}X \longrightarrow\!\!\!\!\rightarrow Y$ is *upper-hemicontinuous* iff, for every sequence $(x(j),y(j))_{j=1}^{\infty}$, if $x(j) \in X$ and $y(j) \in G(x(j))$ for every j, and the sequence $(x(j))_{j=1}^{\infty}$ converges to some point $\bar{x}$, and the sequence $(y(j))_{j=1}^{\infty}$ converges to some point $\bar{y}$, then $\bar{y} \in G(\bar{x})$. Thus, $G{:}X \longrightarrow\!\!\!\!\rightarrow Y$ is upper-hemicontinuous iff the set $\{(x,y)|\ x \in X,\ y \in G(x)\}$ is a closed subset of $X \times Y$.

Notice that most of the length of the definition of upper-hemicontinuity comes from the three hypotheses, which have the effect of limiting the cases in which the conclusion ($\bar{y} \in G(\bar{x})$) is required to hold. Thus, this definition is really rather weak. In particular, if $g{:}X \to Y$ is a continuous function from X to Y and $G(x) = \{g(x)\}$ for every x in X, then $G{:}X \longrightarrow\!\!\!\!\rightarrow Y$ is an upper-hemicontinuous point-to-set correspondence. So upper-hemicontinuous correspondences can be viewed as a generalization of continuous functions.

A *fixed point* of a correspondence $F{:}S \longrightarrow\!\!\!\!\rightarrow S$ is any x in S such that $x \in F(x)$.

KAKUTANI FIXED-POINT THEOREM.   *Let S be any nonempty, convex, bounded, and closed subset of a finite-dimensional vector space. Let $F{:}S \longrightarrow\!\!\!\!\rightarrow S$ be any upper-hemicontinuous point-to-set correspondence such that, for every x in S, $F(x)$ is a nonempty convex subset of S. Then there exists some $\bar{x}$ in S such that $\bar{x} \in F(\bar{x})$.*

Proofs of the Kakutani fixed-point theorem and other related theorems can be found in Burger (1963), Franklin (1980), and Border (1985). Scarf (1973) provides a constructive proof by developing an algorithm for computing the fixed point whose existence is implied by the theorem. (For a generalization of the Kakutani fixed-point theorem, see Glicksberg, 1952.)

To understand the role of the various assumptions in the Kakutani fixed-point theorem, let

$$S = [0,1] = \{x \in \mathbf{R} | 0 \le x \le 1\},$$

and let $F_1 : S \twoheadrightarrow S$ be

$$F_1(x) = \{1\} \text{ if } 0 \le x \le 0.5,$$
$$= \{0\} \text{ if } 0.5 < x \le 1.$$

Then F_1 has no fixed points, and it satisfies all of the assumptions of the Kakutani fixed-point theorem except for upper-hemicontinuity, because the set $\{(x,y) | x \in S, y \in F_1(x)\}$ is not a closed set at $(0.5, 0)$. To satisfy upper-hemicontinuity, we must extend this correspondence to $F_2 : S \twoheadrightarrow S$, where

$$F_2(x) = \{1\} \quad \text{if } 0 \le x < 0.5,$$
$$= \{0,1\} \text{ if } x = 0.5,$$
$$= \{0\} \quad \text{if } 0.5 < x \le 1.$$

F_2 also has no fixed points, and it satisfies all of the assumptions of the Kakutani fixed-point theorem except convex-valuedness, because $F_2(0.5)$ is not a convex set. To satisfy convex-valuedness, we must extend the correspondence to $F_3 : S \twoheadrightarrow S$, where

$$F_3(x) = \{1\} \quad \text{if } 0 \le x < 0.5,$$
$$= [0,1] \text{ if } x = 0.5,$$
$$= \{0\} \quad \text{if } 0.5 < x \le 1.$$

This correspondence F_3 satisfies all the assumptions of the Kakutani fixed-point theorem and has a fixed point, because $0.5 \in F_3(0.5)$.

We can now prove the general existence theorem for Nash equilibria of finite games.

Proof of Theorem 3.1. Let Γ be any finite game in strategic form, where

$$\Gamma = (N, (C_i)_{i \in N}, (u_i)_{i \in N}).$$

The set of randomized-strategy profiles $\times_{i \in N} \Delta(C_i)$ is a nonempty, convex, closed, and bounded subset of a finite-dimensional vector space.

(This set satisfies the above definition of boundedness with $K = |N|$, and it is a subset of $\mathbf{R}^M$, where $M = \cup_{i \in N} C_i$.)

For any σ in $\times_{i \in N} \Delta(C_i)$ and any player j in N, let

$$R_j(\sigma_{-j}) = \underset{\tau_j \in \Delta(C_j)}{\text{argmax}} \; u_j(\sigma_{-j}, \tau_j).$$

That is, $R_j(\sigma_{-j})$ is the set of j's best responses in $\Delta(C_j)$ to the combination of independently randomized strategies σ_{-j} for the other players. By Lemma 3.1 in Section 3.2, $R_j(\sigma_{-j})$ is the set of all probability distributions ρ_j over C_j such that

$$\rho_j(c_j) = 0 \text{ for every } c_j \text{ such that } c_j \notin \underset{d_j \in C_j}{\text{argmax}} \; u_j(\sigma_{-j}, [d_j]).$$

Thus, $R_j(\sigma_{-j})$ is convex, because it is a subset of $\Delta(C_j)$ that is defined by a collection of linear equalities. Furthermore, $R_j(\sigma_{-j})$ is nonempty, because it includes $[c_j]$ for every c_j in $\text{argmax}_{d_j \in C_j} u_j(\sigma_{-j}, [d_j])$, which is nonempty.

Let $R: \times_{i \in N} \Delta(C_i) \longrightarrow \times_{i \in N} \Delta(C_i)$ be the point-to-set correspondence such that

$$R(\sigma) = \underset{j \in N}{\times} R_j(\sigma_{-j}), \quad \forall \sigma \in \underset{i \in N}{\times} \Delta(C_i).$$

That is, $\tau \in R(\sigma)$ iff $\tau_j \in R_j(\sigma_{-j})$ for every j in N. For each σ, $R(\sigma)$ is nonempty and convex, because it is the Cartesian product of nonempty convex sets.

To show that R is upper-hemicontinuous, suppose that $(\sigma^k)_{k=1}^{\infty}$ and $(\tau^k)_{k=1}^{\infty}$ are convergent sequences,

$$\sigma^k \in \underset{i \in N}{\times} \Delta(C_i), \quad \forall k \in \{1,2,3, \ldots \},$$

$$\tau^k \in R(\sigma^k), \quad \forall k \in \{1,2,3, \ldots \},$$

$$\bar{\sigma} = \lim_{k \to \infty} \sigma^k, \text{ and}$$

$$\bar{\tau} = \lim_{k \to \infty} \tau^k.$$

To prove upper-hemicontinuity, we need to show that these conditions imply that $\bar{\tau}$ is in $R(\bar{\sigma})$. These conditions imply that, for every player j in N and every ρ_j in $\Delta(C_j)$,

$$u_j(\sigma^k_{-j}, \tau^k_j) \geq u_j(\sigma^k_{-j}, \rho_j), \quad \forall k \in \{1,2,3, \ldots \}.$$

By continuity of the expected utility function u_j on $\times_{i \in N} \Delta(C_i)$, this in turn implies that, for every j in N and ρ_j in $\Delta(C_j)$,

$$u_j(\overline{\sigma}_{-j}, \overline{\tau}_j) \geq u_j(\overline{\sigma}_{-j}, \rho_j).$$

So $\overline{\tau}_j \in R_j(\overline{\sigma}_{-j})$ for every j in N, and so $\overline{\tau} \in R(\overline{\sigma})$. Thus, $R: \times_{i \in N} \Delta(C_i) \longrightarrow \times_{i \in N} \Delta(C_i)$ is an upper-hemicontinuous correspondence.

By the Kakutani fixed-point theorem, there exists some randomized-strategy profile σ in $\times_{i \in N} \Delta(C_i)$ such that $\sigma \in R(\sigma)$. That is, $\sigma_j \in R_j(\sigma_{-j})$ for every j in N, and so σ is a Nash equilibrium of Γ.   ∎

3.13 Infinite Strategy Sets

We have mainly emphasized finite games, to minimize mathematical difficulties as we explore the fundamental principles of game-theoretic analysis. In this section, however, we broaden our scope to consider a more general class of games, in which players may have infinitely many strategies. In particular, we want to include the possibility that the set of pure strategies for a player may be a bounded interval on the real number line, such as [0,1]. (This set includes all rational and irrational numbers between 0 and 1, so it is an uncountably infinite set.)

To state results in greater generality, all results in this section are based only on the assumption that, for each player i, the strategy set C_i is a *compact metric space*. A compact metric space is a general mathematical structure for representing infinite sets that can be well approximated by large finite sets. One important fact is that, in a compact metric space, any infinite sequence has a convergent subsequence. The reader who is unfamiliar with metric spaces is referred to any mathematics text on real analysis, for example Royden (1968) or Kolmogorov and Fomin (1970). Any closed and bounded subset of a finite-dimensional vector space is an example of a compact metric space. More specifically, any closed and bounded interval of real numbers is an example of a compact metric space, where the distance between two points x and y is $|x - y|$. We will not need to refer to any examples more complicated than these.

(For completeness, we summarize the basic definitions here. A *metric space* is a set M together with a function $\delta: M \times M \to \mathbf{R}$, which defines the *distance* $\delta(x,y)$ between any two points x and y in the set and which satisfies the following properties, for every x, y, and z in M:

$$\delta(x,y) = \delta(y,x) \geq 0,$$

$$\delta(x,y) = 0 \text{ iff } x = y,$$

$$\delta(x,y) + \delta(y,z) \geq \delta(x,z).$$

In a metric space, a point y is the limit of a sequence of points $(x(k))_{k=1}^{\infty}$ iff the distance $\delta(x(k),y)$ goes to 0 as k goes to infinity. An *open ball* of radius ε around a point x, denoted $B(x,\varepsilon)$, is the set of all points in the metric space that have a distance less than ε from x; that is,

$$B(x,\varepsilon) = \{y \mid \delta(x,y) < \varepsilon\}.$$

A set S is an *open* subset of a metric space M iff, for every x in S, there is some positive number ε such that $B(x,\varepsilon) \subseteq S$. A metric space M is *compact* iff every collection of open sets that covers M, in the sense that their union includes all of M, has a finite subcollection that also covers M.)

Let N denote the finite set of players,

$$N = \{1,2, \ldots ,n\}.$$

For each player i in N, let C_i be a compact metric space that denotes the set of all pure strategies available to player i. As usual, let $C = \times_{i \in N} C_i = C_1 \times \cdots \times C_n$. As a finite Cartesian product of compact metric spaces, C is also a compact metric space (with $\delta(c,\hat{c}) = \Sigma_{i \in N} \delta(c_i,\hat{c}_i)$).

When there are infinitely many actions in the set C_i, a randomized strategy for player i can no longer be described by just listing the probability of each individual action. For example, suppose that C_i is the interval $[0,1]$. If player i selected his action from a uniform distribution over the interval $[0,1]$, then each individual action in $[0,1]$ would have zero probability; but the same would be true if he selected his action from a uniform distribution over the interval from 0.5 to 1. To describe a probability distribution over C_i, we must list the probabilities of subsets of C_i. Unfortunately, for technical reasons, it may be mathematically impossible to consistently assign probabilities to all subsets of an infinite set, so some weak restriction is needed on the class of subsets whose probabilities can be meaningfully defined. These are called the *measurable sets*. Here, we let the measurable subsets of C and of each set C_i be the smallest class of subsets that includes all open subsets, all closed subsets, and all finite or countably infinite unions and intersections of sets in the class. These are the *Borel subsets* (and they include virtually all subsets that could be defined without the use of very sophisticated mathematics). Let $\mathcal{B}_i$ denote the set of such measurable or Borel subsets of C_i.

Let $\Delta(C_i)$ denote the set of probability distributions over C_i. That is, $\sigma_i \in \Delta(C_i)$ iff σ_i is a function that assigns a nonnegative number $\sigma_i(Q)$ to each Q that is a Borel subset of C_i, $\sigma_i(C_i) = 1$, and, for any countable collection $(Q_k)_{k=1}^{\infty}$ of pairwise-disjoint Borel subsets of C_i,

$$\sigma_i \left(\bigcup_{k=1}^{\infty} Q_k \right) = \sum_{k=1}^{\infty} \sigma_i(Q_k).$$

We define convergence for randomized strategies by assigning the *weak topology* to $\Delta(C_i)$. This topology is such that a sequence $(\sigma_i^k)_{k=1}^{\infty}$ of probability distributions in $\Delta(C_i)$ converges to a probability distribution σ_i iff, for every bounded continuous function $f:C_i \to \mathbf{R}$,

$$\lim_{k \to \infty} \int_{c_i \in C_i} f(c_i) d\sigma_i^k(c_i) = \int_{c_i \in C_i} f(c_i) d\sigma_i(c_i).$$

This equation asserts that, if $\tilde{c}_i^k$ is a random strategy drawn from C_i according to the σ_i^k distribution, and $\tilde{c}_i$ is a random strategy drawn according to the σ_i distribution, then the expected value of $f(\tilde{c}_i^k)$ must converge to the expected value of $f(\tilde{c}_i)$ as k goes to infinity. With this topology, the set of probability distributions in $\Delta(C_i)$ is itself a compact metric space (with a suitably defined distance function, called the *Prohorov metric*; see Billingsley, 1968).

A function $g:C \to \mathbf{R}$ is *(Borel) measurable* iff, for every number x in $\mathbf{R}$, the set $\{c \in C | g(c) \geq x\}$ is a Borel subset of C. A function $g:C \to \mathbf{R}$ is *bounded* iff there exists some number K such that $|g(c)| \leq K$ for every c in C. To be able to define expected utilities, we must require that players' utility functions are measurable and bounded in this sense.

Let $u = (u_1, \ldots, u_n)$ and $\hat{u} = (\hat{u}_1, \ldots, \hat{u}_n)$ be any two profiles of utility functions defined on C, such that, for each i in N, $u_i:C \to \mathbf{R}$ and $\hat{u}_i:C \to \mathbf{R}$ are bounded measurable functions, to be interpreted as possible payoff functions for player i. We may define the *distance* between the utility-function profiles u and $\hat{u}$ to be

$$\max_{i \in N} \sup_{c \in C} |u_i(c) - \hat{u}_i(c)|.$$

(The *supremum*, or *sup*, of a set of numbers or values of a function is the least upper bound of the set. If a maximum exists in the set, then the supremum and the maximum are the same.) As usual, we may extend any utility function to the set of all randomized-strategy profiles $\times_{j\in N} \Delta(C_j)$, so

$$u_i(\sigma) = \int_{c_n \in C_n} \cdots \int_{c_1 \in C_1} u_i(c) d\sigma_1(c_1) \ldots d\sigma_n(c_n), \quad \forall \sigma \in \underset{j\in N}{\times} \Delta(C_j),$$

where $\sigma = (\sigma_1, \ldots, \sigma_n)$ and $c = (c_1, \ldots, c_n)$.

The utility-function profiles u and $\hat{u}$ each complete the definition of a game in strategic form, denoted by Γ and $\hat{\Gamma}$ respectively. That is,

$$\Gamma = (N, (C_i)_{i\in N}, (u_i)_{i\in N}), \quad \hat{\Gamma} = (N, (C_i)_{i\in N}, (\hat{u}_i)_{i\in N}).$$

If $\sigma = (\sigma_i)_{i\in N}$ is an equilibrium of Γ, then σ is, of course, not necessarily an equilibrium of $\hat{\Gamma}$. Even if u and $\hat{u}$ are very close, the equilibria of Γ and $\hat{\Gamma}$ may be very far apart. For example, if there is only one player, $C_1 = [0,1]$, $u_1(c_1) = \varepsilon c_1$, and $\hat{u}_1(c_1) = -\varepsilon c_1$, where ε is any small positive number, then the unique equilibrium of Γ is 1 and the unique equilibrium of $\hat{\Gamma}$ is 0, even though the distance between u and $\hat{u}$ is only 2ε.

However, if u and $\hat{u}$ are very close, there is a sense in which the equilibria of Γ are "almost" equilibria of $\hat{\Gamma}$. For any nonnegative number ε, Radner (1980) defined an ε-*equilibrium* (or an ε-*approximate equilibrium*) of any strategic-form game to be a combination of randomized strategies such that no player could expect to gain more than ε by switching to any of his feasible strategies, instead of following the randomized strategy specified for him. That is, σ is an ε-equilibrium of Γ iff

$$u_i(\sigma_{-i}, [c_i]) - u_i(\sigma) \le \varepsilon, \quad \forall i \in N, \quad \forall c_i \in C_i.$$

Obviously, when $\varepsilon = 0$, an ε-equilibrium is just a Nash equilibrium in the usual sense. The following result, due to Fudenberg and Levine (1986), shows that such ε-equilibria have a kind of continuity across games.

THEOREM 3.3. *Let Γ and $\hat{\Gamma}$ be as above. Let α be the distance between the utility-function profiles u and $\hat{u}$, and suppose that σ is an ε-equilibrium of $\hat{\Gamma}$. Then σ is an $(\varepsilon + 2\alpha)$-equilibrium of Γ.*

Proof. For any player i and any action c_i,

$$u_i(\sigma_{-i},[c_i]) - u_i(\sigma)$$

$$= (u_i(\sigma_{-i},[c_i]) - \hat{u}_i(\sigma_{-i},[c_i])) + (\hat{u}_i(\sigma_{-i},[c_i]) - \hat{u}_i(\sigma)) +$$

$$(\hat{u}_i(\sigma) - u_i(\sigma))$$

$$\leq \alpha + \varepsilon + \alpha = 2\alpha + \varepsilon. \quad \blacksquare$$

For any game with continuous utility functions, the ε-equilibrium concept also satisfies a related form of continuity, as stated by the following theorem.

THEOREM 3.4. *Suppose that $u_i:C \to \mathbf{R}$ is continuous, for every player i in N. Let $(\sigma^k)_{k=1}^{\infty}$ be a convergent sequence of randomized-strategy profiles in $\times_{i \in N} \Delta(C_i)$, and let $(\varepsilon_k)_{k=1}^{\infty}$ be a convergent sequence of nonnegative numbers such that, for every k, σ^k is an ε_k-equilibrium of Γ. Let $\bar{\sigma} = \lim_{k \to \infty} \sigma^k$ and let $\bar{\varepsilon} = \lim_{k \to \infty} \varepsilon_k$. Then $\bar{\sigma}$ is an $\bar{\varepsilon}$-equilibrium of Γ. In particular, if $\bar{\varepsilon}$ equals 0, then $\bar{\sigma}$ is an equilibrium of Γ.*

Proof. For any player i and any c_i in C_i,

$$u_i(\bar{\sigma}_{-i},[c_i]) - u_i(\bar{\sigma}) = \lim_{k \to \infty} (u_i(\sigma^k_{-i},[c_i]) - u_i(\sigma^k))$$

$$\leq \lim_{k \to \infty} \varepsilon_k = \bar{\varepsilon}. \quad \blacksquare$$

It is important to be able to relate infinite games to the conceptually simpler class of finite games. A function f_i from C_i into a finite set D_i is *(Borel) measurable* iff, for every b_i in D_i, $\{c_i \in C_i |\ f_i(c_i) = b_i\}$ is a Borel-measurable set. We may say that the game $\hat{\Gamma}$ is *essentially finite* iff there exists some finite strategic-form game $\Gamma^* = (N, (D_i)_{i \in N}, (v_i)_{i \in N})$ and measurable functions $(f_i:C_i \to D_i)_{i \in N}$ such that,

$$\hat{u}_i(c) = v_i(f_1(c_1), \ldots, f_n(c_n)), \quad \forall i \in N, \quad \forall c \in C.$$

The following finite-approximation theorem can be easily proved, using the fact that the sets of strategies are compact.

THEOREM 3.5. *Let α be any strictly positive real number. Suppose that, for every player i, the utility function $u_i:C \to \mathbf{R}$ is continuous on the compact* *domain C. Then there exists an essentially finite game* $\hat{\Gamma} = (N, (C_i)_{i \in N}, (\hat{u}_i)_{i \in N})$ *such that the distance between u and û is less than* α.

These results enable us to prove that there exists an equilibrium of Γ in randomized strategies, if the utility functions $(u_i)_{i \in N}$ are continuous. The existence theorem for finite games immediately implies that a randomized equilibrium exists for any essentially finite game. Then Theorems 3.5 and 3.3 imply that there exists a 2α-equilibrium of Γ, for any positive number α. By compactness of $\times_{i \in N} \Delta(C_i)$, any sequence of randomized-strategy profiles has a convergent subsequence. Thus, letting $\alpha = \varepsilon_k/2$ and letting $\varepsilon_k \rightarrow 0$, we can construct a convergent sequence of randomized-strategy profiles that satisfies the hypotheses of Theorem 3.4, with $\bar{\varepsilon} = 0$. Then the limit $\bar{\sigma}$ is a Nash equilibrium of Γ in randomized strategies.

Simon and Zame (1990) have applied a similar line of argument to prove a more general equilibrium existence theorem for games in which the utility functions may be discontinuous (see also Dasgupta and Maskin, 1986). However, Simon and Zame's formulation uses a weaker concept of a game, in which a range of possible payoff allocations may be allowed for some combinations of actions. Such a modification of the concept of a game is needed because, for strategic-form games as conventionally defined, equilibria need not exist when payoff functions are discontinuous. For a simple example, let $N = \{1\}$, $C_1 = [0,1]$, $u_1(c_1) = c_1$ if $c_1 < 0.5$, $u_1(c_1) = 0$ if $c_1 \geq 0.5$. Then player 1 wants to choose the highest number that is strictly less than 0.5, but no such number exists. In a sense, the problem is that we chose the wrong way to define the payoff at the point of discontinuity. If we just changed the payoff there and let $u_1(0.5) = 0.5$, then an equilibrium would exist.

Following Simon and Zame (1990), a *game with an endogenous-sharing rule*, or an *endogenous-sharing game* is any

$$\Gamma^s = (N, (C_i)_{i \in N}, U),$$

where N and $(C_i)_{i \in N}$ are as above, and $U:C \rightarrow\rightarrow \mathbf{R}^N$ is a correspondence that specifies a set of utility payoff allocations in $\mathbf{R}^N$ for every combination of players' strategies. That is, for any c in C, if $w \in U(c)$ then $w = (w_i)_{i \in N}$ is one possible allocation of utility to the players that could occur if they chose the strategy profile c. Suppose that N is finite and each C_i is a compact subset of a metric space, and U is bounded (that is, there exists some number K such that, for every c in C, every w in

$U(c)$, and every i in N, $|w_i| \leq K$). Suppose also that $U{:}C \twoheadrightarrow \mathbf{R}^N$ is upper-hemicontinuous and, for each c in C, $U(c)$ is a nonempty convex subset of $\mathbf{R}^N$. Using these assumptions, Simon and Zame (1990) have proved the following general existence theorem.

THEOREM 3.6. *Let Γ^s be an endogenous-sharing game as defined above. Then there exists some profile of utility functions $(\hat{u}_i)_{i \in N}$ such that each $\hat{u}_i{:}C \to \mathbf{R}$ is a measurable function, $(\hat{u}_1(c), \ldots, \hat{u}_n(c)) \in U(c)$ for every c in C, and the resulting strategic-form game $(N, (C_i)_{i \in N}, (\hat{u}_i)_{i \in N})$ has at least one Nash equilibrium in $\times_{i \in N} \Delta(C_i)$.*

If each function $u_i{:}C \to \mathbf{R}$ is continuous, then letting

$$U(c) = \{(u_1(c), \ldots, u_n(c))\}, \quad \forall c \in C,$$

defines a correspondence $U(\cdot)$ that satisfies all of the properties that we have assumed.

More generally, for any profile of bounded utility functions $u = (u_i)_{i \in N}$, where some $u_i{:}C \to \mathbf{R}$ may be discontinuous, there exists a unique minimal correspondence $U{:}C \twoheadrightarrow \mathbf{R}^N$ such that U is upper-hemicontinuous and, for every c in C, $U(c)$ is a convex set that includes the vector $(u_i(c))_{i \in N}$. When U is this minimal correspondence, $U(c)$ is the set of all limits of expected utility allocations that can be achieved by randomizing over strategy profiles that are arbitrarily close to c. To be more precise, let $B(c,\varepsilon)$ denote the set of all strategy profiles e such that, for every player i, the distance between e_i and c_i is less than ε. Then, for any c in C, a payoff vector w is in $U(c)$ iff, for every positive number ε, there exists some probability distribution μ over the set $B(c,\varepsilon)$ such that, for every player i,

$$\left| w_i - \int_{e \in B(c,\varepsilon)} u_i(e) d\mu(e) \right| \leq \varepsilon.$$

Applying Theorem 3.6 to this minimal correspondence U, we find that, for any profile of discontinuous utility functions u, there exists some profile of utility functions $\hat{u}$ such that $\hat{u}$ differs from u only at points of discontinuity and an equilibrium of the resulting game $\hat{\Gamma}$ exists.

For example, consider a two-player game where $C_1 = [0,1]$, $C_2 = [0,\frac{1}{4}]$,

$$u_1(c_1,c_2) = 1 - u_2(c_1,c_2) = (c_1 + c_2)/2 \qquad \text{if } c_1 < c_2,$$
$$= \tfrac{1}{2} \qquad\qquad\qquad \text{if } c_1 = c_2,$$
$$= 1 - (c_1 + c_2)/2 \quad \text{if } c_1 > c_2.$$

(This game is a modified version of a classic example of Hotelling, 1929.) Following Simon and Zame, we can interpret this game by supposing that the players are two psychiatrists choosing where to locate their offices on a long road, represented by the interval from 0 to 1. The portion of the road represented by the interval from 0 to $\tfrac{1}{4}$ is in Oregon, and the portion represented by the interval from $\tfrac{1}{4}$ to 1 is in California. Player 1 is licensed to practice in both California and Oregon, but player 2 is licensed to practice only in Oregon. Patients are distributed uniformly along this road, each patient will go to the psychiatrist whose office is closer. The payoff to each psychiatrist is the fraction of the patients that go to him.

The payoff function for this game is discontinuous when the players choose the same point in the interval $[0,\tfrac{1}{4}]$ (that is, when they locate across the road from each other in Oregon). We have, quite naturally, assumed that they share the patients equally in this case. However, there is no equilibrium of this game when we define payoffs at the discontinuity in this way. To see why, notice that player 1 can guarantee himself a payoff arbitrarily close to $\tfrac{3}{4}$ by choosing c_1 slightly larger than $\tfrac{1}{4}$, and player 2 can guarantee herself a payoff of at least $\tfrac{1}{4}$ by choosing c_2 equal to $\tfrac{1}{4}$. However, when c_1 and c_2 both equal $\tfrac{1}{4}$, player 1's payoff drops to $\tfrac{1}{2}$ in this model.

The smallest upper-hemicontinuous convex-valued endogenous-sharing rule U that contains the given profile of utility functions u is

$$U(c_1,c_2) = \{((c_1 + c_2)/2, 1 - (c_1 + c_2)/2)\} \text{ if } c_1 < c_2,$$
$$U(c_1,c_2) = \{(\alpha, 1 - \alpha) \mid c_1 \le \alpha \le 1 - c_1\} \text{ if } c_1 = c_2,$$
$$U(c_1,c_2) = \{(1 - (c_1 + c_2)/2, (c_1 + c_2)/2)\} \text{ if } c_1 > c_2.$$

To get an equilibrium, we may use any selection $\hat{u}$ such that

$$\hat{u}_1(\tfrac{1}{4},\tfrac{1}{4}) = \tfrac{3}{4}, \quad \hat{u}_2(\tfrac{1}{4},\tfrac{1}{4}) = \tfrac{1}{4}.$$

The resulting game $\hat{\Gamma}$ has an equilibrium (in pure strategies) at (c_1,c_2) $= (\tfrac{1}{4},\tfrac{1}{4})$, with payoffs $(\tfrac{3}{4},\tfrac{1}{4})$. That is, to get an equilibrium, we must adjust the payoff functions at the points of discontinuity so that, when both players locate at the California–Oregon border, player 1 gets all the California patients and player 2 gets all the Oregon patients.

If we approximated the original game by a finite game in which c_1 and c_2 are required to be integer multiples of $1/K$, where K is any large positive integer that is divisible by 4, then there would be an equilibrium in pure strategies at $(\frac{1}{4} + 1/K, \frac{1}{4})$, with payoffs $(\frac{3}{4} - 1/(2K), \frac{1}{4} + 1/(2K))$. As K goes to infinity, this equilibrium and payoff allocation converge to the equilibrium and payoff allocation that we got for $\hat{\Gamma}$.

Exercises

Exercise 3.1. Suppose that Γ^1 is derived from Γ by eliminating pure strategies that are strongly dominated in Γ. Show that σ is an equilibrium of Γ if and only if σ is an equilibrium of Γ^1. (NOTE: If $\hat{C}_i$ is a subset of C_i, then any probability distribution ρ_i in $\Delta(\hat{C}_i)$ may be identified with the probability distribution in $\Delta(C_i)$ that gives the same probabilities as ρ_i to the pure strategies in $\hat{C}_i$, and gives probability 0 to the pure strategies that are in C_i but not in $\hat{C}_i$.)

Exercise 3.2. Suppose that Γ^2 is derived from Γ by eliminating pure strategies that are weakly dominated in Γ. Show that, if σ is an equilibrium of Γ^2, then σ is an equilibrium of Γ.

Exercise 3.3. For each of the following two-player games, find all equilibria. As usual, player 1 chooses the row and player 2 chooses the column. (HINT: Each game here has an odd number of equilibria.)

a.

	x_2	y_2
x_1	2,1	1,2
y_1	1,5	2,1

b.

	x_2	y_2
x_1	3,7	6,6
y_1	2,2	7,3

c.

	x_2	y_2
x_1	7,3	6,6
y_1	2,2	3,7

d.

	x_2	y_2	z_2
x_1	0,4	5,6	8,7
y_1	2,9	6,5	5,1

e.

	x_2	y_2	z_2
x_1	0,0	5,4	4,5
y_1	4,5	0,0	5,4
z_1	5,4	4,5	0,0

Exercise 3.4. Find all equilibria of the three-player game (due to B. O'Neill) shown in Table 3.15, where $C_i = \{x_i, y_i\}$ for each i in $\{1,2,3\}$. That is, each player gets 0 unless exactly one player i chooses y_i, in which case this player i gets 5, the player before i in the $1 \to 2 \to 3 \to 1$ cycle gets 6, and the player after i in this cycle gets 4. (HINTS: Notice the cyclical symmetry of this game. The number of equilibria is even, which is rather unusual.)

Exercise 3.5. a. Consider a Bayesian game with incomplete information in which player 1 may be either type α or type β, where type α has probability .9 and type β has probability .1 (as assessed by player 2). Player 2 has no private information. Depending on player 1's types, the payoffs to the two players depend on their actions in $C_1 = \{x_1, y_1\}$ and $C_2 = \{x_2, y_2\}$ as shown in Table 3.16. Show that there exists a Bayesian equilibrium in which player 2 chooses x_2.

b. (A version of the *electronic mail* game of Rubinstein, 1989.) Now suppose that the information structure is altered by the following pre-play communication process. If player 1 is type β, then he sends no letter to player 2. Otherwise, player 1 sends to player 2 a letter saying, "I am not type β." Thereafter, each time either player receives a message from the other, he sends back a letter saying, "This is to confirm the receipt of your most recent letter." Suppose that every letter has a probability $\frac{1}{10}$ of being lost in the mail, and the players continue exchanging letters until one is lost. Thus, when player 1 chooses his

Table 3.15 A three-player game in strategic form

	C_2 and C_3				
	x_3			y_3	
C_1	x_2	y_2		x_2	y_2
x_1	0,0,0	6,5,4		4,6,5	0,0,0
y_1	5,4,6	0,0,0		0,0,0	0,0,0

Table 3.16   Expected payoffs for all types and action profiles in a Bayesian
game

$t_1 = \alpha$

C_1	C_2	
	x_2	y_2
x_1	2,2	$-2,0$
y_1	0,-2	0,0

$t_1 = \beta$

C_1	C_2	
	x_2	y_2
x_1	0,2	1,0
y_1	1,-2	2,0

action in $\{x_1,y_1\}$, he knows whether he is β or α; and if he is α, then he also knows how many letters he got from player 2. When player 2 chooses her action in $\{x_2,y_2\}$, she knows how many letters she got from player 1.

After the players have stopped sending letters, what probability would player 2 assign to the event that a letter sent by player 1 was lost in the mail (so the total number of letters sent by 1 and 2 together is odd)?

If player 1 is not type β, then what probability would player 1 assign to the event that a letter sent by player 1 was lost in the mail?

Show that there is no Bayesian equilibrium of this game in which player 2 ever chooses x_2.

Exercise 3.6.   Consider the perturbed version of the Battle of the Sexes game shown in Table 3.17. Here $0 < \varepsilon < 1$, and t_1 and t_2 are independent random variables drawn from a uniform distribution over the interval [0,1]. Each player i observes only t_i before choosing his action in $C_i = \{f_i, s_i\}$. Find three Bayesian equilibria of this game. (HINT: There is an equilibrium in which each player i has some types that choose s_i and some other types that choose f_i.)

Exercise 3.7.   Consider a first-price sealed-bid auction between two bidders with independent *private values* for the object being sold. Before

Table 3.17   A perturbed version of the Battle of the Sexes game

	C_2	
C_1	f_2	s_2
f_1	$3 + \varepsilon t_1, 1$	$\varepsilon t_1, \varepsilon t_2$
s_1	$0, 0$	$1, 3 + \varepsilon t_2$

this auction, each bidder i (for $i = 1,2$) observes a random variable t_i that is independently drawn from a uniform distribution over the interval $[0,1]$. Then the value of the object to bidder i is $v_i = t_i + 0.5$. Each bidder i submits a sealed bid b_i that can be any nonnegative number, and his payoff is

$$u_i(b_1,b_2,t_1,t_2) = (t_i + 0.5) - b_i \quad \text{if } b_i > b_{-i},$$
$$= 0 \quad \text{if } b_i < b_{-i},$$
$$= ((t_i + 0.5) - b_i)/2 \text{ if } b_1 = b_2.$$

Find an equilibrium in which each bidder uses a strategy of the form $b_i = \alpha t_i + \beta$. In this equilibrium, what is the conditionally expected payoff to bidder i, given his type t_i?

Exercise 3.8.   Now consider a first-price sealed-bid auction between two bidders with a *common value* for the object being sold. As in Exercise 3.7, before this auction, each bidder i (for $i = 1,2$) observes a random variable t_i that is independently drawn from a uniform distribution over the interval from 0 to 1, but in this situation the actual value of the object to each bidder i is $v_i = t_1 + t_2$. Thus, given only his own information t_i, the expected value of the object to bidder i is $t_i + 0.5$, as in Exercise 3.7. In the auction, each bidder i submits a sealed bid b_i that can be any nonnegative number, and his payoff is

$$u_i(b_1,b_2,t_1,t_2) = (t_1 + t_2) - b_i \quad \text{if } b_i > b_{-i},$$
$$= 0 \quad \text{if } b_i < b_{-i},$$
$$= ((t_1 + t_2) - b_i)/2 \text{ if } b_1 = b_2.$$

Find an equilibrium in which each bidder uses a strategy of the form $b_i = \alpha t_i + \beta$. In this equilibrium, what is the conditionally expected payoff to bidder i, given his type t_i? Show that, for each value of t_i, the equilibrium bid in this auction is lower than the equilibrium bid in the private-values example in Exercise 3.7.

Exercise 3.9.   a. Verify the equilibrium of the perturbed game described in Section 3.6.

b. How would the equilibrium change if it were common knowledge that player 1 is an intelligent rational decision-maker but that there is still a probability .1 that player 2 may be replaced by a naive agent who would choose a bid according to the uniform distribution on the interval $[0,1]$?

Exercise 3.10.   Let $N = \{1, \ldots, n\}$ be a fixed set of players, and let C_1, $\ldots, C_n$ be fixed finite strategy sets for the n players. For any sequence of games $(\Gamma^k)_{k=1}^{\infty}$, where each $\Gamma^k = (N, (C_i)_{i \in N}, (u_i^k)_{i \in N})$, and for any $\Gamma = (N, (C_i)_{i \in N}, (u_i)_{i \in N})$, we may say that $\lim_{k \to \infty} \Gamma^k = \Gamma$ iff, for every i in N and every c in C, $\lim_{k \to \infty} u_i^k(c) = u_i(c)$. For any such game Γ, let $E(\Gamma)$ denote the set of all Nash equilibria of Γ, and let $D(\Gamma)$ denote the set of all Nash equilibria of Γ that assign zero probability to all weakly dominated pure strategies.

a. Is $E(\cdot)$ upper-hemicontinuous? Justify your answer by a proof or a counterexample.

b. Is $D(\cdot)$ upper-hemicontinuous? Justify your answer by a proof or a counterexample.

Exercise 3.11.   Let Γ be a two-person zero-sum game in strategic form. Show that $\{\sigma_1 | \sigma$ is an equilibrium of $\Gamma\}$ is a convex subset of the set of randomized strategies for player 1.

Exercise 3.12.   Suppose that Y and Z are nonempty compact (closed and bounded) convex subsets of finite-dimensional vector spaces. Let $g: Y \times Z \to \mathbf{R}$ be a continuous function such that, for every z in Z, $g(\cdot, z)$ is an affine function on Y. That is, for any z in Z, and any x and y in Y, and any λ between 0 and 1,

$$g(\lambda x + (1 - \lambda)y, z) = \lambda g(x, z) + (1 - \lambda)g(y, z).$$

For any z in Z, let $F(z) = \text{argmax}_{y \in Y} \, g(y, z)$. Prove that $F: Z \to\to Y$ is upper-hemicontinuous and nonempty-convex-valued. (HINT: The image of a compact set under a continuous function is a compact set; and a compact subset of the real number line is closed and bounded.)

Exercise 3.13.   Let Γ be the two-person game in which the pure strategy sets are $C_1 = C_2 = [0,1]$, and the utility function of each player i is $u_i(c_1,c_2) = c_1c_2 - 0.1(c_i)^2$. For any number ε such that $0 < \varepsilon < \frac{1}{2}$, let $\hat{\Gamma}_\varepsilon$ be the game in which the strategy sets are the same as in Γ, but the utility function of each player i is $\hat{u}_i(c_1,c_2) = u_i(\max\{\varepsilon,c_1\}, \max\{\varepsilon,c_2\})$. (In effect, all strategies between 0 and ε are rounded up to ε.)

a. Show that (0,0) and (1,1) are the only equilibria of Γ.

b. Show that (1,1) is the unique equilibrium of $\hat{\Gamma}_\varepsilon$.

c. Show that (0,0) is an ε-equilibrium of $\hat{\Gamma}_\varepsilon$.

d. Show that the distance between the utility-function profiles (u_1,u_2) and $(\hat{u}_1,\hat{u}_2)$, as defined in Section 3.13, is less than ε.

Exercise 3.14.   A *strong Nash equilibrium* is a Nash equilibrium such that there is no nonempty set of players who could all gain by deviating together to some other combination of strategies that is jointly feasible for them, when the other players who are not in this set are expected to stay with their equilibrium strategies. (See Aumann, 1959, and Bernheim, Peleg, and Whinston, 1987.) Here, to formalize this definition, we can say that a randomized strategy profile σ is a strong Nash equilibrium iff, for each randomized strategy profile τ that is different from σ, there exists some player i such that $\tau_i \neq \sigma_i$ and $u_i(\tau) \leq u_i(\sigma)$.

a. Show an example of a game that has strong Nash equilibria and show another example of a game that does not have any strong Nash equilibria. (HINT: It suffices to consider some of the examples in Section 3.2.)

b. Recall the game in Exercise 3.4. Show that, for each equilibrium of this game, there exists a set of two or three players who could all do better by changing their strategies in such a way that the new strategy profile (including the original equilibrium strategy of the player who is outside this set, if it contains only two players) is still an equilibrium.

4

Sequential Equilibria of Extensive-Form Games

4.1 Mixed Strategies and Behavioral Strategies

Since the work of Selten (1975) and Kreps and Wilson (1982), there has been an increasing appreciation of the importance of studying equilibria of games in extensive form, not just in strategic form. Analysis of *equilibria in behavioral strategies* and of *sequential equilibria* of an extensive-form game can provide insights beyond those generated by analysis of equilibria of any one representation of the game in strategic form. Before we can define these two solution concepts for extensive-form games, however, we need to formulate the more basic concepts of randomized strategies for extensive-form games.

We begin by recalling some notation developed in Chapter 2. Throughout this chapter, we let Γ^e denote an extensive-form game as defined in Section 2.1. The set of players in Γ^e is N. For any player i in N, we let S_i denote the set of all possible information states for player i in Γ^e. We assume that players' information states are labeled such that S_i and S_j are disjoint sets whenever $i \neq j$, and we let S^* denote the union of all these sets; that is

$$S^* = \bigcup_{i \in N} S_i.$$

For any player i and any information state s in S_i, we let Y_s denote the set of nodes belonging to player i with information state s, and we let D_s denote the set of all moves available to i at information state s. That is, D_s is the set of all move-labels of alternative branches following any node in Y_s. A (pure) strategy for a player in Γ^e is a function that specifies a move for the player at each of his possible information states in the game. We let C_i denote the set of such pure strategies for any player i in Γ^e, so

$$C_i = \underset{s \in S_i}{\times} D_s.$$

In Chapter 3 we defined randomized-strategy profiles and equilibria for strategic-form games. In Chapter 2 we defined two different ways of representing an extensive-form game by a strategic-form game: the normal representation and the multiagent representation. These two representations give us two different ways to define randomized-strategy profiles for any given extensive-form game Γ^e. A *mixed-strategy profile* of Γ^e is defined to be any randomized-strategy profile for the normal representation of Γ^e. That is, the set of mixed-strategy profiles of Γ^e is

$$\underset{i \in N}{\times} \Delta(C_i).$$

A *behavioral-strategy profile* of Γ^e is defined to be any randomized-strategy profile for the multiagent representation of Γ^e. Thus, the set of behavioral-strategy profiles of Γ^e is

$$\underset{s \in S^*}{\times} \Delta(D_s) = \underset{i \in N}{\times} \underset{s \in S_i}{\times} \Delta(D_s).$$

In the language of Section 2.1, a mixed-strategy profile specifies, for each player, a probability distribution over his set of overall strategies for playing Γ^e, whereas a behavioral-strategy profile specifies a probability distribution over the set of possible moves for each possible information state of each player. A behavioral-strategy profile can also be called a *scenario* for the game, because it specifies a probability for every alternative branch at every decision node in the game tree.

In general, if σ is a behavioral-strategy profile in $\times_{i \in N} \times_{s \in S_i} \Delta(D_s)$, then we can write

$$\sigma = (\sigma_i)_{i \in N} = (\sigma_{i.s})_{s \in S_i, i \in N}$$

where $\sigma_i = (\sigma_{i.s})_{s \in S_i} \in \underset{s \in S_i}{\times} \Delta(D_s)$

and $\sigma_{i.s} = (\sigma_{i.s}(d_s))_{d_s \in D_s} \in \Delta(D_s)$.

Here the number $\sigma_{i.s}(d_s)$ can be called a *move probability* because it denotes the conditional probability (under scenario σ) that player i would choose move d_s if the path of play reached a node that is controlled by player i with information state s. Any σ_i in $\times_{s \in S_i} \Delta(D_s)$, which specifies a probability for every possible move at every possible information state of player i, is called a *behavioral strategy* for player i.

On the other hand, a *mixed strategy* for player i is any τ_i in $\Delta(C_i)$ that specifies a probability $\tau_i(c_i)$ for every (nonrandomized) plan c_i that player i could use to determine his moves at every information state.

To appreciate the distinction between mixed strategies and behavioral strategies, consider the extensive-form game in Figure 4.1. The normal representation of this game (which is also the reduced normal representation, because there are no redundant strategies) is shown in Table 4.1.

For any numbers α and β between 0 and $\frac{1}{2}$, the mixed-strategy profile

$$(\alpha[w_1y_1] + \alpha[x_1z_1] + (.5 - \alpha)[w_1z_1] + (.5 - \alpha)[x_1y_1],$$

$$\beta[w_2y_2] + \beta[x_2z_2] + (.5 - \beta)[w_2z_2] + (.5 - \beta)[x_2y_2])$$

is an equilibrium of this normal representation in strategic form. However, all of these mixed-strategy profiles should be considered equivalent in the context of the original extensive-form game. In all of them, each player plans at each of his information states to use each of his two possible moves with probability 0.5. The only differences among these equilibria arise from the different ways that a player can correlate his planned moves at different information states. Because each player will actually move at only one of his two information states when the game is played, the correlation between the move that he chooses at one information state and the move that he would have chosen at any other information state has no significance. All that matters are the marginal distributions over moves that the randomized strategies generate at the various information states. Furthermore, these marginal distributions are all that are needed to describe a behavioral-strategy profile for the extensive-form game. That is, all of these mixed-strategy profiles correspond to the behavioral-strategy profile

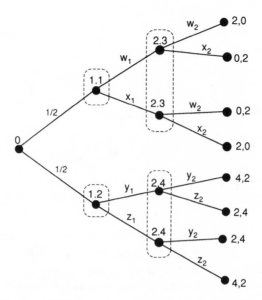

Figure 4.1

Table 4.1 A game in strategic form, the normal representation of Figure 4.1

C_1	C_2			
	w_2y_2	w_2z_2	x_2y_2	x_2z_2
w_1y_1	3,1	2,2	2,2	1,3
w_1z_1	2,2	3,1	1,3	2,2
x_1y_1	2,2	1,3	3,1	2,2
x_1z_1	1,3	2,2	2,2	3,1

$$(.5[w_1] + .5[x_1], .5[y_1] + 0.5[z_1], .5[w_2] + .5[x_2], .5[y_2] + .5[z_2]).$$

Formally defining equivalence relationships between mixed-strategy profiles and behavioral-strategy profiles requires some technical care. For example, consider the game in Figure 4.2, and consider the mixed strategy $.5[a_1y_1] + .5[b_1z_1]$ for player 1. It might at first seem that this mixed strategy should correspond to the behavioral strategy $(.5[a_1] + .5[b_1], .5[y_1] + .5[z_1])$ for player 1, but this would be wrong. The only way that player 1 would plan to use z_1 at a 1.3 node under the mixed

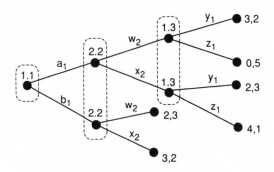

Figure 4.2

strategy $.5[a_1y_1] + .5[b_1z_1]$ is if he chooses the pure strategy b_1z_1, in which case the path of play could not possibly reach a 1.3 node. In fact, under the mixed strategy $.5[a_1y_1] + .5[b_1z_1]$, player 1 would choose y_1 with conditional probability 1 if the path of play reached a 1.3 node, because a_1y_1 is the only strategy that has positive probability and is compatible with information state 3 of player 1. So the behavioral strategy for player 1 that corresponds to $.5[a_1y_1] + .5[b_1z_1]$ is $(.5[a_1] + .5[b_1], [y_1])$.

In general, for any player i, any pure strategy c_i in C_i, and any information state s in S_i, we say that s and c_i are *compatible* iff there exists at least one combination of pure strategies for the other players such that a node at which player i moves with information state s could occur with positive probability when i implements strategy c_i. That is, s and c_i are compatible iff there exists some $c_{-i} = (c_j)_{j \in N-i}$ in $C_{-i} = \times_{j \in N-i} C_j$ such that

$$\sum_{x \in Y_s} P(x|c) > 0,$$

where $c = (c_{-i}, c_i)$ and $P(\cdot|\cdot)$ is as defined in Section 2.2. We let $C_i^*(s)$ denote the set of all pure strategies in C_i that are compatible with i's information state s. For any τ_i in $\Delta(C_i)$ and any s in S_i, we say that s is *compatible* with τ_i iff there exists some c_i in $C_i^*(s)$ such that $\tau_i(c_i) > 0$.

For any s in S_i and any d_s in D_s, let $C_i^{**}(d_s,s)$ denote the set of pure strategies that are compatible with s and designate d_s as i's move at information state s. That is,

$$C_i^{**}(d_s,s) = \{c_i \in C_i^*(s) | c_i(s) = d_s\}.$$

For example, in the game in Figure 4.2, $C_1^*(3) = \{a_1y_1, a_1z_1\}$ and $C_1^{**}(y_1,3) = \{a_1y_1\}$.

A behavioral strategy $\sigma_i = (\sigma_{i.s})_{s \in S_i}$ for player i is a *behavioral representation* of a mixed strategy τ_i in $\Delta(C_i)$ iff, for every s in S_i and every d_s in D_s,

$$(4.1) \qquad \sigma_{i.s}(d_s) \left(\sum_{e_i \in C_1^*(s)} \tau_i(e_i) \right) = \sum_{c_i \in C_1^{**}(d_s,s)} \tau_i(c_i).$$

That is, σ_i is a behavioral representation of τ_i iff, for every move d_s at every information state s in S_i that is compatible with τ_i, $\sigma_{i.s}(d_s)$ is the conditional probability that i would plan to choose d_s at s given that he chose a pure strategy that is compatible with s. It is straightforward to check that any randomized strategy τ_i in $\Delta(C_i)$ has at least one behavioral representation in $\times_{s \in S_i} \Delta(D_s)$. If some information state s in S_i is not compatible with τ_i, then any probability distribution $\sigma_{i.s}(\cdot)$ in $\Delta(D_s)$ would satisfy condition (4.1), because both sides are 0. Thus, τ_i can have more than one behavioral representation. A behavioral-strategy profile $\sigma = (\sigma_i)_{i \in N}$ is a behavioral representation of a mixed-strategy profile τ in $\times_{i \in N} \Delta(C_i)$ iff, for every player i, σ_i is a behavioral representation of τ_i.

Given any behavioral strategy $\sigma_i = (\sigma_{i.s})_{s \in S_i}$ in $\times_{s \in S_i} \Delta(D_s)$, the *mixed representation* of σ_i is defined to be τ_i in $\Delta(C_i)$ such that

$$\tau_i(c_i) = \prod_{s \in S_i} \sigma_{i.s}(c_i(s)), \quad \forall c_i \in C_i.$$

That is, the mixed representation of a behavioral strategy σ_i is the mixed strategy in $\Delta(C_i)$ in which i's planned move at each information state s has the marginal probability distribution $\sigma_{i.s}$ and is determined independently of his moves at all other information states. The mixed representation of a behavioral-strategy profile $\sigma = (\sigma_i)_{i \in N}$ is the mixed-strategy profile τ such that, for every player i, τ_i is the mixed representation of σ_i. It is straightforward to show that, if τ_i is the mixed representation of a behavioral strategy σ_i for player i, then σ_i is a behavioral representation of τ_i.

Two mixed strategies in $\Delta(C_i)$ are *behaviorally equivalent* iff they share a common behavioral representation. Thus, in the example in Figure 4.1, player 1's mixed strategies $.5[w_1y_1] + .5[x_1z_1]$ and $.5[w_1z_1] + .5[x_1y_1]$ are behaviorally equivalent, because both have the behavioral representation $(.5[w_1] + .5[x_1], .5[y_1] + .5[z_1])$, for which the mixed representation is $.25[w_1y_1] + .25[x_1z_1] + .25[w_1z_1] + .25[x_1y_1]$. Similarly, in the game in Figure 4.2, $.5[a_1y_1] + .5[b_1z_1]$ and $.5[a_1y_1] + .5[b_1y_1]$ are behaviorally equivalent, because both have the behavioral representation $(.5[w_1] + .5[x_1], [y_1])$, for which $.5[a_1y_1] + .5[b_1y_1]$ is the mixed representation. Two mixed strategy profiles τ and $\hat{\tau}$ in $\times_{i \in N} \Delta(C_i)$ are *behaviorally equivalent* iff, for every player i, τ_i and $\hat{\tau}_i$ are behaviorally equivalent.

We say that two mixed strategies τ_i and ρ_i in $\Delta(C_i)$ are *payoff equivalent* iff, for every player j and every τ_{-i} in $\times_{l \in N-i} \Delta(C_l)$,

$$u_j(\tau_{-i},\tau_i) = u_j(\tau_{-i},\rho_i),$$

where $u_j(\cdot)$ is j's utility function in the normal representation of Γ^e. That is, τ_i and ρ_i are payoff equivalent if no player's expected utility depends on which of these two randomized strategies is used by player i.

The following theorem, due to Kuhn (1953), is proved in Section 4.11.

THEOREM 4.1. *If Γ^e is a game with perfect recall (see Section 2.1), then any two mixed strategies in $\Delta(C_i)$ that are behaviorally equivalent are also payoff equivalent.*

Figure 4.3 shows that the technical assumption of perfect recall is needed to prove Kuhn's Theorem. This game does not have perfect recall, because player 1 at information state 3 cannot recall whether he chose x_1 or y_1 at his information state 1. Player 1's mixed strategies

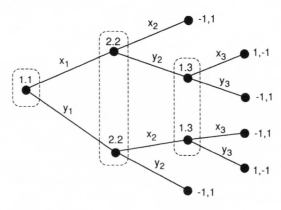

Figure 4.3

$.5[x_1x_3] + .5[y_1y_3]$ and $.5[x_1y_3] + .5[y_1x_3]$ are behaviorally equivalent, because they have the same behavioral representation $(.5[x_1] + .5[y_1],$ $.5[x_3] + .5[y_3])$. However, $.5[x_1x_3] + .5[y_1y_3]$ and $.5[x_1y_3] + .5[y_1x_3]$ are not payoff equivalent; against $[x_2]$, for example, $.5[x_1x_3] + .5[y_1y_3]$ gives player 1 an expected payoff of 0, but $.5[x_1y_3] + .5[y_1x_3]$ gives player 1 an expected payoff of -1.

4.2 Equilibria in Behavioral Strategies

If we defined an equilibrium of an extensive-form game to be an equilibrium of its normal representation in strategic form, then games like Figure 4.1 would have an enormous multiplicity of equilibria that are all behaviorally equivalent.

To avoid this problem, we might consider defining an equilibrium of an extensive-form game to be an equilibrium of its multiagent representation, but such a definition could create a more serious problem of nonsensical equilibria in which a player fails to coordinate his moves across different information states. For example, $([b_1],[w_2],[z_1])$ is an equilibrium of the multiagent representation of Figure 4.2. In this equilibrium, given that player 2 is expected to choose w_2, the agent for 1's first information state anticipates that switching from b_1 to a_1 would reduce his payoff from 2 to 0, because the agent for 1's last information state is planning to do z_1. Furthermore, given the expectation that 1's first agent will choose b_1, 1's last agent figures that his planned move would never be carried out and so planning to choose z_1 is as good as planning to choose y_1. However, if player 1 really expected player 2 to choose w_2, then player 1's actual best response would be to choose a_1 at his first information state and y_1 at his last information state. The nonsensical equilibria arise in the multiagent representation because the possibility of a coordinated change that involves more than one information state is not considered.

To avoid both of these problems, a *(Nash) equilibrium* of an extensive-form game, or an *equilibrium in behavioral strategies*, is defined to be any equilibrium σ of the multiagent representation such that the mixed representation of σ is also an equilibrium of the normal representation. That is, an equilibrium of Γ^e is defined to be a behavioral strategy profile that both is an equilibrium of the multiagent representation and corresponds to an equilibrium of the normal representation. Thus, for example, the unique equilibrium in behavioral strategies of the game in Figure 4.1 is

$$(.5[w_1] + .5[x_1], .5[y_1] + .5[z_1], .5[w_2] + .5[x_2], .5[y_2] + .5[z_2]),$$

and the unique equilibrium of the extensive-form game in Figure 4.2 is

$$(.5[a_1] + .5[b_1], .5[w_2] + .5[x_2], [y_1]).$$

Let Γ^e be an extensive-form game with perfect recall, and let Γ be its normal representation in strategic form. By Kuhn's Theorem, if τ is a Nash equilibrium of Γ, and $\hat{\tau}$ is any other mixed-strategy profile that is behaviorally equivalent to τ, then $\hat{\tau}$ is also a Nash equilibrium of Γ (because switching any player's strategy in an equilibrium to a payoff-equivalent strategy will not violate the property of being an equilibrium). So for any behavioral-strategy profile σ, the mixed representation of σ is an equilibrium of Γ if and only if every mixed-strategy profile that has σ as a behavioral representation is an equilibrium of Γ.

The following theorem tells us that, to find an equilibrium of Γ^e, it suffices to find an equilibrium of the normal representation of Γ^e, because any equilibrium of the normal representation corresponds to an equilibrium of the multiagent representation. The proof is in Section 4.11.

THEOREM 4.2. *If Γ^e is an extensive-form game with perfect recall and τ is an equilibrium of the normal representation of Γ^e, then any behavioral representation of τ is an equilibrium of the multiagent representation of Γ^e.*

Theorem 4.2, together with the existence theorem for strategic-form games (Theorem 3.1), immediately implies the following general existence theorem for extensive-form games.

THEOREM 4.3. *For any extensive-form game Γ^e with perfect recall, a Nash equilibrium in behavioral strategies exists.*

The perfect recall assumption is needed for this existence theorem. For example, the game without perfect recall in Figure 4.3 does not have any equilibrium in behavioral strategies. The unique equilibrium of the normal representation in strategic form is $(.5[x_1x_3] + .5[y_1y_3], .5[x_2] + .5[y_2])$, which gives each player an expected payoff of 0, but this is not the mixed representation of any equilibrium of the multiagent representation. The behavioral representation of this mixed-strategy profile is $(.5[x_1] + .5[y_1], .5[x_2] + .5[y_2], .5[x_3] + .5[y_3])$, which happens to be an equilibrium of the multiagent representation, but it gives player 1 an expected payoff of $-\frac{1}{2}$, instead of 0. Such games without perfect recall are highly counterintuitive and have few (if any) real applications, so the nonexistence of equilibria in behavioral strategies for such games is not problematic. This example is discussed here mainly to illustrate the importance of the perfect recall assumption.

4.3 Sequential Rationality at Information States with Positive Probability

We say that a strategy for player i is *sequentially rational* for player i at information state s in S_i if i would actually want to do what this strategy specifies for him at s when information state s actually occurred.

In Section 2.2, we justified the procedure of studying an extensive-form game through its normal representation in strategic form, by assuming that at the beginning of the game each player chooses an expected-utility-maximizing strategy that tells him what to do at every possible information state and thereafter he mechanically applies this strategy throughout the game. This prior-planning assumption was theoretically convenient, but it is artificial and unrealistic in most circumstances. So let us now drop the prior-planning assumption and ask whether equilibrium strategies are sequentially rational.

For any extensive-form game Γ^e, the given structure of the game specifies a *chance probability* for every branch immediately following a chance node. A behavioral-strategy profile σ specifies a *move probability* for every branch immediately following a decision node, where $\sigma_{i.s}(m)$ is the move probability of a branch with move label m immediately following a node controlled by player i with information state s.

For any behavioral-strategy profile σ and any two nodes x and y, if y follows x then let $\overline{P}(y|\sigma,x)$ denote the multiplicative product of all the chance probabilities and move probabilities specified by σ for the branches on the path from x to y. If y does not follow x, then let $\overline{P}(y|\sigma,x)$ $= 0$. So $\overline{P}(y|\sigma,x)$ is the conditional probability that the path of play would go through y after x if all players chose their moves according to scenario σ and if the play of the game started at x.

Let Ω denote the set of all terminal nodes in the game Γ^e, and let

$$U_i(\sigma|x) = \sum_{y \in \Omega} \bar{P}(y|\sigma,x)w_i(y),$$

where $w_i(y)$ denotes the payoff to player i at terminal node y. That is, $U_i(\sigma|x)$ is the expected utility payoff to player i if the play of the game began in node x (instead of the root) and all players thereafter chose their moves as specified by σ. Notice that $U_i(\sigma|x)$ depends only on the components of σ that are applied at nodes that follow the node x in the tree.

Suppose that player i has an information state s that occurs at only one node x, so $Y_s = \{x\}$. Then the behavioral-strategy profile σ is sequentially rational for i at s iff

$$\sigma_{i.s} \in \underset{\rho_s \in \Delta(D_s)}{\text{argmax}} \; U_i(\sigma_{-i.s},\rho_s|x).$$

(Here $(\sigma_{-i.s},\rho_s)$ is the behavioral strategy that differs from σ only in that player i would behave according to ρ_s at state s.) That is, σ is sequentially rational for i at state s iff $\sigma_{i.s}$ would maximize player i's expected utility when the node x occurred in the game, if player i anticipated that all moves after x would be determined according to σ.

Sequential rationality is somewhat more subtle to define at information states that occur at more than one node. At such an information state, a player's expected utility would depend on his beliefs about which of the possible nodes had actually occurred in the path of play. So these beliefs may have an important role in our analysis.

For any information state s of any player i, a *belief-probability* distribution for i at s is a probability distribution over Y_s, the set of nodes labeled "*i.s.*" That is, for any s in S_i, a belief-probability distribution for i at s is a vector $\pi_{i.s}$ in $\Delta(Y_s)$. If we assert that $\pi_{i.s}$ is player i's belief-probability distribution at state s, then we mean that, for each node x in Y_s, $\pi_{i.s}(x)$ would be the conditional probability that i would assign to the event that he was making a move at node x, when he knew that he was making a move at some node in Y_s.

A *beliefs vector* is any vector $\pi = (\pi_{i.s})_{i \in N, s \in S_i}$ in the set $\underset{i \in N}{\times} \underset{s \in S_i}{\times} \Delta(Y_s)$ $= \underset{s \in S*}{\times} \Delta(Y_s)$. That is, a beliefs vector specifies a belief-probability distribution for each information state of each player.

Given a beliefs vector π, which specifies $\pi_{i.s}$ as player i's belief-probability distribution at an information state s in S_i, a behavioral-strategy profile σ is *sequentially rational* for i at s with beliefs π iff

(4.2) $\sigma_{i.s} \in \underset{\rho_s \in \Delta(D_s)}{\mathrm{argmax}} \sum_{x \in Y_s} \pi_{i.s}(x) U_i(\sigma_{-i.s}, \rho_s | x).$

That is, σ is sequentially rational for player i at state s iff $\sigma_{i.s}$ would maximize player i's expected payoff when a node in Y_s occurred in the path of play, given the belief probabilities that i would assign to the various nodes in Y_s when he learned that one of them had occurred, and assuming that all moves after this node would be determined according to σ.

Equivalently (by Lemma 3.1), σ is sequentially rational for player i at his information state s with beliefs π iff, for any move d_s in D_s,

$$\text{if } \sigma_{i.s}(d_s) > 0, \text{ then } d_s \in \underset{e_s \in D_s}{\mathrm{argmax}} \sum_{x \in Y_s} \pi_{i.s}(x) U_i(\sigma_{-i.s}, [e_s] | x).$$

For any move e_s in D_s, where $s \in S_i$, player i's conditionally expected payoff

$$\sum_{x \in Y_s} \pi_{i.s}(x) U_i(\sigma_{-i.s}, [e_s] | x)$$

is the *sequential value* of the move e_s for player i at state s, relative to the scenario σ and beliefs π. Thus, if σ is not sequentially rational for i at s with π, then there must exist some move b_s in D_s such that the move probability $\sigma_{i.s}(b_s)$ is strictly positive but the sequential value of b_s for i at s (relative to σ and π) is strictly less than the sequential value of some other move that is also available to i at s. Any such move b_s is an *irrational move* in the scenario σ, with beliefs π.

The above definition of sequential rationality begs the question (first addressed by Kreps and Wilson, 1982) of how should a rational intelligent player determine his belief probabilities? Belief probabilities are conditional probabilities given information that may become available to a player during the play of the game. Thus, these belief probabilities must be related by Bayes's formula to what this player would believe at the beginning of the game.

For any node y and any behavioral-strategy profile σ, we let $\overline{P}(y|\sigma)$ denote the probability that the path of play, starting at the root, will reach node y when all players choose their moves according to the scenario σ. That is, $\overline{P}(y|\sigma) = \overline{P}(y|\sigma, x^0)$, where x^0 denotes the root or initial node of the tree, so $\overline{P}(y|\sigma)$ is the multiplicative product of all the chance probabilities and move probabilities specified by σ for the branches on the path from the root to y. We refer to $\overline{P}(y|\sigma)$ as the *prior probability* of node y under the scenario σ.

Now suppose that σ is a behavioral-strategy profile that describes the behavior that an intelligent individual would anticipate in the play of Γ^e. Then, by Bayes's formula, for any player i and any information state s in S_i, player i's belief-probability distribution $\pi_{i,s}$ at state s should satisfy the following equation, for every node x in Y_s:

$$(4.3) \qquad \pi_{i,s}(x) \sum_{y \in Y_s} \overline{P}(y|\sigma) = \overline{P}(x|\sigma).$$

That is, the conditional probability of node x given state s multiplied by the total prior probability of the $i.s$ nodes must equal the prior probability of node x, under the scenario σ. We say that a beliefs vector π is *weakly consistent* with a scenario σ iff π satisfies condition (4.3), for every player i, every information state s in S_i, and every node x in Y_s.

If the probability of the path of play going through a node in Y_s is strictly positive, so that

$$(4.4) \qquad \sum_{y \in Y_s} \overline{P}(y|\sigma) > 0,$$

then equation (4.3) completely characterizes i's belief probabilities at s, because it is equivalent to

$$(4.5) \qquad \pi_{i,s}(x) = \frac{\overline{P}(x|\sigma)}{\displaystyle\sum_{y \in Y_s} \overline{P}(y|\sigma)}, \quad \forall x \in Y_s.$$

On the other hand, if condition (4.4) is violated, then state s has zero probability under σ, or s is said to be *off the path* of σ. Unfortunately, equation (4.3) cannot be used to compute the belief probabilities at an information state s that has zero probability in the scenario σ, because both sides of the equation would be 0 no matter what $\pi_{i,s}(x)$ may be.

At information states that can occur with positive probability in an equilibrium, the equilibrium strategies are always sequentially rational.

THEOREM 4.4. *Suppose that σ is an equilibrium in behavioral strategies of an extensive-form game with perfect recall. Let s be an information state in S_i that occurs with positive probability under σ, so condition (4.4) is satisfied. Let π be a beliefs vector that is weakly consistent with σ. Then σ is sequential rational for player i at state s with beliefs π.*

The proof of Theorem 4.4 is given in Section 4.11. This theorem should not be surprising because, as we saw in Chapter 1, Bayes's formula itself can be derived using an axiom that optimal decisions after observing an event should be consistent with optimal contingent plans that a decision-maker would formulate before observing this event. Thus, without the prior planning assumption, equilibrium strategies are still rational to implement at all information states that have positive probability.

To illustrate these ideas, consider again the simple card game discussed in Section 2.1. In Section 3.2, we showed that the unique equilibrium of this game is

$$(\tau_1, \tau_2) = (\tfrac{1}{3}[Rr] + \tfrac{2}{3}[Rf], \tfrac{2}{3}[M] + \tfrac{1}{3}[P]).$$

In behavioral strategies, this equilibrium is

$$(\sigma_{1.a}, \sigma_{1.b}, \sigma_{2.0}) = ([R], \tfrac{1}{3}[r] + \tfrac{2}{3}[f], \tfrac{2}{3}[M] + \tfrac{1}{3}[P]).$$

That is, player 1 should raise for sure at 1.a, when he has a red card; player 1 should raise with probability $\tfrac{1}{3}$ and fold with probability $\tfrac{2}{3}$ at 1.b, when he has a black card; and player 2 should meet with probability $\tfrac{2}{3}$ and pass with probability $\tfrac{1}{3}$, when she has seen player 1 raise.

To verify sequential rationality, we must check whether each player would actually be willing to implement these equilibrium strategies when his moves are not planned in advance. It is easy to see that player 1 should raise when he holds a winning red card, since he has nothing to gain by folding. When player 1 raises with a black card, he gets $+1$ if player 2 passes, and he gets -2 if player 2 meets. Thus, under the assumption that player 2 would implement her equilibrium strategy and meet with probability $\tfrac{2}{3}$, the sequential value of raising for player 1 at 1.b is $-2 \times \tfrac{2}{3} + 1 \times \tfrac{1}{3} = -1$. On the other hand, when player 1 folds with a black card, he gets a payoff of -1 also, so player 1 should be willing to randomize between raising and folding when he has a black card, as his equilibrium strategy specifies.

The decision problem for player 2 is a bit more subtle at the two nodes labeled 2.0. If she passes, then she gets -1 at either node. If she raises, then she gets -2 at her upper node (where player 1 has a red card), but she gets $+2$ at her lower node (where player 1 has a black card). So her optimal move depends on her belief probabilities for these two nodes. To compute these probabilities, we apply Bayes's formula. Let a_0 denote her upper node and let b_0 denote her lower node. Under the behavioral equilibrium σ, the probability of reaching the upper node is

$$\overline{P}(a_0|\sigma) = .5 \times \sigma_{1.a}(R) = .5 \times 1 = .5,$$

and the probability of her lower node is

$$\overline{P}(b_0|\sigma) = .5 \times \sigma_{1.b}(r) = .5 \times \tfrac{1}{3} = \tfrac{1}{6}.$$

Thus, the conditional probabilities of the two nodes given that player 1 has raised are

$$\pi_{2.0}(a_0) = \frac{.5}{.5 + \tfrac{1}{6}} = 0.75,$$

$$\pi_{2.0}(b_0) = \left(\frac{\tfrac{1}{6}}{.5 + \tfrac{1}{6}}\right) = 0.25.$$

With these belief probabilities, the sequential value of meeting for player 2 at her state 0, after player 1 raises, is $.75 \times (-2) + .25 \times 2 = -1$. On the other hand, if player 2 passes after a raise, then her sequential value is -1, which would be her payoff for sure. So the beliefs $\pi_{2.0}$ do indeed make player 2 willing to randomize between meeting and passing, as her equilibrium strategy specifies.

When we exhibit an equilibrium diagrammatically on an extensive-form game, we should show both the move probabilities of the behavioral-strategy profile and the belief probabilities of a beliefs vector that is consistent with the equilibrium. To facilitate reading, let us follow the convention of writing move probabilities in parentheses and belief probabilities in angle brackets ("<·>"). In Figure 4.4, we show the equilibrium and consistent beliefs of our simple card game in extensive form. At each branch that immediately follows a player's decision node, we show the move probability (a component of the behavioral-strategy profile σ) in parentheses. At each decision node, we show the belief probability (a component of the beliefs vector π) in angle brackets.

4.4 Consistent Beliefs and Sequential Rationality at All Information States

Theorem 4.4 looks like a complete resolution of the question of sequential rationality, but in fact it is not. It says nothing about behavior at information states that have zero probability in equilibrium. To guar-

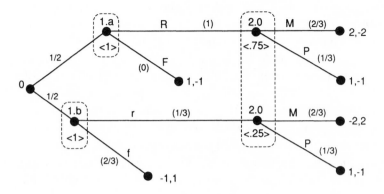

Figure 4.4

antee sequential rationality at all information states, including states with zero probability, we must refine our concept of equilibrium in a substantial and important way.

At first, it might seem that rational behavior at states of zero probability might be a minor technical issue of little interest. After all, if something will almost surely not occur, why worry about it? The problem is that the probability of an information state is determined endogenously by the equilibrium itself. If we allow that players might behave irrationally in an event that endogenously gets zero probability, then we might find that such an event gets zero probability only because players are taking pains to ensure they never create a situation in which someone would behave so irrationally. That is, such a possibility of irrational behavior can have a major impact on players' behavior at states that occur with positive probability. Thus, if we want to study the behavior of rational intelligent players who do not make binding prior plans, we must be careful to apply the criterion of sequential rationality to behavior at all information states, not just at the states of positive probability.

Notice that a player's expected payoff, as evaluated at the beginning of the game, is not affected at all by the move that he plans to use at an information state that is supposed to occur in equilibrium with zero probability. Thus, in formulating a strategy that maximizes his expected payoff given only what he knows at the beginning of the game, there is no restriction at all on what moves a player might plan to use at information states that are supposed to have zero probability. For example, in the game in Figure 4.5, suppose that player 1 is expected to choose y_1. Then player 2 would expect a payoff of 9 no matter what strategy she planned to use. Thus, the strategy to choose y_2 if she gets a chance to move would be as good as any other strategy for player 2, by the criterion of maximizing 2's expected utility given her information at the beginning of the game. Because 1 prefers to choose y_1 if 2 is planning to choose y_2, $([y_1],[y_2])$ is a Nash equilibrium of this game. However, it would be irrational for player 2 to choose y_2 if she actually got the opportunity to move, because it gives her a payoff of 2 when she could get 3 by choosing x_2. Thus, $([y_1],[y_2])$ is not sequentially rational at player 2's node, even though it satisfies the definition of a Nash equilibrium for this game. (The other equilibrium of this game $([x_1],[x_2])$ is sequentially rational, however.)

The unreasonable equilibrium in Figure 4.5 involves a weakly dominated strategy (y_2), but it is easy to construct similar examples in which no strategies are weakly dominated. For example, consider Figure 4.6. No pure strategies are weakly dominated in this game, and $([y_1], [y_2], .5[x_3] + .5[y_3], [y_4])$ is an equilibrium in behavioral strategies, but there is no consistent beliefs vector that would make this equilibrium sequentially rational for player 3. The move x_3 is irrational for player 3 in this scenario, because player 4 would be expected to choose y_4 after player 3 moves, so the sequential value of x_3 would be less than the sequential value of y_3 for player 3 $(0 < 2)$.

In general, a *weak sequential equilibrium* of Γ^e is any (σ,π) such that σ is a behavioral-strategy profile in $\times_{s \in S*}\Delta(D_s)$, π is a beliefs vector in $\times_{s \in S*}\Delta(Y_s)$, σ is sequentially rational for every player at every information state with beliefs π, and π is weakly consistent with σ. (This term has been used by Hillas, 1987.) That is, (σ,π) is a weak sequential equilibrium iff it satisfies conditions (4.2) and (4.3) for every player i

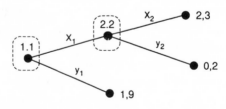

Figure 4.5

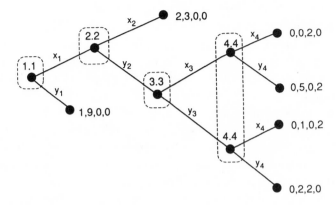

Figure 4.6

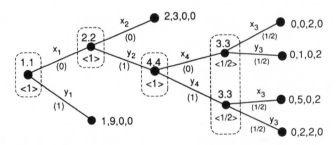

Figure 4.7

and every information state s. We say that a behavioral-strategy profile is a *weak sequential-equilibrium scenario* iff there exists some beliefs vector π such that (σ, π) is a weak sequential equilibrium.

So the concept of weak sequential equilibrium can exclude some unreasonable equilibria. As we have seen, $([y_1], [y_2], .5[x_3] + .5[y_3], [y_4])$ is an equilibrium of the game in Figure 4.6, but it is not a weak sequential-equilibrium scenario. However, it is easy to construct other games in which equally unreasonable equilibria are not excluded by the concept of weak sequential equilibrium.

For example, consider Figure 4.7, which is derived from Figure 4.6 by just switching the order in which players 3 and 4 make their moves. Because neither player observes the other's move, the order should not matter, and indeed these two extensive-form games have the same representations (normal or multiagent) in strategic form. The behavioral-strategy profile $([y_1], [y_2], .5[x_3] + .5[y_3], [y_4])$ is a weak sequential-equilibrium scenario of the game in Figure 4.7. To make this scenario a weak sequential equilibrium, we let the belief probabilities for the two decision nodes of player 3 both equal $\frac{1}{2}$, as shown in the figure. Such beliefs are weakly consistent with this scenario because both of the 3.3 nodes have prior probabilities equal to 0 under this scenario (because both 3.3 nodes follow the x_1 branch, which has move probability 0); so Bayes's formula (4.3) cannot be used to determine the belief probabilities $\pi_{3.3}(x)$. When player 3 thinks that his two decision nodes are equally likely, he is willing to randomize between x_3 and y_3, each of which would give him a conditionally expected payoff of $\frac{1}{2} \times 2 + \frac{1}{2} \times 0 = 1$, when he knows only that some 3.3 node has been reached by the path of play. So (working backward through Figure 4.7), player 4 would be willing to choose y_4 at her 4.4 node if she expected player 3 to randomly choose x_3 and y_3 with equal probability; player 2 would be willing to choose y_2 at her 2.2 node if she expected players 3 and 4 to follow this scenario thereafter (because $\frac{1}{2} \times 5 + \frac{1}{2} \times 2 > 3$); and player 1 would prefer to choose y_1 if he expected player 2 to choose y_2.

The problem with this weak sequential equilibrium is that the belief probabilities specified for player 3 are unreasonable. If player 3 intelligently understood the given scenario $([y_1], [y_2], .5[x_3] + .5[y_3], [y_4])$, then he would expect player 4 to choose y_4 if the path of play reached the 4.4 node; so he would believe that the path of play would have to be at the lower 3.3 node, with belief probability 1, if a 3.3 decision node were reached by the path of play. (With such beliefs, $.5[x_3] + .5[y_3]$ would not be sequentially rational for player 3, because x_3 would be an irrational move.) To develop a more sophisticated equilibrium concept that excludes such scenarios, we need some stronger concept of consistency for determining belief probabilities when all nodes with an information state have zero probability under a given scenario. It is not enough to simply say that any belief probabilities π are "consistent" when zero prior probabilities make both sides of equation (4.3) equal to 0.

To formulate a stronger definition of consistency of beliefs, let us try to think more systematically about the problem of how beliefs should be defined for all information states. We begin by making one simplifying assumption about the game Γ^e (to be relaxed in Section 4.8). Let us assume that every branch that immediately follows a chance node is assigned a chance probability in Γ^e that is strictly positive.

With this assumption, any behavioral-strategy profile σ that assigns a strictly positive move probability to every move at every information state will generate a strictly positive prior probability $\overline{P}(y|\sigma)$ for every node y in the tree. The set of all behavioral-strategy profiles in which all move probabilities are positive is $\times_{s \in S*} \Delta^0(D_s)$. (Recall the definition of Δ^0 in equation 1.6.) So if $\sigma \in \times_{s \in S*} \Delta^0(D_s)$, then $\overline{P}(y|\sigma) > 0$ for every node y, and there is a unique beliefs vector π that satisfies Bayes's formula (4.3) or (4.5). Let Ψ^0 denote the set of all pairs (σ, π) such that $\sigma \in \times_{s \in S*} \Delta^0(D_s)$ and π satisfies Bayes's formula (4.5) for every player i and every information state s in S_i.

Given the extensive-form game Γ^e, any stronger definition of "consistent" beliefs can be characterized mathematically by a set Ψ such that

$$\Psi \subseteq \left(\underset{s \in S*}{\times} \Delta(D_s) \right) \times \left(\underset{s \in S*}{\times} \Delta(Y_s) \right),$$

where Ψ denotes the set of all pairs (σ, π) such that the beliefs vector π is consistent with the scenario σ. If consistency is supposed to be about satisfying all the implications of Bayes's formula, then Ψ^0 should be a subset of this set Ψ. That is, if (σ, π) is in Ψ^0, then the beliefs vector π should be considered consistent with σ in any stronger sense, because Bayes's formula would allow no other beliefs vector with the scenario σ in $\times_{s \in S*} \Delta^0(D_s)$. For natural technical reasons, we can also suppose that this set Ψ should be a closed subset of $(\times_{s \in S*} \Delta(D_s)) \times (\times_{s \in S*} \Delta(Y_s))$, so that scenario-belief pairs that can be arbitrarily closely approximated by consistent pairs are themselves consistent. (For example, the set of pairs (σ, π) such that π is weakly consistent with σ is a closed set, and it contains Ψ^0 as a subset.) Thus, the strongest reasonable definition of consistency that we can consider is to let Ψ be the smallest closed set that contains Ψ^0.

So, following Kreps and Wilson (1982), we say that a beliefs vector π is *fully consistent* with a behavioral-strategy profile σ in the game Γ^e iff there exists some sequence $(\hat{\sigma}^k)_{k=1}^{\infty}$ such that

(4.6) $\hat{\sigma}^k \in \underset{s \in S*}{\times} \Delta^0(D_s), \quad \forall k \in \{1,2,3,\ldots\},$

(4.7) $\sigma_{i.s}(d_s) = \lim_{k \to \infty} \hat{\sigma}_{i.s}^k(d_s), \quad \forall i \in N, \quad \forall s \in S_i, \quad \forall d_s \in D_s,$

(4.8) $$\pi_s(x) = \lim_{k \to \infty} \frac{\overline{P}(x|\hat{\sigma}^k)}{\sum\limits_{y \in Y_s} \overline{P}(y|\hat{\sigma}^k)}, \quad \forall s \in S^*, \quad \forall x \in Y_s.$$

That is, π is fully consistent with σ iff there exist behavioral-strategy profiles that are arbitrarily close to σ and that assign strictly positive probability to every move, such that the beliefs vectors that satisfy Bayes's formula for these strictly positive behavioral-strategy profiles are arbitrarily close to π. With this definition of full consistency, if we let Ψ denote the set of all pairs (σ, π) such that π is fully consistent with σ, then the set Ψ is exactly the smallest closed subset of $(\times_{s \in S^*} \Delta(D_s)) \times (\times_{s \in S^*} \Delta(Y_s))$ that contains Ψ^0. It is easy to show that full consistency implies weak consistency; that is, if π is fully consistent with σ, then π is also weakly consistent with σ. Hereafter, when we refer to beliefs being *consistent* with a scenario, we always mean that they are fully consistent, unless indicated otherwise.

Recall from Section 1.6 that, when the number of possible states is finite, all Bayesian conditional-probability systems can be characterized as limits of conditional-probability systems that are generated by sequences of probability distributions in which all events have positive probability, just as conditions (4.6)–(4.8) require. The only additional feature in (4.6)–(4.8) is that the sequences of probability distributions are required to treat the moves at different information states as stochastically independent random variables, to express the assumption that players choose their strategies without communication. It can be shown that, for any behavioral-strategy profile σ in $\times_{s \in S^*} \Delta(D_s)$, there exists at least one beliefs vector that is fully consistent with σ, although there may exist more than one fully consistent beliefs vector when σ is not in $\times_{s \in S^*} \Delta^0(D_s)$.

To illustrate the implications of full consistency, consider the game in Figure 4.8, and consider the scenario $([z_1],[y_2],[y_3])$, which is an equilibrium of this game. If we perturb this scenario slightly so that all move probabilities are positive, then we get a scenario

$$(\varepsilon_0[x_1] + \varepsilon_1[y_1] + (1 - \varepsilon_0 - \varepsilon_1)[z_1], \varepsilon_2[x_2] + (1 - \varepsilon_2)[y_2],$$
$$\varepsilon_3[x_3] + (1 - \varepsilon_3)[y_3])$$

as shown in parentheses in Figure 4.8, where ε_0, ε_1, ε_2, and ε_3 are very small positive numbers. Then Bayes's formula (4.5) implies that the belief probabilities shown in Figure 4.8 must satisfy

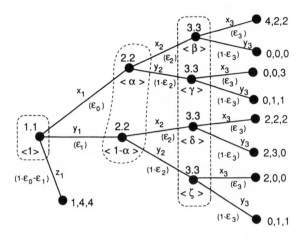

Figure 4.8

$$\alpha = \frac{\varepsilon_0}{\varepsilon_0 + \varepsilon_1}$$

$$\beta = \frac{\varepsilon_0\varepsilon_2}{\varepsilon_0\varepsilon_2 + \varepsilon_0(1 - \varepsilon_2) + \varepsilon_1\varepsilon_2 + \varepsilon_1(1 - \varepsilon_2)}$$

$$= \frac{\varepsilon_2\varepsilon_0}{\varepsilon_0 + \varepsilon_1} = \varepsilon_2\alpha,$$

$$\gamma = \frac{\varepsilon_0(1 - \varepsilon_2)}{\varepsilon_0\varepsilon_2 + \varepsilon_0(1 - \varepsilon_2) + \varepsilon_1\varepsilon_2 + \varepsilon_1(1 - \varepsilon_2)}$$
$$= (1 - \varepsilon_2)\alpha,$$

$$\delta = \frac{\varepsilon_1\varepsilon_2}{\varepsilon_0\varepsilon_2 + \varepsilon_0(1 - \varepsilon_2) + \varepsilon_1\varepsilon_2 + \varepsilon_1(1 - \varepsilon_2)}$$

$$= \frac{\varepsilon_2\varepsilon_1}{\varepsilon_0 + \varepsilon_1} = \varepsilon_2(1 - \alpha),$$

$$\zeta = \frac{\varepsilon_1(1 - \varepsilon_2)}{\varepsilon_0\varepsilon_2 + \varepsilon_0(1 - \varepsilon_2) + \varepsilon_1\varepsilon_2 + \varepsilon_1(1 - \varepsilon_2)}$$
$$= (1 - \varepsilon_2)(1 - \alpha).$$

As these positive numbers ε_0, ε_1, ε_2, and ε_3 go to 0, these consistent beliefs must satisfy

$$\beta = 0, \ \gamma = \alpha, \ \delta = 0, \ \zeta = 1 - \alpha,$$

where α may be any number in the interval $[0,1]$. So there is a one-parameter family of beliefs vectors that are fully consistent with the scenario $([z_1],[y_2],[y_3])$. The belief probability α, which is the conditional probability that player 2 would assign to the event that player 1 chose x_1 if the path of play reached a decision node controlled by player 2 (which is a zero-probability event under this scenario), can be any number between 0 and 1. However, full consistency implies that the sum of belief probabilities $\beta + \gamma$, which is the conditional probability that player 3 would assign to the event that player 1 chose x_1 if the path of play reached a decision node controlled by player 3, must be equal to this same number α. That is, players 2 and 3 must agree about the conditional probability of x_1, conditional on z_1 not being chosen. Also, the sum of belief probabilities $\beta + \delta$, which is the conditional probability that player 3 would assign to the event that player 2 chose x_2 if the path of play reached a decision node controlled by player 3, must equal 0, because the move probability of x_2 is 0 in this scenario.

As defined by Kreps and Wilson (1982), a *sequential equilibrium* (or a *full sequential equilibrium*) of Γ^e is any (σ,π) in $(\times_{s \in S*} \Delta(D_s)) \times (\times_{s \in S*} \Delta(Y_s))$ such that the beliefs vector π is fully consistent with σ and, with beliefs π, the scenario σ is sequentially rational for every player at every information state. We can say that a behavioral-strategy profile σ is a *sequential-equilibrium scenario*, or that σ can be *extended* to a sequential equilibrium, iff there exists some π such that (σ,π) is a sequential equilibrium.

The unique sequential equilibrium of our simple card game is shown in Figure 4.4. For the game in Figure 4.5, the unique sequential-equilibrium scenario is $([x_1],[x_2])$. For the games in Figures 4.6 and 4.7, the unique sequential-equilibrium scenario is $([x_1], [x_2], .5[x_3] + .5[y_3], .5[x_4] + .5[y_4])$; to extend this scenario to a sequential equilibrium, we assign belief probabilities $\frac{1}{2}$ to each of player 4's nodes in Figure 4.6, or to each of player 3's nodes in Figure 4.7. For the game in Figure 4.8, the unique sequential-equilibrium scenario is $([x_1],[x_2],[x_3])$.

To see why the equilibrium $([z_1],[y_2],[y_3])$ cannot be extended to a sequential equilibrium of Figure 4.8, recall that full consistency with this scenario would require that the belief probabilities shown in the figure must satisfy $\beta = \delta = 0$, $\gamma = \alpha$, and $\zeta = 1 - \alpha$. Given that player 3 would choose y_3 at information state 3, player 2's expected payoff at information state 2 (with these belief probabilities) would be $0\alpha + 3(1 - \alpha)$ if she chose x_2, or would be $1\alpha + 1(1 - \alpha) = 1$ if she chose y_2. So choosing y_2 would be sequentially rational for player 2 at information state 2 iff $0\alpha + 3(1 - \alpha) \le 1$, so $\alpha \ge \frac{2}{3}$. On the other hand, with these belief probabilities, player 3's expected payoff at information state 3 would be $2\beta + 3\gamma + 2\delta + 0\zeta = 3\alpha$ if he chose x_3, or would be $0\beta + 1\gamma + 0\delta + 1\zeta = 1$ if he chose y_3. So choosing y_3 would be sequentially rational for player 3 at information state 3 iff $1 \ge 3\alpha$, so $\alpha \le \frac{1}{3}$. Thus (because α cannot be both less than $\frac{1}{3}$ and greater than $\frac{2}{3}$), there is no way to assign belief probabilities that are fully consistent with $([z_1],[y_2],[y_3])$ and make $([z_1],[y_2],[y_3])$ sequentially rational for both players 2 and 3.

We can now state two basic theorems that are necessary to confirm that this definition of sequential equilibrium is a reasonable one.

THEOREM 4.5. *If (σ,π) is a sequential equilibrium of an extensive-form game with perfect recall, then σ is an equilibrium in behavioral strategies.*

THEOREM 4.6. *For any finite extensive-form game, the set of sequential equilibria is nonempty.*

The proof of Theorem 4.5 is deferred to Section 4.11. The proof of Theorem 4.6 is given in Chapter 5, after our development of the concept of perfect equilibrium.

To illustrate Theorem 4.5, recall the example in Figure 4.2. The scenario $([b_1],[w_2],[z_1])$ is an equilibrium of the multiagent representation of this game, but it is not an equilibrium in behavioral strategies. With this scenario, fully consistent beliefs must assign belief probability 1 to the upper 1.3 node that follows w_2. With such beliefs, however, z_1 would be an irrational move for player 1 at information state 3. Thus, $([b_1],[w_2],[z_1])$ is not a sequential-equilibrium scenario, so Theorem 4.5 is satisfied. (Notice that this argument requires full consistency. In fact, $([b_1],[w_2],[z_1])$ is a weak sequential-equilibrium scenario.)

4.5 Computing Sequential Equilibria

Consider the game shown in Figure 4.9, adapted from Rosenthal (1981). This game can be interpreted as follows. After the upper chance event, which occurs with probability .95, the players alternate choosing be-

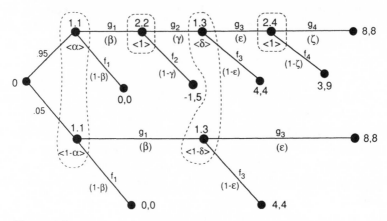

Figure 4.9

tween generous (g_k, for $k = 1,2,3,4$) and selfish (f_k) actions until someone is selfish or until both have been generous twice. Each player loses $1 each time he is generous, but gains $5 each time the other player is generous. Everything is the same after the lower chance event, which occurs with probability .05, except that 2 is then incapable of being selfish (perhaps because she is the kind of person whose natural integrity would compel her to reciprocate any act of generosity). Player 1 does not directly observe the chance outcome.

The move probabilities and belief probabilities that make up a sequential equilibrium are shown in Figure 4.9 in parentheses and angle brackets. To characterize the beliefs, let α denote the probability that player 1 would assign to the event that player 2 is capable of selfishness at the beginning of the game, and let δ denote the probability that 1 would assign to the event that 2 is capable of selfishness if he had observed her being generous once. To characterize the behavioral strategies, let β be the probability that 1 would be generous in his first move; let γ be the probability that 2 would be generous in her first move, if she is capable of selfishness; let ε be the probability that 1 would be generous in his second move; and let ζ be the probability that 2 would be generous at her second move, if she is capable of selfishness. We now show how to solve for these variables, to find a sequential equilibrium of this game.

Two of these variables are easy to determine. Obviously, $\alpha = .95$, because that is the probability of the upper alternative at the chance node, the outcome of which is unobservable to player 1. Also, it is easy to see that $\zeta = 0$, because player 2 would have no incentive to be generous at the last move of the game, if she is capable of selfishness.

To organize the task of solving for the other components of the sequential equilibrium, we use the concept of *support*, introduced in Section 3.3. At any information state, the support of a sequential equilibrium is the set of moves that are used with positive probability at this information state, according to the behavioral-strategy profile in the sequential equilibrium. We select any information state and try to guess what might be the support of the sequential equilibrium at this state. Working with this guess, we can either construct a sequential equilibrium or show that none exists with this support and so go on to try another guess.

It is often best to work through games like this backward. We already know what player 2 would do at information state 4, so let us consider now player 1's information state 3, where he makes his second move. There are three possible supports to consider: $\{g_3\}$, $\{f_3\}$, and $\{g_3,f_3\}$. Because $\zeta = 0$, player 1's expected payoff (or sequential value) from choosing g_3 at state 3 is $3\delta + 8(1-\delta)$; whereas player 1's expected payoff from choosing f_3 at state 3 is $4\delta + 4(1-\delta) = 4$. By Bayes's formula (4.5),

$$\delta = \frac{.95\beta\gamma}{.95\beta\gamma + .05\beta} = \frac{19\gamma}{19\gamma + 1}.$$

Even if $\beta = 0$, full consistency requires that $\delta = 19\gamma/(19\gamma + 1)$, because this equation would hold for any perturbed behavioral strategies in which β was strictly positive.

Let us try first the hypothesis that the support of the sequential equilibrium at 1's state 3 is $\{g_3\}$; so $\varepsilon = 1$. Then sequential rationality for player 1 at state 3 requires that $3\delta + 8(1-\delta) \geq 4$. This inequality implies that $0.8 \geq \delta = 19\gamma/(19\gamma + 1)$; so $\gamma \leq 4/19$. But $\varepsilon = 1$ implies that player 2's expected payoff from choosing g_2 at state 2 (her first move) would be 9, whereas her expected payoff from choosing f_2 at state 2 would be 5. Thus, sequential rationality for player 2 at state 2 requires that $\gamma = 1$, when $\varepsilon = 1$. Since $\gamma = 1$ and $\gamma \leq 4/19$ cannot both be satisfied, there can be no sequential equilibrium in which the support at 1's state 3 is $\{g_3\}$.

Let us try next the hypothesis that the support of the sequential equilibrium at 1's state 3 is $\{f_3\}$; so $\varepsilon = 0$. Then sequential rationality for player 1 at state 3 requires that $3\delta + 8(1-\delta) \leq 4$. This inequality implies that $0.8 \leq \delta = 19\gamma/(19\gamma + 1)$; so $\gamma \geq 4/19$. But $\varepsilon = 0$ implies that player 2's expected payoff from choosing g_2 at state 2 would be 4, whereas her expected payoff from choosing f_2 at state 2 would be 5. Thus, sequential rationality for player 2 at state 2 requires that $\gamma = 0$ when $\varepsilon = 0$. Because $\gamma = 0$ and $\gamma \geq 4/19$ cannot both be satisfied, there can be no sequential equilibrium in which the support at 1's state 3 is $\{f_3\}$.

The only remaining possibility is that the support of the sequential equilibrium at state 3 is $\{g_3, f_3\}$; so $0 < \varepsilon < 1$. Then sequential rationality for player 1 at state 3 requires that $3\delta + 8(1 - \delta) = 4$, or else player 1 would not be willing to randomize between g_3 and f_3. With consistent beliefs, this implies that $0.8 = \delta = 19\gamma/(19\gamma + 1)$; so $\gamma = 4/19$. Thus, player 2 must be expected to randomize between g_2 and f_2 at state 2 (her first move). Player 2's expected payoff from choosing g_2 at state 2 is $9\varepsilon + 4(1-\varepsilon)$, whereas her expected payoff from choosing f_2 at state 2 is 5. Thus, sequential rationality for player 2 at state 2 requires that $5 = 9\varepsilon + 4(1-\varepsilon)$; so $\varepsilon = 0.2$. It only remains now to determine 1's move at state 1. If he chooses s_1, then he gets 0; but if he chooses g_1, then his expected payoff is

$$.95 \times (4/19) \times .2 \times 3 + .95 \times (4/19) \times .8 \times 4 + .95 \times (15/19)$$
$$\times (-1) + .05 \times .2 \times 8 + .05 \times .8 \times 4 = 0.25,$$

when $\alpha = .95$, $\gamma = 4/19$, $\varepsilon = .2$, and $\zeta = 0$. Because $0.25 > 0$, sequential rationality for player 1 at state 1 requires that $\beta = 1$. That is, in the unique sequential equilibrium of this game, player 1 should be generous at his first move.

Consider now the scenario $([f_1], [f_2], [f_3], [f_4])$, in which each player would always be selfish at any information state; so $\beta = \gamma = \varepsilon = \zeta = 0$. If the chance node and the nodes and branches following the lower .05-probability branch were eliminated from this game, then this would be the unique sequential-equilibrium scenario. That is, if it were common knowledge that player 2 would choose between selfishness and generosity only on the basis of her own expected payoffs, then no player would ever be generous in a sequential equilibrium. Furthermore, $([f_1], [f_2], [f_3], [f_4])$ is an equilibrium in behavioral strategies of the actual game given in Figure 4.9 and can even be extended to a weak sequential equilibrium of this game, but it cannot be extended to a full sequential equilibrium of this game. For example, we could make this scenario a

weak sequential equilibrium, satisfying sequential rationality at all information states, by letting the belief probabilities α and δ both equal
.95. However, the belief probability $\delta = .95$ would not be consistent (in
the full sense) with this scenario because, if player 2 would be expected
to be selfish at state 2 ($\gamma = 0$), then player 1 would infer at state 3, after
he chose generosity and did not get a selfish response from player 2,
that player 2 must be incapable of selfishness, so δ must equal 0. But
with $\delta = 0$, selfishness (f_3) would be an irrational move for player 1 at
state 3, because he would expect generosity (g_3) to get him a payoff of
8, with probability $1 - \delta = 1$.

This example illustrates the fact that small initial doubts may have a
major impact on rational players' behavior in multistage games. If player
1 had no doubt about player 2's capacity for selfishness, then perpetual
selfishness would be the unique sequential equilibrium. But when it is
common knowledge at the beginning of the game that player 1 assigns
even a small positive probability of .05 to the event that player 2 may
be the kind of person whose natural integrity would compel her to
reciprocate any act of generosity, then player 1 must be generous at
least once with probability 1 in the unique sequential equilibrium. The
essential idea is that, even if player 2 does not have such natural integrity, she still might reciprocate generosity so as to encourage player 1
to assign a higher probability to the event that she may continue to be
generous in the future. Her incentive to do so, however, depends on
the assumption that player 1 may have at least some initial uncertainty
about player 2's type in this regard. The crucial role of such small initial
uncertainties in long-term relationships has been studied in other examples of economic importance (see Kreps, Milgrom, Roberts, and
Wilson, 1982).

For a second example, consider the game in Figure 4.10. To characterize the sequential equilibria of this game, let α denote the belief
probability that player 2 would assign in information state 3 to the
upper 2.3 node, which follows player 1's x_1 move. With these beliefs at
information state 3, player 2's conditionally expected payoff would be
$8\alpha + 0(1 - \alpha)$ if she chose e_3, $7\alpha + 7(1 - \alpha) = 7$ if she chose f_3, or $6\alpha
+ 9(1 - \alpha) = 9 - 3\alpha$ if she chose g_3. So move e_3 would be optimal for
player 2 at state 3 when both $8\alpha \geq 7$ and $8\alpha \geq 9 - 3\alpha$, that is, when $\alpha
\geq 7/8$. Move f_3 would be optimal for 2 when both $7 \geq 8\alpha$ and $7 \geq 9 -
3\alpha$, that is, when $2/3 \leq \alpha \leq 7/8$. Move g_3 would be optimal for 2 when
both $9 - 3\alpha \geq 8\alpha$ and $9 - 3\alpha \geq 7$, that is, when $\alpha \leq 2/3$. Notice that

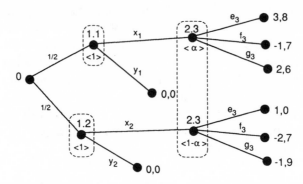

Figure 4.10

there is no value of α for which e_3 and g_3 both would be optimal for player 2 at state 3, so the only possible supports at information state 3 for a sequential equilibrium are $\{e_3\}$, $\{f_3\}$, $\{g_3\}$, $\{e_3,f_3\}$, and $\{f_3,g_3\}$. We consider these possibilities in this order.

If player 2 were to use e_3 for sure at state 3, then player 1 would want to choose x_1 at state 1 and x_2 at state 2. Then consistency with $([x_1],[x_2])$ would require that $\alpha = \frac{1}{2}$, but e_3 would be irrational for 2 at state 3 if $\alpha = \frac{1}{2}$ (because g_3 would be better for her). So there is no sequential equilibrium with support $\{e_3\}$ at state 3.

If player 2 were to use f_3 for sure at state 3, then player 1 would want to choose y_1 at state 1 and y_2 at state 2. Then consistency with $([y_1],[y_2])$ would impose no restriction on the belief probability α, because 2.3 nodes would have prior probability 0. However, sequential rationality for player 2 at information state 3 would require $\frac{2}{3} \leq \alpha \leq \frac{7}{8}$, or else player 2 would not be willing to choose f_3. So the behavioral-strategy profile $([y_1],[y_2],[f_3])$ with any beliefs vector such that $\frac{2}{3} \leq \alpha \leq \frac{7}{8}$ together form a sequential equilibrium.

If player 2 were to use g_3 for sure at state 3, then player 1 would want to choose x_1 at state 1 and y_2 at state 2. Then consistency with $([x_1],[y_2])$ would require that $\alpha = 1$, in which case g_3 would be an irrational move for player 2 at state 3 (e_3 would be better for her). So there is no sequential equilibrium with support $\{g_3\}$ at state 3.

For a randomization between e_3 and f_3 to be sequentially rational for player 2 at state 3, the belief probability α must equal $\frac{7}{8}$ (so that $8\alpha = 7$). There are two ways to construct a sequentially rational scenario, in which player 2 randomizes between e_3 and f_3, and with which such beliefs would be consistent. One way is to have player 1 choose y_1 at state 1 and y_2 at state 2; then the 2.3 nodes have zero probability. To make such moves rational for player 1, the move probability of e_3 cannot be greater than $\frac{1}{4}$, or else x_1 would be better than y_1 for player 1 at state 1. So, for any β such that $0 \le \beta \le \frac{1}{4}$, the behavioral-strategy profile $([y_1], [y_2], \beta[e_3] + (1 - \beta)[f_3])$ with the belief probability $\alpha = \frac{7}{8}$ together form a sequential equilibrium.

The other way to make $\alpha = \frac{7}{8}$ consistent is to have player 1 randomize at state 2. To make player 1 willing to randomize between x_2 and y_2 at state 2, when player 2 is expected to randomize between e_3 and f_3 at state 3, the move probability of e_3 must be exactly $\frac{2}{3}$. With this randomized strategy for player 2, player 1 would want to choose x_1 for sure at state 1. Then $\alpha = \frac{7}{8}$ is consistent if the move probability of x_2 at state 2 is $\frac{1}{7}$. So the behavioral-strategy profile $([x_1], (\frac{1}{7})[x_2] + (\frac{6}{7})[y_2], (\frac{2}{3})[e_3] + (\frac{1}{3})[f_3])$ with the belief probability $\alpha = \frac{7}{8}$ together form a sequential equilibrium.

For player 2 to randomize f_3 and g_3, the belief probability α must equal $\frac{2}{3}$ (so $7 = 6\alpha + 9(1 - \alpha)$). If player 2 would randomize between f_3 and g_3, then player 1 would want to choose y_2 at state 2. So the only way to make $\alpha = \frac{2}{3}$ consistent is to have player 1 also choose y_1 at state 1; then the 2.3 nodes have zero prior probability. Player 1 will be willing to choose y_1 at state 1 if player 2 uses a strategy $(1 - \gamma)[f_3] + \gamma[g_3]$ where $\gamma \le \frac{1}{3}$ (so $0 \ge -1(1 - \gamma) + 2\gamma$). So, for any γ such that $0 \le \gamma \le \frac{1}{3}$, the behavioral-strategy profile $([y_1], [y_2], (1 - \gamma)[f_3] + \gamma[g_3])$ with the belief probability $\alpha = \frac{2}{3}$ together form a sequential equilibrium of the game in Figure 4.10.

4.6 Subgame-Perfect Equilibria

A concept of subgame-perfect equilibrium was defined for extensive-form games by Selten (1965, 1975, 1978). This concept is older and weaker than the concept of sequential equilibrium, but it still has some important applications (especially for games with infinite pure-strategy sets, for which a definition of sequential equilibrium has not yet been formalized).

For any node x in Γ^e, let $F(x)$ be the set of all nodes and branches that follow x, including the node x itself. The node x is a *subroot* if, for every s in S^*, either $Y_s \cap F(x) = \varnothing$ or $Y_s \subseteq F(x)$. That is, if x is a subroot, then any player who moves at x or thereafter will know that the node x has occurred. A *subgame* of Γ^e is any game that can be derived from Γ^e by deleting all nodes and branches that do not follow some subroot x and making that node x the root of the subgame.

Let Γ_x^e be a subgame that begins at a subroot x in Γ^e. If the node x occurred in the play of the game Γ^e, then it would be common knowledge among the players who move thereafter that they were playing in the subgame that begins at x. That is, a game theorist who was modeling this game after node x occurred could describe the commonly known structure of the situation by the extensive-form game Γ_x^e and could try to predict players' behavior just by analyzing this game. Rational behavior for the players in Γ^e at node x and thereafter must also appear rational when viewed within the context of the subgame Γ_x^e. Thus, Selten (1965, 1975) defined a *subgame-perfect equilibrium* of Γ^e to be any equilibrium in behavioral strategies of Γ^e such that, for every subgame of Γ^e, the restriction of these behavioral strategies to this subgame is also an equilibrium in behavioral strategies of the subgame.

To understand why we only consider subgames, recall Figure 2.7 (Chapter 2). This figure shows part of the extensive-form representation of our simple card game, the portion that follows the 1.b node, where player 1 moves with a black card. In any equilibrium of the game in Figure 2.7, the move probability of a raise by player 1 would have to be 0, because player 2 would do better by meeting a raise. On the other hand, a raise by player 1 in state b has move probability $\frac{1}{3}$ in the unique equilibrium of the simple card game, as shown in Figure 4.4. However, the sequential-equilibrium scenario of the simple card game is a subgame-perfect equilibrium. The portion of the game tree that follows the 1.b node is not a subgame, because player 2's information state 0 occurs at two nodes, of which only one is included in Figure 2.7. Indeed, as we discussed in Section 2.7, there is no reason to suppose that a solution for Figure 4.4 should apply to the game in Figure 2.7, because Figure 2.7 represents a game in which player 2 knows that 1's card is black, whereas 2's uncertainty about the color of 1's card is crucial to the analysis of the simple card game in Figure 4.4. Thus, in the definition of subgame-perfect equilibria, we only require that an equilibrium should remain an equilibrium when it is restricted to a subgame, which represents a portion of the game that would be common knowledge if it occurred.

It is straightforward to check that, if (σ, π) is any sequential equilibrium of Γ^e, then σ is a subgame-perfect equilibrium of Γ^e. On the other

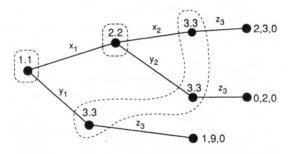

Figure 4.11

hand, the set of subgame-perfect equilibria may be much larger than the set of sequential-equilibrium scenarios.

For the simple example in Figure 4.5, the equilibrium $([y_1],[y_2])$ is not subgame perfect, because its restriction to the subgame beginning at player 2's node is $[y_2]$, which is not an equilibrium of this subgame. On the other hand, consider Figure 4.11, which differs from Figure 4.5 only by the addition of a dummy player 3, who has no nontrivial information or strategic alternatives. For this game, there are no proper subgames (other than the terminal nodes, considered as one-node subgames, and the original game itself), so $([y_1],[y_2],[z_3])$ is a subgame-perfect equilibrium of Figure 4.11. The concept of sequential equilibrium is not sensitive to the addition of such a dummy, however; and $([y_1],[y_2],[z_3])$ is not a sequential-equilibrium scenario of Figure 4.11.

4.7 Games with Perfect Information

A game with *perfect information* is any extensive-form game in which each information state of each player occurs at exactly one decision node. That is, Γ^e is a game with perfect information iff, for every player i and every information state s in S_i, the set Y_s (the set of nodes labeled "$i.s$" in Γ^e) contains exactly one node. Such games with perfect information represent situations in which individuals move one at a time (never simultaneously) and, whenever a player moves, he knows everything that every player observed and did at every past decision node. Many recreational games, such as chess and checkers, are games with perfect information. (Most card games, on the other hand, do not have perfect information.)

If Γ^e is a game with perfect information, then a beliefs vector π is in $\times_{s \in S^*} \Delta(Y_s)$ iff $\pi_s(x) = 1$ for every s in S^* and x in Y_s (because, with perfect information, $x \in Y_s$ implies that $Y_s = \{x\}$). That is, the only possible beliefs vector π is the vector of all ones.

With perfect information, every decision node in the game is a subroot, so every subgame-perfect equilibrium is a sequential-equilibrium scenario. That is, the concepts of subgame-perfect equilibrium and sequential equilibrium coincide for games with perfect information.

Given any extensive-form game Γ^e, we can generate a game that differs from Γ^e only in the payoffs by choosing payoff functions in $\mathbf{R}^{N \times \Omega}$, where Ω is the set of terminal nodes in Γ^e. We may say that a property holds *for all generic games with perfect information* iff, for every extensive-form game Γ^e with perfect information, there is an open and dense set of payoff functions in $\mathbf{R}^{N \times \Omega}$ that generate games, differing from Γ^e only in the payoffs, which all satisfy this property. (A set is *dense* in $\mathbf{R}^{N \times \Omega}$ iff every vector in $\mathbf{R}^{N \times \Omega}$ is the limit of some sequence of vectors in this set.)

Zermelo (1913) showed that a game with perfect information must have a sequential equilibrium in pure strategies, without randomization. Furthermore, a game with perfect information "almost always" has a unique sequential equilibrium, as the following version of Zermelo's theorem asserts. (The proof is in Section 4.11. Note that we are assuming that an extensive-form game has a finite number of nodes.)

THEOREM 4.7. *If Γ^e is an extensive-form game with perfect information, then there exists at least one sequential equilibrium of Γ^e in pure strategies (so every move probability is either 0 or 1). Furthermore, for all generic games with perfect information, there is exactly one sequential equilibrium.*

With rules against repeating any situation more than a fixed number of times, so that it can be finitely described in extensive form, chess is an example of a game with perfect information to which Theorem 4.7 can be applied. So there exists an equilibrium in pure strategies of chess. Then, because chess is a two-person zero-sum game, we can apply Theorem 3.2 to show that exactly one of the following three statements must be true for chess. Either (1) there is a pure strategy for the white player that can guarantee that white will win, no matter what black does; or (2) there is a pure strategy for the black player that can guarantee that black will win, no matter what white does; or (3) each player has a

Table 4.2   A game in strategic form

$\hat{C}_1$	$\hat{C}_2$	
	x_2	y_2
x_1	2,1	3,2
y_1	1,4	4,3

pure strategy that will guarantee himself at least a stalemate, no matter what the other player does. From a game-theoretic point of view, it is just a question of (astronomically complex!) computation to determine which of these statements is true for chess and to characterize the equilibrium strategies.

Given any strategic-form game $\hat{\Gamma} = (N, (\hat{C}_i)_{i \in N}, (u_i)_{i \in N})$, a *Stackelberg solution* of $\hat{\Gamma}$ is defined to be any subgame-perfect equilibrium of an extensive-form game with perfect information that is derived from $\hat{\Gamma}$ by putting the players in some order and making each player i choose a move in $\hat{C}_i$ after observing the moves of all players who precede him in this order. The player who goes first in the order is called the *Stackelberg leader*. In general, there are many Stackelberg solutions for any given strategic-form game, one for each way of ordering the players, but Theorem 4.7 guarantees that each ordering has at least one Stackelberg solution in pure strategies.

For example, consider the two-player game in Table 4.2. In the Stackelberg solution when player 1 is the leader, player 2's strategy is to choose y_2 if player 1 chooses x_1 and to choose x_2 if player 1 chooses y_1; so player 1's strategy is to choose x_1, so that the expected payoff allocation is (3,2). In the Stackelberg solution, when player 2 is the leader, player 1's strategy is to choose x_1 if player 2 chooses x_2 and to choose y_1 if player 2 chooses y_2; so player 2's strategy is to choose y_2, so that the expected payoff allocation is (4,3). It is not necessarily an advantage to be a Stackelberg leader. For this game, player 2 would prefer to be a Stackelberg leader, but player 1 would prefer that player 2 be a Stackelberg leader rather than lead himself.

4.8 Adding Chance Events with Small Probability

Recall again the simple example in Figure 4.5. In this game, $([y_1],[y_2])$ is an equilibrium but cannot be extended to a sequential equilibrium.

This fact is robust, in the sense that it remains true for all games derived from Figure 4.3 by small perturbations in the payoffs (as long as no terminal payoff is changed by more than 0.5). However, as Fudenberg, Kreps, and Levine (1988) have shown, the set of sequential equilibria may change very dramatically when the game is perturbed or transformed by adding chance events with arbitrarily low probability.

For example, let ε be any small positive number and consider Figure 4.12. In this example $([y_1],[y_2])$ can be extended to a sequential equilibrium, with the beliefs vector in which player 2 assigns probability 1 to the lower 2.2 node if she gets an opportunity to move. Thus, by adding a small probability event in which players' moves do not matter, we can transform the game in Figure 4.5 into a game in which all equilibria are sequential. In general, transformations of this form can be used to enlarge the set of sequential equilibria of any game so that it will include any or all of the equilibria of the original game.

Thus, the concept of sequential equilibrium is very sensitive to the inclusion of new low-probability chance events in the structure of the game. This sensitivity occurs because the concept of sequential equilibrium requires that players' beliefs after events of zero probability should be restricted to those that are consistent with the given structure of the game. Adding new branches to the game tree, even if they are chance events with very low probability, may significantly enlarge the set of beliefs that can pass this consistency test.

To put this another way, when we model a conflict situation as a game in extensive form, we might, for simplicity, omit from our model some possible chance events that seem to have very small probability. When we make this modeling omission, we should be aware that it may have a major impact on the set of sequential equilibria. Omitting a small-

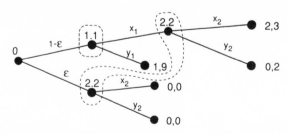

Figure 4.12

probability chance event is equivalent (under the solution concept of sequential equilibrium) to assuming that players would never revise their beliefs to assign a significant probability to this event, as long as there was any other way to explain their observations within the remaining structure of the game (i.e., by inferring that other players had made irrational mistakes).

Throughout this chapter we have assumed that every alternative is assigned strictly positive probability at each chance node. In light of the above remarks, it might be worth considering briefly how sequential equilibria should be defined when this assumption is dropped. For example, in Figure 4.12, when $\varepsilon = 0$, should we admit $([y_1],[y_2])$ with belief probability 1 on the lower 2.2 node as a sequential equilibrium? If we do, then we are treating Figure 4.12 with $\varepsilon = 0$ as a different game from Figure 4.5, although they differ only by a zero-probability chance event. On the other hand, if we do not, then we have a lower-discontinuity in the set of sequential equilibria as ε goes to 0.

One natural way to extend the set of sequential equilibria to games with zero-probability alternatives at chance nodes may be defined as follows. First, revise the game by making "chance" another player who gets payoff 0 at every terminal node, and let every chance node be assigned a different information state. (So "chance" has perfect information whenever it moves.) Then a sequential equilibrium of the original game may be defined to be any sequential equilibrium of this revised game in which chance's behavioral strategy is to randomize according to the probabilities given at each chance node in the original game. With this definition, it can be shown that, when we specify everything about an extensive-form game except its terminal payoffs and the chance probabilities, then the set of sequential equilibria will depend upper-hemicontinuously on these payoffs and the chance probabilities.

So $([y_1],[y_2])$ with belief probability 1 on the lower 2.2 node is a sequential equilibrium of Figure 4.12 even when $\varepsilon = 0$. This distinction between Figure 4.12 with $\varepsilon = 0$ and Figure 4.5 can be justified by distinguishing between an *impossible* event and a *zero-probability* event. A chance event is not impossible if a rational individual might infer that this event had positive probability after some players made moves that supposedly had zero probability. In Figure 4.12 with $\varepsilon = 0$, the event that player 2 may have an opportunity to make a payoff-irrelevant move is a zero-probability event but is not impossible, whereas it is an impossible event in the game described by Figure 4.5.

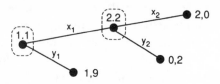

Figure 4.13

This distinction between impossible events and zero-probability possible events is actually fundamental to the whole concept of sequential equilibrium. To see why, consider the simple example in Figure 4.13. The unique sequential-equilibrium scenario of this game is $([y_1],[y_2])$. There are other equilibria where player 2 may use a randomized strategy (as long as the move probability of x_2 is not more than ½), but there is no equilibrium in which player 1 uses x_1 with positive probability. Thus, if we accept sequential equilibrium or Nash equilibrium as an upper solution concept, then we must conclude that the event of player 1 choosing x_1 has zero probability. But can we go on to conclude that it is an impossible event? That is, can we assert that the statement "player 1 chooses x_1 in this game" is definitely false? If so, then the statement "if player 1 chooses x_1 in this game then player 2 will choose x_2" would be true, because (by the basic rules of logic) any implication is true if its hypothesis is false. However, player 1 would rationally choose x_1 if his doing so would imply that player 2 would choose x_2. (See Bicchieri, 1989, for related examples and philosophical discussion.)

To avoid this logical contradiction, we must allow that, before player 1 chooses his move, the event that he will irrationally choose x_1 is not impossible but is merely a zero-probability possible event. In general, before any rational player moves, we must suppose that the event that he will make an irrational mistake is a possible event that has zero probability. Building on this idea, further refinements of the equilibrium concept are developed in Chapter 5 by first considering perturbed games where the probability of such irrational mistakes is small but positive and then taking the limit as these mistake probabilities go to 0.

4.9 Forward Induction

Sequential rationality and subgame-perfectness are *backward induction* principles for the analysis of extensive-form games, because they require that any predictions that can be made about the behavior of players at the end of a game are supposed to be anticipated by the players earlier in the game. In contrast, Kohlberg and Mertens (1986) have suggested a different kind of *forward induction* principle that may also be used in the analysis of extensive-form games.

A forward induction principle would assert that the behavior of rational intelligent players in a subgame may depend on the options that were available to them in the earlier part of the game, before the subgame. For example, consider the game in Figure 4.14. This game has three sequential-equilibrium scenarios: $([a_1],[y_1],[y_2])$, $([a_1], .8[x_1] + .2[y_1], (\frac{3}{8})[x_2] + (\frac{5}{8})[y_2])$, and $([b_1],[x_1],[x_2])$. However, a forward-induction argument may eliminate the first two of these equilibria. Among the three equilibria of the subgame beginning at the 1.1 node, there is only one equilibrium that would give player 1 a payoff that is not less than he could have gotten by choosing a_1 before the subgame. So, Kohlberg and Mertens suggest, player 2 should infer that player 1 would only have chosen b_1 and caused the subgame to occur if he were expecting to play the (x_1,x_2) equilibrium in the subgame; therefore, player 2 should play x_2 (her unique best response to x_1) in the subgame. If player 2 can be expected to reason this way, then player 1 should use the strategy b_1x_1.

An essential element of this argument is that other players may make inferences about the move at the 1.1 node from the fact that the person deciding between x_1 and y_1 is the same person who just chose b_1 over a_1. Thus, to analytically formulate such forward-induction arguments in general, we should consider the normal representation, rather than the multiagent representation. When we do so for this example, we find

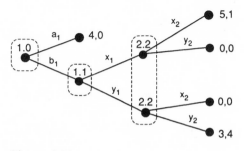

Figure 4.14

that b_1y_1 is a strongly dominated strategy for player 1 in the normal representation, because it can only give him less than the payoff that he could have gotten from a_1. There are no dominated strategies in the multiagent representation of this game.

Unfortunately, some natural forward-induction arguments may be incompatible with other natural backward-induction arguments. For example, consider the game in Figure 4.15. What should player 2 be expected to do at the 2.2 decision node? A forward-induction argument might suggest that, because b_1y_1 is a weakly dominated strategy for player 1, player 2 would expect player 1 to choose x_1 at state 3 if he chose b_1 at state 1; so player 2 should choose a_2. On the other hand, backward induction determines a unique subgame-perfect equilibrium $([a_1],[b_2],[y_1],[y_2])$ in which player 2 would choose b_2 at the 2.2 node. In fact, the reduced normal representation (in Table 4.3) of this game is an example of a game in which iterative elimination of weakly dominated strategies can lead to different results, depending on the order of elimination. Backward induction in the extensive-form game corresponds in the normal representation to iteratively eliminating weakly

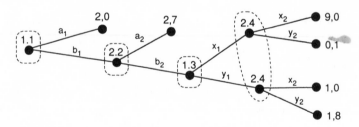

Figure 4.15

Table 4.3   A game in strategic form, the reduced normal representation of Figure 4.15

C_1	C_2		
	a_2·	b_2x_2	b_2y_2
a_1·	2,0	2,0	2,0
b_1x_1	2,7	9,0	0,1
b_1y_1	2,7	1,0	1,8

dominated strategies in the order: first b_2x_2 for player 2, then b_1x_1 for player 1, then a_2· for player 2, and then b_1y_1 for player 1, leaving the equilibrium $(a_1·,b_2y_2)$. Forward induction corresponds to iteratively eliminating weakly dominated strategies in the order: first b_1y_1 for player 1, then b_2x_2 and b_2y_2 for player 2, leaving the equilibria $(a_1·,a_2·)$ and $(b_1x_1,a_2·)$.

A number of examples where forward-induction arguments lead to striking conclusions have been discussed by Glazer and Weiss (1990), van Damme (1989), and Ben-Porath and Dekel (1987). Much of this discussion uses Kohlberg and Mertens's (1986) concept of stable sets of equilibria (see Section 5.6), but some examples can be analyzed using just iterative elimination of weakly dominated strategies.

For example, van Damme (1989) suggested the following variation of the Battle of the Sexes game (Table 3.5): before either player decides whether to go to the shopping center or the football match, player 1 has the opportunity to publicly burn one dollar (or something worth one unit of his utility). This game is shown in extensive form in Figure 4.16, where b_1 denotes the move of burning a dollar, and a_1 denotes the move of not burning a dollar. Then (suppressing the irrelevant move for player 1 if he did not do as his strategy specifies in the first stage, and naming player 2's strategies so that the move that 2 will use

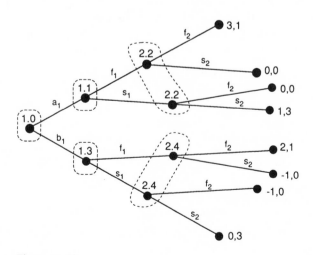

Figure 4.16

if 1 does not burn a dollar comes first), we can write the purely reduced normal representation of this game as shown in Table 4.4.

This game has many Nash equilibria, both in extensive and strategic form. However, iterative elimination of weakly dominated strategies in the normal representation eliminates all equilibria except $(a_1 f_1, f_2 f_2)$, the best equilibrium for player 1. The strategy $b_1 s_1$ is strongly dominated for player 1. But once this strategy is eliminated, $f_2 s_2$ and $s_2 s_2$ become weakly dominated for player 2. Once these strategies are eliminated, $a_1 s_1$ becomes strongly dominated for player 1. Then, when $a_1 s_1$ is also eliminated, $s_2 f_2$ becomes weakly dominated for player 2. Eliminating $s_2 f_2$ leaves only the strategy $f_2 f_2$ for player 2, for which $a_1 f_2$ is the unique best response for player 1.

The first two steps of this iterative elimination correspond to the following forward-induction argument. If player 1 burned a dollar, then player 2 would search for a rational explanation of his action. Burning a dollar could only be optimal for player 1 if he thought that this move would steer player 2 to go to the football match (in the strategy $s_2 f_2$), in which case player 1 could only benefit by going to the football match himself, following the strategy $b_1 f_1$. So, after eliminating the strategy $b_1 s_1$, player 2 would have to take the burning of a dollar as a signal that player 1 will go to the football match. But if player 1 knows that player 2 would interpret the burning of the dollar according to this forward-induction argument, then it would be irrational for player 1 to not burn the dollar and go shopping; so player 2 should not expect player 1 to go shopping even if he does not burn the dollar.

This conclusion, that player 1's unused option to publicly burn a dollar enables him to get the outcome that he likes best, is so remarkable that it may call into question the validity, as a solution concept, of iterative elimination of weakly dominated strategies in the normal rep-

Table 4.4   Battle of the Sexes game with option to burn a dollar for player 1, in strategic form

C_1	C_2			
	$f_2 f_2$	$f_2 s_2$	$s_2 f_2$	$s_2 s_2$
$a_1 f_1$	3,1	3,1	0,0	0,0
$a_1 s_1$	0,0	0,0	1,3	1,3
$b_1 f_1$	2,1	−1,0	2,1	−1,0
$b_1 s_1$	−1,0	0,3	−1,0	0,3

resentation. For comparison, let us consider the equilibrium (a_1s_1, s_2s_2), which is best for player 2 and which this forward-induction argument eliminates.

This equilibrium (a_1s_1, s_2s_2) corresponds to a sequential equilibrium of the extensive-form game, in which player 2 would go shopping whether player 1 burned a dollar or not; so player 1 goes shopping without burning a dollar, but player 1 would go shopping also even if he did burn a dollar. In the context of this sequential equilibrium, if player 1 burned a dollar, then this move would just be interpreted by player 2 as an irrational mistake, with no intrinsic meaning or significance for predicting 1's future behavior. In any sequential equilibrium, each player implicitly applies the conservative assumption that the other players will behave rationally in the future, with probability 1, even when they have made mistakes in the past. So, in this sequential equilibrium, player 2 may still expect player 1 to go shopping with her, even if he has burned a dollar, because (s_1, s_2) is an equilibrium in the remaining subgame. Of course, if 2 would respond to 1's burning move this way, then burning a dollar would indeed be an irrational and meaningless mistake for player 1, and it has zero probability in this equilibrium; so this equilibrium is sequentially rational.

At any given point in the game, forward induction stipulates that a player should adopt beliefs about other players' strategies so that their observed earlier moves would be rational, if any such beliefs exist. So the forward-induction argument against the (a_1s_1, s_2s_2) is that, if player 1 did burn a dollar, then player 2 would try to find some rational explanation for his observed behavior, and the only possible explanation would be that he burned the dollar in an effort to get her to go to the football match, as happens under the equilibrium $(b_1 f_1, s_2 f_2)$. However, a second application of forward induction would also eliminate such an equilibrium; because if player 1 made the mistake of *not* burning the dollar in the $(b_1 f_1, s_2 f_2)$ equilibrium, then player 2 would have to explain this behavior by supposing that he was not burning because he expected her to go to the football match without any burning, as a part of an equilibrium like $(a_1 f_1, f_2 f_2)$. Thus, we must question the basic assumption that player 2 should be able to explain or rationalize player 1's behavior in all parts of the game. Although a_1 and b_1 can each be rationalized by various subgame-perfect equilibria of this game, in every subgame-perfect equilibrium there is at least one zero-probability move by player 1 (a_1 or b_1) that would have to be interpreted by player 2 as an irrational mistake if it occurred.

4.10 Voting and Binary Agendas

The analysis of sequential voting games is an important area of application for the ideas of subgame-perfectness and iterative elimination of weakly dominated strategies. In this section, we give a brief introduction to some basic results in the analysis of voting and agendas. No results from this section are used elsewhere in this book, so this section may be skipped without loss of continuity.

For a simple introductory example, suppose that a committee consisting of nine individuals must choose one of four options, which are numbered 1, 2, 3, and 4. Three of the individuals think that 1 is the best option, 2 is second-best, 4 is third-best, and 3 is worst. Another three of the individuals think that 2 is the best option, 3 is second-best, 4 is third-best, and 1 is worst. Another three of the individuals think that 3 is the best option, 1 is second-best, 4 is third-best, and 2 is worst.

Suppose that, within this committee, any question must be decided by a majority vote. For any two options α and β, let us say that α *can beat* β and write $\alpha > \beta$ iff a majority of the committee would prefer α over β. Thus, for this example,

$$1 > 2, 2 > 3, 3 > 1, 1 > 4, 2 > 4, 3 > 4.$$

Notice that option 4 can be beaten by each of the other options; but options 1, 2, and 3 form a cycle in which each can be beaten by one of the other two.

Now suppose that the chairman of the committee plans to organize the committee's decision-making process by asking for votes on a sequence of questions, each of which has two possible responses. We assume that the chairman's plan for asking questions, called his *agenda*, is common knowledge among the individuals in the committee at the beginning of the meeting.

For example, the agenda might be as follows. The first question will be whether to eliminate 1 or 2 from the list of feasible options under consideration. Then, if α denotes the option in $\{1,2\}$ that was not eliminated at the first stage, the second question will be whether to eliminate this option α or option 3. Then the third and final question will be to choose between option 4 and the other remaining option (α or 3).

Let us suppose that the individuals' preferences are common knowledge. With this agenda, the individual voters should recognize that the option that survives the second stage will beat option 4 at the third stage and thus will be the finally chosen option. So if the second vote is between options 2 and 3, then option 3 will be eliminated and option 2 will be chosen, because $2 > 3$. If the second vote is between 1 and 3, then 1 will be eliminated and 3 will be chosen, because $3 > 1$. Thus, if 1 is eliminated at the first stage, then the second vote will be between 2 and 3; so 2 will be chosen. But if 2 is eliminated at the first stage, then the second vote will be between 1 and 3; so 3 will be chosen. Because $2 > 3$, the majority should vote at the first stage to eliminate option 1, so the finally chosen option will be option 2 (rather than option 3). Notice that a majority of the individuals vote to eliminate 1 rather than 2 at the first stage, in spite of the fact that a majority prefers 1 over 2, because we are assuming that the voters are sophisticated or intelligent enough to understand that eliminating 2 would lead to 3 being the ultimate outcome, not 1.

This result, that the committee will choose option 2, depends crucially on the particular agenda. For example, if the order in which options are to be considered is reversed (so that the first question is whether to eliminate 4 or 3, the second question is whether to eliminate 2 or the survivor from the first vote, and the third question is whether to choose 1 or the survivor of the second vote), then the finally chosen outcome from sophisticated voting will be option 3. There are other agendas under which option 1 would be the finally chosen outcome. However, with the given preferences, there is no agenda that would lead to the final choice of option 4, because any other option would beat 4 at the last stage.

Let us now define a general model of voting in committee. Let K denote a nonempty finite set of social options, among which a committee must ultimately choose one. We write $\alpha > \beta$ iff there is a majority that prefers α over β. Suppose that the number of individuals in the committee is odd, and each individual has a strict preference ordering over the options in K. So for any pair of different options α and β in K,

either $\alpha > \beta$ or $\beta > \alpha$, but not both.

(Notice that this majority preference relation $>$ is not necessarily a transitive ordering, as the above example illustrates.) We assume that this majority preference relation is common knowledge among all the individuals in the committee.

A *binary agenda* for K can be defined (using the graph-theoretic terminology of Section 2.1) to be a finite rooted tree together with a function G that assigns a nonempty set $G(x)$ to each node in the tree, such that the following three properties are satisfied. First, letting x_0 denote the root of the tree, $G(x_0)$ must equal K. Second, each nonterminal node x must be immediately followed by exactly two nodes in the tree; and if y and z denote the two nodes that immediately follow node x, then $G(x) = G(y) \cup G(z)$. (The sets $G(y)$ and $G(z)$ do not have to be disjoint.) Third, for each terminal node w, $G(w)$ must be a set that consists of exactly one option in K.

Successive votes of the committee will determine a path through such an agenda tree starting at the root. When the path reaches a node x, $G(x)$ denotes the set of options that are still under consideration for choice by the committee. If y and z are the nodes that immediately follow node x, then the next question to be decided by a vote at node x is, "Should the set of options under consideration now be reduced to $G(y)$ or $G(z)$?" We require that the union of $G(y)$ and $G(z)$ must equal $G(x)$, because this question should not be phrased in such a way as to automatically eliminate some options; options can be eliminated from consideration only as a result of a majority vote. A succession of votes continues until a terminal node is reached, where only one option remains under consideration. We assume that the agenda planned by the chairman of the committee must be common knowledge among all the individuals in the committee at the start of the voting process.

A *sophisticated voting solution* for such an agenda is a mapping ϕ that selects an option $\phi(x)$ for each node x in the agenda tree, such that the following two properties are satisfied. First, for any node x, $\phi(x) \in G(x)$. Second, for any nonterminal node x, if y and z denote the two nodes that immediately follow x, and $\phi(y) > \phi(z)$ or $\phi(y) = \phi(z)$, then $\phi(x) = \phi(y)$. That is, a sophisticated voting solution tells us which option $\phi(x)$ would be ultimately chosen if the voting began at node x in the agenda. At each round of voting, a majority will vote to move to the immediately following node that has the solution they prefer.

Given the majority preference relation $>$, there is a unique sophisticated voting solution to any such binary agenda. We can easily construct this sophisticated voting solution ϕ by working backward through the tree, starting with the terminal nodes (where $G(x) = \{\phi(x)\}$) and using the fact that, if y and z immediately follow x, then $\phi(x)$ must be the option that a majority prefers among $\{\phi(y), \phi(z)\}$. The option that is selected by the sophisticated voting solution at the root of the agenda tree is the *sophisticated outcome* of the agenda.

Suppose that the individuals in the committee all vote simultaneously and independently on any question that is put to them, after everyone has learned the results of all previous votes. Thus, each node in a binary agenda actually represents a whole set of nodes in an underlying extensive-form game. In this game, each individual player has a different information state for each node x in the agenda tree, and he has two possible moves at this information state, where each move is a vote for one of the two alternatives that immediately follow node x in the agenda tree. The next move after node x in the agenda tree will be to the alternative designated by a majority of the voters.

With this identification, it is straightforward to show that a sophisticated voting solution corresponds to a subgame-perfect (and sequential) equilibrium of the underlying extensive-form game. However, there are other subgame-perfect equilibria as well. In the above example, there is a subgame-perfect equilibrium in which every player always votes for option 1. Given that the other eight individuals are all expected to vote for option 1, each individual cannot change the outcome by his own vote; so voting for 1 would be an optimal move for him even if 1 is his least-preferred option!

However, it can be shown (see Farquharson, 1969; Moulin, 1979) that all equilibria other than the sophisticated voting solution can be eliminated by iterative elimination of weakly dominated strategies, in both the normal and multiagent representations. To see why, suppose first that x is a nonterminal node that is immediately followed by terminal nodes y and z, where $G(y) = \{\alpha\}$ and $G(z) = \{\beta\}$. Consider an individual who strictly prefers α over β. If his vote at x would make a difference in the outcome, then voting for β could only make him worse off, by changing the outcome from α to β. So any strategy in which he would vote for β at node x would be weakly dominated by a strategy in which he would vote for α at node x (leaving all other moves the same). When all such weakly dominated strategies are eliminated, the result at node x must be to choose the alternative $\phi(x)$ in $\{\alpha,\beta\}$ that is preferred by a majority, because every individual must vote at x for the option in $\{\alpha,\beta\}$ that he actually prefers. But then, for the purposes of analyzing the game that remains, we can treat node x as if it were a terminal node where the option $\phi(x)$ is chosen, and the argument continues by backward induction.

Given the majority preference relation $>$ on K, the *top cycle* of K (or the *Condorcet set*) is defined to be the smallest set K^* such that K^* is a nonempty subset of K and, for every α in K^* and every β in $K \setminus K^*$, $\alpha > \beta$ (see Miller, 1977, 1980). Equivalently, an option α is in the top cycle K^* iff, for every option β in K, if $\beta \neq \alpha$, then there exists some finite sequence of options $(\gamma_1, \ldots, \gamma_m)$ such that

$$\gamma_m = \beta, \alpha > \gamma_1, \text{ and } \gamma_j > \gamma_{j+1}, \quad \forall j \in \{1, \ldots, m-1\}.$$

We refer to such a sequence $(\gamma_1, \ldots, \gamma_m)$ as a *chain* from α to β. That is, a chain from α to β is a finite sequence of options in K such that α can beat the first option in the sequence, β is the last option in the sequence, and each option in the sequence can beat the next option in the sequence. (If α can beat β itself, then this definition can be satisfied by letting m equal 1.) Thus, for the example discussed at the beginning of this section, options 1, 2, and 3 are in the top cycle, but option 4 is not. In this example, option 1 can beat options 2 and 4 directly, and there is a chain of length two (2,3) from option 1 to option 3.

The sophisticated voting solution is unique, once the agenda is specified, but there is a wide range of possible agendas that could be implemented. In fact, the set of sophisticated outcomes of all possible binary agendas is exactly the top cycle. That is, we have the following theorem, due to Miller (1977) (see also Moulin, 1986, 1988).

THEOREM 4.8. *Given the set of options K and the majority preference relation $>$, for any option α in K, there exists a binary agenda for which α is the sophisticated outcome if and only if α is in the top cycle of K.*

Proof. We show first that the sophisticated outcome of a binary agenda must be in the top cycle. If a node x is immediately followed by nodes y and z, and $\phi(y)$ is in the top cycle, then $\phi(x)$ must be in the top cycle, because if $\phi(x)$ is not equal to $\phi(y)$, then $\phi(x) = \phi(z) > \phi(y)$, which is possible only if $\phi(z)$ is also in the top cycle. (Recall that any option in the top cycle can beat any option that is not in the top cycle.) Thus, if $\phi(y)$ is in the top cycle, then the sophisticated voting solution must select an option in the top cycle at every node that precedes y in the agenda tree. Furthermore, there must exist at least one terminal node that is assigned to an outcome in the top cycle, because the top cycle is a nonempty subset of K and K is the set of options under consideration at the root. Thus, the sophisticated outcome at the root of the agenda must be in the top cycle.

We now prove that any option in the top cycle is the sophisticated outcome of some binary agenda, using induction in the number of options in the set under consideration. If there are only two options, then the result holds trivially, because the top cycle just consists of the option that is preferred by a majority. So suppose inductively that the theorem holds when there are l or fewer options under consideration, and let K be any set of $l + 1$ options. Let α be any option in the top cycle of K. Let us define the *distance* from α to any other option β to be the length of the shortest chain from α to β. (If α itself can beat β, then the distance from α to β is one.) Let β be any option whose distance from α is maximal, among all options in K other than α. Let $(\gamma_1, \ldots, \gamma_m)$ be a chain of minimal length from α to β, such that

$$\alpha > \gamma_1 > \cdots > \gamma_m = \beta.$$

Let x_0 denote the root of our agenda tree, and let y_0 and z_0 denote the two nodes that immediately follow x_0. We must have $K = G(x_0) = G(y_0) \cup G(z_0)$. Let

$$G(y_0) = K \setminus \{\beta\}, \quad G(z_0) = \{\gamma_1, \ldots, \gamma_m\}.$$

Then $G(y_0) \cup G(z_0)$ equals K because $\gamma_m = \beta$. Option α must still be in the top cycle of $K \setminus \{\beta\}$, because there is no option whose shortest chain from α includes β (because there is no option in K whose distance from α is greater than β). Thus, by induction, using the fact that $K \setminus \{\beta\}$ is a set with only l members, we know that there exists an agenda such that the sophisticated voting solution at y_0 is $\phi(y_0) = \alpha$. Because γ_1 is in the top cycle of $\{\gamma_1, \ldots, \gamma_m\}$ and $m \leq l$ (due to the fact that $\{\gamma_1, \ldots, \gamma_m\} \subseteq K \setminus \{\alpha\}$), the induction hypothesis also implies that we can choose the agenda so that the sophisticated voting solution at z_0 is $\phi(z_0) = \gamma_1$. Thus, the sophisticated outcome of the overall agenda is $\phi(x_0) = \alpha$, because $\alpha > \gamma_1$. ∎

Smaller sets of options can be generated as sophisticated outcomes if we restrict the class of binary agendas, for example, to agendas in which at most one option is eliminated at a time (see Miller, 1980; Shepsle and Weingast, 1984; Moulin, 1986, 1988).

4.11 Technical Proofs

Proof of Theorem 4.1. For any node x in Γ^e, let $p(x)$ denote the product of all of the probabilities assigned to alternatives at chance nodes that are on the path from the root to node x. (If there are no such chance nodes, then let $p(x)$ equal 1.) For each player i, let $B_i(x)$ denote the set of all strategies for i where he is planning to make all the moves necessary for the play of the game to reach node x. That is, $B_i(x)$ is the subset of C_i such that $c_i \in B_i(x)$ iff, for any information state s in S_i and any move d_s in D_s, if there is any branch on the path from the root to node x that is an alternative with move-label d_s at a decision node that belongs to player i with information state s, then $c_i(s)$ must equal d_s. Let $B(x) = \times_{i \in N} B_i(x)$.

With these definitions, it can be shown that, for any node x and any strategy combination c in C,

$$P(x|c) = p(x) \ \text{if} \ c \in B(x),$$

$$P(x|c) = 0 \quad \text{if} \ c \notin B(x).$$

(This result can be proved from the definition of P in Section 2.2 by induction. It is clearly true if x is the root. So one can show that, if it holds for node x, then it holds for all nodes that immediately follow x.)

Let τ be a mixed-strategy profile in $\times_{i \in N} \Delta(C_i)$, and let σ be a behavioral-strategy profile in $\times_{i \in N} \times_{s \in S_i} \Delta(D_s)$ that is a behavioral representation of τ. For any node x, let $\hat{P}(x|\tau)$ be the probability of reaching node x in the play of the game when all players randomize according to τ; that is,

$$\hat{P}(x|\tau) = \sum_{c \in C} \left(\prod_{i \in N} \tau_i(c_i) \right) P(x|c) = p(x) \prod_{i \in N} \left(\sum_{c_i \in B_i(x)} \tau_i(c_i) \right).$$

We now prove that, for any node x,

(4.9) $\hat{P}(x|\tau) = \overline{P}(x|\sigma).$

(Recall from Section 4.3 that $\overline{P}(x|\sigma)$ is the multiplicative product of all chance probabilities and move probabilities specified by σ on the path from the root to node x.)

Equation (4.9) is clearly true if x is the root, because both sides are equal to 1. Now, suppose inductively that $\hat{P}(y|\tau) = \overline{P}(y|\sigma)$ is true at node

y, and let x be a node that immediately follows y. We need to show that (4.9) holds under this assumption.

First suppose that y is a chance node and the alternative branch from y to x has probability q. By definition of $\overline{P}$, $\overline{P}(x|\sigma) = q\,\overline{P}(y|\sigma)$. Also,

$$\hat{P}(x|\tau) = \sum_{c \in C} \left(\prod_{i \in N} \tau_i(c_i) \right) P(x|c)$$

$$= \sum_{c \in C} \left(\prod_{i \in N} \tau_i(c_i) \right) qP(y|c) = q\hat{P}(y|\tau).$$

So (4.9) holds in this case.

Now suppose that y is a node belonging to player j with information state r, and the alternative branch from y to x has move-label d_r. By definition of $\overline{P}$, $\overline{P}(x|\sigma) = \sigma_{j,r}(d_r)\,\overline{P}(y|\sigma)$. Notice that $p(x) = p(y)$ and, for any player i other than j, $B_i(x) = B_i(y)$. Furthermore, because Γ^e has perfect recall, $B_j(y) = C_j^*(r)$ and $B_j(x) = C_j^{**}(d_r,r)$. Thus,

$$\hat{P}(x|\tau) = p(x) \prod_{i \in N} \left(\sum_{c_i \in B_i(x)} \tau_i(c_i) \right)$$

$$= p(y)\,\sigma_{j,r}(d_r) \prod_{i \in N} \left(\sum_{c_i \in B_i(y)} \tau_i(c_i) \right)$$

$$= \sigma_{j,r}(d_r)\,\hat{P}(y|\tau),$$

because

$$\sigma_{j,r}(d_r) \left(\sum_{c_j \in C_j^*(r)} \tau_j(c_j) \right) = \sum_{c_j \in C_j^{**}(d_r,r)} \tau_j(c_j).$$

So (4.9) holds in this case as well.

Thus, by induction, (4.9) holds for every node x. By (4.9), the probability distribution over outcomes of the game depends only on the behavioral representations of the players' mixed strategies. Thus, the expected payoffs generated by behaviorally equivalent strategies must be the same.

For any player i and any behavioral-strategy profile σ in $\times_{i \in N} \times_{s \in S_i} \Delta(D_s)$, let us define $v_i(\sigma)$ to be the expected payoff to player i when every player j uses his behavioral strategy σ_j. Then

$$v_i(\sigma) = \sum_{x \in \Omega} \overline{P}(x|\sigma)w_i(x),$$

where Ω denotes the set of terminal nodes in Γ^e and $w_i(x)$ denotes i's payoff at any terminal node x. For any mixed-strategy profile τ in $\times_{i \in N} \Delta(C_i)$, the expected payoff $u_i(\tau)$ in the normal representation satisfies

$$u_i(\tau) = \sum_{x \in \Omega} \hat{P}(x|\tau) w_i(x).$$

Thus, we can write the conclusion of this proof as follows: if σ is a behavioral representation of τ, then $u_i(\tau) = v_i(\sigma)$. ∎

Proof of Theorem 4.2. Let τ be an equilibrium of the normal representation of Γ^e and let σ be a behavioral representation of τ. If, contrary to the theorem, σ is not an equilibrium of the multiagent representation of Γ^e, then there exists some player i, some state r in S_i, and some ρ_r in $\Delta(D_r)$ such that

$$v_i(\sigma_{-i.r}, \rho_r) > v_i(\sigma),$$

where $v_i : \times_{s \in S*} \Delta(D_s) \to \mathbf{R}$ is the utility function for each agent for player i in the multiagent representation of Γ^e. Let λ_i be the mixed representation of i's behavioral strategy that differs from $\sigma_i = (\sigma_{i.s})_{s \in S_i}$ by substituting ρ_r for $\sigma_{i.r}$. Then, by the proof of Theorem 4.1 (which uses the assumption of perfect recall),

$$u_i(\tau_{-i}, \lambda_i) = v_i(\sigma_{-i.r}, \rho_r) > v_i(\sigma) = u_i(\tau),$$

which contradicts the assumption that τ is an equilibrium of the normal representation. So σ must be an equilibrium of the multiagent representation. ∎

Proof of Theorem 4.4. For any s in $S*$, let $X(s)$ denote the set of all terminal nodes that do not follow any node in Y_s. For any player i, any state s in S_i, any ρ_s in $\Delta(D_s)$, and any behavioral-strategy profile σ, i's utility function in the multiagent representation satisfies the following equation

$$v_i(\sigma_{-i.s}, \rho_s)$$

$$= \sum_{y \in Y_s} \overline{P}(y|\sigma_{-i.s}, \rho_s) U_i(\sigma_{-i.s}, \rho_s|y) + \sum_{x \in X(s)} \overline{P}(x|\sigma_{-i.s}, \rho_s) w_i(x),$$

where $w_i(x)$ denotes the payoff to player i at node x. However, for any y in Y_s, $\overline{P}(y|\sigma_{-i.s}, \rho_s) = \overline{P}(y|\sigma)$, because the probability of the node y

occurring depends only on the moves at information states that occur before i's state s; so it does not depend on the behavior of player i at state s. Also, for any node x in $X(s)$, $\overline{P}(x|\sigma_{-i.s},\rho_s) = \overline{P}(x|\sigma)$, because there are no alternatives chosen by i with state s on the path to such a node x. So when π is weakly consistent with σ,

$$v_i(\sigma_{-i.s},\rho_s) = \sum_{y \in Y_s} \overline{P}(y|\sigma) \, U_i(\sigma_{-i.s},\rho_s|y) + \sum_{x \in X(s)} \overline{P}(x|\sigma)w_i(x)$$

$$= \left(\sum_{y \in Y_s} \pi_{i.s}(y)U_i(\sigma_{-i.s},\rho_s|y) \right)\left(\sum_{z \in Y_s} \overline{P}(z|\sigma) \right)$$

$$+ \sum_{x \in X(s)} \overline{P}(x|\sigma)w_i(x).$$

So if $\Sigma_{z \in Y_s}\overline{P}(z|\sigma) > 0$ but σ is not sequentially rational for i at s, then there exists some ρ_s such that $v_i(\sigma_{-i.s},\rho_s) > v_i(\sigma_{-i.s},\sigma_{i.s}) = v_i(\sigma)$, which is impossible if σ is an equilibrium in behavioral strategies. ∎

Proof of Theorem 4.5. Suppose that (σ,π) is a sequential equilibrium, but σ is not an equilibrium in behavioral strategies. Then there exists some behavioral strategy θ_i in $\times_{s \in S_i}\Delta(D_s)$ such that

$$v_i(\sigma_{-i},\theta_i) > v_i(\sigma).$$

Let us choose θ_i such that the above inequality is satisfied and such that, for any other behavioral strategy ϕ_i, if $v_i(\sigma_{-i},\phi_i) > v_i(\sigma)$, then $|\{s \in S_i|\phi_{i.s} \neq \sigma_{i.s}\}| \geq |\{s \in S_i|\theta_{i.s} \neq \sigma_{i.s}\}|$.

Let r be an information state for player i such that $\theta_{i.r} \neq \sigma_{i.r}$ and, for every s in S_i such that nodes in Y_s follow nodes in Y_r, $\theta_{i.s} = \sigma_{i.s}$. (Such an r can be found because, by the perfect-recall assumption, states in S_i can be ordered in such a way that "later" states never occur before "earlier" states.) Let $\hat{\theta}_i$ be any behavioral strategy for player i such that $\hat{\theta}_{i.s} = \theta_{i.s}$ for every s other than r in S_i.

Because $\hat{\theta}_{i.s} = \theta_{i.s} = \sigma_{i.s}$ for every s in S_i that occurs in nodes that follow nodes in Y_r, $U_i(\sigma_{-i},\hat{\theta}_i|y) = U_i(\sigma_{-i.r},\hat{\theta}_{i.r}|y)$ for every y in Y_r. Also, $\overline{P}(x|\sigma_{-i},\hat{\theta}_i) = \overline{P}(x|\sigma_{-i},\theta_i)$ for every node x in $X(r)$ or in Y_r, because no nodes in Y_r occur on the path from the root to such nodes x. (For nodes in Y_r, this result relies on perfect recall, which implies that no information state can occur twice on a path from the root.) Thus,

$$v_i(\sigma_{-i},\hat{\theta}_i) = \sum_{y \in Y_r} \overline{P}(y|\sigma_{-i},\theta_i)U_i(\sigma_{-i.r},\hat{\theta}_{i.r}|y) + \sum_{x \in X(r)} \overline{P}(x|\sigma_{-i},\theta_i)w_i(x).$$

By perfect recall, the information states of player i and moves controlled by player i that occur on the path from the root to a node y in Y_r must be the same for every node y in Y_r. Thus, for any node y in Y_r, the ratio

$$\frac{\overline{P}(y|\rho)}{\sum_{z \in Y_r} \overline{P}(z|\rho)}$$

does not depend on i's behavioral strategy ρ_i, as long as the denominator is nonzero, because ρ_i contributes the same multiplicative factors to $\overline{P}(y|\rho)$ as to $\overline{P}(z|\rho)$ for every other node z in Y_r. So the fully consistent belief probabilities for player i at nodes in Y_r do not depend on player i's behavioral strategy, and so

$$\overline{P}(y|\sigma_{-i},\theta_i) = \pi_{i,r}(y) \left(\sum_{z \in Y_r} \overline{P}(z|\sigma_{-i},\theta_i) \right), \quad \forall y \in Y_r.$$

This implies that

$$v_i(\sigma_{-i},\hat{\theta}_i) = \left(\sum_{y \in Y_r} \pi_{i,r}(y) U_i(\sigma_{-i,r},\hat{\theta}_{i,r}|y) \right) \left(\sum_{z \in Y_r} \overline{P}(z|\sigma_{-i},\theta_i) \right)$$

$$+ \sum_{x \in X(r)} \overline{P}(x|\sigma_{-i},\theta_i) w_i(x).$$

By definition of a sequential equilibrium, this formula is maximized by letting $\hat{\theta}_{i,r} = \sigma_{i,r}$. But then $v_i(\sigma_{-i},\hat{\theta}_i) > v_i(\sigma)$ and $|\{s \in S_i | \hat{\theta}_{i,s} \neq \sigma_{i,s}\}| < |\{s \in S_i | \theta_{i,s} \neq \sigma_{i,s}\}|$, which violates the way that θ_i was constructed. So no such θ_i exists, and σ is an equilibrium in behavioral strategies. ∎

Proof of Theorem 4.7. Given that Γ^e has perfect information, a behavioral-strategy profile σ (with the beliefs vector of all 1's) is a sequential equilibrium of Γ^e iff, for each player i in N and each state s in S_i,

(4.10) $\sigma_{i,s} \in \underset{\rho_s \in \Delta(D_s)}{\text{argmax}}\ U_i(\sigma_{-i,s},\rho_s|x), \quad \text{where } \{x\} = Y_s.$

By Lemma 3.1, (4.10) would be satisfied if

(4.11) $\sigma_{i,s} = [d_s], \quad \text{where } d_s \in \underset{e_s \in D_s}{\text{argmax}}\ U_i(\sigma_{-i,s},[e_s]|x), \ \{x\} = Y_s.$

Notice that, in (4.10) and (4.11), $U_i(\sigma_{-i,s}, \rho_s | x)$ and $U_i(\sigma_{-i,s}, [e_s] | x)$ depend only on the components of σ for moves at nodes that follow node x.

For any node x, let $\iota(x)$ be the number of decision nodes in the subgame beginning at node x. For any s in S^*, let $\iota(s) = \iota(x)$ where $Y_s = \{x\}$.

Suppose first that $\iota(s) = 1$. Then $U_i(\sigma_{-i,s}, \rho_s | x)$ depends only on ρ_s, because there are no decision nodes after x. So $\text{argmax}_{e_s \in D_s} U_i(\sigma_{-i,s}, [e_s] | x)$ is independent of σ; so we can begin our construction of the nonrandomized equilibrium σ by letting $\sigma_{i,s}$ satisfy (4.11).

Now suppose inductively that, for some number k, $\sigma_{j,r}$ has been defined at all j and r such that $r \in S_j$ and $\iota(r) < k$, and suppose (4.11) is satisfied for all s such that $\iota(s) < k$. Then, for any (i,s,x) such that $s \in S_i$, $\{x\} = Y_s$, and $\iota(x) = k$, $U_i(\sigma_{-i,s}, [e_s] | x)$ is well defined for every e_s in D_s; so we can construct $\sigma_{i,s}$ to satisfy (4.11) for this (i,s,x) as well.

Thus, arguing by induction on k in the preceding paragraph, we can construct σ so that (4.11) is satisfied at all i and s such that $i \in N$ and $s \in S_i$. Such a behavioral-strategy profile σ is then a sequential equilibrium in pure (unrandomized) strategies.

Now let us consider choosing some vector of payoffs in $\mathbf{R}^{N \times \Omega}$ to generate a game $\tilde{\Gamma}^e$ that differs from Γ^e only in the payoffs. For any player i, any node x, and any pure-strategy profile d in $\times_{s \in S^*} D_s$, let $\tilde{U}_i(d | x)$ denote the conditionally expected payoff to player i in the game $\tilde{\Gamma}^e$, if the play of the game started at node x and everyone implemented the move specified by d at every node thereafter. If $\{x\} = Y_s$, $s \in S_i$, $d \in D = \times_{r \in S^*} D_r$, and $d_s \neq e_s \in D_s$, then the difference $\tilde{U}_i(d | x) - \tilde{U}_i(d_{-i,s}, e_s | x)$ is a nonconstant linear function of the terminal payoffs in $\tilde{\Gamma}^e$. Any nonconstant linear function is not equal to 0 on a dense open set. The intersection of finitely many open dense sets is also open and dense. Thus, there is an open dense set of payoff vectors in $\mathbf{R}^{N \times \Omega}$ such that the game $\tilde{\Gamma}^e$ will satisfy

(4.12) $U_i(d | x) \neq U_i(d_{-i,s}, e_s | x)$,

$$\forall i \in N, \quad \forall s \in S_i, \quad \forall d \in D, \quad \forall e_s \in D_s \setminus \{d_s\}.$$

Now, suppose that condition (4.12) is satisfied. Then there is only one sequential equilibrium of $\tilde{\Gamma}^e$, which must therefore be in pure strategies (since we know that at least one sequential equilibrium in pure strategies exists). To see why this is so, suppose that $\tilde{\Gamma}^e$ has two distinct sequential equilibria, σ and $\hat{\sigma}$. Let x be a decision node where player i moves with information state s such that either $\sigma_{i.s} \neq \hat{\sigma}_{i.s}$ or $\sigma_{i.s}$ assigns positive probability to more than one move in D_s. Furthermore, we can select this node x so that, at every decision node that follows x, σ and $\hat{\sigma}$ select the same move without randomization. Then, by (4.12), no two moves at x would give player i the same expected payoff, when σ or $\hat{\sigma}$ is implemented thereafter. That is, for any two moves d_s and e_s in D_s,

$$\tilde{U}_i(\sigma_{-i.s},[d_s]|x) = \tilde{U}_i(\hat{\sigma}_{-i.s},[d_s]|x) \neq \tilde{U}_i(\sigma_{-i.s},[e_s]|x) = \tilde{U}_i(\hat{\sigma}_{-i.s},[e_s]|x).$$

This result implies that there is some move d_s such that

$$\operatorname*{argmax}_{e_s \in D_s} \tilde{U}_i(\sigma_{-i.s},[e_s]|x) = \operatorname*{argmax}_{e_s \in D_s} \tilde{U}_i(\hat{\sigma}_{-i.s},[e_s]|x) = \{d_s\};$$

so we must have $\sigma_{i.s} = \hat{\sigma}_{i.s} = [d_s]$. This contradiction of the way that x was defined implies that there cannot be two distinct sequential equilibria of $\tilde{\Gamma}^e$ when condition (4.12) is satisfied. ∎

Exercises

Exercise 4.1. Consider the extensive-form game and the scenario shown in Figure 4.17.

a. Compute belief probabilities that are consistent with this scenario.

b. For each player, at each of his information states, compute the sequential value of each of his possible moves, relative to this scenario with these consistent beliefs.

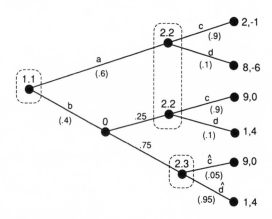

Figure 4.17

c. Identify all irrational moves in this scenario with these consistent beliefs. That is, for each information state of each player, identify every move that has a positive move probability in this scenario but does not maximize the sequential value for this player at this information state.

d. Find a sequential equilibrium of this game.

Exercise 4.2.   Consider the game in Exercise 2.1 (Chapter 2).

a. Show that the normal representation of this game has two pure-strategy equilibria. Show that each of these Nash equilibria corresponds to a sequential equilibrium of the extensive-form game.

b. For each of the sequential equilibria that you found in (a), characterize the connected component of the set of all sequential equilibria in which it lies. That is, for each sequential equilibrium that you identified in (a), show all other sequential equilibria that could be constructed by continuously varying the move probabilities and belief probabilities without leaving the set of sequential equilibria.

Exercise 4.3.   Consider the game shown in Figure 4.18.

a. Let $x = 1$. Find the three pure-strategy equilibria of this game.

b. Suppose that π is a beliefs vector that is fully consistent with some behavioral-strategy vector σ for this game. Let α denote the lower node in Y_2, where player 2 has information state 2 and player 1 has chosen

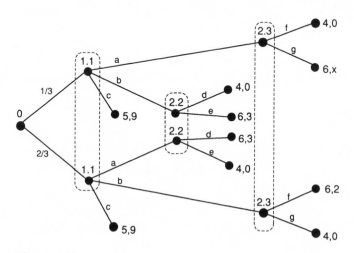

Figure 4.18

a. Let β denote the lower node in Y_3, where player 2 has information state 3 and player 1 has chosen b. Show that there is a formula for expressing $\pi_{2.3}(\beta)$ in terms of $\pi_{2.2}(\alpha)$.

c. With $x = 1$, show that two of the three pure-strategy equilibria from part (a) are (full) sequential-equilibrium scenarios, but the other equilibrium is not. Characterize all beliefs vectors that would make this other equilibrium sequentially rational, and show that they are weakly consistent with it but are not fully consistent. (So it is a weak sequential-equilibrium scenario.)

d. What is the lowest value of x such that all three pure-strategy equilibria are (full) sequential equilibria? For the equilibrium that was not sequential in part (c), show the fully consistent beliefs that make it a sequential equilibrium at this new value of x.

Exercise 4.4. In Exercise 2.3 (Chapter 2), we considered a revised version of the simple card game in which player 1, after looking at his card, can raise $1.00, raise $0.75, or pass. Show that this game has sequential equilibria in which the probability of a $0.75 raise is 0. In such a sequential equilibrium, what probability would player 2 assign to the event that player 1 has a red (winning) card if he raised $0.75?

Exercise 4.5. Find all the sequential equilibria of the game in Figure 2.8 (Exercise 2.4 in Chapter 2). Be sure to specify all move probabilities and all belief probabilities for each sequential equilibrium.

Exercise 4.6. Find all the sequential equilibria of the game shown in Figure 4.19. (This game differs from Figure 2.8 only in that the payoffs after the z_1 branch have been changed from (2,6) to (7,6).)

Exercise 4.7. Consider the game shown in Figure 4.20.
    a. Find the unique subgame-perfect equilibrium of this game.
    b. Now revise the game by eliminating player 1's "g" move at information state 3. Find the unique subgame-perfect equilibrium of this revised game.

Exercise 4.8. Recall the revised card game described in Exercise 2.2 (Chapter 2), where player 1's card is winning for him iff it is a 10 or higher. Find the unique sequential equilibrium of this game. Before the

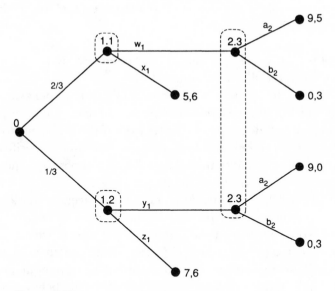

Figure 4.19

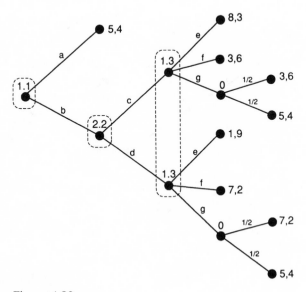

Figure 4.20

card is drawn, what is the expected payoff for each player, if they play according to this equilibrium?

Exercise 4.9. (A game adapted from Fudenberg and Kreps, 1988.) Consider the game shown in Figure 4.21.

a. Show that this game has a unique equilibrium in behavioral strategies, in which the probability that player 3 moves (because player 1 does x_1 or player 2 does x_2) is 1. (HINT: Even if the probability of player 3 moving were 0, an equilibrium is defined to include a specification of the conditional probability that player 3 would choose x_3 if he moved. The other two players, being intelligent, are supposed to know this conditional probability. Thus, in equilibrium, players 1 and 2 are supposed to agree about what player 3 would do if he moved.)

b. Now, consider instead the (nonequilibrium) case where players 1 and 2 might disagree and have different beliefs about what player 3 would do if he moved. Show that different beliefs about player 3's strategy might induce players 1 and 2 to rationally behave in such a way that the probability of player 3 moving is 0.

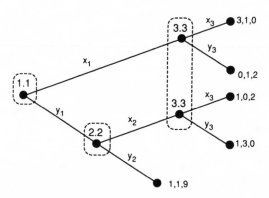

Figure 4.21

5

Refinements of Equilibrium in Strategic Form

5.1 Introduction

We have seen that any extensive-form game can be represented in strategic form, using the normal representation, the reduced normal representation, or the multiagent representation. These representations raised the hope that we might be able to develop all the principles of game-theoretic analysis in the context of the conceptually simpler strategic form. However, it is easy to construct extensive-form games that are the same in strategic form (under any representation) but have different sets of sequential equilibria. For example, $([y_1],[y_2])$ is not a sequential-equilibrium scenario of the game in Figure 4.5, but it is a sequential-equilibrium scenario of the game in Figure 5.1, and both games have the same normal representation in strategic form, which is also the reduced normal and multiagent representation. Thus, if sequential equilibrium is accepted as an exact solution that characterizes rational behavior, then games must be analyzed in the extensive form, instead of the strategic form.

However, the sequential equilibrium shown in Figure 5.1 involves the use by player 2 of a strategy y_2 that is weakly dominated in the strategic-form game. Thus, it is questionable whether $([y_1],[y_2])$ should be considered a reasonable pair of strategies for rational and intelligent players to choose in this game, even though it satisfies the definition of a sequential equilibrium. That is, sequential equilibrium may still be too weak to serve as an exact characterization of rational behavior in games. To more accurately characterize rationality in games, we can look for other ways to refine the concept of equilibrium for games in extensive or strategic form.

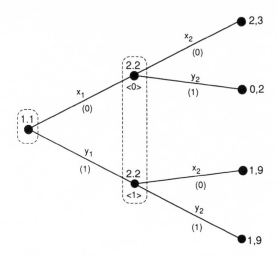

Figure 5.1

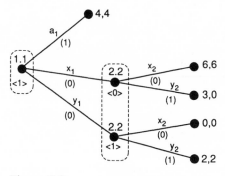

Figure 5.2

Consider the games shown in Figures 5.2 and 5.3, suggested by Kohlberg and Mertens (1986). These two games have the same purely reduced normal representation (up to a relabeling of the strategies). The game in Figure 5.2 has a sequential equilibrium in which player 1 uses move a_1 with probability 1. However, there is no such sequential (or even subgame-perfect) equilibrium of the game in Figure 5.3. The difference between the two games is that, in Figure 5.3, player 2 knows that player 1's choice between x_1 and y_1 could only be made in a situation

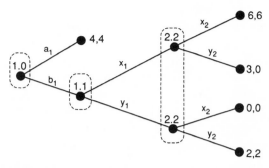

Figure 5.3

where a_1 was no longer an available option for him. Thus, in any sequential equilibrium of the game in Figure 5.3, player 2 must be confident that, if player 1 chose x_1 or y_1, he would choose the move that was better for him, and that is surely x_1. So player 2 must assign probability 1 to her upper node (following x_1) in any sequential equilibrium of the game in Figure 5.3; so she should choose x_2 and player 1 should choose b_1 and x_1. Nonetheless, because these two games have the same purely reduced normal representation, arguments that they should have the same set of solutions have substantial intuitive appeal.

The natural way to avoid distinguishing between games like those in Figures 5.2 and 5.3, or in Figures 4.5 and 5.1, is to develop refinements of the Nash equilibrium concept that can be applied directly to the strategic form. By developing a solution theory that is based on such a refinement and then applying it to extensive-form games via one of the representations that we have studied (normal, purely reduced normal, fully reduced normal, or multiagent) we can guarantee that extensive-form games that share the same strategic-form representation (in the designated sense) will have the same solutions.

It may be helpful to make a list of some of the properties that would be theoretically desirable. from a refinement of the Nash equilibrium concept. The refined equilibrium concept should select a nonempty set of equilibria for any finite game, because it is hard to take a solution concept seriously if it is sometimes empty. (What is supposed to happen in games that have no solutions?) When our intuition seems to clearly identify one equilibrium of a game as being unreasonable, in some sense, then the refined equilibrium concept should formalize our intuition and exclude this equilibrium. For example, an equilibrium in which some player chooses a dominated strategy with positive probability may naturally be considered unreasonable. In an extensive-form game, an equilibrium that cannot be extended by any beliefs vector to a sequential equilibrium may be considered unreasonable. When we work on solution concepts for the strategic form, this criterion suggests that, once we have selected some way of representing extensive-form games in strategic form (multiagent, normal, etc.), then, for any given game in strategic form, our refined solution concept should select only equilibria that correspond to sequential equilibria in all of the extensive-form games that the given strategic-form game represents. Selten (1975) defined a concept of perfect equilibrium that has all of these properties when we use the multiagent representation to map extensive-form games into strategic form.

5.2 Perfect Equilibria

Let $\Gamma = (N, (C_i)_{i \in N}, (u_i)_{i \in N})$ be any finite game in strategic form. A randomized-strategy profile σ in $\times_{i \in N} \Delta(C_i)$ is a *perfect equilibrium* of Γ iff there exists a sequence $(\hat{\sigma}^k)_{k=1}^{\infty}$ such that

$$(5.1) \qquad \hat{\sigma}^k \in \underset{i \in N}{\times} \Delta^0(C_i), \quad \forall k \in \{1,2,3,\ldots\},$$

$$(5.2) \qquad \lim_{k \to \infty} \hat{\sigma}_i^k(c_i) = \sigma_i(c_i), \quad \forall i \in N, \quad \forall c_i \in C_i, \quad \text{and}$$

$$(5.3) \qquad \sigma_i \in \underset{\tau_i \in \Delta(C_i)}{\text{argmax}} \; u_i(\hat{\sigma}_{-i}^k, \tau_i), \quad \forall i \in N.$$

(Here $\sigma = (\sigma_i)_{i \in N}$, $\hat{\sigma}^k = (\hat{\sigma}_i^k)_{i \in N}$, $\hat{\sigma}_{-i}^k = (\hat{\sigma}_j^k)_{j \in N-i}$.) Condition (5.1) asserts that every pure strategy of every player gets strictly positive probability in each $\hat{\sigma}^k$. (Recall the definition of $\Delta^0(\cdot)$ in Section 1.6.) Condition (5.2) asserts that, for large k, $\hat{\sigma}^k$ is a randomized-strategy profile that is very close to σ in all components. Condition (5.3) asserts that every player's strategy in σ is a best response to the other players' strategies in each $\hat{\sigma}^k$.

It is easy to show that any perfect equilibrium of Γ is indeed a Nash equilibrium of Γ. This condition holds because the best-response correspondence is upper-hemicontinuous (as shown in the proof of Theorem 3.1), so conditions (5.2) and (5.3) imply that

$$\sigma_i \in \underset{\tau_i \in \Delta(C_i)}{\text{argmax}}\ u_i(\sigma_{-i}, \tau_i), \quad \forall i \in N.$$

In a Nash equilibrium, each player's equilibrium strategy is a best response to the other players' equilibrium strategies. In a perfect equilibrium, there must also be arbitrarily small perturbations of all players' strategies such that every pure strategy gets strictly positive probability and each player's equilibrium strategy is still a best response to the other players' perturbed strategies. Thus, in a perfect equilibrium, the optimality of a player's strategy choice does not depend on an assumption that some pure strategies are getting zero probability in the equilibrium.

The close logical relationship between perfect and sequential equilibria can be seen by comparing conditions (5.1)–(5.3) with conditions (4.2) and (4.6)–(4.8) in Chapter 4. In both solution concepts, the equilibrium σ is tested by a comparison with some perturbed strategies in which every move or pure strategy gets positive probability. In the definition of sequential equilibrium, the perturbed strategies are used to generate beliefs at events of zero probability (by conditions 4.6–4.8), and these beliefs are then applied in the sequential-rationality or optimality condition (4.2). In the definition of perfect equilibrium, the perturbed strategies are used directly in the optimality condition (5.3). In fact, when Γ is the multiagent representation of Γ^e, condition (5.3) is a stronger requirement than conditions (4.2) and (4.8) together. Thus, as the following theorem asserts, every perfect equilibrium of the multiagent representation of Γ^e corresponds to a sequential equilibrium of Γ^e.

THEOREM 5.1. *Suppose that Γ^e is an extensive-form game with perfect recall and σ is a perfect equilibrium of the multiagent representation of Γ^e. Then there exists a vector of belief probabilities π such that (σ, π) is a sequential equilibrium of Γ^e.*

Proof. In this proof, we use the notation developed in Chapter 4. Let i be any player in Γ^e and let s be any information state in S_i. Let $X(s)$ denote the set of all terminal nodes that do not follow any node in Y_s. Let $(\hat{\sigma}^k)_{k=1}^{\infty}$ be a sequence of behavioral-strategy profiles in $\times_{r \in S^*} \Delta^0(D_r)$ that satisfy the conditions (5.1)–(5.3) of a perfect equilibrium for σ

(when Γ is the multiagent representation of Γ^e). For any k and any node y in Y_s, let

$$\hat{\pi}_s^k(y) = \frac{\overline{P}(y|\hat{\sigma}^k)}{\sum\limits_{z \in Y_s} \overline{P}(z|\hat{\sigma}^k)} \; .$$

Notice that $\overline{P}(y|\hat{\sigma}^k) > 0$ for every y in Y_s, so $\Sigma_{z \in Y_s}\overline{P}(z|\hat{\sigma}^k) > 0$, because $\hat{\sigma}^k$ assigns positive probability to all combinations of pure strategies and because every alternative at every chance node is assumed to have positive probability in the game Γ^e. Let

$$\pi_s(y) = \lim_{k \to \infty} \hat{\pi}_s^k(y).$$

Then by (5.1) and (5.2), π is a beliefs vector that is fully consistent with σ.

For any terminal node x, we let $w_i(x)$ denote the payoff to i at node x. We let $v_s(\cdot)$ denote the utility function of i's agent for state s in the multiagent representation, as defined in Section 2.6. When this agent uses the randomized strategy ρ_s in $\Delta(D_s)$ against the strategies specified by $\hat{\sigma}^k$ for all other agents, his expected payoff is

$$v_s(\hat{\sigma}^k_{-i.s}, \rho_s)$$

$$= \sum_{y \in Y_s} \overline{P}(y|\hat{\sigma}^k_{-i.s}, \rho_s)U_i(\hat{\sigma}^k_{-i.s}, \rho_s|y) + \sum_{x \in X(s)} \overline{P}(x|\hat{\sigma}^k_{-i.s}, \rho_s)w_i(x).$$

However, for any y in Y_s, $\overline{P}(y|\hat{\sigma}^k_{-i.s}, \rho_s) = \overline{P}(y|\hat{\sigma}^k)$, because the probability of the node y occurring depends only on the strategies of agents who may move before state s occurs; so it does not depend on the behavior of player i at state s. Thus,

$$v_s(\hat{\sigma}^k_{-i.s}, \rho_s)$$

$$= \sum_{y \in Y_s} \overline{P}(y|\hat{\sigma}^k)U_i(\hat{\sigma}^k_{-i.s}, \rho_s|y) + \sum_{x \in X(s)} \overline{P}(x|\hat{\sigma}^k)w_i(x)$$

$$= \left(\sum_{y \in Y_s} \hat{\pi}_s^k(y)U_i(\hat{\sigma}^k_{-i.s}, \rho_s|y) \right)\left(\sum_{z \in Y_s} \overline{P}(z|\hat{\sigma}^k) \right) + \sum_{x \in X(s)} \overline{P}(x|\hat{\sigma}^k)w_i(x).$$

The fact that $(\hat{\sigma}^k)_{k=1}^{\infty}$ supports σ as a perfect equilibrium of the multiagent representation of Γ^e implies that

$$\sigma_s \in \underset{\rho_s \in \Delta(D_s)}{\text{argmax}} \; v_s(\hat{\sigma}^k_{-i.s}, \rho_s),$$

which in turn implies that

$$\sigma_s \in \underset{\rho_s \in \Delta(D_s)}{\text{argmax}} \; \sum_{y \in Y_s} \hat{\pi}^k_s(y) U_i(\hat{\sigma}^k_{-i.s}, \rho_s | y),$$

because these two objectives differ by a strictly increasing affine transformation whose coefficients are independent of ρ_s. Then, by upper-hemicontinuity of the best-response mapping,

$$\sigma_s \in \underset{\rho_s \in \Delta(D_s)}{\text{argmax}} \; \sum_{y \in Y_s} \pi_s(y) U_i(\sigma_{-i.s}, \rho_s | y).$$

This inclusion is the sequential-rationality condition for sequential equilibrium, so (σ, π) is a sequential equilibrium of Γ^e. ∎

Selten (1975) defined a *perfect equilibrium* of an extensive-form game to be any perfect equilibrium of its multiagent representation. In contrast, recall that a Nash equilibrium of an extensive-form game is defined to be any Nash equilibrium of its multiagent representation for which the mixed representation is also a Nash equilibrium of the normal representation. Figure 4.2 was used to illustrate an argument for not using the multiagent representation alone to define Nash equilibria of extensive-form games, and it is helpful to see why this argument does not apply to the concept of perfect equilibrium. Recall that the multiagent representation of Figure 4.2 has a Nash equilibrium at $([b_1], [w_2], [z_1])$ that is rather perverse and should not be considered to be an equilibrium of the given extensive-form game. This perverse equilibrium would not be a perfect equilibrium of the multiagent representation, however. To see why, suppose that, for large k, $\hat{\sigma}^k_{1.1} = \varepsilon_1[a_1] + (1 - \varepsilon_1)[b_1]$ and let $\hat{\sigma}^k_{2.2} = (1 - \varepsilon_2)[w_2] + \varepsilon_2[x_2]$, where $\varepsilon_1(k)$ and $\varepsilon_2(k)$ are small positive numbers (each less than 0.5). When we perturb $[b_1]$ and $[w_2]$ to $\hat{\sigma}^k_{1.1}$ and $\hat{\sigma}^k_{2.2}$, respectively, the unique best response for 1's agent at state 3 would be y_1, not z_1, because the probability of (a_1, w_2) occurring is positive and much larger than the probability of (a_1, x_2). In fact, the unique perfect equilibrium of the multiagent representation of this game is $(.5[a_1] + .5[b_1], .5[w_2] + .5[x_2], [y_1])$. This behavioral-strategy profile, with belief probabilities ½ at each of the 2.2 and 1.3 nodes, is also the unique sequential equilibrium of this game.

On the other hand, perfect equilibria of the normal representation may not correspond to sequential equilibria. To see why, consider again the example in Figure 5.3, for which the normal representation is shown in Table 5.1. The strategy profile $([a_1x_1],[y_2])$ is a perfect equilibrium of this normal representation in strategic form, even though it does not correspond to a sequential equilibrium of the extensive-form game in Figure 5.3. To show that $([a_1x_1],[y_2])$ is a perfect equilibrium of Table 5.1, consider the perturbed randomized-strategy profile

$$\hat{\sigma}^k = ((1 - \varepsilon)[a_1x_1] + .1\varepsilon[a_1y_1] + .1\varepsilon[b_1x_1] + .8\varepsilon[b_1y_1],$$

$$\varepsilon[x_2] + (1 - \varepsilon)[y_2]),$$

where $\varepsilon = 1/(k + 2)$. When $\varepsilon \leq \frac{1}{3}$, $[a_1x_1]$ is a best response to $\varepsilon[x_2] + (1 - \varepsilon)[y_2]$ for player 1, because $4 \geq 6\varepsilon + 3(1 - \varepsilon)$. On the other hand, when the probability of b_1y_1 is positive and more than three times larger than the probability of b_1x_1, then y_2 is the best response for player 2. So this sequence $\{\hat{\sigma}^k\}_{k=1}^{\infty}$ satisfies (5.1)–(5.3) for $([a_1x_1],[y_2])$ as a perfect equilibrium of the normal representation.

However, there is no perfect equilibrium of the multiagent representation of Figure 5.3 in which 1 chooses a_1 or 2 chooses y_2 with positive probability. In the multiagent representation, given any behavioral-strategy profile in which 1's agent for the 1.0 node assigns any positive probability to the move b_1, x_1 must be the best unique response for 1's agent at the 1.1 node. So the perfect equilibrium strategy of 1's agent for the 1.1 node must be $[x_1]$. Thus, the perturbed behavioral-strategy profiles $\hat{\sigma}^k$ that support a perfect equilibrium of the multiagent representation must satisfy $\lim_{k \to \infty} \hat{\sigma}_{1.1}^k(x_1) = 1$ and $\lim_{k \to \infty} \hat{\sigma}_{1.1}^k(y_1) = 0$. So under $\hat{\sigma}^k$, when k is large, the probability of the moves (b_1,x_1) occurring must be positive and much larger than the probability of the

Table 5.1   A game in strategic form, the normal representation of Figure 5.3

C_1	C_2	
	x_2	y_2
a_1x_1	4,4	4,4
a_1y_1	4,4	4,4
b_1x_1	6,6	3,0
b_1y_1	0,0	2,2

moves (b_1, y_1) occurring. (These probabilities are $\hat{\sigma}_{1.0}^k(b_1)\hat{\sigma}_{1.1}^k(x_1)$ and $\hat{\sigma}_{1.0}^k(b_1)\hat{\sigma}_{1.1}^k(y_1)$, respectively.) So for all large k, x_2 is player 2's unique best response to $\hat{\sigma}^k$, and the perfect equilibrium strategy of 2's agent must be $[x_2]$. This result in turn implies that $\lim_{k \to \infty} \hat{\sigma}_{2.2}^k(x_2) = 1$ and $\lim_{k \to \infty} \hat{\sigma}_{2.2}^k(y_2) = 0$. So for all large k, the best response to $\hat{\sigma}^k$ for 1's agent at the 1.0 node must be b_1, a response that gives him a payoff close to 6 when x_1 and x_2 have high probability. Thus, the unique perfect equilibrium of the multiagent representation is $([b_1], [x_1], [x_2])$.

5.3 Existence of Perfect and Sequential Equilibria

THEOREM 5.2. *For any finite game in strategic form, there exists at least one perfect equilibrium.*

Proof. Let $\Gamma = (N, (C_i)_{i \in N}, (u_i)_{i \in N})$ be any finite game in strategic form. Let λ be any randomized-strategy profile in $\times_{i \in N} \Delta^0(C_i)$. (For example, we could let $\lambda_i(c_i) = 1/|C_i|$ for every i and every c_i in C_i.) For any number k such that $k \geq 1$, we define a mapping $\delta^k : \times_{i \in N} \Delta(C_i) \to \times_{i \in N} \Delta^0(C_i)$ such that

$$\delta^k(\sigma) = \tau \text{ iff, } \forall i \text{ and } \forall c_i, \tau_i(c_i) = (1 - (1/k))\sigma_i(c_i) + (1/k)\lambda_i(c_i).$$

Similarly, for any c in $\times_{i \in N} C_i$, we define $\delta^k(c)$ such that

$$\delta^k(c) = \tau \text{ iff, } \forall i, \tau_i = (1 - (1/k))[c_i] + (1/k)\lambda_i.$$

For any c in $\times_{i \in N} C_i$, we let

$$\hat{u}_i^k(c) = u_i(\delta^k(c)).$$

Let $\hat{\Gamma}^k = (N, (C_i)_{i \in N}, (\hat{u}_i^k)_{i \in N})$.

Then $\hat{\Gamma}^k$ is a finite game in strategic form. To interpret this game, suppose that each player has an independent probability $1/k$ of irrationally trembling or losing rational control of his actions. Suppose also that, if player i trembled, then his behavior would be randomly determined, according to the randomized strategy λ_i, no matter what rational choices he might have wanted to make. Thus, if the randomized strategy σ_i in $\Delta(C_i)$ would describe player i's behavior when he is not trembling, then $\delta^k(\sigma_i)$ would describe his behavior before it is learned whether he is trembling or not. When such trembling is introduced into the game Γ, the result is the game $\hat{\Gamma}^k$. Of course, as k becomes very large, the probability of anyone trembling becomes very small, and the game $\hat{\Gamma}^k$ becomes a slight perturbation of the game Γ.

Let σ^k be a Nash equilibrium of $\hat{\Gamma}^k$, which exists by Theorem 3.1. Choosing a subsequence if necessary, we assume that $\{\sigma^k\}$ is a convergent sequence in the compact set $\times_{i\in N} \Delta(C_i)$. Choosing a further subsequence if necessary, we can also assume that, for every player i, the set of pure strategies in C_i that get zero probability in σ^k is the same for all k. (This can be done because there are only finitely many subsets of C_i.) Let $\sigma = \lim_{k\to\infty} \sigma^k$. For each k, let $\hat{\sigma}^k = \delta^k(\sigma^k)$. Then

$$\hat{\sigma}^k \in \underset{i\in N}{\times} \Delta^0(C_i), \quad \forall k;$$

$$\lim_{k\to\infty} \hat{\sigma}^k = \lim_{k\to\infty} \sigma^k = \sigma;$$

and, for any k, any player i, and any c_i in C_i,

$$\text{if } c_i \notin \underset{d_i\in C_i}{\text{argmax }} u_i(\hat{\sigma}^k_{-i},[d_i]) = \underset{d_i\in C_i}{\text{argmax }} \hat{u}^k_i(\sigma^k_{-i},[d_i])$$

$$\text{then } \sigma^k_i(c_i) = 0; \text{ and } \sigma_i(c_i) = 0.$$

This last condition implies that, for all k,

$$\sigma_i \in \underset{\tau_i\in\Delta(C_i)}{\text{argmax }} u_i(\hat{\sigma}^k_{-i},\tau_i).$$

Thus, σ satisfies the conditions (5.1)–(5.3) for a perfect equilibrium.   ■

Theorems 5.1 and 5.2 immediately imply Theorem 4.6, the existence theorem for sequential equilibria.

5.4 Proper Equilibria

Because the concept of perfect equilibrium is applied to extensive-form games via the multiagent representation, it still distinguishes between the games in Figures 5.2 and 5.3, which have the same purely reduced normal representation. Specifically, $([a_1],[y_2])$ and $([x_1],[x_2])$ are both perfect equilibria of the multiagent representation of Figure 5.2, but $([b_1],[x_1],[x_2])$ is the unique perfect equilibrium of the multiagent representation of Figure 5.3. To avoid distinguishing between such games and yet stay within the set of sequential equilibria, we need a solution concept that identifies equilibria that can always be extended to sequential equilibria, even when we apply the solution concept to the normal representation. Proper equilibria, which form a subset of the perfect equilibria, are such a solution concept.

We begin with an equivalent characterization of perfect equilibria. Given a finite game $\Gamma = (N, (C_i)_{i \in N}, (u_i)_{i \in N})$ in strategic form, and given any positive number ε, a randomized-strategy profile σ is an ε-perfect equilibrium iff

$$\sigma \in \underset{i \in N}{\times} \Delta^0(C_i)$$

and, for each player i and each pure strategy c_i in C_i,

$$\text{if } c_i \notin \underset{e_i \in C_i}{\text{argmax }} u_i(\sigma_{-i},[e_i]), \text{ then } \sigma_i(c_i) < \varepsilon.$$

It can be shown that a randomized-strategy profile $\bar{\sigma}$ is a perfect equilibrium of Γ iff there exists a sequence $(\varepsilon(k), \hat{\sigma}^k)_{k=1}^{\infty}$ such that,

$$\lim_{k \to \infty} \varepsilon(k) = 0, \quad \lim_{k \to \infty} \hat{\sigma}_i^k(c_i) = \bar{\sigma}_i(c_i), \quad \forall i \in N, \quad \forall c_i \in C_i,$$

and, for every k, $\hat{\sigma}^k$ is an $\varepsilon(k)$-perfect equilibrium.

This characterization of perfect equilibria can be used to motivate an analogous but stronger concept of equilibrium. We say that a randomized-strategy profile σ is an ε-*proper equilibrium* iff

$$\sigma \in \underset{i \in N}{\times} \Delta^0(C_i)$$

(so all pure strategies get strictly positive probability) and, for every player i and every pair of two pure strategies c_i and e_i in C_i,

$$\text{if } u_i(\sigma_{-i},[c_i]) < u_i(\sigma_{-i},[e_i]), \text{ then } \sigma_i(c_i) \leq \varepsilon \, \sigma_i(e_i).$$

We then say that a randomized-strategy profile $\bar{\sigma}$ is a *proper equilibrium* iff there exists a sequence $(\varepsilon(k), \sigma^k)_{k=1}^{\infty}$ such that

$$\lim_{k \to \infty} \varepsilon(k) = 0, \quad \lim_{k \to \infty} \sigma_i^k(c_i) = \bar{\sigma}_i(c_i), \quad \forall i \in N, \quad \forall c_i \in C_i,$$

and, for every k, σ^k is an $\varepsilon(k)$-proper equilibrium.

Thus, a perfect equilibrium is an equilibrium that can be approximated by randomized-strategy profiles in which every pure strategy is assigned positive probability, but any pure strategy that would be a mistake for a player is assigned an arbitrarily small probability. A proper equilibrium is an equilibrium that can be approximated by randomized-strategy profiles in which every pure strategy is assigned positive probability, but any pure strategy that would be a mistake for a player is assigned much less probability (by a multiplicative factor arbitrarily close to 0) than any other pure strategy that would be either a best response or a less costly mistake for him. It is straightforward to show that any proper equilibrium is a perfect equilibrium.

THEOREM 5.3. *For any finite game in strategic form, the set of proper equilibria is a nonempty subset of the set of perfect equilibria.*

Proof. A full proof of existence is given in Myerson (1978a), but I sketch here a more intuitive argument due to Kohlberg and Mertens (1986). Given $\Gamma = (N, (C_i)_{i \in N}, (u_i)_{i \in N})$, for each positive number ε less than 1, let $\hat{\Gamma}_\varepsilon$ be a game in which each player i in N has a pure strategy set that is the set of all strict orderings of C_i and for which the outcome is determined as follows. After the players choose their orderings, a pure strategy in C_i for each player i is independently selected according to the probability distribution in which the first strategy in i's chosen order has the highest probability and each of i's subsequent strategies has probability ε times the probability of the strategy that immediately precedes it in the ordering. Then each player's payoff in $\hat{\Gamma}_\varepsilon$ is his expected payoff in Γ from the pure-strategy profile in C that is randomly selected in this way.

By Theorem 3.1, we can find an equilibrium in randomized strategies of $\hat{\Gamma}_\varepsilon$, for any ε. For each ε, let $\hat{\sigma}^\varepsilon$ be the randomized-strategy profile in $\times_{i \in N} \Delta^0(C_i)$ such that, for each strategy c_i of each player i, $\hat{\sigma}^\varepsilon_i(c_i)$ is the probability of c_i being selected for i in $\hat{\Gamma}_\varepsilon$ when the players follow the equilibrium of $\hat{\Gamma}_\varepsilon$ that we found. It can be shown that $\hat{\sigma}^\varepsilon$ is an ε-proper equilibrium of Γ. By compactness of $\times_{i \in N} \Delta(C_i)$, we can find a subsequence of $(\hat{\sigma}^{\varepsilon(k)})_{k=1}^\infty$ that converges to some randomized-strategy profile σ and such that $\varepsilon(k) \to 0$ as $k \to \infty$. Then σ is a proper equilibrium. ∎

Just as perfect equilibria can be viewed as a way of identifying sequential equilibria in the multiagent representation, so proper equilibria can be used to identify sequential equilibria in the normal representation. The following result, analogous to Theorem 5.1, has been proved by Kohlberg and Mertens (1986) and van Damme (1984).

THEOREM 5.4. *Suppose that Γ^e is an extensive-form game with perfect recall and τ is a proper equilibrium of the normal representation of Γ^e. Then there exists a vector of belief probabilities π and a behavioral-strategy profile σ such that (σ, π) is a sequential equilibrium of Γ^e and σ is a behavioral representation of τ.*

Proof. Let $(\hat{\tau}^k, \varepsilon(k))_{k=1}^{\infty}$ be chosen so that each $\hat{\tau}^k$ is an $\varepsilon(k)$-proper equilibrium of the normal representation of Γ^e, $\lim_{k\to\infty} \hat{\tau}^k = \tau$, and $\lim_{k\to\infty} \varepsilon(k) = 0$. For each k, every node in Γ^e has positive probability under $\hat{\tau}^k$, because every pure-strategy profile has positive probability, so $\hat{\tau}^k$ has a unique behavioral representation $\hat{\sigma}^k$, and there is a unique beliefs vector $\hat{\pi}^k$ that is consistent with $\hat{\sigma}^k$. Choosing a subsequence if necessary, we can guarantee (because the sets of behavioral-strategy profiles and beliefs vectors are compact) that there exists a behavioral-strategy profile σ and a beliefs vector π such that $\lim_{k\to\infty} \hat{\sigma}^k = \sigma$ and $\lim_{k\to\infty} \hat{\pi}^k = \pi$. Then σ is a behavioral representation of τ, and π is fully consistent with σ. Choosing a further subsequence if necessary, we can assume also that, for each player i and each pair of strategies b_i and c_i in C_i, if there is any k in the sequence such that $u_i(\hat{\tau}_{-i}^k, [b_i]) > u_i(\hat{\tau}_{-i}^k, [c_i])$, then $u_i(\hat{\tau}_{-i}^k, [b_i]) > u_i(\hat{\tau}_{-i}^k, [c_i])$ for every k. (Notice there are only finitely many ways to order the players' strategy sets, so at least one must occur infinitely often in the sequence.)

For each player i and each information state s in S_i, let $C_i^*(s)$ denote the set of strategies that are compatible with state s and let $A_i^*(s)$ denote the set of strategies in $C_i^*(s)$ that maximize i's expected payoff against $\hat{\tau}_{-i}^k$. (By the construction of the $(\hat{\tau}^k)_{k=1}^{\infty}$ sequence, this set of maximizers is independent of k.) That is,

$$A_i^*(s) = \underset{c_i \in C_i^*(s)}{\operatorname{argmax}}\ u_i(\hat{\tau}_{-i}^k, [c_i]).$$

By ε-properness,

$$\lim_{k\to\infty} \frac{\sum_{b_i \in A_i^*(s)} \hat{\tau}_i^k(b_i)}{\sum_{c_i \in C_i^*(s)} \hat{\tau}_i^k(c_i)} = 1.$$

So by equation (4.1), if $\sigma_{i.s}(d_s) > 0$ in the limiting behavioral representation σ, then there must exist some strategy b_i in C_i such that $b_i \in$
$A_i^*(s)$ and $b_i(s) = d_s$.

Using perfect recall and the fact that each state s has positive prior
probability under each $\hat{\tau}^k$ we can show that a strategy in $C_i^*(s)$ that
achieves the maximum of $u_i(\hat{\tau}^k_{-i},[c_i])$ over all c_i in $C_i^*(s)$ must also achieve
the maximum of $\Sigma_{x \in Y_s} \hat{\pi}^k_{i.s}(x) U_i(\hat{\sigma}^k_{-i},[c_i]|x)$ over all c_i in $C_i^*(s)$. (Recall sim-
ilar arguments in the proofs of Theorems 5.1 and 4.4.) So

$$A_i^*(s) = \underset{c_i \in C_i^*(s)}{\mathrm{argmax}} \sum_{x \in Y_s} \hat{\pi}^k_{i.s}(x) U_i(\hat{\sigma}^k_{-i},[c_i]|x).$$

By continuity of the objective, this equation implies

$$A_i^*(s) \subseteq \underset{c_i \in C_i^*(s)}{\mathrm{argmax}} \sum_{x \in Y_s} \pi_{i.s}(x) U_i(\sigma_{-i},[c_i]|x).$$

We need to show that the behavioral-strategy profile σ is sequentially
rational with beliefs π. So suppose that, contrary to the theorem, there
is some information state s of some player i, and some move d_s in D_s
such that

$$\sigma_{i.s}(d_s) > 0 \text{ but } d_s \notin \underset{e_s \in D_s}{\mathrm{argmax}} \sum_{x \in Y_s} \pi_{i.s}(x) U_i(\sigma_{-i.s},[e_s]|x).$$

By perfect recall (which induces a partial order on each player's information states), we can choose the state s so that there is no such failure
of sequential rationality at any information state of player i that could
occur after s in the game. (That is, we pick s as close as possible to the
end of the game among all information states at which sequential
rationality fails.) However, because $\sigma_{i.s}(d_s) > 0$, there exists some strategy b_i

$$b_i(s) = d_s \text{ and } b_i \in \underset{c_i \in C_i^*(s)}{\mathrm{argmax}} \sum_{x \in Y_s} \pi_{i.s}(x) U_i(\sigma_{-i},[c_i]|x).$$

Thus, using move d_s would not be optimal for player i at state s if he
planned to use the behavioral strategy σ_i at all subsequent information
states, but using d_s would be optimal for player i at state s if he planned
to use the strategy b_i at all subsequent information states. However, this
result implies a contradiction of our assumption that the behavioral-
strategy profile σ is sequentially rational for player i (with the consistent

beliefs π) at all information states after s. So there cannot be any information state where sequential rationality fails. ∎

For example, in the game shown in Table 5.1, the strategy profile

$$\hat{\sigma} = ((1 - \varepsilon)[a_1 x_1] + .1\varepsilon[a_1 y_1] + .1\varepsilon[b_1 x_1] + .8\varepsilon[b_1 y_1],$$

$$\varepsilon[x_2] + (1 - \varepsilon)[y_2]),$$

is ε-perfect, for any ε less than $\frac{1}{3}$. So, as we have seen, $([a_1 x_1],[y_2])$ (the limit of $\hat{\sigma}$ as $\varepsilon \to 0$) is a perfect equilibrium of this game, even though it does not correspond to any sequential equilibrium of the game in Figure 5.3. However, $\hat{\sigma}$ is not an ε-proper equilibrium. The strategy $b_1 y_1$ would be a worse mistake than $b_1 x_1$ for player 1 against $\hat{\sigma}_2$, because $0\varepsilon + 2(1 - \varepsilon) < 6\varepsilon + 3(1 - \varepsilon)$. Thus, ε-properness requires that the ratio $\hat{\sigma}_1(b_1 y_1)/\hat{\sigma}_1(b_1 x_1)$ must not be more than ε, whereas this ratio of probabilities is 8 in $\hat{\sigma}$.

In fact, for any ε less than 1, the probability assigned to $b_1 y_1$ must be less than the probability assigned to $b_1 x_1$ in an ε-proper equilibrium, so x_2 must be a best response for player 2. So player 2 must use x_2 for sure in any proper equilibrium; and player 1 must use $b_1 x_1$, which is his best response to x_2. Thus, the unique proper equilibrium of the strategic-form game in Table 5.1 is $(b_1 x_1, x_2)$. It can be justified as a proper equilibrium by ε-proper equilibria of the form

$$((1 - \varepsilon - .5\varepsilon^2)[b_1 x_1] + .5\varepsilon^2[b_1 y_1] + .5\varepsilon[a_1 x_1] + .5\varepsilon[a_1 y_1],$$

$$(1 - \varepsilon)[x_2] + \varepsilon[y_2]).$$

This proper equilibrium corresponds to the unique sequential equilibrium of the extensive-form game in Figure 5.3.

Similarly, the unique proper equilibrium of the normal representation of the game in Figure 5.2 (which differs from Table 5.1 only in that player 1's two payoff-equivalent strategies are merged into one and the labels of his three pure strategies are changed to a_1, x_1, and y_1) is $([x_1],[x_2])$. In general, adding or eliminating a player's pure strategy that is payoff equivalent (for all players) to another pure strategy of the same player makes no essential change in the set of proper equilibria.

However, adding a pure strategy that is randomly redundant may change the set of proper equilibria. For example, consider the games in Tables 5.2 and 5.3, which we also considered in Section 2.4. The strategy $b_1 z_1$ in Table 5.2 is randomly redundant, because it is payoff

Table 5.2   A game in strategic form, the purely reduced representation of
Figure 2.5

	C_2	
C_1	x_2	y_2
$a_1.$	6,0	6,0
b_1x_1	8,0	0,8
b_1y_1	0,8	8,0
b_1z_1	3,4	7,0

Table 5.3   A game in strategic form, the fully reduced normal representation
of Figure 2.5

	C_2	
C_1	x_2	y_2
$a_1.$	6,0	6,0
b_1x_1	8,0	0,8
b_1y_1	0,8	8,0

equivalent to the randomized strategy $.5[a_1.] + .5[b_1y_1]$. So Table 5.3
differs from Table 5.2 only by the elimination of a randomly redundant
strategy. However, the proper equilibria of these two games are different. The unique proper equilibrium of Table 5.2 is $([a_1.], (7/12)[x_2] +
(5/12)[y_2])$, but the unique proper equilibrium of Table 5.3 is $([a_1.], .5[x_2]
+ .5[y_2])$. So eliminating a randomly redundant strategy for player 1
changes the unique proper equilibrium strategy for player 2.

To understand why player 2 must randomize differently proper equilibria of these two games, notice first that $a_1.$ is a best response for player
1 only if the probability of x_2 is between ¼ and ¾. So to justify a proper
equilibrium in which player 1 uses $a_1.$, player 2 must be willing to
randomize between x_2 and y_2 in any ε-proper equilibrium $\hat{\sigma}$ that approximates the proper equilibrium. For Table 5.2, this willingness im-
plies that

(5.4) $8\hat{\sigma}_1(b_1y_1) + 4\hat{\sigma}_1(b_1z_1) = 8\hat{\sigma}_1(b_1x_1).$

So b_1x_1 must be tied for second best for player 1 (after $a_1.$) with either
b_1y_1 or b_1z_1 (or else one side of this equation would have to be of order

ε times the other side). For Table 5.3, without $b_1 z_1$, the corresponding equation is

$$8\hat{\sigma}_1(b_1 y_1) = 8\hat{\sigma}_1(b_1 x_1).$$

Thus, player 2 must use $.5[x_2] + .5[y_2]$ in Table 5.3, so that player 1 will be willing to assign comparable probabilities to $b_1 y_1$ and $b_1 x_1$; so Table 5.3 has ε-proper equilibria of the form

$$((1 - \varepsilon)[a_1 \cdot] + .5\varepsilon[b_1 x_1] + .5\varepsilon[b_1 y_1], .5[x_2] + .5[y_2]).$$

But if player 2 used $.5[x_2] + .5[y_2]$ in Table 5.2, then $b_1 z_1$ would be better for player 1 than either $b_1 x_1$ or $b_1 y_1$ (because $\frac{3}{2} + \frac{7}{2} > \frac{0}{2} + \frac{8}{2}$), and so $\hat{\sigma}_1(b_1 z_1)$ would have to be much larger than $\hat{\sigma}_1(b_1 x_1)$ in an ε-proper equilibrium $\hat{\sigma}$, so that equation (5.4) could not be satisfied. For player 1 to be willing to satisfy (5.4) in an ε-proper equilibrium, player 2 must make player 1 indifferent between $b_1 x_1$ and $b_1 z_1$ by using $(\frac{7}{12})[x_2] + (\frac{5}{12})[y_2]$. So, for example, Table 5.2 has ε-proper equilibria of the form

$$((1 - \varepsilon - 2\varepsilon^2/3)[a_1 \cdot] + ((\varepsilon + \varepsilon^2)/3)[b_1 x_1] + (\varepsilon^2/3)[b_1 y_1]$$
$$+ (2\varepsilon/3)[b_1 z_1], (\frac{7}{12})[x_2] + (\frac{5}{12})[y_2]).$$

Why should the addition of a randomly redundant strategy for player 1 change player 2's rational behavior? Kohlberg and Mertens (1986) showed that this lack of invariance is an inevitable consequence of Theorem 5.4. In particular, Table 5.2 is the purely reduced normal representation of the extensive-form game in Figure 5.4, whereas Table 5.3 is the purely reduced normal representation of the extensive-form game in Figure 5.5. Each of these two extensive-form games has a unique subgame-perfect and sequential equilibrium, shown in Figures 5.4 and 5.5. In particular, subgame-perfectness requires that player 2 use $(\frac{7}{12})[x_2] + (\frac{5}{12})[y_2]$ in Figure 5.4 but $.5[x_2] + .5[y_2]$ in Figure 5.5. (In each figure, the subgame beginning with player 2's decision node has a unique equilibrium, but these equilibria are different.) So the sensitivity of proper equilibria to the addition of randomly redundant strategies is an inevitable result of the fact that two extensive-form games that have the same fully reduced normal representation may have disjoint sets of sequential-equilibrium scenarios.

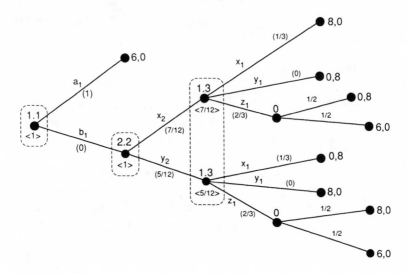

Figure 5.4

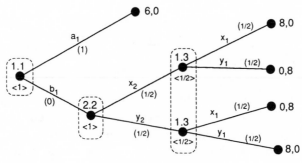

Figure 5.5

5.5 Persistent Equilibria

Consider again, in Table 5.4, the Battle of the Sexes game that we saw in Chapter 3. This game has three equilibria: $([f_1],[f_2])$, $([s_1],[s_2])$, and $(.75[f_1] + .25[s_1], .25[s_2] + .75[f_2])$. All three of these equilibria are perfect and proper.

The randomized equilibrium of this game is worse for both players than either of the pure-strategy equilibria. It represents a lack of co-ordination that neither player can unilaterally overcome. In a game like

Table 5.4   Battle of the Sexes game in strategic form

C_1	C_2	
	f_2	s_2
f_1	3,1	0,0
s_1	0,0	1,3

this, as Schelling (1960) has argued, both players should be groping for some focal factor that would steer them away from this equilibrium and toward $([f_1],[f_2])$ or $([s_1],[s_2])$. Thus, the randomized equilibrium may seem unstable or unreasonable as a solution to this game. Kalai and Samet (1984) developed a concept of persistent equilibrium that excludes this randomized equilibrium for this example.

Given $\Gamma = (N, (C_i)_{i \in N}, (u_i)_{i \in N})$, a finite game in strategic form, a *retract* is a set Θ of the form $\Theta = \times_{i \in N} \Theta_i$ such that, for every i, Θ_i is a nonempty closed convex subset of $\Delta(C_i)$. For any positive number ε, a randomized-strategy profile σ is in the ε-*neighborhood* of a set of randomized-strategy profiles Θ iff there exists some $\hat{\sigma}$ in Θ such that

$$|\hat{\sigma}_i(c_i) - \sigma_i(c_i)| < \varepsilon, \quad \forall i \in N, \quad \forall c_i \in C_i.$$

A retract Θ is *absorbing* iff there is some positive number ε such that, for every randomized-strategy profile σ in the ε-neighborhood of Θ, there exists some ρ in Θ such that

$$\rho_i \in \underset{\tau_i \in \Delta(C_i)}{\mathrm{argmax}}\ u_i(\sigma_{-i}, \tau_i), \quad \forall i \in N.$$

That is, a retract is absorbing if, for any randomized-strategy profile that is sufficiently close to the retract, there is a best response for each player such that the randomized-strategy profile composed of these best responses is in the retract. In a sense, then, if we only assumed that players would use strategies that were close to an absorbing retract, then rational responses could lead the players into the retract.

A *persistent* retract is a minimal absorbing retract. That is, a retract is persistent iff it is absorbing and there does not exist any other absorbing retract $\hat{\Theta} = \times_{i \in N} \hat{\Theta}_i$ such that $\hat{\Theta}_i \subseteq \Theta_i$ for every i. A *persistent equilibrium* is any equilibrium that is contained in a persistent retract.

Table 5.5   A three-player game in strategic form

C_1	C_2 and C_3			
	x_3		y_3	
	x_2	y_2	x_2	y_2
x_1	1,1,1	0,0,0	0,0,0	0,0,0
y_1	0,0,0	0,0,0	0,0,0	1,1,0

Kalai and Samet (1984) have proved the following general existence theorem.

THEOREM 5.5.   *For any finite game in strategic form, the set of persistent equilibria is nonempty, and it includes at least one proper equilibrium.*

In the Battle of the Sexes game, there are three absorbing retracts: $\{[f_1]\} \times \{[f_2]\}$, $\{[s_1]\} \times \{[s_2]\}$, and $\Delta(\{f_1,s_1\}) \times \Delta(\{f_2,s_2\})$. Thus $\{[f_1]\} \times \{[f_2]\}$ and $\{[s_1]\} \times \{[s_2]\}$ are the only persistent retracts; so $([f_1],[f_2])$ and $([s_1],[s_2])$ are the only persistent equilibria.

Kalai and Samet (1984) also discuss the three-player game shown in Table 5.5, where $C_i = \{x_i,y_i\}$ for each player i.

The equilibrium (x_1,x_2,x_3) is perfect, proper and persistent. (Perfectness and properness always coincide for games in which no player has more than two pure strategies.) However, there is another, less intuitive equilibrium (y_1,y_2,x_3) that is also perfect and proper but is not persistent. (Notice that

$$(.5\varepsilon[x_1] + (1 - .5\varepsilon)[y_1], .5\varepsilon[x_2] + (1 - .5\varepsilon)[y_2], (1 - \varepsilon)[x_3] + \varepsilon[y_3])$$

is ε-perfect and ε-proper.) Iterative elimination of weakly dominated strategies also eliminates the (y_1,y_2,x_3) equilibrium.

5.6 Stable Sets of Equilibria

Kohlberg and Mertens (1986) sought to develop a solution theory for strategic-form games that, when applied to extensive-form games via their fully reduced normal representations, would always identify equilibria that could be extended (with appropriate beliefs vectors) to sequential equilibria. However, as the games in Figures 5.4 and 5.5 illustrate, two games can have the same fully reduced normal representation and yet have sets of sequential equilibria that are disjoint (when projected into the set of randomized-strategy profiles of the reduced normal representation). Thus Kohlberg and Mertens broadened their scope to consider sets of equilibria, rather than individual equilibria, as candidates for "solutions" to a game.

Let $\Gamma = (N, (C_i)_{i \in N}, (u_i)_{i \in N})$ be a finite game in strategic form. For any vector e in $\mathbf{R}^N$ such that

$$0 < e_i \leq 1, \quad \forall i \in N,$$

and for any randomized strategy profile λ in $\times_{i \in N} \Delta^0(C_i)$, let $\delta_{e,\lambda} : \times_{i \in N} \Delta(C_i) \to \times_{i \in N} \Delta^0(C_i)$ be defined by

$$\delta_{e,\lambda}(\sigma) = \hat{\sigma} \quad \text{iff,} \quad \forall i \in N, \ \hat{\sigma}_i = (1 - e_i)\sigma_i + e_i\lambda_i.$$

Let $\hat{\Gamma}_{e,\lambda} = \big(N, (C_i)_{i \in N}, (u_i(\delta_{e,\lambda}([\cdot])))_{i \in N}\big)$. That is, $\hat{\Gamma}_{e,\lambda}$ is the perturbation of Γ in which each player i has an independent probability e_i of accidentally implementing the randomized strategy λ_i. We can say that a set Θ is *prestable* iff it is a closed subset of the set of equilibria of Γ and, for any positive number ε_0, there exists some positive number ε_1 such that, for any e in $\mathbf{R}^N$ and any λ in $\times_{i \in N} \Delta^0(C_i)$, if $0 < e_i < \varepsilon_1$ for every player i, then there exists at least one Nash equilibrium of $\hat{\Gamma}_{e,\lambda}$ that is in an ε_0-neighborhood of Θ. That is, a prestable set contains at least one equilibrium of Γ that is close to an equilibrium of every perturbed game $\hat{\Gamma}_{e,\lambda}$, for every e with sufficiently small components and every λ that assigns positive probability to all pure strategies. Kohlberg and Mertens (1986) then define a *stable set* of equilibria of Γ to be any prestable set that contains no other prestable subsets. That is, a stable set is a minimal prestable set.

In proving the existence of perfect equilibria, we constructed a perfect equilibrium that was a limit of equilibria of games of the form $\hat{\Gamma}_{e,\lambda}$ (where $e = (1/k, \ldots, 1/k)$ and $k \to \infty$). However, a stable set of equilibria must contain limits of equilibria of all sequences of games of this form when the components of the e vectors are going to 0.

Kohlberg and Mertens have proved the following two theorems.

THEOREM 5.6. *For any finite strategic-form game Γ, there is at least one connected set of equilibria that contains a stable subset.*

THEOREM 5.7. *If Θ is any stable set of the game Γ, then Θ contains a stable set of any game that can be obtained from Γ by eliminating any pure strategy that is weakly dominated or that is an inferior response (that is, not a best response) to every equilibrium in Θ.*

Kohlberg and Mertens (1986) show an example of an extensive-form game with a unique sequential equilibrium and a stable set of its reduced normal representation that does not contain the unique sequential equilibrium. However, they also defined a concept of *fully stable* sets of equilibria, which may be larger than stable sets and which always contain proper equilibria that correspond to sequential equilibria in every extensive-form game that has Γ as its normal or reduced normal representation. However, fully stable sets of equilibria can include equilibria that assign positive probability to weakly dominated strategies.

To appreciate the power of Theorems 5.6 and 5.7, consider the extensive-form game in Figure 4.10. The normal representation of this game is in Table 5.6. The set of equilibria of this strategic-form game has two connected components. First, there is a connected set consisting of the single equilibrium

$$((\tfrac{1}{7})[x_1x_2] + (\tfrac{6}{7})[x_1y_2], (\tfrac{2}{3})[e_3] + (\tfrac{1}{3})[f_3]).$$

Second, there is a connected set consisting of equilibria of the form

$$([y_1y_2], \beta[e_3] + (1 - \beta - \gamma)[f_3] + \gamma[g_3]), \quad \text{where } 4\beta + 3\gamma \leq 1.$$

(This second connected set contains the union of the equilibria generated in three separate cases considered in Section 4.5.) Using Theorem 5.7, we can show that this second connected set of equilibria does not contain any stable subset. For every equilibrium in this second set, it is never a best response for player 1 to choose x_2 at his 1.2 decision node.

Table 5.6 A game in strategic form, the normal representation of Figure 4.10

	C_2		
C_1	e_3	f_3	g_3
x_1x_2	2.0,4.0	−1.5,7.0	0.5,7.5
x_1y_2	1.5,4.0	−0.5,3.5	1.0,3.0
y_1x_2	0.5,0	−1.0,3.5	−0.5,4.5
y_1y_2	0,0	0,0	0,0

(On the other hand, there are equilibria in this set where choosing x_1 at 1.1 would be a best response.) So any stable subset of the second connected set must contain a stable subset of the game where the strategies $x_1 x_2$ and $y_1 x_2$ for player 1 are eliminated, by Theorem 5.7. (Weak domination alone could eliminate $y_1 x_2$ but not $x_1 x_2$.) However, when these two strategies are eliminated, the strategies f_3 and g_3 are weakly dominated for player 2 in the game that remains; so another application of Theorem 5.7 implies that a stable subset of the second connected set must contain a stable set of the game where the strategies $x_1 x_2$, $y_1 x_2$, f_3, and g_3 are all eliminated. But eliminating these strategies eliminates all of the equilibria in the second connected set (and the empty set is not stable!), so there can be no stable subset of the second connected set of equilibria. Then, by the existence theorem for stable sets, the first connected set, consisting of just one equilibrium, must be a stable set.

For another example, consider the *Beer-Quiche game* of Cho and Kreps (1987). In this two-player game, player 1's type may be "surly" or "wimpy," with probabilities .9 or .1, respectively. Knowing his type, player 1 enters a tavern and must order either beer or quiche. He prefers beer if he is surly, but he prefers quiche if he is wimpy. Player 2 does not know 1's type but, after seeing what player 1 has ordered, player 2 must decide whether to fight with player 1 or just leave him alone and go away. Player 2's payoff is 0 if she goes away, $+1$ if she fights and player 1 is wimpy, or -1 if she fights and player 1 is surly. Player 1's payoff is the sum $x + y$, where $x = 1$ if he orders the item that he prefers between beer and quiche, and $x = 0$ otherwise, and $y = 2$ if he does not fight with player 2, and $y = 0$ otherwise. This game has two connected sets of sequential equilibria.

In one set of equilibria, shown in Figure 5.6, player 1 always orders beer, no matter what his type is; player 2 does not fight when he orders beer; but player 2 would fight with a probability β that is at least $\frac{1}{2}$ if player 1 ordered quiche. To make such a scenario a sequential equilibrium, player 2's belief probability α of the event that player 1 is wimpy after he orders quiche must be at least $\frac{1}{2}$. Formally, the parameters α and β must satisfy the following conditions:

$$\text{either } \tfrac{1}{2} \le \alpha \le 1 \text{ and } \beta = 1, \text{ or } \alpha = \tfrac{1}{2} \text{ and } \tfrac{1}{2} \le \beta < 1.$$

In the other set of equilibria, shown in Figure 5.7, player 1 always orders quiche, no matter what his type is; and player 2 does not fight when he orders quiche, but player 2 would fight with a probability β

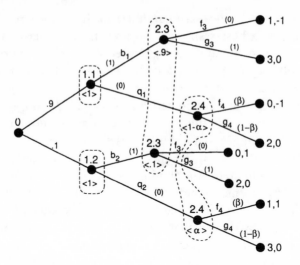

Figure 5.6

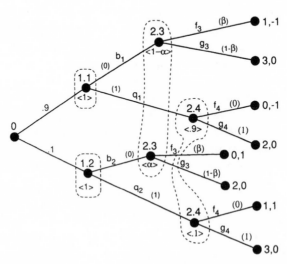

Figure 5.7

that is at least ½ if player 1 ordered beer. To make such a scenario a sequential equilibrium, player 2's belief probability α of the event that player 1 is wimpy after he orders beer must be at least ½. Again, the parameters α and β must satisfy the conditions:

$$\text{either } \tfrac{1}{2} \le \alpha \le 1 \text{ and } \beta = 1, \text{ or } \alpha = \tfrac{1}{2} \text{ and } \tfrac{1}{2} \le \beta < 1.$$

There may seem to be something counterintuitive about the equilibria in Figure 5.7, where player 1 orders quiche even if he is surly. These equilibria rely on the supposition that ordering beer would be taken as evidence of wimpiness, but a wimpy type would be less inclined to order beer than a surly type. In particular, the set of randomized strategies for player 2 against which ordering beer would be optimal for the wimpy type of player 1 is a strict subset of the set of strategies for player 2 against which ordering beer would be optimal for the surly type of player 1. Thus, it is natural to suppose that the belief probability that player 2 would assign to the event that player 1 is wimpy should not be larger when player 1 has ordered beer than when he has ordered quiche. This criterion eliminates the sequential equilibria in Figure 5.7 and leaves only the sequential equilibria in Figure 5.6, where player 1 orders beer. We can formalize this intuition by using Theorem 5.7 to show that there is no stable subset of the equilibria where player 1 buys quiche.

Consider first the normal representation, shown in Table 5.7. Here player 1's information state 1 is the surly type, 1's state 2 is the wimpy type, player 2's state 3 is the state of having seen beer ordered, and 2's state 4 is the state of having seen quiche ordered. We can verify that the counterintuitive equilibrium $([q_1q_2],[f_3g_4])$ is perfect and proper, because it can be approximated by ε-proper equilibria in which the

Table 5.7 Beer-Quiche game in strategic form, the normal representation

C_1	C_2			
	f_3f_4	f_3g_4	g_3f_4	g_3g_4
b_1b_2	0.9,−0.8	0.9,−0.8	2.9,0	2.9,0
b_1q_2	1.0,−0.8	1.2,−0.9	2.8,0.1	3.0,0
q_1b_2	0,−0.8	1.8,0.1	0.2,−0.9	2.0,0
q_1q_2	0.1,−0.8	2.1,0	0.1,−0.8	2.1,0

probability of q_1b_2 is small but much greater than the probabilities of b_1q_2 and b_1b_2. However, the strategy q_1b_2 (order quiche if surly, order beer if wimpy) is strongly dominated for player 1 (by $(\frac{1}{9})[b_1q_2]$ + $(\frac{8}{9})[q_1q_2]$, which has the same probability .1 of ordering beer but orders it only when surly). After q_1b_2 is eliminated, f_3f_4 and f_3g_4 and g_3g_4 are all weakly dominated by g_3f_4 for player 2, and b_1b_2 is the unique best response for player 1 to g_3f_4. So all equilibria except (b_1b_2,g_3f_4) can be eliminated by iterative elimination of weakly dominated strategies. Thus, by Theorem 5.7, any stable set of equilibria must include this equilibrium, in which player 1 always drinks beer and player 2 goes away when he drinks beer, but player 2 would fight if player 1 ordered quiche.

In the multiagent representation, there are no weakly dominated strategies, but Theorem 5.7 can still be applied to show that there is no stable set of the equilibria where player 1 orders quiche. In the set of equilibria where player 1 orders quiche, there is one equilibrium (with $\alpha = \frac{1}{2}$ and $\beta = \frac{1}{2}$) where ordering beer would be a best response for the surly agent 1.1, but there are no equilibria where ordering beer would be a best response for the wimpy agent 1.2. So a stable set of equilibria in which player 1 orders quiche would necessarily have a nonempty subset in the modified game where we eliminate b_2, the wimpy agent's option to order beer. Then buying beer would be proof of surliness in this modified game; so f_3, fighting when beer is ordered, would be weakly dominated for agent 2.3. But when f_3 is eliminated, the surly agent would never choose to buy quiche in any equilibrium of the game that remains. So by Theorem 5.7, the multiagent representation of the original game has no stable subset of the equilibria where player 1 always orders quiche.

This argument does not eliminate the equilibria in Figure 5.6, where player 1 orders beer, because 1's only move that is never a best response to these equilibria is q_1, quiche for the surly type. (Quiche for the wimpy type is a best response when $\beta = \frac{1}{2}$.) Eliminating the option of buying quiche for the surly type guarantees that player 2 would want to fight if player 1 bought quiche, but that still leaves the equilibrium in which $\beta = 1$.

Cho and Kreps (1987) and Banks and Sobel (1987) have developed other solution criteria related to Kohlberg-Mertens stability, with applications to signaling games of economic interest.

5.7 Generic Properties

Once we specify a finite set of players N and finite pure-strategy sets C_i for every player i, the set of possible payoff functions $(u_i)_{i \in N}$ is $\mathbf{R}^{N \times C}$, where $C = \times_{i \in N} C_i$. This set $\mathbf{R}^{N \times C}$ is a finite-dimensional vector space. We can say that a property is satisfied *for all generic games in strategic form* iff, for every nonempty finite N and $(C_i)_{i \in N}$, there is an open and dense set of payoff functions $(u_i)_{i \in N}$ in $\mathbf{R}^{N \times C}$ that generate games $(N,$ $(C_i)_{i \in N}, (u_i)_{i \in N})$ that all satisfy this property. On the other hand, we can say that the property holds *for some generic games in strategic form* iff there exist some nonempty finite sets N and $(C_i)_{i \in N}$ and a nonempty open set of payoff functions in $\mathbf{R}^{N \times C}$ that generate games that satisfy this property. That is, if a property is satisfied for some generic games, then there exists a game that satisfies the property *robustly*, in the sense that any sufficiently small perturbation of the payoffs would generate a new game that also satisfies the property. If a property is satisfied for all generic games, then small perturbations of the payoffs in any game can generate a new game that satisfies this property robustly, even if the original game does not. (For further discussion of genericity, see Mas-Colell, 1985, pages 316–319.)

Given any extensive-form game Γ^e, we can generate a game that differs from Γ^e only in the payoffs by choosing payoff functions in $\mathbf{R}^{N \times \Omega}$, where Ω is the set of terminal nodes and N is the set of players in Γ^e. We may say that a property holds *for all generic games in extensive form* iff, for every extensive-form game Γ^e, there exists an open and dense set of payoff functions in $\mathbf{R}^{N \times \Omega}$ that generate games, differing from Γ^e only in the payoffs, that all satisfy this property. A property holds *for some generic games in extensive form* iff there exists some extensive-form game Γ^e and a nonempty open set of payoff functions in $\mathbf{R}^{N \times \Omega}$ that generate games, differing from Γ^e only in the payoffs, that all satisfy this property.

Van Damme (1987) has shown (see his Theorems 2.4.7, 2.5.5, and 2.6.1) that, for all generic games in strategic form, all equilibria are perfect and proper. This result might at first seem to suggest that perfectness and properness usually have no power to eliminate unreasonable equilibria. However, for some generic games in extensive form, there exist nonempty sets of imperfect and improper equilibria of both the normal and multiagent representations. For example, recall the game in Figure 4.5; $([y_1],[y_2])$ is an imperfect and improper equilibrium of the normal representation of every game that can be derived from Figure 4.5 by changing payoffs by less than ± 0.5. To understand how this distinction between strategic-form genericity and extensive-form genericity can arise, notice that, for all generic games in strategic form, no two payoffs are identical. But player 2 gets from (y_1,x_2) the same payoff she gets from (y_1,y_2) in the normal representation of every game that is generated from Figure 4.5 by changing the terminal payoffs.

Thus, the essential significance and power of the equilibrium refinements like perfectness and properness is really evident only when we apply them to games in extensive form, even though these refinements are defined in terms of games in strategic form. On the other hand, for some generic games in strategic form, persistent equilibria can be guaranteed to exclude at least one equilibrium. For example, after any small perturbation of payoffs in the Battle of the Sexes game in Table 5.4 (with every payoff changed by less than ± 0.5), there will exist a randomized equilibrium that is not persistent.

5.8 Conclusions

The discussion in the preceding two chapters indicates only a part of the rich literature on refinements of Nash equilibrium (see also McLennan, 1985; Reny, 1987; Fudenberg and Tirole, 1988; Farrell, 1988; Grossman and Perry, 1986; Okuno-Fujiwara, Postlewaite, and Mailath, 1991; and Kohlberg, 1989). There are a variety of questions that we can use to evaluate these various refinements. Does the proposed solution concept satisfy a general existence theorem? Is the set of solutions invariant when we transform a game in a way that seems irrelevant? Does the solution concept exclude all equilibria that seem intuitively unreasonable? Does it include all the equilibria that seem intuitively reasonable? Does the intrinsic logic of the solution concept seem compelling as a characterization of rational behavior?

Unfortunately, we now have no solution concept for which an unqualified "Yes" can be given in answer to all of these questions. For example, we have no compelling argument to prove that rational intelligent players could not possibly behave according to an equilibrium that is not proper; nor do we have any convincing argument that every proper equilibrium must be considered as a possible way that rational intelligent players might behave in a game. Thus, it is hard to argue that proper equilibria should be accepted as either an upper solution or as a lower solution, in the sense of Section 3.4. The basic rationale for considering perturbed models with small probabilities of mistakes, as we do when we test for perfectness and properness, is that they give us a way to check that the justification of an equilibrium does not depend on the (unreasonable) assumption that players completely ignore the possible outcomes of the game that have zero probability in the equilibrium. Thus, we can only argue that testing for perfectness and properness of equilibria is a useful way to formalize part of our intuitive criteria about how rational intelligent players might behave in a game. Similar comments can be made about other solution concepts in the literature.

On the basis of the logical derivation of the concept of sequential equilibrium, one can reasonably argue that sequential equilibrium might be an upper solution for extensive-form games, so being a sequential equilibrium might be a necessary (even if not sufficient) condition for a reasonable characterization of rational intelligent behavior. However, the case for requiring full consistency, as defined in Section 4.4, was not completely compelling, so there might be some other reasonable consistency requirement that allows more beliefs vectors than full consistency but less than weak consistency. Such intermediate consistency conditions have been used by Fudenberg and Tirole (1988) in their definition of *perfect Bayesian equilibrium*.

We should draw a distinction between *refinements* of Nash equilibrium and criteria for *selection* among the set of equilibria. A refinement of Nash equilibrium is a solution concept that is intended to offer a more accurate characterization of rational intelligent behavior in games. However, the ultimate refinement that exactly characterizes rational behavior can still include multiple equilibria for many games (e.g., the Battle of the Sexes). That is, a game can have many different equilibria such that, if players in some given culture or environment expected each other to behave according to any one of these equilibria, then they could rationally and intelligently fulfill these expectations.

A selection criterion is then any objective standard, defined in terms of the given structure of the mathematical game, that can be used to determine the focal equilibrium that everyone expects. As such, a selection criterion would only be valid if there is something in the cultural environment of the players that would predispose them to focus attention on this selection criterion. Equity and efficiency criteria of cooperative game theory can be understood as such selection criteria. Other solution concepts that have been offered are based on the assumption that players share a rich natural language for communication (Farrell, 1988; Myerson, 1989); and these solution concepts also can be considered as equilibrium-selection criteria, rather than as equilibrium refinements that are supposed to be valid whenever players are rational and intelligent.

Exercises

Exercise 5.1.   Consider the two-person game in strategic form shown in Table 5.8.

a. Find all Nash equilibria of this game. Which of these equilibria are (trembling-hand) perfect? Which of these equilibria are proper?

b. Show that player 1's strategy bg is randomly redundant.

c. Consider now the reduced game after the redundant strategy bg has been eliminated. Which equilibria of this reduced game are perfect? Which equilibria of this reduced game are proper?

d. Interpret the answers to parts (a) and (c) in terms of Exercise 4.7 (Chapter 4).

Exercise 5.2.   Consider the two-person game in strategic form shown in Table 5.9.

a. Characterize the set of all Nash equilibria of this game.

b. Show that $(.5[x_1] + .5[z_1], [x_2])$ is not a perfect equilibrium of this game.

c. Which Nash equilibria are perfect?

d. Find a proper equilibrium of this game.

Table 5.8   A game in strategic form

	C_2	
C_1	c	d
a·	5,4	5,4
be	8,3	1,9
bf	3,6	7,2
bg	4,5	6,3

Table 5.9   A game in strategic form

C_1	C_2		
	x_2	y_2	z_2
x_1	1,2	3,0	0,3
y_1	1,1	2,2	2,0
z_1	1,2	0,3	3,0

Exercise 5.3.   Consider the game shown in Figure 4.19.

a. Show that the set of equilibria in which player 1 would choose z_1 at his information state 2 does not contain any stable subset.

b. Explain why your argument in part (a) would not apply to the game in Figure 2.8.

6

Games with Communication

6.1 Contracts and Correlated Strategies

There are many games, like the Prisoners' Dilemma (Table 3.4), in which the Nash equilibria yield very low payoffs for the players, relative to other nonequilibrium outcomes. In such situations, the players would want to transform the game, if possible, to extend the set of equilibria to include better outcomes. Players might seek to transform a game by trying to communicate with each other and coordinate their moves, perhaps even by formulating contractual agreements. (Another way to extend the set of equilibria, creating a long-term relationship or repeated game, is studied in Chapter 7.)

We argued in Chapter 2 that there is no loss of generality in assuming that each player chooses his strategy simultaneously and independently. In principle, we argued, anything that a player can do to communicate and coordinate with other players could be described by moves in an extensive-form game, so planning these communication moves would become part of his strategy choice itself.

Although this perspective can be fully general in principle, it is not necessarily the most fruitful way to think about all games. There are many situations where the possibilities for communication are so rich that to follow this modeling program rigorously would require us to consider enormously complicated games. For example, to formally model player 1's opportunity to say just one word to player 2, player 1 must have a move and player 2 must have an information state for every word in the dictionary! In such situations, it might be more useful to leave communication and coordination possibilities out of our explicit game model. If we do so, then we must instead use solution concepts that express an assumption that players have implicit communication opportunities, in addition to the strategic options explicitly described in the game model. In this chapter we investigate such solution concepts and show that they can indeed offer important analytical insights into many situations.

Let us begin by considering the most extreme kind of implicit-coordination assumption: that players can not only communicate with each other but can even sign jointly binding contracts to coordinate their strategies. Consider the example in Table 6.1 (a modified version of the Prisoners' Dilemma). In this game, the unique equilibrium is (y_1,y_2), which gives a payoff allocation of $(1,1)$. Now suppose that some lawyer or other outside intervener presented the two players with a contract that says: "We, the undersigned, promise to choose x_1 and x_2 if this contract is signed by both players. If it is signed by only one player, then he or she will choose y_1 or y_2." The option for each player i to sign the contract can be introduced into the game description, as the strategy s_i, shown in Table 6.2. This transformed game has an equilibrium at (s_1,s_2), which is the unique perfect equilibrium and gives the payoff allocation $(2,2)$.

Even better expected payoffs could be achieved if the contract committed the players to a correlated or jointly randomized strategy. For

Table 6.1   A game in strategic form

C_1	C_2	
	x_2	y_2
x_1	2,2	0,6
y_1	6,0	1,1

Table 6.2   A game in strategic form, including one contract-signing strategy for each player

C_1	C_2		
	x_2	y_2	s_2
x_1	2,2	0,6	0,6
y_1	6,0	1,1	1,1
s_1	6,0	1,1	2,2

example, another contract might specify that, if both players sign the contract, then a coin will be tossed after which occurs they will implement (x_1,y_2) if Heads and (y_1,x_2) if Tails. Then the expected payoff allocation when both players sign this second contract is (3,3). Assuming that the players must make their signing decisions independently and that each player can sign only one contract, and letting $\hat{s}_i$ denote player i's option to sign the second contract, the transformed game becomes as shown in Table 6.3. In this game, there are three perfect equilibria: one at (s_1,s_2), one at $(\hat{s}_1,\hat{s}_2)$, and a third where each player i randomizes between s_i and $\hat{s}_i$ (with probabilities $\frac{2}{3}$ and $\frac{1}{3}$, respectively).

In realistic situations where players can sign contracts, there are usually a very wide range of contracts that could be considered. Furthermore, in such situations, contract-signing is usually only the last move in a potentially long bargaining process. If we were to try to list all strategies for bargaining and signing contracts as explicit options in a strategic-form game, it would become unmanageably complicated. So, as noted earlier, a fruitful analytical approach is to take the contract-signing options out of the explicit structure of the game and instead express the impact of these options by developing a new solution concept that takes them into account.

So let us introduce the concept of a game with contracts. When we say that a given game is *with contracts*, we mean that, in addition to the options that are given to the players in the formal structure of the game, the players also have very wide options to bargain with each other and to sign contracts, where each contract binds the players who sign it to some correlated strategy that may depend on the set of players who sign. After saying that a game is with contracts, we can then leave the

Table 6.3   A game in strategic form, including two contract-signing strategies for each player

C_1	C_2			
	x_2	y_2	s_2	$\hat{s}_2$
x_1	2,2	0,6	0,6	0,6
y_1	6,0	1,1	1,1	1,1
s_1	6,0	1,1	2,2	1,1
$\hat{s}_1$	6,0	1,1	1,1	3,3

actual structure of these contract-signing options implicit and take them into account by an appropriate solution concept.

Formally, given any strategic-form game $\Gamma = (N, (C_i)_{i \in N}, (u_i)_{i \in N})$, a *correlated strategy* for a set of players is any probability distribution over the set of possible combinations of pure strategies that these players can choose in Γ. That is, given any $S \subseteq N$, a correlated strategy for S is any probability distribution in $\Delta(C_S)$, where

$$C_S = \underset{i \in S}{\times} C_i.$$

In this notation, we also can write

$$C = C_N, \text{ and } C_{-i} = C_{N-i}.$$

A set of players S might implement a correlated strategy τ_S by having a trustworthy mediator (or a computer with a random-number generator) randomly designate a profile of pure strategies in C_S in such a way that the probability of designating any $c_S = (c_i)_{i \in S}$ in C_S is $\tau_S(c_S)$. Then the mediator would tell each player i in S to implement the strategy c_i that is the i-component of the designated profile c_S.

Given any correlated strategy μ in $\Delta(C)$ for all players, for each i, let $U_i(\mu)$ denote the expected payoff to player i when μ is implemented in the game Γ. That is,

$$U_i(\mu) = \sum_{c \in C} \mu(c) u_i(c).$$

Let $U(\mu) = (U_i(\mu))_{i \in N}$ denote the expected payoff allocation to the players in N that would result from implementing μ. Here, we define an *allocation* to be any vector that specifies a payoff for each player.

With this notation, a *contract* can be represented mathematically by any vector $\tau = (\tau_S)_{S \subseteq N}$ in $\times_{S \subseteq N} \Delta(C_S)$. For any such contract τ, τ_S represents the correlated strategy that would be implemented by the players in S if S were the set of players to sign the contract. So for any allocation in the set

$$\{U(\mu) | \mu \in \Delta(C)\},$$

there exists a contract such that, if the players all signed this contract, then they would get this expected payoff allocation. This set of possible expected payoff allocations is a closed and convex subset of $\mathbf{R}^N$.

However, not all such contracts could actually be signed by everyone in an equilibrium of the implicit contract-signing game. For example, in the game shown in Table 6.1, player 1 would refuse to sign a contract that committed the players to implement (x_1, y_2). Such a contract would give him a payoff of 0 if player 2 signed it, but player 1 could guarantee himself a payoff of 1 by signing nothing and choosing y_1.

For any player i, the *minimax value* (or *security level*) for player i in game Γ is

$$(6.1) \qquad v_i = \min_{\tau_{N-i} \in \Delta(C_{N-i})} \left(\max_{c_i \in C_i} \sum_{c_{N-i} \in C_{N-i}} \tau_{N-i}(c_{N-i}) u_i(c_{N-i}, c_i) \right).$$

That is, the minimax value for player i is the best expected payoff that player i could get against the worst (for him) correlated strategy that the other players could use against him. A *minimax strategy* against player i is any correlated strategy in $\Delta(C_{N-i})$ that achieves the minimum in (6.1). The theory of two-person zero-sum games (Section 3.8) tells us that the minimax value also satisfies

$$v_i = \max_{\tau_i \in \Delta(C_i)} \left(\min_{c_{N-i} \in C_{N-i}} \sum_{c_i \in C_i} \tau_i(c_i) u_i(c_{N-i}, c_i) \right).$$

So player i has a randomized strategy that achieves the above maximum and guarantees him an expected payoff that is not less than his minimax value, no matter what the other players may do.

We may say that a correlated strategy μ in $\Delta(C)$ for the set of all players is *individually rational* iff

$$(6.2) \qquad U_i(\mu) \geq v_i, \quad \forall i \in N.$$

The inequalities in condition (6.2) are called *participation constraints* or *individual-rationality constraints* on the correlated strategy μ.

Suppose now that players make their decisions about which contract to sign independently. For any such individually rational correlated strategy μ, there exists a contract τ with $\tau_N = \mu$ such that all players signing this contract is an equilibrium of the implicit contract-signing game. To prove this fact, for each player i, let τ_{N-i} be a minimax strategy against player i. Then no player could do better than his minimax value if he were the only player to not sign the contract. Thus, each player would want to sign if he expected all other players to sign. (If condition 6.2 is satisfied with all strict inequalities, then this equilibrium of the contract-signing game is also perfect, proper, and persistent.)

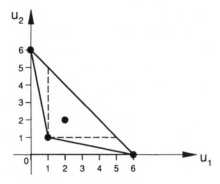

Figure 6.1

Conversely, no equilibrium of any contract-signing game can generate an expected payoff allocation in which some player gets strictly less than his minimax value, because he could then do better by signing nothing and using the strategy that guarantees him his minimax value. So

$$\{U(\mu) \mid \mu \in \Delta(C) \text{ and } U_i(\mu) \geq v_i, \quad \forall i \in N\}$$

is exactly the set of payoff allocations that can be achieved in equilibria of the implicit contract-signing game, when every player i has the option to sign nothing and choose an action in C_i. This set of all possible expected payoff allocations that are individually rational is also closed and convex.

For the game in Table 6.1, the set of possible expected payoff allocations is the closed convex set that has extreme points $(0,6)$, $(6,0)$, and $(1,1)$. (The allocation $(2,2)$ is in the interior of this set.) The set of possible expected payoff allocations satisfying individual rationality is the smaller triangle with extreme points (corners) at $(1,1)$, $(5,1)$, and $(1,5)$, as shown in Figure 6.1.

6.2 Correlated Equilibria

There are many situations in which players cannot commit themselves to binding contracts. Players' strategies might be unobservable to the legal enforcers of contracts; or sanctions to guarantee compliance with contracts might be inadequate; or some players' strategies might involve inalienable rights (such as the inalienable right of a worker to quit a job). In such situations, it might still be possible for the players to communicate and coordinate with each other. We say that a game is *with communication* if, in addition to the strategy options explicitly specified in the structure of the game, the players have a very wide range of implicit options to communicate with each other. Aumann (1974) showed that the solutions for games with communication may have remarkable properties, even without contracts.

Consider the example in Table 6.4. Without communication, there are three equilibria of this game: (x_1,x_2), which gives the payoff allocation (5,1); (y_1,y_2), which gives the payoff allocation (1,5); and a randomized equilibrium, which gives the expected payoff allocation (2.5,2.5). The best symmetric payoff allocation (4,4) cannot be achieved by the players without binding contracts, because (y_1,x_2) is not an equilibrium. However, even without contracts, communication can allow the players to achieve an expected payoff allocation that is better for both than (2.5,2.5). For example, the players might plan to toss a coin and choose (x_1,x_2) if Heads occurs or (y_1,y_2) if Tails occurs, and thus implement the correlated strategy $0.5[x_1,x_2] + 0.5[y_1,y_2]$. Even though the coin toss has no binding force on the players, such a plan is *self-enforcing*, in the sense that neither player could gain by unilaterally deviating from this plan.

With the help of a *mediator* (that is, a person or machine that can help the players communicate and share information), there is a self-enforcing plan that generates an even better expected payoff allocation $(3\frac{1}{3}, 3\frac{1}{3})$. To be specific, suppose that a mediator randomly recommends strategies to the two players in such a way that each of the pairs (x_1,x_2), (y_1,y_2), and (y_1,x_2) is recommended with probability $\frac{1}{3}$. Suppose also that each player learns only his own recommended strategy from the mediator. Then, even though the mediator's recommendation has no binding force, there is a Nash equilibrium of the transformed game with mediated communication in which both players plan to always obey

Table 6.4 A game in strategic form

C_1	C_2	
	x_2	y_2
x_1	5,1	0,0
y_1	4,4	1,5

the mediator's recommendations. If player 1 heard a recommendation "y_1" from the mediator, then he would think that player 2 was told to do x_2 or y_2 with equal probability, in which case his expected payoff from y_1 would be as good as from x_1 (2.5 from either strategy). If player 1 heard a recommendation "x_1" from the mediator, then he would know that player 2 was told to do x_2, in which case his best response is x_1. So player 1 would always be willing to obey the mediator if he expected player 2 to obey the mediator, and a similar argument applies to player 2. That is, the players can reach a self-enforcing understanding to each obey the mediator's recommendation when he plans to randomize in this way. Randomizing between (x_1,x_2), (y_1,y_2), and (y_1,x_2) with equal probability gives the expected payoff allocation

$$\tfrac{1}{3}(5,1) + \tfrac{1}{3}(4,4) + \tfrac{1}{3}(1,5) = (3\tfrac{1}{3}, 3\tfrac{1}{3}).$$

Notice that the implementation of this correlated strategy $\tfrac{1}{3}[x_1,x_2] + \tfrac{1}{3}[y_1,y_2] + \tfrac{1}{3}[y_1,x_2]$ without contracts required that each player get different partial information about the outcome of the mediator's randomization. If player 1 knew when player 2 was told to choose x_2, then player 1 would be unwilling to choose y_1 when it was also recommended to him. So this correlated strategy could not be implemented without some kind of mediation or noisy communication. With only direct unmediated communication in which all players observe anyone's statements or the outcomes of any randomizations, the only self-enforcing plans that the players could implement without contracts would be randomizations among the Nash equilibria of the original game (without communication), like the correlated strategy $.5[x_1,x_2] + .5[y_1,y_2]$ that we discussed earlier. (However, Barany, 1987, and Forges, 1990, have shown that, in any strategic-form and Bayesian game with four or more players, a system of direct unmediated communication between pairs of players can simulate any centralized communication system with a mediator, provided the communication between any pair of players is not directly observable by the other players.)

Consider now what the players could do if they had a bent coin for which player 1 thought that the probability of Heads was 0.9 while player 2 thought that the probability of Heads was 0.1, and these assessments were common knowledge. With this coin, it would be possible to give each player an expected payoff of 4.6 by the following self-enforcing plan: toss the coin and then implement the (x_1,x_2) equilibrium if Heads occurs, and implement the (y_1,y_2) equilibrium otherwise.

However, the players' beliefs about this coin would be *inconsistent*, in the sense of Section 2.8. That is, there is no way to define a prior probability distribution for the outcome of the coin toss and two other random variables such that player 1's beliefs are derived by Bayes's formula from the prior and his observation of one of the random variables, player 2's beliefs are derived by Bayes's formula from the prior and her observation of the other random variable, and their beliefs about the probability of Heads are common knowledge. (See Aumann, 1976, for a general proof of this fact.)

The existence of such a coin, about which the players have inconsistent beliefs, would be very remarkable and extraordinary. With such a coin, the players could make bets with each other that would have arbitrarily large positive expected monetary value to both! Thus, as a pragmatic convention, let us insist that the existence of any random variables about which the players may have such inconsistent beliefs should be explicitly listed in the structure of the game and should not be swept into the implicit meaning of the phrase "game with communication." (These random variables could be explicitly modeled either by a Bayesian game with beliefs that are not consistent with any common prior or by a game in generalized extensive-form, where a distinct subjective probability distribution for each player could be assigned to the set of alternatives at each chance node.) Similarly (but more obviously), any player's opportunities to observe random variables that affect payoffs must be explicitly modeled in the game. Thus, when we say that a particular game is played "with communication," we mean that players and mediators have wide but implicit opportunities to observe and exchange information about payoff-irrelevant random variables that are all sampled from objective probability distributions about which everyone agrees.

In general, for any finite strategic-form game $\Gamma = (N, (C_i)_{i \in N}, (u_i)_{i \in N})$, a mediator who was trying to help coordinate the players' actions would (at least) need to let each player i know which strategy in C_i was recommended for him. When the mediator can communicate separately and confidentially with each player, however, no player would have to be told the recommendations for any other players. Without contracts, player i would then be free to choose any strategy in C_i, after hearing the mediator's recommendation. So in the game with mediated communication, each player i would actually have an enlarged set of communication strategies that would include all mappings from C_i into C_i, each of which represents a possible rule for choosing an element of C_i as a function of the mediator's recommendation in C_i.

Now, suppose that it is common knowledge that the mediator will determine his recommendations according to the probability distribution μ in $\Delta(C)$; so $\mu(c)$ denotes the probability that any given pure-strategy profile $c = (c_i)_{i \in N}$ would be recommended by the mediator. Then it would be an equilibrium for all players to obey the mediator's recommendations iff

$$(6.3) \qquad U_i(\mu) \geq \sum_{c \in C} \mu(c) u_i(c_{-i}, \delta_i(c_i)), \quad \forall i \in N, \; \forall \delta_i : C_i \to C_i$$

(where $U_i(\mu)$ is as defined in Section 6.1). Following Aumann (1974, 1987a), we say that μ is a *correlated equilibrium* of Γ iff $\mu \in \Delta(C)$ and μ satisfies condition (6.3). That is, a correlated equilibrium is any correlated strategy for the players in Γ that could be self-enforcingly implemented with the help of a mediator who can make nonbinding confidential recommendations to each player.

It can be shown that condition (6.3) is equivalent to the following system of inequalities:

$$(6.4) \qquad \sum_{c_{-i} \in C_{-i}} \mu(c)(u_i(c) - u_i(c_{-i}, e_i)) \geq 0, \quad \forall i \in N, \; \forall c_i \in C_i, \; \forall e_i \in C_i.$$

(Here $c = (c_{-i}, c_i)$.) To interpret this inequality, notice that, given any c_i, dividing the left-hand side by $\sum_{c_{-i} \in C_{-i}} \mu(c)$, would make it equal to the difference between player i's conditionally expected payoff from obeying the mediator's recommendation and his conditionally expected payoff from using the action e_i, given that the mediator has recommended c_i, when all other players are expected to obey the mediator's recommendations. Thus, condition (6.4) asserts that no player i could expect to increase his expected payoff by using some disobedient action e_i after getting any recommendation c_i from the mediator. These inequalities (6.3) and (6.4) can be called *strategic incentive constraints*, because they represent the mathematical inequalities that a mediator's correlated strategy must satisfy to guarantee that all players could rationally obey his recommendations.

The set of correlated equilibria is a compact and convex set, for any finite game in strategic form. Furthermore, it can be characterized by a finite collection of linear inequalities, because a vector μ in $\mathbf{R}^N$ is a

correlated equilibrium iff it satisfies the strategic incentive constraints
(6.3 or 6.4) and the following *probability constraints*

$$\sum_{e \in C} \mu(e) = 1 \text{ and } \mu(c) \geq 0, \quad \forall c \in C.$$

Thus, for example, if we want to find the correlated equilibrium that
maximizes the sum of the players' expected payoffs in Γ, we have a
problem of maximizing a linear objective $(\Sigma_{i \in N} U_i(\mu))$ subject to linear
constraints. This is a linear programming problem, which can be solved
by any one of many widely available computer programs.

For the game in Table 6.4, the problem of finding the correlated
equilibrium that maximizes the expected sum of the players' payoffs
can be formulated as follows:

maximize $6\mu(x_1,x_2) + 0\mu(x_1,y_2) + 8\mu(y_1,x_2) + 6\mu(y_1,y_2)$

subject to

$(5 - 4)\mu(x_1,x_2) + (0 - 1)\mu(x_1,y_2) \geq 0,$

$(4 - 5)\mu(y_1,x_2) + (1 - 0)\mu(y_1,y_2) \geq 0,$

$(1 - 0)\mu(x_1,x_2) + (4 - 5)\mu(y_1,x_2) \geq 0,$

$(0 - 1)\mu(x_1,y_2) + (5 - 4)\mu(y_1,y_2) \geq 0,$

$\mu(x_1,x_2) + \mu(x_1,y_2) + \mu(y_1,x_2) + \mu(y_1,y_2) = 1,$

$\mu(x_1,x_2) \geq 0, \quad \mu(x_1,y_2) \geq 0, \quad \mu(y_1,x_2) \geq 0, \quad \text{and} \quad \mu(y_1,y_2) \geq 0.$

Using standard techniques of Lagrangean analysis (see Theorem 2 of
Myerson, 1985a) or the simplex algorithm for linear programming, we
can show that the optimal solution to this problem is the correlated
equilibrium where

$$\mu(x_1,x_2) = \mu(y_1,x_2) = \mu(y_1,y_2) = \frac{1}{3} \text{ and } \mu(x_1,y_2) = 0.$$

That is, among all correlated equilibria, $\frac{1}{3}[x_1,x_2] + \frac{1}{3}[y_1,y_2] +$
$\frac{1}{3}[y_1,x_2]$ maximizes the sum of the players' expected payoffs. So the
strategic incentive constraints imply that the players' expected sum of
payoffs cannot be higher than $3\frac{1}{3} + 3\frac{1}{3} = 6\frac{2}{3}$.

(This result can also be derived by elementary methods that exploit
the symmetry of the example. If $\hat{\mu}$ is any correlated equilibrium of this
game and $\mu(x_1,x_2) = \mu(y_1,y_2) = .5\hat{\mu}(x_1,x_2) + .5\hat{\mu}(y_1,y_2)$, $\mu(x_1,y_2) = \hat{\mu}(x_1,y_2)$,
and $\mu(y_1,x_2) = \hat{\mu}(y_1,x_2)$, then μ is also a correlated equilibrium and

generates the same expected sum of players' payoffs as $\hat{\mu}$ does. So there is no loss of generality in considering only correlated equilibria such that $\mu(x_1,x_2) = \mu(y_1,y_2)$. With this condition, the objective is to maximize $12\mu(y_1,y_2) + 8\mu(y_1,x_2)$, while the probability constraints and the second strategic incentive constraint imply that $1 - 2\mu(y_1,y_2) \geq \mu(y_1,x_2)$ and $\mu(y_1,y_2) \geq \mu(y_1,x_2)$. But the maximum of

$$12\mu(y_1,y_2) + 8\min\{\mu(y_1,y_2),\ 1 - 2\mu(y_1,y_2)\}$$

is $6\frac{2}{3}$, achieved when $\mu(y_1,y_2) = \frac{1}{3}$.)

It may be natural to ask why we have been focusing our attention on mediated communication systems in which it is rational for all players to obey the mediator. The reason for this focus is that such communication systems can simulate any equilibrium of any game that can be generated from any given strategic-form game by adding any communication system. To see why, let us try to formalize a general framework for describing communication systems that might be added to a given strategic-form game Γ as above. Given any communication system, let R_i denote the set of all strategies that player i could use to determine the reports that he sends out into the communication system, and let M_i denote the set of all messages that player i could receive from the communication system. For any $r = (r_i)_{i \in N}$ in $R = \times_{i \in N} R_i$ and any $m = (m_i)_{i \in N}$ in $M = \times_{i \in N} M_i$, let $v(m|r)$ denote the conditional probability that m would be the messages received by the various players if each player i were sending reports according to r_i. This function $v{:}R \rightarrow \Delta(M)$ is our basic mathematical characterization of the communication system. (If all communication is directly between players, without noise or mediation, then every player's message would be composed directly of other players' reports to him, and so $v(\cdot|r)$ would always put probability 1 on some vector m; but noisy communication or randomized mediation allows $0 < v(m|r) < 1$.)

Given such a communication system, the set of pure communication strategies that player i can use for determining the reports that he sends and the action in C_i that he ultimately implements (as a function of the messages he receives) is

$$B_i = \{(r_i,\delta_i)|\ r_i \in R_i,\ \delta_i{:}M_i \rightarrow C_i\}.$$

Player i's expected payoff depends on the communication strategies of all players according to the function $\overline{u}_i$, where

$$\bar{u}_i((r_j, \delta_j)_{j \in N}) = \sum_{m \in M} v(m \,|\, r) u_i((\delta_j(m_j))_{j \in N}).$$

Thus, the communication system $v : R \to \Delta(M)$ generates a *communication game* Γ_v, where

$$\Gamma_v = (N, (B_i)_{i \in N}, (\bar{u}_i)_{i \in N}).$$

This game Γ_v is the appropriate game in strategic form to describe the structure of decision-making and payoffs when the game Γ has been transformed by allowing the players to communicate through the communication system v before choosing their ultimate payoff-relevant actions. To characterize rational behavior by the players in the game with communication, we should look among the equilibria of Γ_v.

However, any equilibrium of Γ_v is equivalent to a correlated equilibrium of Γ as defined by the strategic incentive constraints (6.3). To see why, let $\sigma = (\sigma_i)_{i \in N}$ be any equilibrium in randomized strategies of this game Γ_v. Let μ be the correlated strategy defined by

$$\mu(c) = \sum_{(r,\delta) \in B} \sum_{m \in \delta^{-1}(c)} \left(\prod_{i \in N} \sigma_i(r_i, \delta_i) \right) v(m \,|\, r), \quad \forall c \in C,$$

where we use the notation:

$$B = \underset{i \in N}{\times} B_i; \quad (r, \delta) = ((r_i, \delta_i)_{i \in N}); \text{ and}$$

$$\delta^{-1}(c) = \{ m \in M \,|\, \delta_i(m_i) = c_i, \ \forall i \in N \}.$$

That is, the probability of any outcome c in C under the correlated strategy μ is just the probability that the players would ultimately choose this outcome after participating in the communication system v, when every player determines his plan for sending reports and choosing actions according to σ. So μ effectively simulates the outcome that results from the equilibrium σ in the communication game Γ_v. Because μ is just simulating the outcomes from using strategies σ in Γ_v, if some player i could have gained by disobeying the mediator's recommendations under μ, when all other players are expected to obey, then he also could have gained by similarly disobeying the recommendations of his own strategy σ_i when applied against σ_{-i} in Γ_v. More precisely, if condition (6.3) were violated for some i and δ_i, then player i could gain by switching from σ_i to $\hat{\sigma}_i$ against σ_{-i} in Γ_v, where

$$\hat{\sigma}_i(r_i, \gamma_i) = \sum_{\zeta_i \in Z(\delta_i, \gamma_i)} \sigma_i(r_i, \zeta_i), \quad \forall (r_i, \gamma_i) \in B_i, \text{ and}$$

$$Z(\delta_i, \gamma_i) = \{\zeta_i \mid \delta_i(\zeta_i(m_i)) = \gamma_i(m_i), \forall m_i \in M_i\}.$$

This conclusion would violate the assumption that σ is an equilibrium. So μ must satisfy the strategic incentive constraints (6.3), or else σ could not be an equilibrium of Γ_ν.

Thus, any equilibrium of any communication game that can be generated from a strategic-form game Γ by adding a system for preplay communication must be equivalent to a correlated equilibrium satisfying the strategic incentive constraints (6.3) or (6.4). This fact is known as the *revelation principle* for strategic-form games.

For any communication system ν, there may be many equilibria of the communication game Γ_ν, and these equilibria may be equivalent to different correlated equilibria. In particular, for any equilibrium $\bar{\sigma}$ of the original game Γ, there are equilibria of the communication game Γ_ν in which every player i chooses a strategy in C_i according to $\bar{\sigma}_i$, independently of the reports that he sends or the messages that he receives. (Such an equilibrium σ of Γ_ν can be constructed as follows: if $\delta_i(m_i) = c_i$ for all m_i in M_i, then $\sigma_i(r_i, \delta_i) = \bar{\sigma}_i(c_i)/|R_i|$ for every r_i in R_i; and if $\delta_i(m_i) \neq \delta_i(\hat{m}_i)$ for some m_i and $\hat{m}_i$, then $\sigma_i(r_i, \delta_i) = 0$.) That is, adding a communication system does not eliminate any of the equilibria of the original game, because there are always equilibria of the communication game in which reports and messages are treated as having no meaning and hence are ignored by all players. (Recall also the discussion of Table 3.9 in Chapter 3.) Such equilibria of the communication game are called *babbling equilibria*.

The set of correlated equilibria of a strategic-form game Γ has simple and tractable mathematical structure, because it is closed and convex and is characterized by a finite system of linear inequalities. On the other hand, the set of Nash equilibria of Γ, or of any specific communication game that can be generated from Γ, does not generally have any such simplicity of structure. So the set of correlated equilibria, which characterizes the union of the sets of equilibria of all communication games that can be generated from Γ, may be easier to analyze than the set of equilibria of any one of these games. This observation demonstrates the analytical power of the revelation principle. That is, the general conceptual approach of accounting for communication possibilities in the solution concept, rather than in the explicit game model, not only simplifies our game models but also generates solutions that are much easier to analyze.

6.3 Bayesian Games with Communication

The revelation principle for strategic-form games asserts that any equilibrium of any communication system can be simulated by a communication system in which the only communication is from a central mediator to the players, without any communication from the players to the mediator. The one-way nature of this communication should not be surprising, because the players have no private information to tell the mediator about, within the structure of the strategic-form game. More generally, however, players in Bayesian games might have private information about their types, and two-way communication would then allow the players' actions to depend on each other's types, as well as on extraneous random variables like coin tosses. Thus, in Bayesian games with communication, there might be a need for players to talk as well as to listen in mediated communication systems (see Forges, 1986).

Let $\Gamma^b = (N, (C_i)_{i \in N}, (T_i)_{i \in N}, (p_i)_{i \in N}, (u_i)_{i \in N})$, be a finite Bayesian game with incomplete information, as defined in Section 2.8. Let us suppose now that Γ^b is a game with communication; so the players have wide opportunities to communicate, after each player i learns his type in T_i but before he chooses his action in C_i.

Consider mediated communication systems of the following form: first, each player is asked to confidentially report his type to the mediator; then, after getting these reports, the mediator confidentially recommends an action to each player. The mediator's recommendations depend on the players' reports in either a deterministic or a random fashion. For any $c = (c_i)_{i \in N}$ in $C = \times_{i \in N} C_i$ and any $t = (t_i)_{i \in N}$ in $T = \times_{i \in N} T_i$, let $\mu(c|t)$ denote the conditional probability that the mediator would recommend to each player i that he should use action c_i, if each player j reported his type to be t_j. Obviously, these numbers $\mu(c|t)$ must satisfy the following *probability constraints*

$$(6.5) \qquad \sum_{c \in C} \mu(c|t) = 1 \quad \text{and} \quad \mu(d|t) \geq 0, \quad \forall d \in C, \ \forall t \in T.$$

In general, any such function $\mu: T \to \Delta(C)$ is called a *mediation plan* or *mechanism* for the game Γ^b with communication.

If every player reports his type honestly to the mediator and obeys the recommendations of the mediator, then the expected utility for type t_i of player i from the plan μ would be

$$U_i(\mu|t_i) = \sum_{t_{-i} \in T_{-i}} \sum_{c \in C} p_i(t_{-i}|t_i)\mu(c|t)u_i(c,t),$$

where $T_{-i} = \times_{j \in N-i} T_j$ and $t = (t_{-i},t_i)$.

We must recognize, however, that each player could lie about his type or disobey the mediator's recommendation. That is, we assume here that the players' types are not verifiable by the mediator and that the choice of an action in C_i can be controlled only by player i. Thus, a mediation plan μ induces a communication game Γ_μ^b in which each player must select his type report and his plan for choosing an action in C_i as a function of the mediator's recommendation. Formally Γ_μ^b is itself a Bayesian game, of the form

$$\Gamma_\mu^b = (N, (B_i)_{i \in N}, (T_i)_{i \in N}, (p_i)_{i \in N}, (\bar{u}_i)_{i \in N})$$

where, for each player i,

$$B_i = \{(s_i,\delta_i)|s_i \in T_i, \delta_i:C_i \to C_i\},$$

and $\bar{u}_i:(\times_{j \in N}B_j) \times T \to \mathbf{R}$ is defined by the equation

$$\bar{u}_i((s_j,\delta_j)_{j \in N},t) = \sum_{c \in C} \mu(c|(s_j)_{j \in N})u_i((\delta_j(c_j))_{j \in N},t).$$

A strategy (s_i,δ_i) in B_i represents a plan for player i to report s_i to the mediator and then to choose his action in C_i as a function of the mediator's recommendation according to δ_i; so he would choose $\delta_i(c_i)$ if the mediator recommended c_i. The action that player i chooses cannot depend on the type-reports or recommended actions of any other player, because each player communicates with the mediator separately and confidentially.

Suppose, for example, that the true type of player i were t_i but that he used the strategy (s_i,δ_i) in the communication game Γ_μ^b. If all other players were honest and obedient to the mediator, then i's expected utility payoff would be

$$U_i^*(\mu,\delta_i,s_i|t_i) = \sum_{t_{-i} \in T_{-i}} \sum_{c \in C} p_i(t_{-i}|t_i)\mu(c|t_{-i},s_i)u_i((c_{-i},\delta_i(c_i)),t).$$

(Here (t_{-i}, s_i) is the type profile in T that differs from $t = (t_{-i}, t_i)$ only in that the i-component is s_i instead of t_i. Similarly, $(c_{-i}, \delta_i(c_i))$ is the action profile in C that differs from c only in that the i-component is $\delta_i(c_i)$ instead of c_i.)

Bayesian equilibrium (defined in Section 3.9) is still an appropriate solution concept for a Bayesian game with communication, except that we must now consider the Bayesian equilibria of the induced communication game Γ_μ^b, rather than just the Bayesian equilibria of Γ. We say that a mediation plan μ is *incentive compatible* iff it is a Bayesian equilibrium for all players to report their types honestly and to obey the mediator's recommendations when he uses the mediation plan μ. Thus, μ is incentive compatible iff it satisfies the following general *incentive constraints*:

(6.6) $U_i(\mu \mid t_i) \geq U_i^*(\mu, \delta_i, s_i \mid t_i),$

$$\forall i \in N, \quad \forall t_i \in T_i, \quad \forall s_i \in T_i, \quad \forall \delta_i : C_i \to C_i.$$

If the mediator uses an incentive-compatible mediation plan and each player communicates independently and confidentially with the mediator, then no player could gain by being the only one to lie to the mediator or disobey his recommendations. Conversely, we cannot expect rational and intelligent players to all participate honestly and obediently in a mediation plan unless it is incentive compatible.

In general, there can be many different Bayesian equilibria of a communication game Γ_μ^b, even if μ is incentive compatible. Furthermore, as in the preceding section, we could consider more general communication systems, in which the reports that player i can send and the messages that player i might receive are respectively in some arbitrary sets R_i and M_i, not necessarily T_i and C_i. However, given any general communication system and any Bayesian equilibrium of the induced communication game, there exists an equivalent incentive-compatible mediation plan, in which every type of every player gets the same expected utility as he would get in the given Bayesian equilibrium of the induced communication game. In this sense, there is no loss of generality in assuming that the players communicate with each other through a mediator who first asks each player to reveal all of his private information and who then reveals to each player only the minimum information needed to guide his action, in such a way that no player has any incentive to lie or disobey. This result is the *revelation principle* for general Bayesian games.

The formal proof of the revelation principle for Bayesian games is almost the same as for strategic-form games. Given a general communication system $v:R \rightarrow \Delta(M)$ and communication strategy sets $(B_i)_{i \in N}$ as defined in Section 6.2, a Bayesian equilibrium of the induced communication game would then be a vector σ that specifies, for each i in N, each (r_i, δ_i) in B_i, and each t_i in T_i, a number $\sigma_i(r_i, \delta_i | t_i)$ that represents the probability that i would report r_i and choose his final action according to δ_i (as a function of the message that he receives) if his actual type were t_i. If σ is such a Bayesian equilibrium of the communication game Γ_v^b induced by the communication system v, then we can construct an equivalent incentive-compatible mediation plan μ by letting

$$\mu(c|t) = \sum_{(r,\delta) \in B} \sum_{m \in \delta^{-1}(c)} \left(\prod_{i \in N} \sigma_i(r_i, \delta_i | t_i) \right) v(m|r), \quad \forall c \in C, \quad \forall t \in T.$$

(See Myerson, 1982.)

This construction can be described more intuitively as follows. The mediator first asks each player to (simultaneously and confidentially) reveal his type. Next the mediator computes (or simulates) the reports that would have been sent by the players, with these revealed types, under the given equilibrium. He computes the recommendations or messages that would have been received by the players, as a function of these reports, under the given communication system or mechanism. Then he computes the actions that would have been chosen by the players, as a function of these messages and the revealed types in the given equilibrium. Finally, the mediator tells each player to do the action computed for him at the last step. Thus, the constructed mediation plan simulates the given equilibrium of the given communication system. To verify that this constructed mediation plan is incentive compatible, notice that any type of any player who could gain by lying to the mediator or disobeying his recommendations under the constructed mediation plan (when everyone else is honest and obedient) could also gain by similarly lying to himself before implementing his equilibrium strategy or disobeying his own recommendations to himself after implementing his equilibrium strategy in the given communication game; but the definition of a Bayesian equilibrium implies that such gains are not possible.

Thus, incentive-compatible mediation plans are the appropriate generalization of the concept of correlated equilibria for Bayesian games with communication. We can synonymously use the terms *communication equilibrium* or *generalized correlated equilibrium* to refer to any incentive-compatible mediation plan of a Bayesian game.

(Notice, however, that the set of incentive-compatible mediation plans is *not* equal to the set of correlated equilibria of the type-agent representation of Γ^b in strategic form, introduced in Section 2.8. The correlated equilibria of the type-agent representation have no clear interpretation in terms of the given Bayesian game with communication, unless one makes the unnatural assumption that a mediator can send each player a message that depends on his actual type while the players can send no reports to the mediator at all!)

Like the set of correlated equilibria, the set of incentive-compatible mediation plans is a closed convex set, characterized by a finite system of inequalities (6.5 and 6.6) that are linear in μ. On the other hand, it is generally a difficult problem to characterize the set of Bayesian equilibria of any given Bayesian game. Thus, by the revelation principle, it may be easier to characterize the set of all equilibria of all games that can be induced from Γ^b with communication than it is to compute the set of equilibria of Γ^b or of any one communication game induced from Γ^b.

For example, consider again the two-player Bayesian game from Table 3.13 in Chapter 3. Suppose that the two players can communicate, either directly or through some mediator, or via some tatonnement (groping) process, before they choose their actions in C_1 and C_2. In the induced communication game, could there ever be a Bayesian equilibrium giving the outcomes (x_1,x_2) if player 2 is type 2.1 and (y_1,y_2) if player 2 is type 2.2, as naive analysis of the two matrices in Table 3.13 would suggest? The answer is "No," by the revelation principle. If there were such a communication game, then there would be an incentive-compatible mediation plan achieving the same outcomes. But this would be the plan satisfying

$$\mu(x_1,x_2|1.0,2.1) = 1, \quad \mu(y_1,y_2|1.0,2.2) = 1,$$

which is not incentive compatible, because player 2 could gain by lying about his type. In fact, there is only one incentive-compatible mediation plan for this example, and it is $\overline{\mu}$, defined by

$$\overline{\mu}(x_1,x_2|1.0,2.1) = 1, \quad \overline{\mu}(x_1,y_2|1.0,2.2) = 1.$$

That is, this game has a unique communication equilibrium, which is equivalent to the unique Bayesian equilibrium of the game without communication that we found in Section 3.9. (This analysis assumes that player 2 cannot choose her action and show it verifiably to player 1 before he chooses his action. She can say whatever she likes to player 1 about her intended action before they actually choose, but there is nothing to prevent her from choosing an action different from the one she promised if she has an incentive to do so.)

In the insurance industry, the inability to get individuals to reveal unfavorable information about their chances of loss is known as adverse selection, and the inability to get fully insured individuals to exert efforts against their insured losses is known as moral hazard. This terminology can be naturally extended to more general game-theoretic models. The need to give players an incentive to report information honestly can be called *adverse selection*. The need to give players an incentive to implement recommended actions can be called *moral hazard*. (See Rasmusen, 1989, for a survey of economic games involving adverse selection and moral hazard.) In this sense, we can say that the general incentive constraints (6.6) are a general mathematical characterization of the effect of adverse selection and moral hazard in Bayesian games.

We can also ask how these incentive constraints might be relaxed if there is a regulator (or contract-enforcer) to eliminate moral hazard, or an auditor to eliminate adverse selection. The participation constraints (6.2) characterize what players in a strategic-form game can achieve when they bargain over contracts without adverse-selection or moral-hazard problems. The strategic incentive constraints (6.3) characterize what players can achieve with rational communication in a case where there is a moral-hazard problem but no adverse selection. The next case to consider is that of adverse selection without moral hazard, in Bayesian collective-choice problems.

6.4 Bayesian Collective-Choice Problems and Bayesian Bargaining Problems

A Bayesian collective-choice problem differs from a Bayesian game in that we are given a set of possible outcomes that are jointly feasible for the players together, rather than a set of actions or strategies for each player separately. That is, a *Bayesian collective-choice problem* is any Γ^c such that

$$\Gamma^c = (N, D, (T_i)_{i \in N}, (p_i)_{i \in N}, (u_i)_{i \in N}),$$

where D is the nonempty set of possible *outcomes* or *social options*. The other components $(N, (T_i)_{i \in N}, (p_i)_{i \in N}, (u_i)_{i \in N})$ of the Bayesian collective-choice problem have the same interpretation and the same mathematical structure as in the Bayesian game Γ^b (as defined in Section 2.8), except that each utility function u_i is now a function from $D \times T$ into $\mathbf{R}$ (where $T = \times_{i \in N} T_i$, as before).

A *collective-choice plan* or *mechanism* for Γ^c is any function $\mu : T \to \Delta(D)$. That is, for every choice d in D and every possible profile of types t in T, a mechanism μ specifies the probability $\mu(d|t)$ that d would be the chosen outcome if t were the combination of types reported by the players. So the components of μ must satisfy the probability constraints

$$\sum_{d \in D} \mu(d|t) = 1, \quad \mu(e|t) \geq 0, \quad \forall e \in D, \quad \forall t \in T.$$

The expected payoff for type t_i of player i if all players report honestly is

$$U_i(\mu|t_i) = \sum_{t_{-i} \in T_{-i}} \sum_{d \in D} p_i(t_{-i}|t_i)\mu(d|t)u_i(d,t);$$

and the expected payoff for type t_i of player i if he reports s_i while the other players are honest is

$$U_i^*(\mu,s_i|t_i) = \sum_{t_{-i} \in T_{-i}} \sum_{d \in D} p_i(t_{-i}|t_i)\mu(d|t_{-i},s_i)u_i(d,t).$$

The mechanism μ is *incentive compatible* iff honest reporting by all players is a Bayesian equilibrium of the game induced by μ. That is, μ is incentive compatible iff it satisfies the following *informational incentive constraints*:

(6.7) $U_i(\mu|t_i) \geq U_i^*(\mu,s_i|t_i), \quad \forall i \in N, \quad \forall t_i \in T_i, \quad \forall s_i \in T_i.$

These definitions are the same as those given in the preceding section, except that there is no longer any question of players disobeying recommended actions. With this simplification, the revelation principle applies equally to Bayesian collective-choice problems and to Bayesian games. That is, any equilibrium of reporting strategies in any game that could be generated from Γ^c by a rule for choosing outcomes as a function of players' reports must be equivalent to some incentive-compatible mechanism that satisfies condition (6.7).

The Bayesian collective-choice problem Γ^c *subsumes* a Bayesian game $\hat{\Gamma}^b$ iff $\hat{\Gamma}^b$ can be written in the form

$$\hat{\Gamma}^b = (N, (\hat{C}_i)_{i \in N}, (T_i)_{i \in N}, (p_i)_{i \in N}, (\hat{u}_i)_{i \in N}),$$

where (letting $\hat{C} = \times_{i \in N} \hat{C}_i$) there exists some function $g : \hat{C} \to D$ such that

$$\hat{u}_i(c,t) = u_i(g(c),t), \quad \forall i \in N, \quad \forall c \in \hat{C}, \quad \forall t \in T.$$

That is, Γ^c subsumes any Bayesian game that has the same sets of players and types as Γ^c, has the same probability functions as Γ^c, and has payoff functions that can be derived from the payoff functions in Γ^c by specifying an outcome in the collective-choice problem for every profile of actions in the Bayesian game. If $\hat{\mu}$ is an incentive-compatible mechanism for a Bayesian game $\hat{\Gamma}^b$ (in the sense of 6.6), $\hat{\Gamma}^b$ is subsumed by the Bayesian collective-choice problem Γ^c, and

$$\mu(d|t) = \sum_{c \in g^{-1}(d)} \hat{\mu}(c|t), \quad \forall d \in D, \quad \forall t \in T$$

(where $g^{-1}(d) = \{c \in C | g(c) = d\}$), then μ must be an incentive-compatible mechanism for Γ^c as well (in the sense of 6.7). In this sense, the incentive-compatible mechanisms of a Bayesian collective-choice problem Γ^c subsume or include the incentive-compatible mechanisms of all the Bayesian games that are subsumed by Γ^c, when outcomes can depend on players' actions in any arbitrary way.

For example, consider a bargaining game where player 1 is the seller and player 2 is a potential buyer of a single individual object. The object is worth \$10 to player 2, but it might be worth any amount between \$0 and \$9 to player 1. Player 1's type is the value of the object to him. Player 2 has no private information. Suppose that an outcome is a pair (k,y), where k is the number of months during which the players bargain before trading, and y is the price (in dollars) at which the object is ultimately sold. Let the cost of time be described by a *discount factor* of 0.99 per month. So the utility payoffs (in present discounted value) from an outcome (k,y) are

$$u_1(k,y,t) = (0.99^k)(y - t), \quad u_2(k,y,t) = (0.99^k)(10 - y),$$

where t is the dollar value of the object to player 1.

With this collective-choice structure, suppose that the players are involved in some bargaining game that has an equilibrium with the following structure: if the value of the object is \$0 to the seller, then trade occurs immediately (with no delay, $k = 0$) at price \$5, otherwise trade occurs after some delay but always at the "fair" price that is halfway between the seller's value and the buyer's value. Without knowing anything more about the rules of the bargaining game, we can tell how long it must take for the two players to trade when the seller's value is $9, because any equilibrium of the bargaining game must correspond to an incentive-compatible mechanism of the collective-choice problem that subsumes it.

So let $K(t)$ denote the number of months by which trading will be delayed in this mechanism if the value of the object to the seller is t. The price depends on the value of the object to the seller according to the formula $Y(t) = (10 + t)/2$ in this mechanism (because we are given that the price is always halfway between the seller's and buyer's values). The informational incentive constraints for the seller imply that

$$(0.99^{K(t)})\left(\frac{10 + t}{2} - t\right) = \max_{s \in [0,9]}(0.99^{K(s)})\left(\frac{10 + s}{2} - t\right), \quad \forall t \in [0,9].$$

So for any t in the interval $[0,9]$, if $K(\cdot)$ is differentiable, then

$$0 = \left(\frac{\partial}{\partial s}\right)\left((0.99^{K(s)})\left(\frac{10 + s}{2} - t\right)\right)\bigg|_{s=t}$$

$$= (0.99^{K(t)})(\log_e(0.99)(10 - t)K'(t) + 1)/2.$$

So $K'(t) = -1/((10 - t)\log_e(0.99))$. Given that $K(0) = 0$, this differential equation has a unique solution, which satisfies

$$0.99^{K(t)} = 1 - t/10.$$

For the highest value of $t = 9$, the delay in bargaining must be $K(9) = \log_e(0.10)/\log_e(0.99) = 229.1$. So the maximum delay before trading must be 229.1 months, or about 19 years, in any equilibrium of this game where the price always equalizes the buyer's and seller's gains from trade.

In many situations, a mechanism cannot be implemented without the prior consent or agreement of the players, so a feasible mechanism must also satisfy some participation constraints. To characterize such participation constraints, our model must specify what will happen if such agreement does not occur. The simplest way to do this is to specify one designated outcome $d*$ in D, called the *disagreement outcome*, that will occur if the players fail to agree on a mechanism. (More general models of disagreement are discussed in Section 6.6.) A Bayesian collective-choice problem together with specification of such a disagreement outcome $d*$ is called a *Bayesian bargaining problem*.

A mechanism μ for such a Bayesian bargaining problem is *individually rational* iff it satisfies the following *participation constraints*:

$$(6.8) \qquad U_i(\mu \,|\, t_i) \geq \sum_{t_{-i} \in T_{-i}} p_i(t_{-i} \,|\, t_i) u_i(d^*, t), \quad \forall i \in N, \quad \forall t_i \in T_i.$$

That is, each player, knowing his type, would agree to participate in the mechanism μ only if μ is at least as good for him as the disagreement outcome. There is no loss of generality in assuming that the players will agree on a mechanism that satisfies condition (6.8) for all types. Given any equilibrium of a mechanism-selection game in which some players would sometimes insist on the disagreement outcome, there is an equivalent individually rational mechanism that would choose the disagreement outcome whenever one or more players would insist on it in the given equilibrium of the mechanism-selection game. Thus, we say that a mechanism μ is *feasible* for a Bayesian bargaining problem iff it is incentive compatible and individually rational, in the sense of conditions (6.7) and (6.8).

It is common to choose utility payoff scales so that every player's payoff from the disagreement outcome is always 0, that is,

$$u_i(d^*, t) = 0, \quad \forall i \in N, \quad \forall t \in T.$$

With such payoff scales, the participation constraints (6.8) reduce to

$$U_i(\mu \,|\, t_i) \geq 0, \quad \forall i \in N, \quad \forall t_i \in T_i.$$

For a simple example, consider a trading situation involving one seller (player 1) and one potential buyer (player 2) of some divisible commodity. The seller has one unit available, and he knows its quality, which may be either good (type 1.a) or bad (type 1.b). If it is of good quality, then it is worth \$40 per unit to the seller and \$50 per unit to the buyer. If it is of bad quality, then it is worth \$20 to the seller and \$30 to the buyer. The buyer thinks that the probability of good quality is .2. Suppose that the buyer cannot verify any claims that the seller might make about the quality, and the seller cannot offer any quality-contingent warranties when he sells the object.

To formulate this example as a Bayesian collective-choice problem, we let $T_1 = \{1.a, 1.b\}$ and $T_2 = \{2.0\}$ (so the buyer's type can be ignored), and we represent the player's beliefs by letting

$$p_2(1.a \,|\, 2.0) = .2, \quad p_2(1.b \,|\, 2.0) = .8, \quad p_1(2.0 \,|\, t_1) = 1, \quad \forall t_1.$$

The set of possible outcomes is

$$D = \{(q,y) \mid 0 \le q \le 1, y \in \mathbf{R}\},$$

where, for each (q,y) in D, q represents the amount of the commodity that the seller delivers to the buyer and y represents the amount of money that the buyer pays to the seller. (For an indivisible object, q could be reinterpreted as the probability that the buyer gets the object.) The utility functions are

$$u_1((q,y), (1.a,2.0)) = y - 40q,$$

$$u_2((q,y), (1.a,2.0)) = 50q - y,$$

$$u_1((q,y), (1.b,2.0)) = y - 20q,$$

$$u_2((q,y), (1.b,2.0)) = 30q - y.$$

Because of the convexity of D and the linearity of the utility functions, we can restrict our attention to *deterministic trading mechanisms*, that is, functions from T to D, instead of functions from T to $\Delta(D)$. (Any mechanism $\mu{:}T \to \Delta(D)$ would be essentially equivalent to the deterministic mechanism $(Q,Y){:}T \to D$, where

$$Q(t) = \int_{(q,y)\in D} q \, d\mu(q,y\mid t), \quad Y(t) = \int_{(q,y)\in D} y \, d\mu(q,y\mid t).)$$

So let $(Q(\cdot),Y(\cdot))$ be a mechanism for determining the outcome as a function of the seller's type, where $Q(t)$ and $Y(t)$ denote respectively the expected quantity sold to the buyer and the expected cash payment from the buyer to the seller when the seller's type is t. Because the seller has only one unit of the commodity to sell, this mechanism must satisfy

$$0 \le Q(1.a) \le 1, \quad 0 \le Q(1.b) \le 1.$$

The expected utilities from this mechanism for each type of each player are

$$U_1(Q,Y \mid 1.a) = Y(1.a) - 40Q(1.a),$$

$$U_1(Q,Y \mid 1.b) = Y(1.b) - 20Q(1.b),$$

$$U_2(Q,Y \mid 2.0) = .2(50Q(1.a) - Y(1.a)) + .8(30Q(1.b) - Y(1.b)).$$

The informational incentive constraints are

$$Y(1.a) - 40Q(1.a) \ge Y(1.b) - 40Q(1.b),$$

$Y(1.b) - 20\,Q(1.b) \geq Y(1.a) - 20Q(1.a)$.

Suppose that each player has the right to not trade, so the disagreement outcome is $(q,y) = (0,0)$. Then the participation constraints are

$Y(1.a) - 40Q(1.a) \geq 0$,

$Y(1.b) - 20Q(1.b) \geq 0$,

$.2(50Q(1.a) - Y(1.a)) + .8(30Q(1.b) - Y(1.b)) \geq 0$.

The commodity is always worth more to the buyer than to the seller in this example, but there is no incentive-compatible and individually rational trading mechanism under which the buyer always gets all of the commodity. In any such mechanism, we would have $Q(1.a) = Q(1.b) = 1$; but then the seller's participation constraints and the informational incentive constraints would require $40 \leq Y(1.a) = Y(1.b)$, a result that would imply

$U_2(Q,Y|2.0) \leq .2 \times 50 + .8 \times 30 - 40 = -6 < 0$.

Thus, there must be a positive probability that the seller keeps some of the commodity in this example, as an unavoidable cost of the incentive constraints.

Many general conclusions about feasible trading mechanisms can be proved for this example, by analyzing the above constraints. From the informational incentive constraints we get

$40(Q(1.b) - Q(1.a)) \geq Y(1.b) - Y(1.a) \geq 20(Q(1.b) - Q(1.a))$,

so

$Q(1.b) \geq Q(1.a)$.

That is, the good-quality seller must sell less than the bad-quality seller, as a way of signaling his need for a higher price. The bad-quality seller cannot expect lower gains from trade than the good-quality seller expects, because

$U_1(Q,Y|1.b) - U_1(Q,Y|1.a)$

$= (Y(1.b) - 20Q(1.b)) - (Y(1.a) - 20Q(1.a)) + 20Q(1.a) \geq 0$.

Expected gains are bounded above by the inequality

(6.9) $0.3\,U_1(Q,Y|1.a) + 0.7\,U_1(Q,Y|1.b) + U_2(Q,Y|2.0) \leq 8$.

To prove this inequality, check that

$$3U_1(Q,Y|1.a) + 7U_1(Q,Y|1.b) + 10U_2(Q,Y|2.0)$$

$$= 80Q(1.b) + ((Y(1.a) - 20\,Q(1.a)) - (Y(1.b) - 20\,Q(1.b))$$

and recall that $Q(1.b) \leq 1$.

We say that an incentive-compatible mechanism is *incentive efficient* iff there is no other incentive-compatible mechanism that would give a higher expected payoff to at least one type of one player without giving a lower expected payoff to some other type of some player. Thus, for this example, an incentive-compatible mechanism must be incentive efficient if

$$0.3U_1(Q,Y|1.a) + 0.7U_1(Q,Y|1.b) + U_2(Q,Y|2.0) = 8.$$

For a specific example of a feasible trading mechanism, suppose that we want to stipulate that the price of the commodity should be \$25 per unit if it is of bad quality and \$45 per unit if it is of good quality. (These numbers are the averages of the values to the buyer and seller in each case.) If everything is sold in the bad-quality case, then we have

$$Q(1.b) = 1, \quad Y(1.b) = 25,$$

and $Y(1.a) = 45Q(1.a)$. For incentive-compatibility, then, $25 - 20 \geq (45 - 20)Q(1.a)$, so $Q(1.a) \leq 0.2$. So let us complete the trading mechanism by specifying

$$Q(1.a) = 0.2, \quad Y(1.b) = 45 \times 0.2 = 9.$$

This mechanism is feasible (incentive compatible and individually rational), and it gives expected payoffs

$$U_1Q,Y|1.a) = 1, \quad U_1(Q,Y|1.b) = 5, \quad U_2(Q,Y|2.0) = 4.2.$$

Notice that $0.3 \times 1 + 0.7 \times 5 + 4.2 = 8$, so condition (6.9) is satisfied with equality. Thus, even though most of the good-quality commodity would not be sold under this trading mechanism, it is incentive efficient. That is, there is no feasible mechanism that would increase the expected payoff for the buyer and both types of the seller.

It may be somewhat upsetting to discover that there must be a positive probability that the seller keeps some of the commodity in this example. After all, if both players know that the commodity is always worth more to the buyer than to the seller, why do they not keep bargaining as long as the seller has anything to sell? To understand the answer to this question, we must consider a more elaborate model, in which trade can occur at different points in time and there is some cost of waiting to trade. (If there were no cost of waiting, then there would be no incentive for the players to ever stop bargaining at any point in time and actually trade.) In such a model, it is possible to create mechanisms in which the buyer is sure to buy all of the seller's commodity eventually, but *costs of delay of trade* will replace the *costs of failure to (fully) trade* in these mechanisms. For example, if the players use a discount factor of 0.99 per month to discount future profits, then there is an incentive-compatible mechanism for the above example in which the seller sells his entire supply for $25 immediately if his type is 1.b, or he sells his entire supply for $45 after a delay of 160.14 months if his type is 1.a. Notice that $0.99^{160.14} = 0.2$, so

$$(25 - 20) = 5 \geq (0.99^{160.14})(45 - 20),$$

$$(0.99^{160.14})(45 - 40) = 1,$$

$$.8(30 - 25) + .2(0.99^{160.14})(50 - 45) = 4.2;$$

so this mechanism gives the same allocation of expected (present-discounted) utility payoffs as the feasible trading mechanism without delay discussed above. In general, if the players use a common discount factor to measure costs of delay, then the set of expected utility allocations that are achievable by incentive-compatible mechanisms cannot be enlarged by allowing delay of trade. This result is proved and discussed further in Section 10.4 of Chapter 10.

6.5 Trading Problems with Linear Utility

A powerful general characterization of incentive-compatible mechanisms can be derived when each player's type set is an interval of the real number line, each player's utility is linear in his own type, and types are independent. So let us consider a Bayesian collective-choice problem or Bayesian bargaining problem in which the set of players is $N = \{1, 2, \ldots, n\}$, and each player i has a type that is a real number in the interval between two given parameters $L(i)$ and $M(i)$, so

$$T_i = [L(i), M(i)].$$

For each i, let f_i be a continuous positive probability density function on the interval $[L(i),M(i)]$, and suppose that every other player j thinks that player i's type is a random variable that has probability density f_i and is independent of all other players' types. Let the set of social choice options D be a convex subset of $\mathbf{R}^N \times \mathbf{R}^N$, with an element of D denoted

$$(q,y) = ((q_i)_{i \in N}, (y_i)_{i \in N}) = (q_1, \ldots, q_n, y_1, \ldots, y_n).$$

In such an outcome (q,y), q_i may be interpreted as the net quantity of some commodity that is delivered to player i, and y_i may be interpreted as the net monetary payment that player i makes to others. Then suppose that each player has utility that is linear in his trades q_i and y_i, and the type of player i is his monetary value for a unit of the commodity, so

$$u_i(q,y,t) = q_i t_i - y_i, \quad \forall i \in N, \quad \forall t \in T, \quad \forall(q,y) \in D.$$

So we are assuming, in this section, that the players have *independent private values* for the commodity being traded.

Because utility is linear, we can again restrict our attention to deterministic mechanisms, mapping from T into D, without loss of generality. So let us denote a mechanism by a pair (Q,Y) such that, for any t in T,

$$(Q,Y)(t) = (Q(t),Y(t)) = (Q_1(t), \ldots, Q_n(t), Y_1(t), \ldots, Y_n(t)) \in D.$$

Any random mechanism $\mu:T \to \Delta(D)$ would be essentially equivalent to the deterministic mechanism (Q,Y) where, for every i and t,

$$Q_i(t) = \int_{(q,y) \in D} q_i \, d\mu(q,y|t), \quad Y_i(t) = \int_{(q,y) \in D} y_i \, d\mu(q,y|t).$$

That is, $Q(t)$ and $Y(t)$ are respectively the expected quantity of the commodity delivered to i and i's expected net monetary payment to others, when t is the profile of reported types. Whether honest or lying, each player's expected payoff under the random mechanism μ could be expressed in terms of (Q,Y), so it suffices to consider only (Q,Y).

Given any mechanism $(Q,Y):T \to D$ and any type t_i of player i, let

$$\overline{Q}_i(t_i) = \int_{t_{-i} \in T_{-i}} Q_i(t) \left(\prod_{j \in N-i} f_j(t_j) \right) dt_{-i},$$

$$\overline{Y}_i(t_i) = \int_{t_{-i} \in T_{-i}} Y_i(t) \left(\prod_{j \in N-i} f_j(t_j) \right) dt_{-i}.$$

(Here $t = (t_{-i}, t_i)$, as usual.) That is, $\overline{Q}_i(t_i)$ and $\overline{Y}_i(t_i)$ respectively denote the expected quantity of the commodity to be delivered to player i and the expected monetary payment to be made by player i, when he reports his type to be t_i and all other players report their types honestly to the mechanism (Q,Y). So if everyone participates honestly, then the expected utility payoff to type t_i of player i under mechanism (Q,Y) would be

$$U_i(Q,Y|t_i) = \overline{Q}_i(t_i)t_i - \overline{Y}_i(t_i).$$

On the other hand, if player i's type was t_i but he dishonestly reported some other type s_i, while everyone else followed the strategy of participating honestly, then the expected payoff to player i would be

$$U_i^*(Q,Y,s_i|t_i) = \overline{Q}_i(s_i)t_i - \overline{Y}_i(s_i).$$

So the informational incentive constraints that an incentive-compatible mechanism (Q,Y) must satisfy are

$$U_i(Q,Y|t_i) = \overline{Q}_i(t_i)t_i - \overline{Y}_i(t_i) \geq \overline{Q}_i(s_i)t_i - \overline{Y}_i(s_i),$$

$$\forall i \in N, \quad \forall t_i \in T_i, \quad \forall s_i \in T_i.$$

Rewriting these constraints with the roles of t_i and s_i reversed, we can easily show that they imply

$$\overline{Q}_i(t_i)(t_i - s_i) \geq U_i(Q,Y|t_i) - U_i(Q,Y|s_i) \geq \overline{Q}_i(s_i)(t_i - s_i).$$

These inequalities imply that $\overline{Q}(t_i)$ is a nondecreasing (and so Riemann integrable) function of t_i and that

$$(6.10) \qquad U_i(Q,Y|t_i) = U_i(Q,Y|\theta_i) + \int_{\theta_i}^{t_i} \overline{Q}_i(s_i)ds_i,$$

$$\forall i \in N, \quad \forall t_i \in T_i, \quad \forall \theta_i \in T_i.$$

Then, before player i's type t_i is determined, the expected value of $\overline{Y}_i(t_i)$ is

$$(6.11) \qquad \int_{L(i)}^{M(i)} \overline{Y}_i(t_i)f_i(t_i)dt_i$$

$$= \int_{L(i)}^{M(i)} (\overline{Q}_i(t_i)t_i - U_i(Q,Y|t_i))f_i(t_i)dt_i$$

$$= \int_{L(i)}^{M(i)} (\overline{Q}_i(t_i)t_i - U_i(Q,Y \mid \theta_i) - \int_{\theta_i}^{t_i} \overline{Q}_i(s_i)ds_i)f_i(t_i)dt_i$$

$$= \int_{L(i)}^{M(i)} \overline{Q}_i(t_i)\psi_i(t_i,\theta_i)f_i(t_i)dt_i - U_i(Q,Y \mid \theta_i)$$

$$= \int_{L(n)}^{M(n)} \cdots \int_{L(1)}^{M(1)} Q_i(t)\psi_i(t_i,\theta_i)\left(\prod_{j \in N} f_j(t_j)\right) dt_1 \cdots dt_n - U_i(Q,Y \mid \theta_i)$$

where

$$\psi_i(t_i,\theta_i) = t_i - \frac{\displaystyle\int_{t_i}^{M(i)} f_i(s_i)ds_i}{f_i(t_i)} \quad \text{if } t_i > \theta_i,$$

$$\psi_i(t_i,\theta_i) = t_i + \frac{\displaystyle\int_{L(i)}^{t_i} f_i(s_i)ds_i}{f_i(t_i)} \quad \text{if } t_i < \theta_i.$$

Equations (6.10) and (6.11) have many important applications. The rest of this section is devoted to a short survey of some simple examples of particular economic interest, to illustrate the power of these equations.

For example, suppose that the players are bidders in an auction for a single individual object and that the type of each bidder is his value for the object. In this case, we may reinterpret $Q_i(t)$ as the probability that player i will get the object (and thus still as his expected quantity of objects received) when the reported type profile is t. For each player i, suppose that the minimum possible value $L(i)$ is 0, and suppose that a player whose true value for the object is 0 will not pay anything in the auction, so

$$U_i(Q,Y \mid 0) = 0.$$

(That is, the individual-rationality constraint is satisfied with equality for the lowest possible type, 0.) Then (6.11) with

$$\theta_i = L(i) = 0$$

implies that the total expected revenue to the auctioneer is

(6.12) $\quad \displaystyle\sum_{i=1}^{n} \int_{0}^{M(i)} \overline{Y}_i(t_i)f_i(t_i)dt_i$

$$= \int_0^{M(n)} \cdots \int_0^{M(1)} \left(\sum_{i=1}^n Q_i(t) \psi_i(t_i, 0) \right) \left(\prod_{j \in N} f_j(t_j) \right) dt_1 \ldots dt_n.$$

This equation implies that the auctioneer's expected revenue depends only on the Q functions, which describe how the probability of each player getting the object depends on the players' (reported) types. For example, any mechanism that always delivers the object to the bidder whose true value for the object is highest must have

$$Q_i(t) = 1 \quad \text{if} \quad \{i\} = \underset{j \in N}{\text{argmax}} \, t_j,$$

$$Q_i(t) = 0 \quad \text{if} \quad i \notin \underset{j \in N}{\text{argmax}} \, t_j;$$

so all such mechanisms must give the same expected revenue to the seller. (The set of type profiles where two or more players are in a tie for the highest type has Lebesque measure 0, so the integral in equation (6.12) does not depend on who gets the object in case of ties.) Notice that, for any equilibrium of any auction game such that (1) the object is always delivered to the highest bidder and (2) there is an increasing function $b: \mathbf{R} \to \mathbf{R}$ such that each bidder in equilibrium will submit a bid $b(t_i)$ if his value is t_i, the equivalent incentive-compatible mechanism (under the revelation principle) is one in which the object is always delivered to the player with the highest true value for the object. Thus, all such equilibria of all such auction games must have the same expected revenue to the auctioneer. This result explains many of the *revenue-equivalence* results for equilibria of various auction games (first-price auctions, second-price auctions, etc.) that have been known since Ortega-Reichert (1968) (see Milgrom, 1987, or McAfee and McMillan, 1987).

Suppose now that we want to find an incentive-compatible mechanism that maximizes the expected revenue to the auctioneer. Suppose we can find an incentive-compatible individually rational mechanism under which, for any type profile t, the object is always delivered to a player for whom the quantity $\psi_i(t_i, 0)$ is maximal, provided this maximum is nonnegative, and otherwise the auctioneer keeps the object. Then, under this mechanism, the probabilities $(Q_i(t))_{i \in N}$ maximize

$$\sum_{i=1}^n Q(t) \, \psi_i(t_i, 0)$$

subject to

$$\sum_{i=1}^{n} Q_i(t) \le 1, \quad Q_j(t) \ge 0, \quad \forall j,$$

for every t, and so this mechanism maximizes the auctioneer's expected revenue among all possible incentive-compatible individually rational mechanisms.

For example, consider the case where there are two bidders, player 1's type is drawn from the uniform distribution on $[0,1]$, and player 2's type is drawn from the uniform distribution on $[0,2]$. Then

$$\psi_1(t_1,0) = t_1 - (1 - t_1) = 2t_1 - 1$$

$$\psi_2(t_2,0) = t_2 - (2 - t_2) = 2t_2 - 2.$$

We want to deliver the object to player 1 if $2t_1 - 1 > \max\{2t_2 - 2, 0\}$, that is, if

$$t_1 \ge \max\{t_2 - 0.5, 0.5\}.$$

Similarly, we want to deliver the object to player 2 if $2t_2 - 2 > \max\{2t_1 - 1, 0\}$, that is, if

$$t_2 \ge \max\{t_1 + 0.5, 1\}.$$

If $t_1 < 0.5$ and $t_2 < 1$, then we want the auctioneer to keep the object. One way to implement this delivery plan incentive compatibly is to specify that, given the type reports of the two players, the player (if any) who gets the object under this plan will pay the infimum of all reports that he could have made and still gotten the object, given the report of the other player. For example, if the types are $t_1 = 0.8$ and $t_2 = 0.9$, then player 1 gets the object and pays 0.5, even though player 2 actually has the higher value for the object. This mechanism maximizes the auctioneer's expected revenue among all incentive-compatible individually rational mechanisms, even though it may sometimes deliver the object to the player with the lower value for it, or may even not deliver the object to any bidder at all. The discrimination against player 2 when her reported value exceeds player 1's reported value by less than 0.5 serves in this optimal auction mechanism to encourage player 2 to honestly report higher types that may have to pay as much as 1.5 for the object. For more on optimal auctions, see Myerson (1981a).

Equations (6.10) and (6.11) can also be applied to bilateral trading problems. Suppose that player 1 is the seller and player 2 is the only potential buyer of a single indivisible object. Because player 1 has the only object and any transfer from one player is necessarily a transfer to the other player in bilateral trading, the set of possible outcomes is

$$D = \{(q_1,q_2,y_1,y_2) \mid q_2 = -q_1 \in [0,1] \text{ and } y_2 = -y_1 \in \mathbf{R}\}.$$

So in any mechanism (Q,Y), we must have

$$Q_2(t) = -Q_1(t) \in [0,1] \text{ and } Y_2(t) = -Y_1(t), \quad \forall t.$$

For each player i, let θ_i be the type least willing to trade, so

$$\theta_1 = M(1), \quad \theta_2 = L(2).$$

Then equation (6.11) implies

(6.13)
$$
\begin{aligned}
0 &= \int_{L(2)}^{M(2)} \int_{L(1)}^{M(1)} (Y_2(t) + Y_1(t))f_1(t_1)f_2(t_2)dt_1dt_2 \\
&= \int_{L(2)}^{M(2)} \overline{Y}_2(t_2)f_2(t_2)dt_2 + \int_{L(1)}^{M(1)} \overline{Y}_1(t_1)f_1(t_1)dt_1 \\
&= \int_{L(2)}^{M(2)} \int_{L(1)}^{M(1)} Q_2(t)\big(\psi_2(t_2,L(2)) - \psi_1(t_1,M(1))\big)f_1(t_1)f_2(t_2)dt_1dt_2 \\
&\quad - U_1(Q,Y\mid M(1)) - U_2(Q,Y\mid L(2))
\end{aligned}
$$

This and other related results have been studied by Myerson and Satterthwaite (1983).

For example, suppose that the seller and buyer both have values for the object that are independent random variables drawn from the interval [0,1]. (This example was first considered by Chatterjee and Samuelson, 1983.) Then

$$\psi_1(t_1,M(1)) = \psi_1(t_1,1) = t_1 + t_1 = 2t_1,$$

$$\psi_2(t_2,L(2)) = \psi_2(t_2,0) = t_2 - (1 - t_2) = 2t_2 - 1.$$

Substituting these formulas into (6.13), with $f_i(t_i) = 1$ for the uniform distribution, we see that any incentive-compatible mechanism (Q,Y) must satisfy

$$\int_0^1 \int_0^1 2(t_2 - t_1 - \tfrac{1}{2})Q_2(t_1,t_2)dt_1dt_2 = U_1(Q,Y\mid 1) + U_2(Q,Y\mid 0).$$

When players have the right to not trade, the participation constraints imply that each $U_i(Q,Y|t_i)$ must be nonnegative for any individually rational mechanism. So this equation implies that, conditionally on trade occurring in an incentive-compatible individually rational mechanism, the expected difference between the buyer's value and the seller's value must be at least ½.

A trading mechanism is *ex post efficient* if it always allocates all of the commodities being traded to the individuals who value them most highly. That is, for ex post efficiency in a bilateral trading problem with one object, the buyer should get the object whenever it is worth more to her than to the seller.

For this bilateral trading example in which the player's private values are independently drawn from the uniform distribution on the interval [0,1], conditionally on the buyer's value being higher than the seller's, the expected difference between the two values would be only ⅓. (It can be shown by calculus that

$$\int_0^1 \int_0^{t_2} (t_2 - t_1 - \tfrac{1}{3})dt_1 dt_2 = 0.)$$

Thus, because ⅓ < ½, there is no ex post efficient incentive-compatible individually rational mechanism for this example. By the revelation principle, any equilibrium of any trading game must be equivalent to some incentive-compatible individually rational mechanism. So when the traders can lie about their types and can choose to not trade, there is no way to design a trading game for this example such that, in equilibrium, trade occurs whenever the object is actually worth more to the buyer than to the seller. This impossibility result has been proved for more general distributions by Myerson and Satterthwaite (1983).

It may be helpful to see how this failure of ex post efficiency occurs in the context of a specific trading game. So suppose that the players simultaneously submit bids, and the buyer gets the object for a price equal to the average of their bids if her bid is higher than the seller's, but otherwise no trade occurs. Chatterjee and Samuelson (1983) showed that this game has an equilibrium in which the seller bids $\min\{2t_1/3 + \tfrac{1}{4}, t_1\}$ and the buyer bids $\max\{2t_2/3 + \tfrac{1}{12}, t_2\}$. This equilibrium of this game is equivalent to the incentive-compatible individually rational mechanism

$$Q_2(t) = 1 \text{ if } 2t_2/3 + \tfrac{1}{12} > 2t_1/3 + \tfrac{1}{4}, \text{ that is, } t_2 > t_1 + \tfrac{1}{4},$$

$Q_2(t) = 0$ if $t_2 < t_1 + \frac{1}{4}$,

$Y_2(t) = Q_2(t)(2t_2/3 + \frac{1}{12} + 2t_1/3 + \frac{1}{4})/2$

$\quad = Q_2(t)(t_1 + t_2 + 0.5)/3$.

So trade will not occur if the buyer's value exceeds the seller's value by less than $\frac{1}{4}$. This differential is needed because, with any given value for the object, the seller would submit a higher bid than the buyer would submit with the same value, because of the seller's incentive to try to increase the price and the buyer's incentive to try to decrease the price. Leininger, Linhart and Radner (1989) have shown that this game has other equilibria, which are equivalent to other incentive-compatible individually rational mechanisms, but none can guarantee that the buyer gets the object whenever it is worth more to her.

There are a variety of factors that can improve the prospects for achieving ex post efficiency in other situations. We consider here two such factors: large numbers of traders, and uncertainty about the direction of trade.

Gresik and Satterthwaite (1989) have studied a general class of models with many buyers and sellers, who have independent private values for the commodity being traded. They have shown that incentive-compatible individually rational mechanisms can be constructed that are arbitrarily close (in a certain technical sense) to ex post efficiency if the number of buyers and sellers is sufficiently large. This result suggests that large Bayesian trading problems where the traders have independent private values may be well approximated by the classical general equilibrium model of economic theory, in which ex post efficiency is always achieved. To see how this might be so, suppose that there are N_1 sellers, N_2 buyers, each seller has an initial endowment of one unit of the commodity, each trader can consume at most one unit of the commodity, and the traders have independent private values for the commodity that are each drawn from the uniform distribution on the interval [0,1]. Consider a mechanism in which buyers and sellers can only trade at some prespecified price x. By the law of large numbers, if N_1 and N_2 are large, then, with probability close to 1, the fraction of the N_1 sellers who are willing to sell at this price is close to x, and the fraction of the N_2 buyers who are willing to buy at this price is close to $(1 - x)$. So let us choose x such that $N_1 x = N_2(1 - x)$, and match for trading as many buyer–seller pairs as we can among those who offer to trade at this price. Any mutually beneficial trade that could occur after the operation of this mechanism would necessarily involve a player whose offer to trade at price x was turned down (due to a shortage of players offering to trade from the other side of the market). If N_1 and N_2 are large, then, with probability almost 1, the number of such traders will be a very small fraction of the total number of traders.

Cramton, Gibbons, and Klemperer (1987) have shown that uncertainty about the direction of trade (that is, uncertainty about whether any given player will be a net buyer or a net seller of the commodity) can help make ex post efficient mechanisms feasible. For example, suppose as before that players 1 and 2 are the only traders, each player has linear utility for money and the commodity, and their private values for a unit of the commodity are independent random variables drawn from the uniform distribution on the interval [0,1]. Now, however, suppose that each player has an initial endowment of one unit of a given commodity. So everything is the same as in the preceding example, except that now the set of possible outcomes is

$$D = \{(q_1, q_2, y_1, y_2) | q_2 = -q_1 \in [-1,1] \text{ and } y_2 = -y_1 \in \mathbf{R}\}.$$

There are incentive-compatible individually rational mechanisms for this example that guarantee that the player who has the higher private value for the commodity will buy all of the other player's supply. For instance, consider the game where the players simultaneously submit sealed bids, and the player with the higher bid buys the other's supply for the average of their two bids. This game has an equilibrium in which each player i submits a bid of $2t_i/3 + \frac{1}{6}$ when his true value is t_i. So this equilibrium is equivalent to the mechanism (Q,Y) where

$$-Q_1(t) = Q_2(t) = 1 \text{ if } 2t_2/3 + \frac{1}{6} > 2t_1/3 + \frac{1}{6}, \text{ that is, } t_2 > t_1,$$

$$-Q_1(t) = Q_2(t) = -1 \text{ if } t_2 < t_1,$$

$$-Y_1(t) = Y_2(t) = Q_2(t)(2t_2/3 + \frac{1}{6} + 2t_1/3 + \frac{1}{6})/2$$
$$= Q_2(t)(t_1 + t_2 + \frac{1}{2})/3,$$

which is incentive compatible, individually rational, and ex post efficient.

In this equilibrium, each player may bid either higher or lower than his true value, depending on whether it is lower or higher than $\frac{1}{2}$, that is, depending on whether he is more likely to sell or buy in the ultimate transaction. In the terminology of Lewis and Sappington (1989), a player's uncertainty about whether he will ultimately buy or sell creates *countervailing incentives* that may reduce his incentive to lie in trading games and so may decrease the cost of satisfying informational incentive constraints.

6.6 General Participation Constraints for Bayesian Games with Contracts

Any given Bayesian game $\Gamma^b = (N, (C_i)_{i \in N}, (T_i)_{i \in N}, (p_i)_{i \in N}, (u_i)_{i \in N})$ can be subsumed by a Bayesian collective-choice problem Γ^c as defined in Section 6.4, when the set of outcomes is identified with the set of profiles of actions, that is,

$$D = C = \underset{i \in N}{\times} C_i .$$

However, the set of incentive-compatible mechanisms of Γ^c (in the sense of condition 6.7) is larger than the set of incentive-compatible mechanisms of Γ^b (in the sense of condition 6.6), because we ignore moral hazard in the Bayesian collective-choice problem. In effect, the general incentive constraints (6.6) that define incentive compatibility for Bayesian games are based on an assumption that each player i has an inalienable right to control his action in C_i himself. Let us now investigate what happens when we change this assumption and suppose instead that this right is alienable. That is, player i can choose to sign a contract that delegates control of his action in C_i to some agent or regulator, who will exercise control according to the mechanism specified by the contract. However, player i also has the option to refuse to sign any such contract and control his action himself.

If player i refused to participate in a mechanism μ, then the other players' actions could still depend on each others' types according to some plan $\tau_{-i} : T_{-i} \to \Delta(C_{-i})$ that may be specified by the contract for this contingency. Of course, without any report from player i, this plan cannot depend on i's actual type in T_i. On the other hand, each player already knows his type at the beginning of the game, when he makes his participation decision. Thus, if it would be an equilibrium for all players to sign a contract that would implement the mechanism $\mu : T \to \Delta(C)$ when everyone signs, then μ must satisfy the following general *participation constraints*:

(6.14) $\forall i \in N$, $\exists \tau_{-i} : T_{-i} \to \Delta(C_{-i})$ such that, $\forall t_i \in T_i$,

$$U_i(\mu | t_i) \geq \max_{c_i \in C_i} \sum_{t_{-i} \in T_{-i}} \sum_{c_{-i} \in C_{-i}} p_i(t_{-i} | t_i) \tau_{-i}(c_{-i} | t_{-i}) u_i(c,t).$$

(Here "$\exists$" means "there exists.") We may say that μ is *individually rational* iff it satisfies these participation constraints. Notice that we are assuming that each player already knows his type when he decides whether to participate in the mechanism.

Conversely, suppose that a mechanism $\mu : T \to \Delta(C)$ satisfies the participation constraints (6.14) and the informational incentive constraints (6.7). Suppose that, when a player signs a delegation contract, he must make a confidential report about his type (to be used only in implementing the contract); and suppose that each player must make his signing decision and his report simultaneously with and independently of every other player. (Perhaps each player sits in a separate room with a copy of the contract and a type-report form.) Finally, suppose that the contract specifies that $\mu : T \to \Delta(C)$ would be implemented if everyone signs it and, for each player i, the threat τ_{-i} that satisfies (6.14) would be implemented if i were the only player who did not sign. Then it would be an equilibrium for every type of every player to sign the contract and report honestly. Under this equilibrium, because each player is sure that all other players are signing when he signs the delegation contract and reports his type, potential gains from lying or not participating under the terms of some τ_{-i} would not affect his incentives when he actually signs the contract and makes his report.

There is really no loss of generality in assuming that the players will all sign a contract to jointly regulate their actions and then make reports about their types. For example, suppose that, in some equilibrium of a contract-signing game, there is a positive probability that some type t_i of some player i would refuse to sign the delegation contract that everyone else is signing. Then this type t_i of player i would be willing to sign a revised version of the contract that specified that, if after signing he reported this type, then the other players would do whatever the original contract specified for them when he did not sign, and that he must do whatever he would have wanted to do after not signing the original contract with this type. Once a player has signed a delegation contract, there is no loss of generality in assuming that he must make some report about his type, because subsequent silence could be interpreted as being synonymous with some specified report under the terms of the contract.

Thus, for a Bayesian game Γ^b with contracts but without verifiable types (that is, without moral hazard but with adverse selection), a feasible mechanism $\mu:T \to \Delta(C)$ is one that satisfies the informational incentive constraints (6.7) and the general participation constraints (6.14). It can be shown that the set of all mechanisms that satisfy (6.14) is convex, so the set of such feasible mechanisms is convex.

6.7 Sender–Receiver Games

Sender–receiver games are a special class of Bayesian games with communication that have been studied, to gain insights into the problems of communication, since Crawford and Sobel (1982). A sender–receiver game is a two-player Bayesian game with communication in which player 1 (the sender) has private information but no choice of actions, and player 2 (the receiver) has a choice of actions but no private information. Thus, sender–receiver games provide a particularly simple class of examples in which both moral hazard and adverse selection are involved.

A general *sender–receiver game* can be characterized by specifying (T_1, C_2, p, u_1, u_2), where T_1 is the set of player 1's possible types, C_2 is the set of player 2's possible actions, p is a probability distribution over T_1 that represents player 2's beliefs about player 1's type, and $u_1:C_2 \times T_1 \to \mathbf{R}$ and $u_2:C_2 \times T_1 \to \mathbf{R}$ are utility functions for player 1 and player 2, respectively. A sender–receiver game is finite iff T_1 and C_2 are both finite sets.

A *mediation plan* or *mechanism* for the sender–receiver game as characterized above is any function $\mu:T_1 \to \Delta(C_2)$. If such a plan μ were implemented honestly and obediently by the players, then the expected payoff to player 2 would be

$$U_2(\mu) = \sum_{t_1 \in T_1} \sum_{c_2 \in C_2} p(t_1)\mu(c_2|t_1)u_2(c_2,t_1)$$

and the conditionally expected payoff to player 1 if he knew that his type was t_1 would be

$$U_1(\mu|t_1) = \sum_{c_2 \in C_2} \mu(c_2|t_1)u_1(c_2,t_1).$$

The general incentive constraints (6.6) can be simplified in sender–receiver games. Because player 1 controls no actions, the incentive constraints on player 1 reduce to purely informational incentive constraints similar to condition (6.7). On the other hand, because player 2 has no private information, the incentive constraints on player 2 reduce to purely strategic incentive constraints similar to condition (6.4). Thus, a mediation plan μ is *incentive compatible* for the sender–receiver game iff

$$(6.15) \quad \sum_{c_2 \in C_2} \mu(c_2|t_1)u_1(c_2,t_1) \geq \sum_{c_2 \in C_2} \mu(c_2|s_1)u_1(c_2,t_1), \quad \forall t_1 \in T_1, \ \forall s_1 \in T_1,$$

and

$$(6.16) \quad \sum_{t_1 \in T_1} p(t_1)(u_2(c_2,t_1) - u_2(e_2,t_1))\mu(c_2|t_1) \geq 0, \quad \forall c_2 \in C_2, \ \forall e_2 \in C_2.$$

The informational incentive constraints (6.15) assert that player 1 should not expect to gain by claiming that his type is s_1 when it is actually t_1, if he expects player 2 to obey the mediator's recommendations. The strategic incentive constraints (6.16) assert that player 2 should not expect to gain by choosing action e_2 when the mediator recommends c_2 to her, if she believes that player 1 was honest with the mediator.

For example, consider a sender–receiver game, due to Farrell (1988), with $C_2 = \{x_2, y_2, z_2\}$ and $T_1 = \{1.a, 1.b\}$, $p(1.a) = .5 = p(1.b)$, and utility payoffs (u_1, u_2) that depend on player 1's type and player 2's action as given in Table 6.5.

Suppose first that there is no mediation, but that player 1 can send player 2 any message drawn from some large alphabet or vocabulary and that player 2 will be sure to observe player 1's message without any error or noise. Under this assumption, Farrell (1988) has shown that, in any equilibrium of the induced communication game, player 2 will choose action y_2 after any message that player 1 may send with positive probability. To see why, notice that player 2 is indifferent between

Table 6.5 Payoffs in a sender–receiver game

Type of player 1	C_2		
	x_2	y_2	z_2
1.a	2,3	0,2	−1,0
1.b	1,0	2,2	0,3

choosing x_2 and z_2 only if she assesses a probability of 1.a of exactly .5, but with this assessment she prefers y_2. Thus, there is no message that can generate beliefs that would make player 2 willing to randomize between x_2 and z_2. For each message that player 1 could send, depending on what player 2 would infer from receiving this message, player 2 might respond either by choosing x_2 for sure, by randomizing between x_2 and y_2, by choosing y_2 for sure, by randomizing between y_2 and z_2, or by choosing z_2 for sure. Notice that, when player 1's type is 1.a, he is not indifferent between any two different responses among these possibilities, because he strictly prefers x_2 over y_2 and y_2 over z_2. Thus, in an equilibrium of the induced communication game, if player 1 had at least two messages (call them "α" and "β") that may be sent with positive probability and to which player 2 would respond differently, then type 1.a would be willing to send only one of these messages (say, "α"); so the other message ("β") would be sent with positive probability only by type 1.b. But then, player 2's best response to this other message ("β") would be z_2, which is the worst outcome for type 1.b of player 1; so type 1.b would not send it with positive probability either. This contradiction implies that player 2 must use the same response to every message that player 1 sends with positive probability. Furthermore, this response must be y_2, because y_2 is player 2's unique best action given her beliefs before she receives any message.

Thus, as long as the players are restricted to perfectly reliable noiseless communication channels, no substantive communication can occur between players 1 and 2 in any equilibrium of this game; see Forges (1985, 1988) for generalizations of this impossibility result. However, substantive communication can occur when noisy communication channels are used. For example, suppose player 1 has a carrier pigeon that he could send to player 2, but the probability of the pigeon arriving, if sent, is only ½. Then there is an equilibrium of the induced communication game in which player 2 chooses x_2 if the pigeon arrives, player 2 chooses y_2 if the pigeon does not arrive, player 1 sends the pigeon if his type is 1.a, and player 1 does not send the pigeon if his type is 1.b. Because of the noise in the communication channel (the possibility of the pigeon getting lost), if player 2 got the message "no pigeon arrives," then she would assign a probability ⅓ (= .5 × 1/(.5 × 1 + .5 × ½)) to the event that player 1's type was 1.a (and he sent a pigeon that got lost); so player 2 would be willing to choose y_2, which is better than x_2 for type 1.b of player 1. In effect, the noise in the communication channel gives type

1.b some protection from exposure when he does not imitate type 1.a, so it reduces the incentive of 1.b to imitate 1.a.

Thus, with a noisy communication channel, there can be an equilibrium in which both players get higher expected payoffs than they can get in any equilibrium with direct noiseless communication. By analyzing the incentive constraints (6.15) and (6.16), we can find other mediation plans $\mu:T_1 \to \Delta(C_2)$ in which they both do even better. The informational incentive constraints (6.15) on player 1 are

$$2\mu(x_2|1.a) - \mu(z_2|1.a) \geq 2\mu(x_2|1.b) - \mu(z_2|1.b),$$

$$\mu(x_2|1.b) + 2\mu(y_2|1.b) \geq \mu(x_2|1.a) + 2\mu(y_2|1.a),$$

and the strategic incentive constraints (6.16) on player 2 are

$$0.5\mu(x_2|1.a) - \mu(x_2|1.b) \geq 0,$$

$$1.5\mu(x_2|1.a) - 1.5\mu(x_2|1.b) \geq 0,$$

$$-0.5\mu(y_2|1.a) + \mu(y_2|1.b) \geq 0,$$

$$\mu(y_2|1.a) - 0.5\mu(y_2|1.b) \geq 0,$$

$$-1.5\mu(z_2|1.a) + 1.5\mu(z_2|1.b) \geq 0,$$

$$-\mu(z_2|1.a) + 0.5\mu(z_2|1.b) \geq 0.$$

(The last of these constraints, for example, asserts that player 2 should not expect to gain by choosing y_2 when z_2 is recommended.) To be a mediation plan, μ must also satisfy the probability constraints

$$\mu(x_2|1.a) + \mu(y_2|1.a) + \mu(z_2|1.a) = 1,$$

$$\mu(x_2|1.b) + \mu(y_2|1.b) + \mu(z_2|1.b) = 1,$$

and all $\mu(c_2|t_1) \geq 0$.

If, for example, we maximize the expected payoff to type 1.a of player 1

$$U_1(\mu|1.a) = 2\mu(x_2|1.a) - \mu(z_2|1.a)$$

subject to these constraints, then we get the mediation plan

$$\mu(x_2|1.a) = 0.8, \quad \mu(y_2|1.a) = 0.2, \quad \mu(z_2|1.a) = 0,$$

$$\mu(x_2 | 1.b) = 0.4, \quad \mu(y_2 | 1.b) = 0.4, \quad \mu(z_2 | 1.b) = 0.2.$$

Honest reporting by player 1 and obedient action by player 2 is an equilibrium when a noisy communication channel or mediator generates recommended-action messages for player 2 as a random function of the type reports sent by player 1 according to this plan μ. Furthermore, no equilibrium of any communication game induced by any communication channel could give a higher expected payoff to type 1.a of player 1 than the expected payoff of $U_1(\mu | 1.a) = 1.6$ that he gets from this plan.

On the other hand, the mechanism that maximizes player 2's expected payoff is

$$\mu(x_2 | 1.a) = \tfrac{2}{3}, \quad \mu(y_2 | 1.a) = \tfrac{1}{3}, \quad \mu(z_2 | 1.a) = 0,$$

$$\mu(x_2 | 1.b) = 0, \quad \mu(y_2 | 1.b) = \tfrac{2}{3}, \quad \mu(z_2 | 1.b) = \tfrac{1}{3}.$$

This gives expected payoffs

$$U_1(\mu | 1.a) = 1\tfrac{1}{3}, \quad U_1(\mu | 1.b) = 1\tfrac{1}{3}, \quad U_2(\mu) = 2\tfrac{1}{2}.$$

Once we have a complete characterization of the set of all incentive-compatible mediation plans, the next natural question is: Which mediation plans or mechanisms should we actually expect to be selected and used by the players? That is, if one or more of the players has the power to choose among all incentive-compatible mechanisms, which mechanisms should we expect to observe?

To avoid questions of equity in bargaining (which will be considered in Chapter 8 and thereafter), let us here consider only cases where the power to select the mediator or design the communication channel belongs to just one of the players. To begin with, suppose that player 2 can select the mediation plan. To be more specific, suppose that player 2 will first select a mediator and direct him to implement some incentive-compatible mediation plan; then player 1 can either accept this mediator and communicate with 2 thereafter only through him, or 1 can reject this mediator and thereafter have no further communication with 2.

It is natural to expect that player 2 will use her power to select a mediator who will implement the incentive-compatible mediation plan that is best for player 2. This plan is worse than y_2 for player 1 if his type is 1.b, so some might think that, if player 1's type is 1.b, then he should reject player 2's proposed mediator and refuse to communicate further. However, there is an equilibrium of this mediator-selection game in which player 1 always accepts player 2's proposal, no matter what his type is. In this equilibrium, if 1 rejected 2's mediator, then 2 might reasonably infer that 1's type was 1.b, in which case 2's rational choice would be z_2 instead of y_2, and z_2 is the worst possible outcome for both of 1's types.

Unfortunately, there is another sequential equilibrium of this mediator-selection game in which player 1 always rejects player 2's mediator, no matter what mediation plan she selects. In this equilibrium, player 2 infers nothing about 1 if he rejects the mediator and so she does y_2; but if he accepted the mediator, then she would infer (in this zero-probability event) that player 1 is type 1.b and she would choose z_2 no matter what the mediator subsequently recommended.

Now consider the mediator-selection game in which the informed player 1 can select the mediator and choose the mediation plan that will be implemented, with the only restriction that player 1 must make the selection after he already knows his own type, and player 2 must know what mediation plan has been selected by player 1. For any incentive-compatible mediation plan μ, there is an equilibrium in which 1 chooses μ for sure, no matter what his type is, and they thereafter play honestly and obediently when μ is implemented. In this equilibrium, if any mediation plan other than μ were selected, then 2 would infer from 1's surprising selection that his type was 1.b (she might think "only 1.b would deviate from μ"), and therefore she would choose z_2 no matter what the mediator might subsequently recommend. Thus, concepts like sequential equilibrium cannot determine the outcome of such a mediator-selection game beyond what we already knew from the revelation principle.

To get a more definite prediction about what mediation plans or mechanisms are likely to be selected by the players, we need to make some assumptions that go beyond traditional noncooperative game theory. An introduction to the theory of cooperative mechanism selection is given in Chapter 10.

6.8 Acceptable and Predominant Correlated Equilibria

In this section, we consider the question of what should be the analogue to the concept of perfect equilibrium for strategic-form games with communication. (The treatment here follows Myerson, 1986a.)

The key idea behind the concept of perfect equilibrium is that players should take some account of all possible outcomes of the games, because there is at least some infinitesimal chance of any outcome occurring, perhaps due to the possibility that some players might make mistakes or "tremble" to some irrational move. However, to preserve the assumption of players' rationality, the probability of any player trembling is always assumed to be infinitesimal at most.

Given a finite game $\Gamma = (N, (C_i)_{i \in N}, (u_i)_{i \in N})$ in strategic form and any positive number ε, an ε-correlated strategy is any probability distribution η in $\Delta(C \times (\cup_{S \subseteq N} C_S))$ such that

(6.17) $\varepsilon \eta(c, e_S) \geq (1 - \varepsilon) \sum_{e_i \in C_i} \eta(c, e_{S \cup \{i\}}),$

$\forall c \in C, \quad \forall i \in N, \quad \forall S \subseteq N - i, \quad \forall e_S \in C_S,$

and

(6.18)   if $\eta(c, e_S) > 0$, then $\eta(c, e_{S \cup \{i\}}) > 0,$

$\forall c \in C, \quad \forall i \in N, \quad \forall S \subseteq N - i, \quad \forall e_{S \cup \{i\}} \in C_{S \cup \{i\}}.$

Here the i-component of $e_{S \cup \{i\}}$ is e_i, and all other components of $e_{S \cup \{i\}}$ together form e_S. As before, we use the notation $C_S = \times_{i \in N} C_i$, where $C_\varnothing = \{\varnothing\}$, and $C = C_N$.

In such an ε-correlated strategy η, we interpret $\eta(c, e_S)$ as the probability that a mediator will recommend the strategies in $c = (c_i)_{i \in N}$ to the various players and all players not in S will then choose their strategies rationally, but the players in S will all tremble and accidentally implement the strategies in $e_S = (e_i)_{i \in S}$. Condition (6.18) asserts that every strategy e_i must have positive probability of being chosen by player i as a result of a tremble, given any combination of mediator's recommendations and trembles by the other players. Condition (6.17) asserts that, given any combination of mediator's recommendations and trembles by other players, the conditional probability of player i trembling is not greater than ε. Thus, in any limit of ε-correlated strategies, as ε goes to 0, the conditional probability of any player i trembling would always be 0, independently of any given information about the mediator's recommendations and other players' trembles. Stronger forms of independence of trembles are not required, because Γ is a game with communication.

Notice that, if η is an ε-correlated strategy and $\hat{\varepsilon} > \varepsilon$, then η is also an $\hat{\varepsilon}$-correlated strategy.

An ε-*correlated equilibrium* is defined to be any ε-correlated strategy that satisfies the following incentive constraints:

$$\sum_{c_{-i} \in C_{N-i}} \sum_{S \subseteq N-i} \sum_{e_S \in C_S} \eta(c, e_S)(u_i(c_{-S}, e_S) - u_i(c_{-(S \cup \{i\})}, e_{S \cup \{i\}})) \geq 0,$$

$$\forall i \in N, \quad \forall c_i \in C_i, \quad \forall e_i \in C_i.$$

That is, player i should not expect to gain by using strategy e_i when he is told to use c_i and he is not irrationally trembling.

An *acceptable correlated equilibrium* of Γ is any μ in $\Delta(C)$ such that there exists some sequence $(\varepsilon_k, \eta_k)_{k=1}^{\infty}$ such that, $\varepsilon_k > 0$ and η_k is an ε_k-correlated equilibrium for each k, and

$$\lim_{k \to \infty} \varepsilon_k = 0, \quad \text{and} \quad \lim_{k \to \infty} \eta_k(c, \varnothing) = \mu(c), \quad \forall c \in C.$$

That is, an acceptable correlated equilibrium is any limit of ε-correlated equilibria as ε goes to 0. (Here $\eta(c, \varnothing)$ is the probability that the mediator recommends c and all players choose rationally.) We can think of an acceptable correlated equilibrium as a correlated equilibrium in which obedient behavior by every player could still be rational when each player has an infinitesimal probability of trembling to any strategy.

We say that c_i is an *acceptable strategy* for player i iff, for every positive number ε, there exists some ε-correlated equilibrium η such that

$$\sum_{c_{-i} \in C_{N-i}} \eta(c, \varnothing) > 0.$$

That is, c_i is acceptable iff it can be used rationally by player i when the probabilities of trembling are arbitrarily small. Let E_i denote the set of acceptable strategies for player i in the game Γ, and let $E = \times_{i \in N} E_i$. Myerson (1986a) showed that the set E_i is always nonempty.

Let $R(\Gamma)$ denote the game that is derived from Γ by eliminating all unacceptable strategies; that is,

$$R(\Gamma) = (N, (E_i)_{i \in N}, (u_i)_{i \in N}).$$

(The utility functions in $R(\Gamma)$ are just the restrictions of the utility functions in Γ to the domain $E \subseteq C$.) This game $R(\Gamma)$ is called the *acceptable residue* of Γ.

Any correlated strategy in $\Delta(E)$, for the game $R(\Gamma)$, can also be interpreted as a correlated strategy in $\Delta(C)$ for the game Γ, simply by assigning probability 0 to every combination of strategies in $C\backslash E$. Conversely, if μ is in $\Delta(C)$ and $\Sigma_{e\in E}\ \mu(e) = 1$, then μ can also be interpreted as a correlated strategy in $\Delta(E)$, for the game $R(\Gamma)$, simply by ignoring the zero components for strategy profiles that are not in E.

With this notation, we can state the main theorem on acceptable correlated equilibria. (For the proof, see Myerson, 1986a.)

THEOREM 6.1. *The set of acceptable correlated equilibria of the finite strategic-form game Γ is a nonempty subset of the set of correlated equilibria of Γ. If μ is any correlated equilibrium of Γ, then μ is an acceptable correlated equilibrium if and only if*

$$\sum_{c\in E} \mu(c) = 1.$$

Furthermore, the set of acceptable correlated equilibria of Γ exactly coincides with the set of correlated equilibria of $R(\Gamma)$ (reinterpreted as correlated strategies in $\Delta(C)$).

Thus, to characterize the acceptable correlated equilibria of Γ, it suffices to identify and eliminate all unacceptable strategies of all players. (For computational procedures to identify unacceptable strategies, see Myerson, 1986a.) Then the correlated equilibria of the residual game that remains are the acceptable correlated equilibria of the original game. It can be shown that all weakly dominated strategies are unacceptable, but some unacceptable strategies may also be undominated. For example, in Table 6.6, there are no dominated strategies, but y_1, z_1, y_2, and z_2 are all unacceptable.

Table 6.6   A game in strategic form

C_1	C_2			
	w_2	x_2	y_2	z_2
w_1	2,2	1,1	0,0	0,0
x_1	1,1	1,1	2,0	2,0
y_1	0,0	0,2	3,0	0,3
z_1	0,0	0,2	0,3	3,0

There might be some strategies in E_i that are unacceptable in the context of the acceptable residue $R(\Gamma)$, even though they are acceptable in Γ. This situation might occur because ε-correlated strategies for the game $R(\Gamma)$ do not include any trembles outside of the strategy sets $(E_i)_{i \in N}$. Thus, we consider the acceptable residue of the acceptable residue of Γ; that is, $R^2(\Gamma) = R(R(\Gamma))$. Continuing inductively, we define

$$R^{k+1}(\Gamma) = R(R^k(\Gamma)), \quad \text{for } k = 2,3,4, \ldots.$$

Taking an acceptable residue of a game means eliminating any unacceptable strategies in the game, and Γ itself has only finitely many strategies for each player. Thus, there must exist some K such that

$$R^K(\Gamma) = R^{K+1}(\Gamma) = R^{K+2}(\Gamma) = \ldots = R^\infty(\Gamma).$$

Let E_i^∞ denote the strategy set for each player i in $R^\infty(\Gamma)$. We call $R^\infty(\Gamma)$ the *predominant residue* of Γ, and the strategies in E_i^∞ are the *iteratively acceptable* or *(weakly) predominant* strategies of player i in Γ. We say that μ is a *predominant correlated equilibrium* of Γ iff it is a correlated equilibrium of Γ and

$$\sum_{c \in E^*} \mu(e) = 1, \quad \text{where } E^\infty = \underset{i \in N}{\times} E_i^\infty.$$

That is, a predominant correlated equilibrium is a correlated equilibrium in which all players are always told to use their predominant strategies. It is straightforward to show that the set of predominant equilibria is nonempty and includes at least one Nash equilibrium that can be implemented without communication (see Myerson, 1986a).

In Table 6.6, the strategies w_1, x_1, w_2, x_2 are all acceptable strategies, but only w_1 and w_2 are predominant. Thus, the only predominant equilibrium is (w_1, w_2). The equilibrium at (x_1, x_2) is acceptable but not predominant.

There may exist acceptable and predominant correlated equilibria that can be implemented without communication (so that they are also Nash equilibria) but are not perfect equilibria. For example, consider the three-player game in Table 6.7, where $C_i = \{x_i, y_i\}$ for $i = 1,2,3$.

Both (x_1, x_2, y_3) and (y_1, x_2, x_3) are pure-strategy Nash equilibria in this game. Of these, (x_1, x_2, y_3) is a perfect equilibrium. But (y_1, x_2, x_3) is not perfect, because, when the players tremble independently with small positive probabilities from the strategies (y_1, x_2, x_3), the probability that player 1 trembles to x_1 while player 3 stays at x_3 is much larger than the

Table 6.7   A three-player game in strategic form

| C_1 | C_2 and C_3 | | | |
| | x_3 | | y_3 | |
	x_2	y_2	x_2	y_2
x_1	1,1,1	4,4,0	1,1,1	0,0,4
y_1	3,2,2	3,2,2	0,0,0	0,0,0

probability that player 1 trembles to x_1 and player 3 trembles to y_3. Knowing this, player 2 would expect to do better by switching from x_2 to y_2. However, all strategies are acceptable in this game, so all equilibria are acceptable and predominant, including (y_1, x_2, x_3).

To see why all strategies are acceptable and (y_1, x_2, x_3) is an acceptable equilibrium, let ε be some very small positive number and consider the following mediation plan. With probability $1 - \varepsilon - \varepsilon^3$, the mediator recommends the strategies (y_1, x_2, x_3); with probability ε, the mediator recommends (x_1, x_2, y_3); and with probability ε^3, the mediator recommends (x_1, y_2, x_3). As usual, we assume that each player is only told the mediator's recommendation for himself. To complete the description of an ε^2-correlated strategy, let us suppose that every player trembles independently with probability ε^2 (given any recommendations by the mediator) and, whenever a player trembles, he is equally likely to use either of his strategies. For any sufficiently small ε, this is an ε^2-correlated equilibrium, and it approaches the equilibrium (y_1, x_2, x_3) as ε gets small.

To check that player 2 is willing to use strategy x_2 when the mediator recommends it in this ε^2-correlated equilibrium, consider what she would believe about the other players when she gets this recommendation. The most likely way that she could get this recommendation is when player 1 is told to do y_1 and player 3 is told to do x_3 and nobody is trembling, but in this case player 2 is indifferent between her strategies. The next most likely way that she could get this recommendation, which happens with prior probability ε, is when player 2 is told to use x_1 and player 3 is told to use y_3 and nobody is trembling; and in this case she strictly prefers to obey the recommendation to use x_2. All other events have prior probabilities of order ε^2 or less, so player 2 is willing to obey a recommendation to choose x_2.

6.9 Communication in Extensive-Form and Multistage Games

The analysis of extensive-form games may also be significantly different when players can communicate with each other and with a mediator. For example, consider the games in Figures 6.2 and 6.3.

Like the games in Figures 5.2 and 5.3, the games in Figures 6.2 and 6.3 have the same reduced normal representations. Kohlberg and Mer-

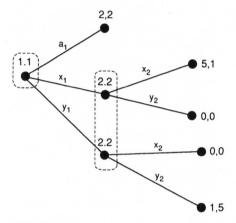

Figure 6.2

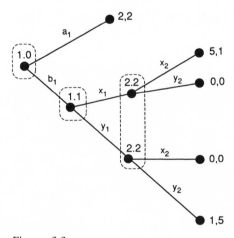

Figure 6.3

tens (1986) have argued that these games should be considered as analytically equivalent. As games with communication, however, they are very different.

In Figure 6.2, the move y_1 is strongly dominated by a_1 for player 1 at the 1.1 node. Thus, even with communication, player 1 would never rationally choose y_1 in Figure 6.2.

On the other hand, consider the following plan that could be implemented by a mediator in Figure 6.3. First, advise player 1 to choose b_1 at his node 1.0. Then, after player 1 has chosen b_1, toss a coin. If the coin comes up Heads, then advise player 1 to choose x_1 and advise player 2 to choose x_2; if the coin comes up Tails, then advise player 1 to choose y_1 and player 2 to choose y_2. When this mediation plan is implemented for the game in Figure 6.3, the probability of player 1 choosing y_1 is ½, and it is an equilibrium for both players to plan to obey the mediator's recommendation. At any point in time, if player 2 is expected to obey the mediator, then player 1 would expect to do strictly worse by disobeying. In particular, player 1's expected payoff from obeying the recommendation to choose b_1 at 1.0 is $.5 \times 5 + .5 \times 1 = 3$, whereas his expected payoff from a_1 would be only 2.

The key to implementing this mediation plan is that player 1 must not learn whether x_1 or y_1 will be recommended for him until it is too late for him to select a_1. In Figure 6.3, there is a point in time (after the 1.0 node but before the 1.1 node) when x_1 and y_1 are still available to player 1 but a_1 is not; however, in Figure 6.2 there is no such point in time. If all communication had to occur before the beginning of the game, then this distinction would not matter. However, under the assumption that the players can communicate with each other and with a mediator at any point in time, Figure 6.3 has a strictly larger set of incentive-compatible mediation plans than does Figure 6.2, even though these games have the same purely reduced normal representations.

Thus, if we want to study extensive-form games as games with communication, the normal representation is not adequate to characterize the set of correlated equilibria or incentive-compatible mediation plans. If we want to assume that, in addition to the given explicit structure of an extensive-form game, the players have implicit opportunities to communicate after they learn private information about their type or information state, then none of the representations in strategic form that we studied in Chapter 2 are adequate to characterize the incentive-compatible mediation plans or communication equilibria of the given game.

(To see the inadequacy of the multiagent representation, consider any sender–receiver game, when the sender's communication possibilities are not included in the explicit structure of the game. In the multiagent or type-agent representation, the agents for the sender have no explicit strategy options, and their access to information is not represented. So, except for the receiver's one agent, all agents appear to be dummies.)

Thus, there is a conceptual trade-off between the revelation principle and the generality of the strategic form. If we want to allow communication opportunities to remain implicit at the modeling stage of our analysis, then we get a mathematically simpler solution concept, because (by the revelation principle) communication equilibria can be characterized by a set of linear incentive constraints. However, if we want to study the game in strategic form, then all communication opportunities must be made an explicit part of the structure of the extensive-form game before we construct the normal or multiagent representation.

Myerson (1986b) studied communication in games that are given in *multistage form*, which is a modification of the extensive form defined by Kuhn (1953). (The multistage form actually resembles more closely the definition of the extensive form used by von Neumann and Morgenstern, 1944.) A multistage game is any game of the form

$$\Gamma^m = \left(N, K, ((C_i^k, T_i^k)_{k=1}^K, u_i)_{i \in N}, (p^k)_{k=1}^K\right),$$

where K is the number of stages in the game; N is the set of players; C_i^k is the set of moves or actions available to player i at the end of stage k; T_i^k is the set of signals or new information that player i could learn at the beginning of stage k; p^k specifies a probability distribution over $\times_{i \in N} T_i^k$, the set of combinations of players' signals at stage k, as a function of the past history of actions and signals (in $\times_{l=1}^{k-1} \times_{i \in N} (C_i^l \times T_i^l)$); and u_i specifies player i's utility payoff as a function of all actions and signals (in $\times_{l=1}^{K} \times_{j \in N} (C_j^l \times T_j^l)$). That is, a multistage game differs from an extensive-form game in that the players' moves are ordered in discrete stages, and we must specify what a player knows at each stage even if he has no alternative moves (that is, even if $|C_i^k| = 1$). Any extensive-form game in which players move in a prespecified order can be represented as a multistage game in the obvious way. (That is, we can identify each player's move with a distinct stage. At each stage k, we specify that a nonmoving player i has no new information and so $|T_i^k| = 1$, unless he moved in the preceding stage, in which case his new information is his recollection of his recent move and $T_i^k = C_i^{k-1}$.)

For multistage games, the revelation principle asserts that any equilibrium of any communication game that can be induced from Γ^m by adding a communication structure is equivalent to some mediation plan of the following form: (1) At the beginning of each stage, the players confidentially report their new information to the mediator. (2) The mediator then determines the recommended actions for the players at this stage, as a function of all reports received at this and all earlier stages, by applying some randomly selected feedback rule. (3) The mediator tells each player confidentially the action that is recommended for him. (4) When all players know the probability distribution that the mediator used to select his feedback rule, it is an equilibrium of the induced communication game for all players to always report their information honestly and choose their actions obediently as the mediator recommends. The distributions over feedback rules that satisfy this incentive-compatibility condition (4) can be characterized by a collection of linear incentive constraints that assert that no player can expect to gain by switching to any manipulative strategy of lying and disobedience, when all other players are expected to be honest and obedient.

Intuitively, the general revelation principle asserts that there is no loss of generality in assuming that, at every stage, the players communicate only with a central mediator to whom they reveal everything they have learned, but who in turn reveals to each player only the minimal information that the player needs to determine his current action at this stage. As the game in Figure 6.3 showed, it may be essential that the mediator should not tell a player what move will be recommended for him at a particular stage before that stage is reached. In general, revealing more information to players makes it harder to prevent them from finding ways to gain by lying or disobeying recommendations.

The possibility of communication, even if it only occurs with infinitesimal probability, can significantly affect our belief-consistency criteria for multistage games and games in extensive form. Even equilibria that are not subgame-perfect may become sequentially rational when communication is allowed. For example, recall the game in Figure 4.6. For this game, $([y_1], [y_2], .5[x_3] + .5[y_3], [y_4])$ is a Nash equilibrium, but it is not a subgame-perfect equilibrium, because player 3 should prefer to choose y_3, in the subgame beginning at his node, when he knows that player 4 would choose y_4. The unique subgame-perfect equilibrium of this game is $([x_1], [x_2], .5[x_3] + .5[y_3], .5[x_4] + .5[y_4])$, which gives the payoff allocation (2,3,0,0).

However, $([y_1], [y_2], .5[x_3] + .5[y_3], [y_4])$ can be extended to a sequential equilibrium if we allow even an infinitesimal probability that the players might be able to communicate. To see how, consider Figure 6.4, where ε is a very small positive number. The game in Figure 6.4 differs from the game in Figure 4.6 only in that players 1, 2, and 4 may observe some payoff-irrelevant signal that occurs with probability ε. According to the sequential equilibrium shown in Figure 6.4, if players 1, 2, and 4 observe this signal, then they will plan to choose the actions (x_1, x_2, x_4); otherwise, with probability $1 - ε$, they will plan to choose the actions (y_1, y_2, y_4). Player 3 can have an opportunity to move only if at least one player made a mistake. Thus, player 3 may consistently believe that, if he gets to move, then, with equal probability, either the signal was not

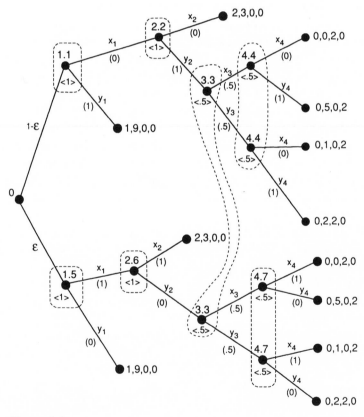

Figure 6.4

observed and player 1 made a mistake choosing x_1 at 1.1, or the signal was observed and player 2 made a mistake choosing y_2 at 2.6. These beliefs can be justified, no matter how small ε may be, by supposing that player 1 is much less likely to make a mistake than player 2. With these beliefs, player 3 would be willing to randomize between x_3 and y_3 as specified in the equilibrium, because he believes that it is equally likely that player 4 may be about to choose x_4 (at 4.7) or y_4 (at 4.4). This sequential equilibrium coincides with the equilibrium ($[y_1]$, $[y_2]$, $.5[x_3] + .5[y_3]$, $[y_4]$) on the upper chance event, which happens almost surely as ε goes to 0.

We observed in Section 4.8 that adding small-probability chance events may significantly change the set of sequential equilibria of a game. Notice, however, that the new chance event added in Figure 6.4 has the special property of not altering the structure of players' alternatives and payoffs given in Figure 4.6. That is, the additional chance event is *payoff irrelevant*, in the sense that its only effect is to create a second copy of the original game tree, which some players can distinguish and others cannot. (The chance event in Figure 4.12 was not payoff irrelevant in this sense.) The assumption that a game is "with communication" implies that such payoff-irrelevant chance events can be considered to be part of the implicit structure of the game, even if they are not explicitly shown.

Thus, the possibility of communication may weaken our notions of sequential rationality. Myerson (1986b) defined a concept of sequential rationality for multistage games with communication and showed that (like acceptability for strategic-form games with communication) it can be characterized by eliminating some unreasonable actions that satisfy a kind of generalized domination criterion.

Exercises

Exercise 6.1.   Consider the game in strategic form shown in Table 6.8.

a. Find the correlated equilibrium that maximizes the expected sum of the two players' payoffs.

b. Find the correlated equilibrium that minimizes the expected sum of the two players' payoffs.

Exercise 6.2.   The two-person game shown in Table 6.9 has a unique Nash equilibrium (see Moulin and Vial, 1978) in which each player's expected payoff is 3.

Table 6.8   A game in strategic form

C_1	C_2	
	a_2	b_2
a_1	4,4	1,6
b_1	6,1	-3,-3

Table 6.9   A game in strategic form

C_1	C_2		
	x_2	y_2	z_2
x_1	0,0	5,4	4,5
y_1	4,5	0,0	5,4
z_1	5,4	4,5	0,0

a. Show that this game has correlated equilibria in which both players get expected payoffs strictly larger than 4.

b. Find the correlated equilibrium that maximizes the expected payoff for player 1.

Exercise 6.3.   Players 1 and 2 are, respectively, seller and buyer for a single indivisible object. The value of the object may be \$0 or \$80 for player 1, and may be \$20 or \$100 for player 2. Ex ante, all four possible combinations of these values are considered to be equally likely. When the players meet to bargain, each player knows his own private value for the object and believes that the other player's private value could be either of the two possible numbers, each with probability ½. We let the type of each player be his private value for the object, so $T_1 = \{0, 80\}$, $T_2 = \{20, 100\}$. Letting $t = (t_1,t_2)$ denote the pair of players' types, the payoffs (net gains from trade) to the two players would be

$$u_1((1,y),t) = y - t_1 \text{ and } u_2((1,y),t) = t_2 - y$$

if the object were sold to the buyer for y dollars, and the payoffs would be

$$u_1((0,0),t) = 0 = u_2((0,0),t)$$

if no trade occurred. No-trade (0,0) is the disagreement outcome in the bargaining problem. Assume that there is no question of delay of trade; the object must be sold today or never.

Let us represent a mechanism for this Bayesian bargaining problem by a pair of functions $Q:T_1 \times T_2 \to [0,1]$ and $Y:T_1 \times T_2 \to \mathbf{R}$, where $Q(t)$ is the probability that the object will be sold to the buyer and $Y(t)$ is the expected net payment from the buyer to the seller if t is the pair of types reported by the players to a mediator.

a. Write down the four informational incentive constraints and the four participation constraints that an incentive-compatible individually rational mechanism must satisfy.

b. The probability that the object is worth more to the buyer than to the seller is ¾. Show that, in any incentive-compatible individually rational mechanism, the probability that the object will be sold to the buyer cannot be greater than ⅔. (HINT: For any positive numbers α and β, the probability of trade cannot be larger than

$$(\tfrac{1}{4})(Q(0,100) + Q(0,20) + Q(80,100) + Q(80,20))$$

$$+ \ \beta U_1(Q,Y|80) + \beta U_2(Q,Y|20)$$

$$+ \ \alpha(U_1(Q,Y|0) - U_1^*(Q,Y,80|0))$$

$$+ \ \alpha(U_2(Q,Y|100) - U_2^*(Q,Y,20|100)).$$

To simplify this formula, try letting $\beta = 2\alpha$ and $\alpha = \tfrac{1}{120}$.)

Exercise 6.4. Player 1 is the plaintiff and player 2 is the defendant in a lawsuit for breach of contract. The expected in-court settlement, to be paid by player 2 to player 1 if they go to court, depends on the strength of both sides' cases as shown in Table 6.10. In addition, if they go to court, then each side must pay trial costs of $10,000. So, for example, if both cases are strong and they go to court, then the expected total cost to player 2 is $90,000, but the expected net benefit to player 1 is only $70,000. Each player knows whether his own case is weak or strong, and each side believes that the other side is equally likely to be weak or strong. Assume that both players are risk neutral. The players could avoid these trial costs by settling out of court. However, in pretrial bargaining, there is nothing to prevent a player from lying about the strength of his case, if he has an incentive to do so. Going to court is the disagreement outcome in any pretrial bargaining.

Table 6.10   Expected in-court settlement for player 1, depending on the
strengths of the cases for both sides

Strength of player 1's case	Settlement based on strength of player 2's case	
	Strong	Weak
Strong	$80,000	$144,000
Weak	$16,000	$80,000

a. Suppose that you have been asked to mediate the pretrial bargaining between these two players. Create an incentive-compatible mediation plan that has the following properties: (1) if they settle out of court, then player 2 will pay player 1 an amount equal to the expected in-court settlement, given their reports to you about the strength or weakness of their cases; (2) if both report to you that they are weak, then they settle out of court with probability 1; and (3) if one side reports that it is weak and the other reports that it is strong, then they settle out of court with a probability q that does not depend on which side reported weakness. Make q as large as possible without violating any informational incentive constraints. (HINT: When you compute expected payoffs, do not forget the net payoff from going to court. Player 1 expects to get money from player 2 even if they do not agree to settle out of court!)

b. Imagine that player 1, knowing that his case is actually strong, gets the following advice from his lawyer. "Given that our case is strong, our expected in-court settlement would be $0.5 \times 80,000 + 0.5 \times 144,000 = $112,000$. So let us tell player 2 that our case is strong and offer to settle out of court for this amount." Show that, even if player 2 were to believe player 1's claim about the strength of 1's case, it would be a mistake for player 1 to make an offer to settle for $112,000 when his case is strong.

Exercise 6.5.   Show that the set of mechanisms μ that satisfy the general participation constraints (6.14) is convex.

Exercise 6.6.   Consider the following generalization of the model in Section 6.5. As before, each player j's type is a random variable that is independent of all other players' types and is drawn from the interval

$T_j = [L(j),M(j)]$ according to the continuous positive probability density function f_j. Let D denote the set of possible outcomes. Suppose that, for some player i, there exist functions $q_i:D \to \mathbf{R}$ and $y_i:D \times T_{-i} \to \mathbf{R}$ such that player i's payoff function can be written

$$u_i(c,t) = q_i(c)t_i - y_i(c,t_{-i}), \ \forall c \in D, \ \forall t = (t_{-i},t_i) \in T.$$

That is, i's payoff is linear in his type t_i, with coefficient $q_i(c)$.

Given any random mechanism $\mu:T \to \Delta(D)$, let

$$Q_i(t) = \int_{c \in D} q_i(c)d\mu(c|t)$$

$$\overline{Q}_i(t_i) = \int_{t_{-i} \in T_{-i}} Q_i(t) \left(\prod_{j \in N-i} f_j(t_j) \right) dt_{-i}.$$

That is, $\overline{Q}_i(t_i)$ is the expected value of $q_i(c)$ for the chosen outcome c, when player i's type is t_i and all players will participate honestly in the mechanism μ.

Show that, if μ is incentive compatible, then $\overline{Q}_i$ is a nondecreasing function and player i's type-contingent expected payoff under μ satisfies condition (6.10),

$$U_i(\mu|t_i) = U_i(\mu|\theta_i) + \int_{\theta_i}^{t_i} \overline{Q}_i(s_i)ds_i, \ \ \forall t_i \in T_i, \ \forall \theta_i \in T_i.$$

Exercise 6.7. Recall the auction situation with independent private values that was described in Exercise 3.7 (Chapter 3). Show that, even if we changed the auction rules, for any incentive-compatible mechanism μ in which the object is always sold to the bidder with the higher type and a bidder of type zero has zero expected gains, the conditionally expected payoff to bidder i, given his type t_i, must be $0.5(t_i)^2$.

Exercise 6.8. Recall the auction situation with common values and independent types that was described in Exercise 3.8 (Chapter 3). Show that, even if we changed the auction rules, for any incentive-compatible mechanism μ in which the object is always sold to the bidder with the higher type and a bidder of type zero has zero expected gains, the conditionally expected payoff to bidder i, given his type t_i, must be $0.5(t_i)^2$. (HINT: You may use the results of Exercise 6.6.)

Exercise 6.9. (An example due to Forges, 1985.) Consider a sender–receiver game with $C_2 = \{x_2, y_2, z_2\}$, $T_1 = \{1.a, 1.b\}$, $p(1.a) = .5 = p(1.b)$, and utility payoffs (u_1, u_2) that depend on player 1's type and player 2's action as shown in Table 6.11.

a. Show that, as long as the players are restricted to perfectly reliable noiseless communication channels, no substantive communication can occur between players 1 and 2 in any equilibrium of this game, so the expected payoff to type 1.a of player 1 must be 0.

b. Now suppose that a mediator can introduce some randomness or noise into the communication channel. Write down the constraints that characterize an incentive-compatible mediation plan μ, and show that there exists an incentive-compatible mediation plan in which type 1.a of player 1 would get an expected payoff of $U_1(\mu | 1.a) = 5$.

Exercise 6.10. Player 1 is the seller of a single indivisible object, and player 2 is the only potential buyer. Depending on the quality of the object, its value to player 1 can be any number between L and M (where $L < M$). Player 1 knows the quality, so we can identify his type t_1 with this value. Player 2 does not know the quality, but the object actually would be worth $g(t_1)$ to player 2 if it were worth t_1 to player 1. Thus, letting $q = 1$ if player 2 buys the object, letting $q = 0$ if player 1 keeps the object, and letting y denote the net payment from player 2 to player 1, we can write the players' payoff functions as $u_1((q,y),t_1) = y - t_1 q$, $u_2((q,y),t_1) = g(t_1)q - y$. The disagreement outcome is no trade $(0,0)$. Player 2 has no private information and assesses a subjective probability distribution for 1's type that has a continuous positive probability density function f_1 on the interval $[L,M]$. For any number r in $[L,M]$, let $F_1(r) = \int_L^r f_1(s)ds$ denote the cumulative probability distribution corresponding to $f_1(\cdot)$.

a. Consider a mechanism (Q,Y) in which, when player 1's type is t_1, the probability of player 2 buying the object is $Q(t_1)$ and the expected

Table 6.11 Payoffs in a sender–receiver game

Type of player 1	C_2		
	x_2	y_2	z_2
1.a	6,6	0,4	9,0
1.b	6,0	9,4	0,6

payment from player 2 to player 1 is $Y(t_1)$. Show that, if (Q,Y) is incentive compatible and individually rational, then $Q(t_1)$ is a nonincreasing function of t_1, and

$$Y(t_1) \geq t_1 Q(t_1) + \int_{t_1}^{M} Q(s)ds, \quad \forall t_1 \in [L,M];$$

so player 2's expected payoff must satisfy

$$U_2(Q,Y) \leq \int_{L}^{M} (g(t_1) - t_1 - F_1(t_1)/f_1(t_1))Q(t_1)f_1(t_1)dt_1.$$

(HINT: Notice that q and $Q(t_1)$ in this model correspond to $-q_1$ and $-\overline{Q}_1(t_1)$ in the notation of Section 6.5. See also Myerson, 1985b.)

b. (The *lemon problem* of Akerlof, 1970.) Suppose now that the distribution of possible seller's types is uniform on the interval $[0,1]$, and the object is always worth 50 percent more to the buyer than to the seller, so

$$L = 0, \quad M = 1, \quad f_1(t_1) = 1, \quad \text{and} \quad g(t_1) = 1.5t_1, \quad \forall t_1 \in [0,1].$$

Show that an incentive-compatible individually rational mechanism must have $Q(t_1) = 0$ for every positive t_1. (Thus, the probability of trade occurring must be 0, even though the object would always be worth more to the buyer than to the seller.)

c. Suppose that the distribution of seller's possible types is uniform on the interval $[L,M]$, and the buyer's value is linear in the seller's value, so there exist parameters A and B such that

$$f_1(t_1) = 1/(M - L) \quad \text{and} \quad g(t_1) = At_1 + B, \quad \forall t_1 \in [L,M].$$

Suppose that player 2 (the buyer) can specify any first and final offer price R. Player 1 will accept the offer and sell at this price if his value is less than R, but player 1 will reject the offer and keep the object if his value is greater than R. As a function of the parameters L, M, A, and B, what is the optimal offer price R for the buyer? (It can be shown that, without loss of generality, we can assume that R is in the interval $[L,M]$.)

Show that no other incentive-compatible individually rational mechanism can offer a higher expected payoff to player 2 than what she gets from her optimal first and final offer price. (Recall that $Q(t_1)$ must be a nonincreasing function of t_1. Notice also that $F_1(t_1)/f_1(t_1) = t_1 - L$ for any t_1 in $[L,M]$.)

For the special case in part (b), where $L = 0$, $M = 1$, $A = 1.5$, and $B = 0$, you should get an optimal offer price of $R = 0$. Explain why no higher offer price would be better for the buyer, with reference to the winner's curse that was discussed in Section 3.11.

Exercise 6.11. Consider the extensive-form game shown in Figure 6.5.

a. Find two pure-strategy equilibria and show that only one of them is a sequential-equilibrium scenario (see Selten, 1975; Binmore, 1987–88).

b. Let ε be an arbitrarily small positive number. Suppose that there is a payoff-irrelevant event that has a probability ε of occurring before this game is played and that, if it occurred, would be observed by players 1 and 3 but not by player 2. Show the extensive-form game where this event is taken into account and show that this revised game has a sequential equilibrium in which, if the ε-probability event does not occur, the players behave according to the equilibrium that, by your answer to part (a), was not a sequential-equilibrium scenario of the original game.

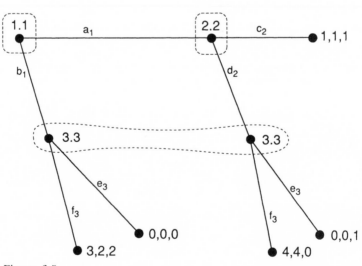

Figure 6.5

Bibliographic Note

The term *incentive compatibility* was first used by Hurwicz (1972). In this book, this term is used in a somewhat different sense, called *Bayesian incentive compatibility* by d'Aspremont and Gerard-Varet (1979).

Gibbard (1973) first expressed the revelation principle for a non-Bayesian dominant-strategy solution concept. The revelation principle for Bayesian equilibria of games with incomplete information was developed by Dasgupta, Hammond, and Maskin (1979); Harris and Townsend (1981); Holmstrom (1977); Myerson (1979); and Rosenthal (1978). Other general formulations of the revelation principle were developed by Myerson (1982, 1986b) for Bayesian games with moral hazard and for multistage games.

7

Repeated Games

7.1 The Repeated Prisoners' Dilemma

People may behave quite differently toward those with whom they expect to have a long-term relationship than toward those with whom they expect no future interaction. To understand how rational and intelligent behavior may be affected by the structure of a long-term relationship, we study repeated games.

In a repeated game, there is an infinite sequence of *rounds*, or points in time, at which players may get information and choose moves. That is, a repeated game has an *infinite time horizon*, unlike the finite game trees that were considered in Chapter 2. In a repeated game, because no move is necessarily the last, a player must always consider the effect that his current move might have on the moves and information of other players in the future. Such considerations may lead players to be more cooperative, or more belligerent, in a repeated game than they would be in a finite game in which they know when their relationship will end.

To introduce the study of repeated games, let us begin with a well-known example, the repeated Prisoners' Dilemma (Table 7.1). Here g_i is i's generous move, and f_i is i's selfish move. As we saw in Chapter 3, (f_1, f_2) is the only equilibrium of this game. Now let us consider what happens if players 1 and 2 expect to repeatedly play this game with each other every day, for a long time.

For example, suppose that the number of times that the game will be played is a random variable, unknown to the players until the game stops, and that this random stopping time is a geometric distribution with expected value 100. That is, the probability of play continuing for

Table 7.1 Prisoners' Dilemma game

D_1	D_2	
	g_2	f_2
g_1	5,5	0,6
f_1	6,0	1,1

exactly k rounds is $(.99^{k-1}) \times .01$. Thus, after each round of play, the probability that players 1 and 2 will meet and play again is .99; and at any time when they are still playing, the expected number of further rounds of play is 100. In this repeated game, generous behavior can be supported in equilibrium. Consider, for example, the strategy of playing g_i every day until someone plays f_1 or f_2, and thereafter playing f_i. If both players plan to use this strategy, then, at any time in the game each player will get an expected total future payoff of

$$\sum_{k=1}^{\infty} (.99^{k-1})(.01)(5k) = 500,$$

as long as no one has deviated and chosen f_i. But if either player i deviated from this strategy and chose f_i on some particular day, then his expected total future payoff from this day onward would be

$$6 + \sum_{k=2}^{\infty} (.99^{k-1})(.01)(1k) = 105.$$

On the other hand, if anyone has ever played f_1 or f_2, then it is easy to see that neither player could gain by choosing g_i when the other player is expected to follow this strategy and so always choose the selfish move hereafter. Thus, at any point in time, with any history of past moves, neither player can expect to gain by any unilateral deviation from this strategy.

The key to this generous equilibrium is that, whenever the players meet, they believe that there is a very high probability that they will play again; so the hope of inducing future generous behavior by the other player can give each player an incentive to be generous. We could not construct such an equilibrium if the players knew in advance when the game would end. For example, if it were common knowledge that the game would be played for exactly 100 rounds, then the resulting

100-round game would have a unique sequential equilibrium, with both always playing (f_1, f_2) at every round. At the last (100th) round, a generous move cannot induce any future generosity by the other player (because there is no future), so there is no reason for either player to be generous. So the players should choose (f_1, f_2) at the 100th round, no matter what the prior history might be. Thus, at the 99th round, the players must know that their moves will have no impact on the 100th-round moves, so they should play (f_1, f_2) on the 99th round as well. Working backward through the game, it is straightforward to verify by induction that the unique sequential-equilibrium scenario is indeed to play f_1 and f_2 at every round.

Thus, rational behavior in a repeated game with a potentially infinite time horizon may be very different from rational behavior in the corresponding game in which the number of rounds is arbitrarily large but finite. So the study of repeated games requires us to use a model that goes beyond the finite extensive form.

7.2. A General Model of Repeated Games

A wide range of models of repeated games have been studied in the game-theory literature. The following model (used by Mertens, Sorin, and Zamir, 1989) provides a general conceptual structure that can include most of these models as special cases. Let N be the nonempty set of players. Let Θ be a nonempty set, denoting the set of possible *states of nature*. For each player i in N, let the nonempty sets D_i and S_i denote respectively the set of *moves* that player i can choose and the set of *signals* that player i may receive, at each round of the game. We use the notation

$$D = \underset{i \in N}{\times} D_i, \quad S = \underset{i \in N}{\times} S_i.$$

Given these sets, we must specify an *initial distribution* q in $\Delta(S \times \Theta)$, a *transition function* $p{:}D \times \Theta \to \Delta(S \times \Theta)$, and, for every player i, a *payoff function* $u_i{:}D \times \Theta \to \mathbf{R}$. These structures

$$\Gamma^r = (N, \Theta, (D_i, S_i, u_i)_{i \in N}, q, p)$$

complete the specification of a general repeated game.

The interpretation of such a general repeated game is as follows. The game is played in an infinite sequence of rounds numbered $1, 2, 3, \ldots$. At each round, some state in Θ is the *(current) state of nature*. Each player

i's new information about the state of nature is summarized by a signal in S_i that he receives at the beginning of the round. As a function of the signals that he has received in the current round and in all previous rounds, each player i must choose a move in D_i. The probability that θ^1 is the state of nature at the first round and that each player i receives the signal s_i^1 at the beginning of the first round is $q(s^1,\theta^1)$. (Here we use the notation $s^k = (s_i^k)_{i \in N}$.) At any round k, if the current state of nature is θ^k and the moves of the players are $d^k = (d_i^k)_{i \in N}$ in D, then $u_i(d^k,\theta^k)$ is the round-k payoff to each player i, and $p(s^{k+1},\theta^{k+1}|d^k,\theta^k)$ is the conditional probability that the signals and the state of nature at round $k+1$ will be (s^{k+1},θ^{k+1}). Notice that this probability is independent of the states and moves at rounds 1 through $k-1$ and the signals at rounds 1 through k.

We say that payoffs are *bounded* in the repeated game Γ^r iff there exists some positive number Ω such that,

$$|u_i(d,\theta)| \leq \Omega, \quad \forall i \in N, \; \forall d \in D, \; \forall \theta \in \Theta.$$

We say that a state θ in Θ is *absorbing* iff

$$\sum_{s \in S} p(s,\theta|d,\theta) = 1, \quad \forall d \in D.$$

Thus, if an absorbing state θ is the current state of nature at round k, then θ will be the state of nature at all rounds after k.

To describe the infinitely repeated Prisoners' Dilemma in the preceding section in the general repeated-game form, let $N = \{1,2\}$, $\Theta = \{0,1\}$, $D_i = \{f_i,g_i\}$, and $S_i = \Theta \cup (D_1 \times D_2)$. Then, for each i in N and d in D, let $u_i(d,1)$ be as shown in Table 7.1, let $u_i(d,0) = 0$, and let

$$q(1,1,1) = 1,$$
$$p(d,d,1|d,1) = 0.99, \quad p(0,0,0|d,1) = 0.01, \quad \text{and}$$
$$p(0,0,0|d,0) = 1.$$

Here, state 1 means that the active play is continuing, and state 0 means that active play has stopped. The equation $q(1,1,1) = 1$ means that, with probability 1, at round 1 players 1 and 2 will each receive a signal "1" and the state of nature will be 1. At each subsequent round $k+1$, if the state of nature is 1, then each player will get a signal telling him the pair of moves chosen in the preceding round. The state 0 is absorbing.

In a general repeated game, we assume that each player at each round k recalls all the signals that he has gotten in rounds 1 through k. To assure that he recalls his own past moves, it then suffices to let one component of his signal in each round equal his own move from the preceding round. Thus, the set of all *pure strategies* C_i for player i in the general repeated game Γ^r is

$$C_i = \{c_i = (c_i^k)_{k=1}^\infty \,|\, c_i^k : (S_i)^k \to D_i, \quad \forall k\}.$$

(Here $(S_i)^k$ denotes the k-fold Cartesian product of the set S_i.) The set of *behavioral strategies* B_i for player i is similarly

$$B_i = \{\sigma_i = (\sigma_i^k)_{k=1}^\infty \,|\, \sigma_i^k : (S_i)^k \to \Delta(D_i), \quad \forall k\}.$$

That is, for any k and any history of signals $(s_i^1, \ldots, s_i^k)$ in $(S_i)^k$, a behavioral strategy σ_i must specify a probability distribution $\sigma_i^k(\cdot|s_i^1, \ldots, s_i^k)$ in $\Delta(D_i)$, which describes how i would randomly determine his move at round k if he had observed this sequence of signals. We let $C = \times_{i \in N} C_i$ and $B = \times_{i \in N} B_i$.

Given any behavioral strategy profile $\sigma = (\sigma_i)_{i \in N}$ in B, let $P^k(d, \theta|\sigma)$ denote the probability that, at round k, θ^k will be the current state of nature and d will be the profile of moves chosen by the players, if every player i uses his behavioral strategy σ_i at all rounds. To make this definition formally precise, we can inductively define $Q^1(\cdot|\sigma)$, $Q^2(\cdot|\sigma)$, ..., and $P^1(\cdot|\sigma)$, $P^2(\cdot|\sigma)$, ... by the following equations

$$Q^1(s^1, \theta^1|\sigma) = q(s^1, \theta^1),$$

$$Q^{k+1}(s^1, \ldots, s^{k+1}, \theta^{k+1}|\sigma) = \sum_{\theta^k \in \Theta} \sum_{d \in D} Q^k(s^1, \ldots, s^k, \theta^k|\sigma)$$

$$\left(\prod_{i \in N} \sigma_i^k(d_i|s_i^1, \ldots, s_i^k) \right) p(s^{k+1}, \theta^{k+1}|d, \theta^k),$$

$$P^k(d, \theta^k|\sigma) = \sum_{(s^1, \ldots, s^k)} Q^k(s^1, \ldots, s^k, \theta^k|\sigma) \prod_{i \in N} \sigma_i^k(d_i|s_i^1, \ldots, s_i^k).$$

(So $Q^k(s^1, \ldots, s^k, \theta^k|\sigma)$ is the probability that, at round k, $(s^1, \ldots, s^k)$ will be the history of profiles of signals received by the various players and θ^k will be the current state of nature, if every player i uses the behavioral strategy σ_i.)

In the analysis of finite games in extensive form, we have assumed that each player's objective is to maximize the expected value of a utility payoff that he gets at the end of the game. For repeated games, to allow an infinite time horizon, we must drop this assumption that all payoffs come at the end of the game. Instead, we suppose that players get payoffs at each round of the game. That is, the payoff outcome to a player in a repeated game is an infinite sequence of payoffs. To describe players' objectives, we must specify some criterion for ranking such payoff sequences.

In the example described above, there was an absorbing state in which the players' payoffs are always 0, and the expected number of rounds until arrival at this absorbing state is always finite. For such examples, we can use the *sum-of-payoffs criterion* to determine players' objectives. That is, we can define each player's objective to be to maximize the expected sum of payoffs that he receives at all rounds of the game. With the sum-of-payoffs criterion, we can define an equilibrium to be a behavioral strategy profile σ such that, for every player i in N and every behavioral strategy $\hat{\sigma}_i$ in B_i,

$$\sum_{k=1}^{\infty} \sum_{d \in D} \sum_{\theta \in \Theta} P^k(d,\theta|\sigma)u_i(d,\theta) \geq \sum_{k=1}^{\infty} \sum_{d \in D} \sum_{\theta \in \Theta} P^k(d,\theta|\sigma_{-i},\hat{\sigma}_i)u_i(d,\theta).$$

Unfortunately, if there is no absorbing state with zero payoffs for which the expected time of arrival is finite, then the expected sum of payoffs may be infinite for all behavioral strategy profiles, so this sum-of-payoffs criterion cannot be applied.

For any number δ such that $0 \leq \delta < 1$, the δ-*discounted average* of a sequence of payoffs $(w_i(1),w_i(2),w_i(3), \ldots)$ is

$$(1 - \delta) \sum_{k=1}^{\infty} (\delta)^{k-1}w_i(k).$$

(In this chapter, superscripts are used both as indexes and as exponents. Where there is danger of confusion, we put the base in parentheses when the superscript denotes exponentiation. Thus, for example, $(\delta)^3 = \delta \times \delta \times \delta$.) Notice that, if $w_i(k)$ is equal to a constant, independent of k, then this δ-discounted average is equal to the same constant, because

$$(1 - \delta) \sum_{k=1}^{\infty} (\delta)^{k-1} = 1.$$

The number δ is called the *discount factor* in this expression. For any general repeated game, if the payoffs are bounded, then the δ-discounted average of each player's sequence of payoffs is a finite number and is bounded in absolute value by the same number that bounds payoffs. So we can more generally apply the δ-discounted average criterion and assume that each player's objective is to maximize the expected δ-discounted average of his sequence of payoffs. (See Epstein, 1983, for an axiomatic derivation of the δ-discounted average criterion and generalizations.) With the δ-discounted average criterion, a behavioral strategy profile σ is an *equilibrium* iff, for every player i and every behavioral strategy $\hat{\sigma}_i$,

$$(1 - \delta) \sum_{k=1}^{\infty} \sum_{d \in D} \sum_{\theta \in \Theta} P^k(d,\theta|\sigma)(\delta)^{k-1}u_i(d,\theta)$$

$$\geq (1 - \delta) \sum_{k=1}^{\infty} \sum_{d \in D} \sum_{\theta \in \Theta} P^k(d,\theta|\sigma_{-i},\hat{\sigma}_i)(\delta)^{k-1}u_i(d,\theta).$$

Given a sequence of payoffs $(w_i(k))_{k=1}^{\infty}$ and a discount factor δ, let $\overline{w}_i(2,\delta)$ denote the δ-discounted average of the (sub)sequence of payoffs that begins at round 2. That is,

$$\overline{w}_i(2,\delta) = (1 - \delta) \sum_{k=1}^{\infty} (\delta)^{k-1}w_i(k+1).$$

Then the following useful *recursion equation* holds:

(7.1) $$(1 - \delta) \sum_{k=1}^{\infty} (\delta)^{k-1}w_i(k) = (1 - \delta)w_i(1) + \delta\overline{w}_i(2,\delta).$$

That is, the δ-discounted average of a sequence of payoffs beginning at round 1 is the weighted average of the payoff at round 1, with weight $(1 - \delta)$, and the δ-discounted average of the sequence of payoffs beginning at round 2, with weight δ.

In the example above, with a geometrically distributed stopping time (arrival time at the absorbing state with zero payoffs), the probability that the players will still be playing for nonzero payoffs at round k is $(\delta)^{k-1}$, when $\delta = .99$. Thus, this example is essentially equivalent to an infinitely repeated Prisoners' Dilemma game with no stopping (no absorbing state "0") but with the .99-discounted average criterion.

In general, the discount factor δ represents a measure of the patience or long-term perspective of the players. If δ is close to 0, then a δ-discounted average criterion puts most weight on the payoffs that players get in the first few rounds of the repeated game, and so is more appropriate for players who are impatient or mainly concerned about their current and near-future payoffs. As δ approaches 1, the ratio of the weights given to the payoffs in any two rounds in the δ-discounted average converges to 1; that is, for any k and l,

$$\lim_{\delta \to 1} \frac{(1-\delta)(\delta)^{k-1}}{(1-\delta)(\delta)^{l-1}} = 1.$$

So if δ is close to 1, then the δ-discounted average criterion approximates a patient criterion in which players are not significantly less concerned with their payoffs in any given future round than with their payoffs in the current round.

Another criterion for ranking payoff sequences is the *limit of average payoffs*. According to this criterion, one sequence of payoffs $w = (w_i(1), w_i(2), \ldots)$ for player i would be preferred by i over another sequence of payoffs $\hat{w} = (\hat{w}_i(1), \hat{w}_i(2), \ldots)$ iff

$$\lim_{M \to \infty} \frac{\sum_{k=1}^{M} w_i(k)}{M} \geq \lim_{M \to \infty} \frac{\sum_{k=1}^{M} \hat{w}_i(k)}{M}.$$

A difficulty with this criterion is that these limits may fail to exist, even when payoffs are bounded. When such limits fail to exist, one option is to weaken the criterion and rank $(w_i(1), w_i(2), \ldots)$ over $(\hat{w}_i(1), \hat{w}_i(2), \ldots)$ iff

$$\liminf_{M \to \infty} \frac{\sum_{k=1}^{M} w_i(k)}{M} \geq \limsup_{M \to \infty} \frac{\sum_{k=1}^{M} \hat{w}_i(k)}{M}.$$

(The *liminf*, or *limit-infimum*, of a sequence is the lowest number that is a limit of some convergent subsequence. The *limsup*, or *limit-supremum*, is the highest number that is a limit of some convergent subsequence.) Another option is to use the *overtaking criterion*, according to which a sequence of payoffs $(w_i(1), w_i(2), \ldots)$ is ranked over another sequence $(\hat{w}_i(1), \hat{w}_i(2), \ldots)$ iff

$$\liminf_{M \to \infty} \sum_{k=1}^{M} (w_i(k) - \hat{w}_i(k)) > 0.$$

Under each of these criteria, however, some payoff sequences may be incomparable. Furthermore, the limit-supremum and limit-infimum are nonlinear operations that do not commute with the expected value operator. A technical way of avoiding these difficulties is to use Banach limits to evaluate payoff sequences in repeated games. The last four paragraphs of this section give a brief introduction to this advanced topic, which is not used subsequently in this book.

Let l_∞ denote the set of bounded sequences of real numbers. A *Banach limit* is defined to be a linear operator $L{:}l_\infty \to \mathbf{R}$ such that, for any sequences $w = (w(1), w(2), \ldots)$ and $x = (x(1), x(2), \ldots)$ in l_∞,

$$L(\alpha w + \beta x) = \alpha L(w) + \beta L(x), \quad \forall \alpha \in \mathbf{R}, \ \forall \beta \in \mathbf{R};$$

$$\limsup_{k \to \infty} w(k) \geq L(w) \geq \liminf_{k \to \infty} w(k);$$

if $w(k+1) = x(k) \ \forall k \in \{1,2,3, \ldots\}$, then $L(w) = L(x)$.

The existence of Banach limits can be proved by using the Hahn-Banach theorem (see Royden, 1968, pp. 186–192), but this proof is nonconstructive. Furthermore, there are many Banach limits, which may lead to different rankings of payoff sequences.

Using ideas from nonstandard analysis (see, for example, Hurd and Loeb, 1985), we can show how to define a class of Banach limits directly from the limit of average payoffs criterion. A *filter* on a set I is defined to be any nonempty collection $\mathcal{F}$ of subsets of I such that

if $A_1 \in \mathcal{F}$ and $A_2 \in \mathcal{F}$, then $A_1 \cap A_2 \in \mathcal{F}$,

if $A_1 \in \mathcal{F}$ and $A_1 \subseteq A_2 \subseteq I$, then $A_2 \in \mathcal{F}$, and

$\varnothing \notin \mathcal{F}$.

An *ultrafilter* on I is any filter $\mathcal{F}$ on I such that, for any A that is a subset of I, if $A \notin \mathcal{F}$ then $I \backslash A \in \mathcal{F}$. That is, among any pair of complementary subsets of I, one subset must be in the ultrafilter. If I is an infinite set, then an ultrafilter $\mathcal{F}$ on I is *free* iff it contains no finite subsets of I.

Now, let I be the set of positive integers. The existence of free ultrafilters on I can be guaranteed by an axiom of choice, because defining a free ultrafilter requires infinitely many choices. Any finite set of integers must be excluded from a free ultrafilter, and any set that contains all but finitely many of the positive integers in I must be included in a free ultrafilter. But for any pair of infinite complementary subsets of I, such as the even numbers and the odd numbers, we must pick one of the two sets to be in our free ultrafilter.

Given any free ultrafilter $\mathcal{F}$ on the set of positive integers, we can define a Banach limit $L_{\mathcal{F}}(\cdot)$ such that, for any w in l_∞, $L_{\mathcal{F}}(w) = z$ iff, for every positive number ε,

$$\left\{ M \,\middle|\, z - \varepsilon \leq \sum_{k=1}^{M} w(k) / M \leq z + \varepsilon \right\} \in \mathcal{F}.$$

It can be shown that, once a free ultrafilter $\mathcal{F}$ is fixed, this limit $L_{\mathcal{F}}(w)$ is uniquely defined for every bounded sequence w, and the operator $L_{\mathcal{F}}:l_\infty \to \mathbf{R}$ satisfies all the conditions of a Banach limit. Thus, this concept of filter-limits enables us to define a player's overall payoff in an infinitely repeated game to be the limit of the average payoffs that he gets in finitely repeated versions of the game; and in making this definition, we do not need to worry about the existence of limits. In effect, a free ultrafilter defines a rule for selecting a convergent subsequence for any bounded sequence that fails to converge.

7.3 Stationary Equilibria of Repeated Games with Complete State Information and Discounting

A general repeated game has *complete state information* iff, at every round, every player knows the current state of nature. That is, for every player i in N and every signal s_i in S_i, there exists some state $\omega(s_i)$ in Θ such that, for every θ and $\hat{\theta}$ in Θ, every d in D, and every s_{-i} in $\times_{j \in N-i} S_j$,

$$p(s, \hat{\theta} | d, \theta) = 0 \text{ if } \hat{\theta} \neq \omega(s_i);$$

so the signal s_i can never occur unless the current state is $\omega(s_i)$. (Here $s = (s_{-i}, s_i)$.)

In a repeated game with complete state information, a behavioral strategy is *stationary* iff the move probabilities depend only on the current state. That is, a behavioral strategy σ_i is stationary iff there exists some $\tau_i: \Theta \to \Delta(D_i)$ such that, for each k and each history of signals $(s_i^1, \ldots, s_i^k)$ in $(S_i)^k$,

$$\sigma_i^k(\cdot | s_i^1, \ldots, s_i^k) = \tau_i(\cdot | \omega(s_i^k)).$$

Because τ_i completely characterizes σ_i when the above equation holds, we may also refer to any such function $\tau_i: \Theta \to \Delta(D_i)$ as a *stationary strategy* for player i. Here $\tau_i(d_i | \theta)$ denotes the probability, at each round when θ is the state of nature, that player i would choose move d_i. We say that a profile of stationary strategies $\tau = (\tau_i)_{i \in N}$ is an equilibrium iff the corresponding behavioral strategies $((\sigma_i)_{i \in N}$ as above) form an equilibrium. For any profile of stationary strategies τ, we can also write

$$\tau_i(\theta) = (\tau_i(d_i | \theta))_{d_i \in D_i} \text{ and } \tau(\theta) = (\tau_j(\theta))_{j \in N} \in \underset{j \in N}{\times} \Delta(D_j).$$

Given a profile of stationary strategies τ as above, let $v_i(\theta)$ denote the expected δ-discounted average payoff for player i in the repeated game if the players plan to always use these stationary strategies and, at round 1, the initial state of nature is θ. Let $v_i = (v_i(\theta))_{\theta \in \Theta}$. If every player plans to implement the stationary strategies τ at every round except that, at round 1 only, player i plans to choose move d_i, and if the current state of nature at round 1 is θ, then the expected δ-discounted average payoff for player i is

$$Y_i(\tau, d_i, v_i, \theta, \delta) =$$

$$\sum_{d_{-i} \in D_{-i}} \left(\prod_{j \in N-i} \tau_j(d_j | \theta) \right) \left((1 - \delta) u_i(d, \theta) + \delta \sum_{\zeta \in \Theta} \sum_{s \in S} p(s, \zeta | d, \theta) v_i(\zeta) \right).$$

(Here $D_{-i} = \times_{j \in N-i} D_j$, $d_i = (d_j)_{j \in N-i}$, and $d = (d_{-i}, d_i)$.) This formula is derived from the basic recursion equation (7.1). When all players use stationary strategies, the conditionally expected δ-discounted average of the sequence of payoffs beginning at round 2, given that ζ was the state of nature at round 2, is the same as what the conditionally expected δ-discounted average of the sequence of payoffs beginning at round 1 would be, if ζ were the given state of nature at round 1. The expected δ-discounted average payoffs, when τ is used at every round, must satisfy

$$(7.2) \qquad v_i(\theta) = \sum_{d_i \in D_i} \tau_i(d_i | \theta) Y_i(\tau, d_i, v_i, \theta, \delta), \quad \forall \theta \in \Theta.$$

When $0 \le \delta < 1$, the system of equations (7.2) always has at most one bounded solution in v_i, given τ. (To prove uniqueness, suppose to the contrary that v_i and $\hat{v}_i$ are two different solutions to the system of equations (7.2). Because the components of the probability distributions $p(\cdot | d, \theta)$ and $\tau_i(\cdot | \theta)$ must sum to 1, (7.2) and the definition of Y_i would imply that

$$0 < \max_{\theta \in \Theta} |v_i(\theta) - \hat{v}_i(\theta)|$$

$$\le \max_{\theta \in \Theta} \max_{d_i \in D_i} |Y_i(\tau, d_i, v_i, \theta, \delta) - Y_i(\tau, d_i, \hat{v}_i, \theta, \delta)|$$

$$\le \delta \max_{\zeta \in \Theta} |v_i(\zeta) - \hat{v}_i(\zeta)| < \max_{\zeta \in \Theta} |v_i(\zeta) - \hat{v}_i(\zeta)|,$$

which is impossible.) Thus, for any profile of stationary strategies τ, the expected δ-discounted average payoff for player i in the repeated game, starting at each state, can always be found by solving (7.2).

For a profile of stationary strategies τ to be an equilibrium of a repeated game with complete state information, no player should expect to do better by deviating to some other move at some round in some state θ that could ever occur in the game. This requirement implies that

$$(7.3) \qquad v_i(\theta) = \max_{d_i \in D_i} Y_i(\tau, d_i, v_i, \theta, \delta), \quad \forall \theta \in \Theta.$$

For each player i, to make (7.3) a necessary condition for an equilibrium, we should restrict the range of θ to be the set of all states that may occur with positive probability when the players use the strategies τ. However, for an equilibrium that is sequentially rational in all states, condition (7.3) is necessary as stated, with θ ranging over all of Θ.

As an application of fundamental results from dynamic programming (see, e.g., Blackwell, 1965; Howard, 1960), it can be shown that, if no player could ever expect to gain by deviating at one round alone from a stationary profile of strategies τ, then no player would ever want to deviate from this profile of strategies to any more complicated strategy. That is, we have the following theorem.

THEOREM 7.1. *Given a repeated game with complete state information and bounded payoffs, and given a profile of stationary strategies τ, if there exists a bounded vector $v = (v_i(\theta))_{\theta \in \Theta, i \in N}$ such that conditions (7.2) and (7.3) are satisfied for every player i, then τ is an equilibrium of the repeated game with the δ-discounted average payoff criterion. In this equilibrium, $v_i(\theta)$ is the expected δ-discounted average payoff for player i in the repeated game when the initial state of nature is θ.*

Proof. Suppose, contrary to the theorem, that τ and v satisfy conditions (7.2) and (7.3) for every player i, but τ is not an equilibrium. Then there exists some positive number ε and some strategy σ_i (which is not necessarily stationary) for some player i such that player i's expected δ-discounted average payoff would increase by more than ε if i changed from τ_i to σ_i while all other players followed the strategies in τ. For any K, let $\hat{\sigma}_i^K$ denote the strategy that coincides with σ_i at all rounds before round K and coincides with the stationary strategy τ_i at round K and thereafter.

Now compare strategies $\hat{\sigma}_i^K$ and $\hat{\sigma}_i^{K+1}$ for player i, when the other players are expected to follow τ. Player i's expected payoffs at all rounds before round K are the same for these two strategies. Furthermore, i's behavior will coincide with τ_i at round $K+1$ and thereafter. So if the state of nature at round $K+1$ were ζ, then $v_i(\zeta)$ would be the expected δ-discounted average of i's sequence of payoffs beginning at round $K+1$. Thus, if the state at round K were θ and player i chose move d_i, then the expected δ-discounted average of i's sequence of payoffs beginning at round K would be $Y_i(\tau,d_i,v_i,\theta,\delta)$, as defined above. But conditions (7.2) and (7.3) imply that

$$\tau_i(d_i|\theta) > 0 \text{ only if } d_i \in \underset{e_i \in D_i}{\operatorname{argmax}} Y_i(\tau,e_i,v_i,\theta,\delta).$$

That is, at the one round where $\hat{\sigma}_i^K$ and $\hat{\sigma}_i^{K+1}$ differ, $\hat{\sigma}_i^K$ always chooses an optimal move for player i. So $\hat{\sigma}_i^K$ must be at least as good as $\hat{\sigma}_i^{K+1}$ for player i against τ_{-i}, for any positive number K. By induction, therefore, no strategy $\hat{\sigma}_i^K$ can be better than $\hat{\sigma}_i^1$, which is identical to τ_i, for player i against τ_{-i}.

Now choose K such that $\Omega(\delta)^{K-1} < \varepsilon/2$, where Ω is the bound on payoffs. Then the expected δ-discounted average of i's sequence of payoffs (beginning at round 1) when he uses $\hat{\sigma}_i^K$ cannot differ by more than $\varepsilon/2$ from the expected δ-discounted average of i's sequence of payoffs when he uses σ_i, because the two strategies do not differ until round K. By our choice of ε, this fact implies that $\hat{\sigma}_i^K$ is better than τ_i for i against τ_{-i}. This contradiction proves the theorem. ∎

To make the most use of Theorem 7.1, we should recognize that many strategies that do not at first appear stationary may become stationary when we consider an equivalent model that has a larger state space. For example, recall from Section 7.1 the strategy where player i in the repeated Prisoners' Dilemma game plays g_i until someone plays f_j at some round, and thereafter plays f_i. This strategy is not stationary as long as there is only one state that represents "active play continuing," as formulated in the model proposed in Section 7.2. However, we can consider an equivalent game that includes two states 1a and 1b in Θ, where 1a represents "active play continuing and no one has ever chosen f_1 or f_2" and 1b represents "active play continuing and someone has chosen f_1 or f_2 at some time in the past." In this model, 1a would be the initial state, a transition from 1a to 1b would occur whenever the moves are not (g_1, g_2), a transition from 1b to 1a would never occur, and payoffs would be the same in both states 1a and 1b. In this model, the strategy described above would be stationary, with

$$\tau_i(g_i | 1a) = 1, \quad \tau_i(f_i | 1b) = 1.$$

In this elaborated-state model, Theorem 7.1 can be applied to show that it is an equilibrium for both players to apply this strategy, if δ is not too small. When both players implement this strategy, conditions (7.2) and (7.3) for each player i become

$$v_i(1a) = (1 - \delta)5 + \delta(.99v_i(1a) + .01v_i(0))$$
$$\geq (1 - \delta)6 + \delta(.99v_i(1b) + .01v_i(0)),$$

$$v_i(1b) = (1 - \delta)1 + \delta(.99v_i(1b) + .01v_i(0))$$
$$\geq (1 - \delta)0 + \delta(.99v_i(1b) + .01v_i(0)),$$

$$v_i(0) = (1 - \delta)0 + \delta v_i(0).$$

For example, the first inequality asserts that, in state 1a, being generous (getting payoff 5) this round and so staying in state 1a next round with probability .99 (unless the transition to state 0 occurs, which has probability .01), is not worse for player i than being selfish (getting payoff 6) this round and so going to state 1b next round with probability .99. The unique solution of the three equations is

$$v_i(1a) = \frac{5(1 - \delta)}{1 - .99\delta},$$

$$v_i(1b) = \frac{1 - \delta}{1 - .99\delta},$$

$$v_i(0) = 0,$$

and the inequalities are satisfied if $\delta \geq 20/99$.

In general, suppose the definition of a "state of nature" has been elaborated sufficiently such that, in a given behavioral strategy profile, all players' strategies are stationary. If no player in any one state could expect to gain by deviating from his given strategy at just one round, then the behavioral strategy profile is an equilibrium (i.e., no player could ever expect to gain by planning to deviate over many different rounds).

For an interesting example, consider the following game, called the *Big Match*, due to Blackwell and Ferguson (1968). Let $N = \{1,2\}$, $\Theta =$

$\{0,1,2\}$, $D_1 = \{x_1, y_1\}$, $D_2 = \{x_2, y_2\}$, $S_1 = S_2 = (D_1 \times D_2) \cup \Theta$. In this game, the initial state is state 1, and each player's initial signal confirms this, so $q(1,1,1) = 1$. As long as the game is in state 1, the payoffs $(u_1(d,1), u_2(d,1))$ are as shown in Table 7.2. The state of nature stays at 1 until player 1 chooses y_1, and then the state changes either to state 0, if player 2 chose x_2 at the round where player 1 first chose y_1, or to state 2, if player 2 chose y_2 at the round where player 1 first chose y_1. States 0 and 2 are absorbing, and the payoffs in these states are always the same as at the last round in state 1, so

$$u_1(d,0) = 0, \ u_2(d,0) = 0, \ u_1(d,2) = 2, \ u_2(d,2) = -2, \quad \forall d \in D.$$

As long as the game stays in state 1, the players observe each other's past moves, so

$$p(d,d,1 \mid d,1) = 1 \text{ if } d_1 = x_1,$$

while

$$p(0,0,0 \mid (y_1,x_2),1) = 1, \ p(2,2,2 \mid (y_1,y_2),1) = 1, \text{ and}$$
$$p(0,0,0 \mid d,0) = 1, \ p(2,2,2 \mid d,2) = 1, \quad \forall d \in D.$$

The interest in this game derives from the fact that player 2 wants to avoid being caught at y_2 in the round where player 1 first chooses y_1; but if player 2 avoids this trap by always choosing x_2, then player 1 can beat her just as badly by always choosing x_1. To find a stationary equilibrium of the Big Match with a δ-discounted average criterion, let $\alpha = \tau_1(x_1 \mid 1)$, $\beta = \tau_2(x_2 \mid 1)$. It is easy to see that there is no pure-strategy equilibrium, so we can assume that $0 < \alpha < 1$ and $0 < \beta < 1$. Now, let $z = \nu_1(1)$ denote the δ-discounted average payoff to player 1 in this game. Because it is a zero-sum game, the δ-discounted average payoff to player 2 in this game must be $-z$. If player 1 chooses y_1 in round 1,

Table 7.2   Payoffs in the Big Match, for all move profiles, at any round when y_1 has not been previously chosen

	D_2	
D_1	x_2	y_2
x_1	$2,-2$	$0,0$
y_1	$0,0$	$2,-2$

then either he gets 0 forever, with probability β, or he gets 2 forever, with probability $1 - \beta$. If player 1 chooses x_1 in round 1, he gets a current expected payoff of $2\beta + 0(1 - \beta)$, and thereafter the game continues in state 1 at round 2. So equations (7.2) and (7.3) imply that

$$z = 0\beta + 2(1 - \beta) = (1 - \delta)2\beta + \delta z.$$

Thus, $\beta = \frac{1}{2}$, and $z = 1$. Similarly, to make player 2 willing to randomize between x_2 and y_2, we must have

$$
\begin{aligned}
-z &= \alpha((1 - \delta)(-2) + \delta(-z)) + (1 - \alpha)(0) \\
&= \alpha((1 - \delta)(0) + \delta(-z)) + (1 - \alpha)(-2);
\end{aligned}
$$

so $\alpha = 1/(2 - \delta)$. Notice that, as δ approaches 1, player 1's equilibrium strategy approaches the strategy in which he always chooses x_1; but that cannot be an equilibrium because player 2 would respond to it by always choosing y_2.

Blackwell and Ferguson (1968) studied the Big Match under the limit of average payoffs criterion and showed that, in this two-person zero-sum game, player 1's minimax value is $+1$ and player 2's minimax value (or security level) is -1, just as it was under the discounted average payoff criterion. Player 2 can guarantee that player 1 cannot expect more than $+1$ by randomizing between x_1 and y_2, each with probability $\frac{1}{2}$, at every round while the state of nature is 1. However, there is no equilibrium under the limit of average payoffs because player 1 has no minimax strategy against player 2. Blackwell and Ferguson showed that, for any positive integer M, player 1 can guarantee himself an expected limit of average payoffs that is not lower than $M/(M + 1)$ (so player 2's limit of average payoffs is not more than $-M/(M + 1)$) by using the following strategy: at each round, choose y_1 with probability $1/(1 + M + x^* - y^*)^2$, where x^* is the number of times that player 2 has chosen x_2 in the past and y^* is the number of times that player 2 has chosen y_2 in the past.

7.4 Repeated Games with Standard Information: Examples

A *repeated game with standard information*, or a *standard repeated game* (also sometimes called a *supergame*), is a repeated game in which there is only one possible state of nature and the players know all of each other's past moves. That is, in a standard repeated game $|\Theta| = 1$, $S_i = \times_{j \in N} D_j$ for every i, and

$$p(d, \ldots, d, \theta \,|\, d, \theta) = 1, \quad \forall d \in D.$$

Repeated games with standard information represent situations in which a group of individuals face exactly the same competitive situation infinitely often and always have complete information about each other's past behavior. Thus, the standard information structure maximizes the players' abilities to respond to each other.

The stationary equilibria of a standard repeated game are just the equilibria of the corresponding one-round game, repeated in every round. However, the nonstationary equilibria of standard repeated games are generally much larger sets, because of the players' abilities to respond to each other and punish each other's deviations from any supposed equilibrium path. In fact, under weak assumptions, almost any feasible payoff allocation that gives each player at least his minimax value can be achieved in an equilibrium of a standard repeated game. A general statement of this result is developed and proved in the next section; in this section, we develop the basic intuition behind this important result by detailed consideration of two examples.

To illustrate the analysis of standard repeated games, consider the repeated game of *Chicken*, where the payoffs to players 1 and 2 are as shown in Table 7.3. Each player i has two moves to choose between at each round: the "cautious" move a_i and the "bold" move b_i. Each player would most prefer to be bold while the other is cautious, but for both to be bold is the worst possible outcome for both. The best symmetric outcome is when both are cautious.

If the players play this game only once, then there are three Nash equilibria of this game, one giving payoffs (6,1), one giving payoffs (1,6), and one randomized equilibrium giving expected payoffs (3,3). When communication is allowed, there is a correlated equilibrium of this game in which the probability of (a_1, a_2) is equal to .5 and the expected payoff allocation is (3.75,3.75), but no higher symmetric ex-

Table 7.3 Payoffs at any round, for all move profiles, in the repeated game of Chicken

	D_2	
D_1	a_2	b_2
a_1	4,4	1,6
b_1	6,1	$-3,-3$

pected payoff allocation can be achieved in a correlated equilibrium of the one-round game. In particular, (a_1, a_2) is not an equilibrium of the one-round game, because each player i would prefer to be bold (choose b_i) if he expected the other player to be cautious (choose a_{-i}).

Suppose now that this game is repeated infinitely, with standard information, and each player uses a δ-discounted average payoff criterion, for some number δ that is between 0 and 1. A strategy for a player in the repeated game is a rule for determining his move at every round as a function of the history of moves that have been used at every preceding round. For example, one celebrated strategy for repeated games like this one (or the repeated Prisoners' Dilemma) is the strategy called *tit-for-tat*. Under the tit-for-tat strategy, a player chooses his cautious move in the first round, and thereafter chooses the same move as his opponent chose in the preceding round.

If both players follow the tit-for-tat strategy, then the actual outcome will be (a_1, a_2) in every round, giving each player a discounted average payoff of 4. However, tit-for-tat should not be confused with the strategy "always choose a_i," under which a player would choose a_i at every round no matter what happened in the past. To see why, suppose first that player 1 is following the strategy of "play a_1 at every round." Then player 2's best response is to always play b_2 and get a discounted average payoff of 6. On the other hand, if player 1 is following the tit-for-tat strategy and the discount factor δ is close to 1, then player 2 will never want to choose b_2, because she will lose more from player 1's reprisal next round than she will gain from choosing b_2 in this round. Thus, the part of a strategy that specifies how to punish unexpected behavior by the opponent may be very important, even if such punishments are not carried out in equilibrium.

Let us check to see how large δ has to be for tit-for-tat to deter bold behavior by one's opponent. Suppose that player 1 is implementing the tit-for-tat strategy and that player 2 is considering whether to be bold at the first round and thereafter go back to being cautious. If she does so, her payoff will be 6 in the first round, 1 in the second round, and 4 at every round thereafter, so her δ-discounted average payoff is

$$(1 - \delta)(6 + 1\delta + \sum_{k=3}^{\infty} 4\delta^{k-1}) = 6 - 5\delta + 3\delta^2.$$

On the other hand, if she is always cautious against player 1's tit-for-tat, she will get the discounted average payoff of 4. So, to deter player 2 from being bold, we need

$$4 \geq (1 - \delta)(6 + 1\delta + \sum_{k=3}^{\infty} 4\delta^{k-1}),$$

or $\delta \geq \frac{2}{3}$. In fact, if $\delta \geq \frac{2}{3}$, then neither player can gain by being the first to deviate from the tit-for-tat strategy, if the other is expected to always use tit-for-tat; so tit-for-tat is an equilibrium.

Although it is an equilibrium for both players to use the tit-for-tat strategy, the equilibrium is not subgame perfect. To be a subgame-perfect equilibrium, the players must always choose sequentially rational moves, even after histories of moves that have zero probability under the given strategies. Consider the situation that would be faced by player 1 at the second round if player 2 accidentally chose b_2 at the first round. Under the assumption that both players will follow tit-for-tat hereafter, the outcome will be (b_1, a_2) in every even-numbered round and (a_1, b_2) in every odd-numbered round; and the discounted average payoff to player 1 will be

$$(1 - \delta)(1 + 6\delta + 1\delta^2 + 6\delta^3 + \ldots) = (1 + 6\delta)/(1 + \delta).$$

On the other hand, if player 1 deviates from tit-for-tat by choosing a_1 at round 2 and both players then follow tit-for-tat thereafter, then the outcome will be (a_1, a_2) at every round after round 1; so the discounted average payoff to player 1 will be

$$(1 - \delta)(1 + 4\delta + 4\delta^2 + 4\delta^3 + \ldots) = 1 + 3\delta.$$

With some straightforward algebra, it can be shown that

$$1 + 3\delta > (1 + 6\delta)/(1 + \delta) \quad \text{when } \delta > \frac{2}{3}.$$

Furthermore, for any δ, either player could gain from unilaterally deviating from the tit-for-tat strategy after both players chose (b_1, b_2), because alternating between payoffs 1 and 6 is better than always getting -3. So it would be irrational for a player to implement the punishment move in tit-for-tat if the other player is also expected to implement tit-for-tat hereafter. That is, the scenario in which both play tit-for-tat is not a subgame-perfect equilibrium.

There is a way to modify tit-for-tat to get around this difficulty, however. Consider the following strategy for player i: at each round, i chooses a_i unless the other player has in the past chosen b_{-i} strictly more times than i has chosen b_i, in which case i chooses b_i. We call this strategy *getting-even*. It can be shown that, as long as δ is greater than $\frac{2}{3}$, it is a subgame-perfect equilibrium for both players to follow the getting-even strategy. For example, if player 2 deviates from the strategy and chooses b_2 at round 1 but is expected to apply the strategy thereafter, then it is rational for player 1 to retaliate and choose b_1 at round 2. Player 1 expects that player 2, under the getting-even strategy, will treat player 1's retaliatory b_1 as a justified response; so player 2 should not reply with a counterretaliatory b_2 in the third round. The condition $\delta \geq \frac{2}{3}$ is needed only to guarantee that neither player wants to choose b_i when the number of past bold moves is equal for the two players.

The distinction between tit-for-tat and getting-even is very fine, because they differ only after a mistake has been made. Let $r_i(k)$ denote the number of rounds at which player i chooses b_i before round k. If player 1 applies the getting-even strategy correctly, then, no matter what strategy player 2 uses, at any round k, $r_2(k) - r_1(k)$ will always equal either 0 or 1 and will equal 1 if and only if player 2 chose b_2 last round. Thus, according to the getting-even strategy, player 1 will choose b_1 if and only if player 2 chose b_1 in the preceding round. The distinction between the two strategies becomes evident only in cases where player 1 himself has made some accidental deviation from his own strategy but then goes back to implementing it.

Neither tit-for-tat nor getting-even can be sustained bilaterally as an equilibrium of this game if $\delta < \frac{2}{3}$. However, as long as $\delta \geq .4$, there are other subgame-perfect equilibria of this repeated game that achieve the outcome (a_1, a_2) in every round with probability 1.

When both players are supposed to choose a_i in equilibrium, player 1 can deter player 2 from choosing b_2 only if he has a credible threat of some retaliation or punishment that would impose in the future a greater cost (in terms of the discounted average) than player 2 would gain from getting 6 instead of 4 in the current round. As the discount factor becomes smaller, however, losses in later payoffs matter less in comparison with gains in current payoffs, so it becomes harder to deter deviations to b_2. To devise a strategy that deters such deviations for the lowest possible discount factor, we need to find the credible threat that is most costly to player 2.

At worst, player 2 could guarantee herself a payoff of $+1$ per round, by choosing a_2 every round. So the discounted average payoff to player 2 while she is being punished for deviating could not be lower than 1. Such a payoff would actually be achieved if player 1 chose b_1 and player 2 chose a_2 forever after the first time that player 2 chose b_2. Furthermore, it would be rational for player 1 to choose b_1 forever and player 2 to choose a_2 forever if each expected the other to do so. Thus, the worst credible punishment against player 2 would be for player 1 to choose b_1 forever while player 2 responds by choosing a_2.

Consider the following *grim strategy* for player i: if there have been any past rounds where one player was bold while the other player was cautious (outcome (b_1, a_2) or (a_1, b_2)) and, at the *first* such round, i was the cautious player, then i chooses b_i now and hereafter; otherwise i chooses a_i. That is, under the grim strategy, i plays cooperatively (that is, "cautiously") until his opponent deviates and is bold, in which case i punishes by being bold forever. As long as $\delta \geq .4$, it is a subgame-perfect equilibrium for both players to follow the grim strategy.

Such infinite unforgiving punishment may seem rather extreme. So we might be interested in finding other strategies that, like tit-for-tat, have only one-round punishments and can sustain (a_1, a_2) forever as the actual outcome in a subgame-perfect equilibrium when the discount factor is smaller than $2/3$. Such a strategy does exist. We call it *mutual punishment*, and it can be described for player i as follows: i chooses a_i if it is the first round, or if the moves were (a_1, a_2) or (b_1, b_2) last round; i chooses b_i if the moves were (b_1, a_2) or (a_1, b_2) last round. If both players are supposed to be following this strategy but one player deviates in a particular round, then in the next round the players are supposed to choose (b_1, b_2), which is the worst possible outcome for both of them (hence the name "mutual punishment"). If player 2 is following this strategy correctly and player 1 deviates to choose b_1, then player 2 will choose b_2 in retaliation the next round and will continue to do so until player 1 again chooses b_1. The idea is that player 2 punishes player 1's deviation until player 1 participates in his own punishment or bares his chest to the lash.

To check whether it is a subgame-perfect equilibrium for both players to follow the mutual-punishment strategy, we must consider two possible deviations: choosing b_i when the strategy calls for a_i, and choosing a_i when the strategy calls for b_i. If the strategy calls for a_i but player i deviates for one round while the other player follows the strategy, then i will get payoff $+6$ this round and -3 the next, whereas he would have gotten $+4$ in both rounds by not deviating. Because the next round is discounted by an extra factor of δ, the deviation is deterred if $4 + 4\delta \geq 6 + -3\delta$, that is, if $\delta \geq 2/7$. On the other hand, if the strategy calls for b_i but i deviates for one round while the other player follows the strategy, then i will get payoff $+1$ this round and -3 the next, whereas he would have gotten -3 in this round and $+4$ in the next by not deviating. So this deviation is deterred if $-3 + 4\delta \geq 1 + -3\delta$, that is, if $\delta \geq 4/7$. Thus, it is a subgame-perfect equilibrium for both players to follow the mutual-punishment strategy as long as the discount factor is at least $4/7$.

Thus far, we have been discussing only equilibria in which the actual outcome is always the cooperative (a_1,a_2) at every round, unless someone makes a mistake or deviates from the equilibrium. There are many other equilibria to this repeated game, however. Any equilibrium of the original one-round game would be (when repeated) an equilibrium of the repeated game. For example, there is an equilibrium in which player 1 always chooses b_1 and player 2 always chooses a_2. This is the best equilibrium for player 1 and the worst for player 2.

There are also equilibria of the repeated game that have very bad welfare properties and are close to being worst for both players. Furthermore, some of these bad equilibria have a natural or logical appeal that may, in some cases, make them the focal equilibria that people actually implement. To see the logic that leads to these bad equilibria, notice that the getting-even and grim equilibria both have the property that the player who was more bold earlier is supposed to be less bold later. Some people might suppose that the opposite principle is more logical: that the player who has been more bold in the past should be the player who is expected to be more bold in the future.

So consider a strategy, which we call the *q-positional* strategy, that may be defined for each player i as follows: i chooses b_i if i has chosen b_i strictly more times than the other player has chosen b_{-i}; i chooses a_i if the other player has chosen b_{-i} strictly more times than i has chosen b_i; and if b_i and b_{-i} have been chosen the same number of times, then i chooses b_i now with probability q and chooses a_i now with probability $1 - q$. The intuitive rationale for this strategy is that the player who has established a stronger reputation for boldness can be bold in the future, whereas the player who has had a more cautious pattern of behavior should conform to the cautious image that he has created. When neither player has the more bold reputation, then they may independently randomize between bold and cautious in some way.

Given any value of the discount factor δ between 0 and 1, there is a value of q such that it is a subgame-perfect equilibrium for both players to follow the q-positional strategy. To compute this q, notice first that, although there are no alternative states of nature intrinsically defined in this repeated game, the players' positional strategies are defined in terms of an implicit state: the difference between the number of past rounds in which player 1 played b_1 and the number of past rounds in which player 2 played b_2. At any round, if this implicit state is positive and both players follow the q-positional strategy, then the payoffs will be (6,1) at every round from now on. Similarly, if the implicit state is negative, then the payoffs will be (1,6) at every round from now on. Let z denote the expected δ-discounted average payoff expected by each player (the same for both, by symmetry) when the implicit state is 0 at the first round (as, of course, it actually is) and both apply the q-positional strategy. Then player 1 is willing to randomize between b_1 and a_1 when the implicit state is 0 if and only if

$$q((1 - \delta)(-3) + \delta z) + (1 - q)((1 - \delta)(6) + \delta(6))$$
$$= q((1 - \delta)(1) + \delta(1)) + (1 - q)((1 - \delta)(4) + \delta z) = z.$$

These equalities assert that, when player 2's probability of being bold is q, player 1's expected δ-discounted average payoff in the truncated game is the same, whether he is bold (first expression) or cautious (second expression) in the first round, and is in fact equal to z (as our definition of z requires). Given δ, these two equations can be solved for q and z. The results are shown in Table 7.4. Thus, if the players have long-term objectives, so δ is close to 1, then the positional equilibrium gives each player an expected δ-discounted average payoff that is only slightly better than the minimum that he could guarantee himself in the worst equilibrium. In effect, the incentive to establish a reputation for boldness pushes the two players into a *war of attrition* that is close to the worst possible equilibrium for both.

The various equilibria discussed here may be taken as examples of the kinds of behavior that can develop in long-term relationships. When people in long-term relationships perceive that their current behavior will have an influence on each other's future behavior, they may ration-

Table 7.4   Expected δ-discounted average payoffs (z) and initial boldness probabilities (q) for four discount factors (δ), in positional equilibrium of repeated Chicken

δ	q	z
.99	.992063	1.0024
.90	.925	1.024
.667	.767	1.273
.40	.582	1.903

Table 7.5   Payoffs at any round, for all move profiles, in a repeated game

D_1	D_2	
	a_2	b_2
a_1	8,8	1,2
b_1	2,1	0,0

ally become more cooperative (as in the getting-even equilibrium) or more belligerent (as in the positional equilibrium), depending on the kind of linkage that is expected between present and future behavior. Qualitatively, the more cooperative equilibria seem to involve a kind of reciprocal linkage (e.g., "expect me tomorrow to do what you do today"), whereas the more belligerent equilibria seem to involve a kind of extrapolative linkage ("expect me tomorrow to do what I do today").

For a simple example in which repeated-game equilibria may be worse for both players than any equilibrium of the corresponding one-round game, consider the game in Table 7.5. It is easy to see that the unique equilibrium of the one-round game is (a_1,a_2), which gives payoffs (8,8). For the repeated game, however, consider the following scenario: if the total number of past rounds when (a_1,a_2) occurred is even, then player 1 chooses a_1 and player 2 chooses b_2; if the total number of past rounds when (a_1,a_2) occurred is odd, then player 1 chooses b_1 and player 2 chooses a_2. When the players implement these strategies, (a_1,a_2) never occurs; so (a_1,b_2) is the outcome at every round (because 0 is even) and payoffs are (1,2). This scenario is a subgame-perfect equilibrium if $\delta \geq$ ⁶⁄₇. For example, if player 2 deviated from this scenario by choosing a_2 at round 1, then her discounted average payoff would be

$$(1 - \delta)(8 + 1\delta + 1\delta^2 + 1\delta^3 + \ldots) = 8 - 7\delta.$$

Notice that $2 \geq 8 - 7\delta$ if $\delta \geq$ ⁶⁄₇.

7.5 General Feasibility Theorems for Standard Repeated Games

The general intuition that we take from the examples in the preceding section is that, in standard repeated games, when players are sufficiently patient, almost any feasible payoff allocation that gives each player at least his minimax security level can be realized in an equilibrium of the repeated game. That is, the payoff allocations that are feasible in equilibria in a standard repeated game may generally coincide with the payoffs that are feasible in the corresponding one-round game with contracts, as studied in Section 6.1. In fact, general feasibility theorems that express this intuition have been formulated and proved under a variety of weak technical assumptions. In the game-theory literature, these feasibility theorems have been referred to as *folk theorems*, because some weak feasibility theorems were understood or believed by many game theorists, as a part of an oral folk tradition, before any rigorous statements were published. We use here the phrase *general feasibility theorem* in place of "folk theorem," because naming a theorem for the fact that it was once unpublished conveys no useful information to the uninitiated reader.

Rubinstein (1979) proved a general feasibility theorem for subgame-perfect equilibria of standard repeated games with the overtaking criterion. Fudenberg and Maskin (1986) proved a general feasibility theorem for subgame-perfect equilibria of standard repeated games with discounting. We state and prove here a version of Fudenberg and Maskin's result, using correlated equilibria instead of Nash equilibria. To strengthen the result, we assume here that there is a mediator or correlating device that, at each round, makes only public announcements that are commonly observed by all players. (That is, there is no confidential private communication with individual players.)

A *correlated strategy* for this game is then any $\mu = (\mu^k)_{k=1}^{\infty}$ such that, for each k, μ^k is a function from $(D)^{2k-2}$ into $\Delta(D)$. (Here $(D)^m$ denotes the m-fold Cartesian product of D, where $D = \times_{i \in N} D_i$ is the set of possible profiles of moves at each round.) For any k and any $(d^1, \ldots, d^{k-1}, c^1, \ldots, c^k)$ in $(D)^{2k-1}$, the number $\mu^k(c^k | d^1, \ldots, d^{k-1}, c^1, \ldots, c^{k-1})$ denotes the conditional probability that $c^k = (c_i^k)_{i \in N}$ would be the profile of moves that would be recommended to the players at round k, by the mediator, if the history of recommendations in previous rounds was $(c^1, \ldots, c^{k-1})$ and if the history of past moves was $(d^1, \ldots, d^{k-1})$. For each k, let $P^k(d^k | \mu)$ be the probability that d^k will be the profile of moves implemented at round k if recommendations are generated according to the correlated strategy μ and everyone always obeys these recommendations. To make this formally precise, we can inductively define

$$Q^k(d^1, \ldots, d^k | \mu) =$$
$$\mu^k(d^k | d^1, \ldots, d^{k-1}, d^1, \ldots, d^{k-1}) Q^{k-1}(d^1, \ldots, d^{k-1} | \mu),$$

where $Q^1(d^1) = \mu^1(d^1)$, and then let

$$P^k(d^k|\mu) = \sum_{(d^1,\ldots,d^{k-1}) \in D^{k-1}} Q^k(d^1,\ldots d^k|\mu).$$

When recommended moves at each round are publicly announced, a *manipulative strategy* for player i is any $e = (e^k)_{k=1}^{\infty}$ where, for each k, e^k is a function from $(D)^{2k-1}$ to D_i. Here $e^k(d^1,\ldots,d^{k-1},c^1,\ldots,c^k)$ represents the move that player i chooses at round k, according to the strategy e, if $(d^1,\ldots,d^{k-1})$ is the history of past moves actually chosen by the players before round k and $(c^1,\ldots,c^k)$ is the history of recommended moves at all rounds up through round k. Let $P_i^k(d^k|\mu,e)$ be the probability that d^k is the profile of moves that actually will be chosen by the players at round k, if recommended moves are determined according to μ, every player except i always obeys the recommendations, but player i uses the manipulative strategy e. To make this definition formally precise, we let

$$Q_i^1(d^1,c^1|\mu,e) = \mu(c^1) \quad \text{if } d^1 = (c_{-i}^1, e^1(c^1)),$$

$$Q_i^1(d^1,c^1|\mu,e) = 0 \qquad \text{if } d^1 \neq (c_{-i}^1, e^1(c^1)),$$

and inductively define, for all $k > 1$,

$$Q_i^k(d^1,\ldots,d^k,c^1,\ldots,c^k|\mu,e) = 0$$

$$\text{if } d^k \neq (c_{-i}^k, e^k(d^1,\ldots,d^{k-1},c^1,\ldots,c^k)),$$

$$Q_i^k(d^1,\ldots,d^k,c^1,\ldots,c^k|\mu,e) =$$

$$\mu^k(c^k|d^1,\ldots,d^{k-1},c^1,\ldots,c^{k-1}) \, Q_i^{k-1}(d^1,\ldots,d^{k-1},c^1,\ldots,c^{k-1}|\mu,e)$$

$$\text{if } d^k = (c_{-i}^k, e^k(d^1,\ldots,d^{k-1},c^1,\ldots,c^k)),$$

and then let

$$P_i^k(d^k|\mu,e) = \sum_{(d^1,\ldots,d^{k-1},c^1,\ldots,c^k) \in D^{2k-1}} Q_i^k(d^1,\ldots,d^k,c^1,\ldots,c^k|\mu,e).$$

To describe expected δ-discounted average payoffs, let

$$U_i(\mu,\delta) = (1 - \delta) \sum_{k=1}^{\infty} (\delta)^{k-1} \sum_{d \in D} P^k(d|\mu)u_i(d),$$

$$U_i^*(\mu,e,\delta) = (1 - \delta) \sum_{k=1}^{\infty} (\delta)^{k-1} \sum_{d \in D} P_i^k(d|\mu,e)u_i(d).$$

The correlated strategy μ is a *publicly correlated equilibrium* of the standard repeated game with the δ-discounted average payoff criterion iff, for every player i and every manipulative strategy e for player i,

$$U_i(\mu,\delta) \geq U_i^*(\mu,e,\delta).$$

For any k and any $(d^1,\ldots,d^k,c^1,\ldots,c^k)$ in $(D)^{2k}$, we define $\mu\backslash(d^1,\ldots,d^k,c^1,\ldots,c^k)$ such that $\mu\backslash(d^1,\ldots,d^k,c^1,\ldots,c^k) = (\hat{\mu}^m)_{m=1}^{\infty}$ iff, for every m and every $(\bar{d}^1,\ldots,\bar{d}^{m-1},\bar{c}^1,\ldots,\bar{c}^m)$ in D^{2m-1},

$$\hat{\mu}^m(c^m|\bar{d}^1,\ldots,\bar{d}^{m-1},\bar{c}^1,\ldots,\bar{c}^{m-1})$$
$$= \mu^{m+k}(c^m|d^1,\ldots,d^k,\bar{d}^1,\ldots,\bar{d}^{m-1},c^1,\ldots,c^k,\bar{c}^1,\ldots,\bar{c}^{m-1}).$$

Thus, if a mediator is using the correlated strategy μ, then $\mu\backslash(d^1,\ldots,d^k,c^1,\ldots,c^k)$ is the correlated strategy that he would be using in the subgame that follows a history of recommendations $(c^1,\ldots,c^k)$ and of actual moves $(d^1,\ldots,d^k)$. A publicly correlated equilibrium μ is *subgame perfect* iff, for every k and every $(d^1,\ldots,d^k,c^1,\ldots,c^k)$ in $(D)^{2k}$, the correlated strategy $\mu\backslash(d^1,\ldots,d^k,c^1,\ldots,c^k)$ is also a publicly correlated equilibrium.

Let $\hat{v}_i$ be the minimax value for player i in the one-round game, when only pure strategies are considered. That is,

(7.4) $$\hat{v}_i = \min_{d_{-i} \in \times_{j \in N-i} D_j} \max_{d_i \in D_i} u_i(d).$$

Let F denote the set of expected payoff allocations that could be achieved by correlated strategies in the corresponding one-round game. That is, $x = (x_i)_{i \in N}$ is in F iff there exists some η in $\Delta(D)$ such that

$$x_i = \sum_{d \in D} \eta(d)u_i(d), \quad \forall i \in N.$$

Then we may state the following general feasibility theorem, due to Fudenberg and Maskin (1986).

THEOREM 7.2. *For any standard repeated game with bounded payoffs, let x be any vector in F such that the set $F \cap \{z \in \mathbf{R}^N | \hat{v}_i \leq z_i \leq x_i \quad \forall i \in N\}$ has a nonempty interior relative to $\mathbf{R}^N$. Then there exists some number $\overline{\delta}$ such that $0 < \overline{\delta} < 1$ and, for every δ such that $\overline{\delta} \leq \delta < 1$, there exists a correlated strategy μ such that μ is a subgame-perfect publicly correlated equilibrium of the repeated game with the δ-discounted average payoff criterion and*

$$U_i(\mu,\delta) = x_i, \quad \forall i \in N.$$

Proof. Because payoffs are bounded, there exists some number Ω such that, for every i and d, $-\Omega < u_i(d) < \Omega$. The assumption that $F \cap \{z \in \mathbf{R}^N | \hat{v}_i \le z_i \le x_i \quad \forall i \in N\}$ has a nonempty interior relative to $\mathbf{R}^N$ implies that there exists some y in F and some positive number ε such that

$$\hat{v}_i < y_i - \varepsilon < y_i \le x_i, \forall i \in N,$$

and, for every j in N, the vector $(y_{-j}, y_j - \varepsilon)$ is also in F.

The construction of correlated strategy μ is best described in terms of $2|N| + 1$ modes of behavior: an initial mode and, for each player i, a mode that we call "harsh punishment against i" and a mode that we call "mild punishment against i." In the initial mode, the mediator recommends moves at each round that are determined according to a probability distribution in $\Delta(D)$ such that each player j gets the expected payoff x_j. When the mode is harsh punishment against i, the mediator recommends that all players other than i should use minimax strategy against player i that achieves the minimum in the above definition of $\hat{v}_i$ (7.4), and that player i should use his best response. When the state is mild punishment against i, the mediator recommends moves at each round that are determined according to a probability distribution in $\Delta(D)$ such that player i gets the expected payoff $y_i - \varepsilon$ and every other player j gets expected payoff y_j.

The mediator stays in the initial mode until some round at which exactly one player disobeys his recommendation. After any round at which one and only one player disobeys the mediator's recommendation, the mediator plans to make recommendations in the mode of harsh punishment against this player at each of the next K rounds and then to make recommendations in the mode of mild punishment against this player at every round thereafter. These plans will only be changed after a round at which a single player disobeys the recommendations, in which case the preceding sentence determines the new plan for the mediator.

At any round, if player i disobeys the mediator and no other players ever disobey, then i's δ-discounted average future payoff is not more than

$$(1 - \delta)(\Omega + (\delta)^1 \hat{v}_i + \cdots + (\delta)^K \hat{v}_i + (\delta)^{K+1}(y_i - \varepsilon) +$$
$$(\delta)^{K+2}(y_i - \varepsilon) + \ldots) = (1 - \delta)\Omega + (\delta - (\delta)^{K+1})\hat{v}_i + (\delta)^{K+1}(y_i - \varepsilon).$$

When the mode is harsh punishment against i, player i cannot gain by unilaterally disobeying, because in the current round his recommended move is his best response to the minimax strategy being used by the other players and disobedience will only delay the time when his planned expected payoff increases from $\hat{v}_i$ to $y_i - \varepsilon$. Player i does better in the initial mode and in the mode of mild punishment against some other player than in the mode of mild punishment against himself, because $x_i \geq y_i > y_i - \varepsilon$. So, to verify that μ is a subgame-perfect publicly correlated equilibrium, it suffices to verify that player i would not want to deviate when the mode is either mild punishment against himself or harsh punishment against some other player.

When the mode is mild punishment against i, the expected δ-discounted average of future payoffs to i is $y_i - \varepsilon$, if no one disobeys. When the mode is harsh punishment against some other player, the expected δ-discounted average of future payoffs is not less than

$$(1 - \delta)(-\Omega - (\delta)^1\Omega - \ldots - (\delta)^{K-1}\Omega + (\delta)^K y_i + (\delta)^{K+1} y_i + \ldots)$$

$$= -\Omega(1 - (\delta)^K) + (\delta)^K y_i.$$

So μ is a subgame-perfect correlated equilibrium if, for every player i,

(7.5) $$y_i - \varepsilon \geq (1 - \delta)\Omega + (\delta - (\delta)^{K+1})\hat{v}_i + (\delta)^{K+1}(y_i - \varepsilon),$$

and

(7.6) $$-\Omega(1 - (\delta)^K) + (\delta)^K y_i \geq (1 - \delta)\Omega + (\delta - (\delta)^{K+1})\hat{v}_i$$
$$+ (\delta)^{K+1}(y_i - \varepsilon).$$

Inequality (7.5) is satisfied when

$$\frac{y_i - \varepsilon - \hat{v}_i}{\Omega - \hat{v}_i} \geq \frac{1 - \delta}{1 - (\delta)^{K+1}}.$$

The left-hand side of this inequality is strictly positive and the right-hand side converges to $1/(K + 1)$ as δ converges to 1. Let us pick K so that

$$\frac{y_i - \varepsilon - \hat{v}_i}{\Omega - \hat{v}_i} > \frac{1}{K + 1}.$$

Then there exists some $\hat{\delta}$ such that (7.5) is satisfied whenever $\hat{\delta} \leq \delta < 1$. Notice also that, as δ converges to 1 with K fixed, the left-hand side of (7.6) converges to y_i and the right-hand side converges to $y_i - \varepsilon$. So there must exist some $\overline{\delta}$ such that, whenever $\overline{\delta} \leq \delta < 1$, (7.6) is satisfied as well as (7.5).   ∎

Notice that a player i may get less than his own minimax value at rounds where he is helping to force some other player j down to j's minimax value, because the act of punishing might hurt the punisher even more than the person being punished. Thus, punishment against a player after he deviates from the equilibrium would not be credible unless the punishers were given some incentive to carry out the punishment. In the equilibrium described in the proof, punishers are rewarded for their efforts in the harsh punishment mode by not being the object of the mild punishment that follows. The assumption that $F \cap \{z \in \mathbf{R}^N | \hat{v}_i \leq z_i \leq x_i \ \forall i \in N\}$ has a nonempty interior, relative to $\mathbf{R}^N$, means that there is enough independence between the different players' payoffs, so any one player can be mildly punished without pushing any other players down to their mild-punishment payoff levels.

The pure-strategy minimax value $\hat{v}_i$, defined in equation (7.4), is generally higher than the randomized-strategy minimax value v_i that was defined in Section 6.1 (equation 6.1). Using the assumption that a mediator can pass recommendations confidentially to the various players (so, at each round, no player knows the currently recommended moves for the other players), we can prove a revised version of Theorem 7.2, in which $\hat{v}_i$ is replaced by v_i and the word "publicly" is deleted. The only change in the proof is that, in the mode of harsh punishment against i, the mediator recommends to the players other than i that they choose moves that are randomly chosen according to a correlated strategy that achieves the minimum in equation (6.1), but the recommended moves for the players other than i are concealed from player i himself.

7.6 Finitely Repeated Games and the Role of Initial Doubt

As noted in Section 7.1, there is a striking contrast between the finitely repeated Prisoners' Dilemma game and the infinitely repeated version of that game. If the Prisoners' Dilemma game is repeated only finitely many times and the finite upper bound on the number of repetitions is common knowledge to the players at the beginning of the game, then the unique equilibrium is for both players to always play selfishly; so their average payoff allocation must be (1,1). But if there is an infinite time horizon and the players' discount factor is close to 1, then any feasible payoff allocation that gives each player more than his minimax value 1 can be achieved in a subgame-perfect equilibrium.

This striking difference between an infinitely repeated game and a long finitely repeated game is actually rather special to the Prisoners' Dilemma, however. Benoit and Krishna (1985) showed that if a strategic-form game has multiple equilibria (when it is not repeated) that give two or more different payoffs to each player, then any payoff allocation vector x that satisfies the conditions of Theorem 7.2 is arbitrarily close to vectors that can be achieved as average payoff allocations in subgame-perfect equilibria of sufficiently long finitely repeated versions of this game.

For example, consider the game shown in Table 7.6. Without repetition, there are three equilibria of this game, giving payoff allocations (1,6), (6,1), and (3,3). The minimax value for each player is 0. To approximate the Pareto-efficient allocation (4,4) in a finitely repeated game, consider the following equilibrium. Each player i plays a_i as long as both have done so, until the last two rounds. If both players have always played (a_1, a_2) before the last two rounds, then at the last two rounds they play the randomized equilibrium that gives expected payoffs (3,3) at each round. On the other hand, if either player ever deviates from (a_1, a_2) in a round *before* the last two rounds, then the players thereafter play the equilibrium that gives payoff 1 at each round to the player who deviated first. (If both deviate first at the same round, then let us say that the players act as if player 1 deviated first.) It is a subgame-perfect equilibrium, with the sum-of-payoffs criterion, for both players to behave according to this scenario in the K-round finitely repeated game, for any positive integer K. Furthermore, this equilibrium gives an expected average payoff per round of $((K - 2)4 + 6)/K$, which converges to 4 as K goes to infinity.

Table 7.6   Payoffs at any round, for all move profiles, in a finitely repeated game

	D_2		
D_1	a_2	b_2	x_2
a_1	4,4	1,6	0,5
b_1	6,1	$-3,-3$	$-4,-4$
x_1	5,0	$-4,-4$	$-5,-5$

There are also equilibria of this finitely repeated game that are worse for both players than any equilibrium of the one-round game (but, of course, are not worse than the minimax values). For example, consider the following scenario, when the game is played $7K + M$ times, for any positive integers K and M, where $2 \leq M \leq 8$. The players implement a seven-round cycle of (b_1,b_2) for four rounds and then (a_1,a_2) for three rounds, until someone deviates or until the last M rounds of the game. Notice that this cycle gives each player an average payoff of 0 per round. If no one ever deviates from this cycle then, during the last M rounds, they play the randomized equilibrium that gives expected payoff allocation (3,3) at each round. If any player deviates from the seven-round cycle of (b_1,b_2) and (a_1,a_2), then, as "punishment" for the player i who deviated first (where, if both deviate first at the same round, we may let i equal 1), the punished player i chooses a_i thereafter and the other (punishing) player j chooses x_j in every round except the last round, when he chooses b_j. Under this punishment scheme, the punished player i gets 0 at every round, except the last, when he gets 1. To give the punishing player j an incentive to use his move x_j (which is dominated in the unrepeated game), we must stipulate that if the punishing player j ever chose b_j when he was supposed to be choosing x_j, then the punishment against i would be terminated and they would switch to playing thereafter according to the equilibrium (b_i,a_j), in which the formerly punishing player j gets 1 and the formerly punished player i gets 6. This scenario is a subgame-perfect equilibrium of the finitely repeated game with the sum-of-payoffs criterion, and it gives each player an expected average payoff per round that converges to the minimax value 0, as K goes to infinity.

In the infinitely repeated Prisoners' Dilemma game (see Table 7.1), one might think that the equilibrium in which both players always choose their selfish moves (f_1,f_2) is somehow "more rational" than the other equilibria, because it is the only equilibrium that corresponds to a sequential equilibrium of finitely repeated versions of the game. However, Kreps, Milgrom, Roberts, and Wilson (1982) have shown that there are other ways of constructing finitely repeated versions of the Prisoners' Dilemma which also have unique equilibria that converge to very different equilibria of the infinitely repeated game, as the finite number of repetitions goes to infinity. Their basic idea is to consider finitely repeated games in which there is a small probability of an alternative state about which some players have incomplete information. They show that, in games that are repeated a large but finite number of times, *small initial doubts* about the state of the game may radically affect the set of sequential equilibria.

Kreps, Milgrom, Roberts, and Wilson (1982) considered a finitely repeated Prisoners' Dilemma game with the sum-of-payoffs criterion, in which there is a small probability ε that player 2 is not really a rational player but is instead a machine that always plays tit-for-tat. Player 1 cannot know that player 2 is not such a machine until she actually deviates from tit-for-tat. Kreps et al. showed that there exists some finite number $M(\varepsilon)$, depending on ε but not on the number of rounds in the game, such that, in every sequential equilibrium, with probability 1, both players will use their generous moves (g_1,g_2) at every round before the last $M(\varepsilon)$ rounds. The essential idea is that, even if player 2 is not a tit-for-tat machine, she has an incentive to cultivate player 1's doubt, because he would always be generous before the last round were he to assign a sufficiently high probability to the event that she is a tit-for-tat machine. In any sequential equilibrium, if player 2 ever deviates from the tit-for-tat strategy, then both players will play selfishly (f_1,f_2) at every round thereafter because, in any subgame, it will be common knowledge that player 2 is rational. So, to preserve player 1's doubt, player 2 always acts like a tit-for-tat machine until the last $M(\varepsilon)$ moves; so player 1 always is generous before the last $M(\varepsilon)$ moves. Before the last $M(\varepsilon)$ rounds, because player 2 is expected to play tit-for-tat no matter what, player 1 learns nothing about her type, and so continues to assign probability ε to the chance that player 2 is a tit-for-tat machine. During the last $M(\varepsilon)$ rounds, player 2 may randomly deviate from the tit-for-tat strategy, if she is not a machine; so the belief probability that player 1 would assign to her being a machine if she has not deviated from tit-for-tat will increase above ε during these last rounds. In a sequential equilibrium, player 2's random deviation probabilities must be just large enough so that player 1's consistent beliefs will make him willing to randomize between g_1 and f_1 in a way such that the rational player 2, if not a machine, would indeed be willing to randomize between continuing to play tit-for-tat and deviating. To make player 1 willing to randomize in the second-to-last round, for example, his belief probability of the event that player 2 is a machine must be $\frac{1}{5}$ in the second-to-last round (for the payoffs given in Table 7.1), if she has not previously deviated from tit-for-tat.

The details of the proofs and the construction of the sequential equilibria in Kreps, Milgrom, Roberts, and Wilson (1982) is quite complicated, but similar results can be more easily derived for the example shown in Figure 4.9 (Chapter 4). In this example, the option of being generous (g_i) or selfish (f_i) alternates between players 1 and 2 until some player is selfish or the finite number of rounds are completed. Any selfish move ends the game (or puts the game in an absorbing state where all payoffs are 0), but a generous move gives -1 (a loss of 1) to the player who is being generous and gives $+5$ to the other player. Let us number the players so that it is player 2 who would make the last move if both were always generous. As discussed in Section 4.5, if it is common knowledge that both players are rational, then every player would be selfish at every node in the unique sequential equilibrium of any finitely repeated version of this game (because the player at the last move would surely be selfish, so the other player at the second-to-last move would have no incentive to be generous, and so on). Rosenthal (1981), who first studied this game, suggested that there may be something wrong with game-theoretic analysis if it leads to the conclusion that players should always be selfish in such a game. However, let us consider instead a modified version of the game in which there is a small probability ε that player 2 is actually a machine that is always generous (as long as active play continues, which means both have always been generous in the past), but player 1 cannot observe whether player 2 is a rational player or a machine (until she is selfish and stops the active play). Suppose that $\frac{1}{5}^M < \varepsilon < \frac{1}{5}^{M-1}$, for some integer M. Then there is a unique sequential equilibrium in which both players are always generous until after player 1's Mth move from the end of the tree (that is, until player i has less than M possible moves remaining). Thereafter, player 1 is selfish with move probability 0.8, and player 2 randomizes in such a way that, for every k less than M, player 1's consistent belief probability of player 2 being a generous machine is $\frac{1}{5}^k$ at player 1's kth move from the end of the tree. The uniqueness of this sequential equilibrium is proved in Section 4.5 for the case of $\varepsilon = 0.05$.

We say that a strategy for a player in a finitely repeated game is *attractive* if, as with tit-for-tat in the Prisoners' Dilemma, introducing a small positive probability of the player being a machine that uses this strategy would substantially change the set of sequential equilibria of the game, by giving this player an incentive to imitate the machine and cultivate the other players' doubts even when he is not the machine.

Not every strategy is attractive, because not every way of perturbing the game with small initial doubt would have such an impact on the set of equilibria. For example, if we supposed instead that player 1 assigned a small positive probability only to the event that player 2 was a machine that always played the generous move, then the unique sequential equilibrium in any finitely repeated version of the Prisoners' Dilemma would be for each player to always be selfish (except, of course, that 2 must be generous if she is a machine). Player 1's best response to the always-generous strategy is to always be selfish; so player 2 has no reason to want to cultivate player 1's belief that she is going to play like an always-generous machine. So always-generous (and, similarly, always-selfish) is not an attractive strategy in the repeated Prisoners' Dilemma.

Thus, when we study games that are repeated a large but finite number of times, the set of expected average payoff allocations that are achievable in sequential equilibria may or may not be smaller than the set that is achievable in equilibria of the corresponding infinitely repeated game; but if it is smaller, then it may be quite sensitive to perturbations involving small probability events. Fudenberg and Maskin (1986) have shown, under some weak assumptions, that any payoff allocation that is achievable in the infinitely repeated game according to Theorem 7.2 may also be approximately achievable as the expected average payoff allocation in a sequential equilibrium of a long finitely repeated version of the game with small-probability perturbations.

7.7 Imperfect Observability of Moves

The general feasibility theorems discussed in Section 7.5 hold for repeated games in which there is only one possible state of nature and the players can observe all of each other's past moves. When all moves are perfectly observable, the players can threaten to punish each other's deviations from an equilibrium and be confident that, in equilibrium, these threats will have zero probability of actually being carried out. When such perfect observability is not assumed, however, punishment threats that deter selfish behavior may actually have to be carried out with positive probability in equilibrium, because a player may appear to have deviated to his selfish move even if he has not. So threats in an equilibrium may have a positive expected cost of being carried out, as well as the expected benefit of deterring selfish behavior. Finding the best equilibria for the players may require a trade-off or balancing between these costs and benefits.

To illustrate this trade-off, let us consider now a class of infinitely repeated games, with discounting and only one possible state of nature, in which players cannot observe each other's moves. The ideas presented in this section have been developed by Abreu (1986); Abreu, Pearce, and Stacchetti (1986); Radner, Myerson, and Maskin (1986); and Abreu, Milgrom, and Pearce (1988). The specific example discussed here is due to Abreu, Milgrom, and Pearce (1988).

Consider the problem of five players, in $N = \{1,2,3,4,5\}$, who are working independently to prevent some kinds of unfortunate accidents from occurring. At each point in time, each player must choose an effort level, which is a number between 0 and 1. In any short interval of time, the probability of an accident occurring is proportional to the length of the interval and a decreasing function of the players' effort levels. Each player loses one unit of payoff every time an accident occurs, and he must also pay an effort cost per unit time that depends only on his own effort level. The players cannot observe one anothers' effort levels, but everyone can observe the accidents whenever they occur. Also, to simplify our analysis, suppose that the players can use publicly observable random variables to correlate their strategies before each round, so they can implement publicly correlated equilibria.

If we think of this game as actually being played in real continuous time, then we have a problem, because we only know how to analyze repeated game models in which time is a discrete variable, measured in "rounds" (see Section 7.10). So we must let each round denote some interval of time of positive duration, say ε years (where ε may be much less than 1). Ideally, this length of time ε should be the length of time that it takes an individual to rethink or change his actions. Let us carry ε as a parameter in our analysis, so that we can study the effect of changing the length of time represented by a round in our discrete-time model.

Let us suppose that, if $e = (e_i)_{i \in N}$ denotes the profile of effort levels in $[0,1]^5$ that are chosen by the five players, then the probability of an accident occurring in a round of length ε is $\varepsilon(6 - \Sigma_{i \in N} e_i)$, and the cost to player i of exerting effort at level e_i for one round is $\varepsilon(e_i + (e_i)^2)/2$. Thus, i's effort cost is increasing and convex in his effort, and is proportional to the length of the interval that we are considering to be a "round." To assure that the accident probability is less than 1, we shall assume that $\varepsilon \le \frac{1}{6}$. We assume that all players use an expected discounted average payoff criterion to determine their optimal strategies in the repeated game. The discount factor δ per round should depend on the length of a round, according to some decreasing exponential function that approaches 1 as the length approaches 0, so let us suppose that the discount factor per round is $\delta = .1^{\varepsilon}$ (so $\varepsilon \le \frac{1}{6}$ implies $\delta > .681$).

Because an accident costs each player one unit of payoff, the expected payoff to each player i, in a round where e denotes the profile of efforts, is

$$-\varepsilon \left(6 - \sum_{j \in N} e_j \right) - \varepsilon \left(\frac{e_i + (e_i)^2}{2} \right) .$$

This expected payoff is maximized over e_i by letting $e_i = \frac{1}{2}$, so the unique stationary equilibrium is for all players to choose effort $\frac{1}{2}$. When the players all choose any common effort level c, the expected payoff to each player is $\varepsilon(4.5c - 6 - 0.5c^2)$, which is increasing in c over the interval from 0 to 1. So the players would like to give one another some incentive to increase his or her effort above the stationary equilibrium level of $\frac{1}{2}$.

Because the players can observe accidents but cannot observe one anothers' efforts, the only way to give one another an incentive to increase his or her effort is to threaten that some punishment may occur when there is an accident. Notice that the probability of an accident depends symmetrically on everyone's effort and is positive even when everyone chooses the maximum effort level 1. So when an accident occurs, there is no way to tell who, if anyone, was not exerting enough effort. Furthermore, the only way to punish (or reward) anyone in this game is by reducing (or increasing) effort to change the probability of accidents, which affects everyone equally. So, at each round in an equilibrium, we can expect all players rationally to be choosing the same effort level; that is, we can expect equilibria to be *symmetric* with respect to the players. So let us assume symmetry and see how to compute the best and the worst symmetric publicly correlated equilibria of this repeated game with imperfect monitoring.

Let y denote the expected discounted average payoff to each player in the best symmetric publicly correlated equilibrium, and let z denote the expected discounted average payoff to each player in the worst symmetric publicly correlated equilibrium. Now consider what happens in the first round of the best symmetric equilibrium. Each player will choose some effort level b. It can be shown that, because of the strict concavity of the cost-of-effort function, players will not randomize their effort choices at any round. Then there will either be an accident or not. The players' behavior at any future round may depend on whether or not there was an accident at round 1, but it cannot depend otherwise on their individual effort choices, which are not commonly observed. So let x_1 denote the expected discounted average of the sequence of payoffs to each player beginning at round 2 if there is an accident at round 1, and let x_2 denote the expected discounted average of the sequence of payoffs to each player beginning at round 2 if there is no accident at round 1. No matter what happens at round 1, the intrinsic structure of the game is the same from round 2 on as from round 1 on, so x_1 and x_2 must be expected discounted averages of payoff sequences from some symmetric publicly correlated equilibrium of the original game. Thus, by definition of y and z, we must have

(7.7) $y \geq x_1 \geq z$ and $y \geq x_2 \geq z$.

In fact, any expected discounted average payoff x_j that satisfies (7.7) can be achieved by publicly randomizing between playing according to the best equilibrium (with probability $(x_j - z)/(y - z)$) and the worst (with probability $(y - x_j)/(y - z)$). The basic recursion formula for y in terms of x_1 and x_2 is

(7.8) $y = \varepsilon(6 - 5b)\big((1 - \delta)(-1 - (b + b^2)/2) + \delta x_1\big)$
$+ (1 - \varepsilon(6 - 5b))\big((1 - \delta)(-(b + b^2)/2) + \delta x_2\big),$

For each player i to be willing to choose the effort level b when the others are doing so in the first round, we must have

(7.9) $b \in \underset{e_i \in [0,1]}{\operatorname{argmax}} \big(\varepsilon(6 - 4b - e_i)\big((1 - \delta)(-1 - (e_i + (e_i)^2)/2) + \delta x_1\big)$
$+ (1 - \varepsilon(6 - 4b - e_i))\big((1 - \delta)(-(e_i + (e_i)^2)/2) + \delta x_2\big)\big).$

Thus, given z, y (which is the best equilibrium's discounted average payoff) must be the largest number that can satisfy equation (7.8) when x_1, x_2, and b satisfy the constraints (7.7) and (7.9).

By a similar argument, given y, z must be the lowest number that satisfies

(7.10) $z = \varepsilon(6 - 5c)\big((1 - \delta)(-1 - (c + c^2)/2) + \delta x_3\big)$

$\qquad + (1 - \varepsilon(6 - 5c))\big((1 - \delta)(-(c + c^2)/2) + \delta x_4\big),$

where c, x_3, and x_4 satisfy the constraints

(7.11) $c \in \underset{e_i \in [0,1]}{\mathrm{argmax}} \big(\varepsilon(6 - 4c - e_i)\big((1 - \delta)(-1 - (e_i + (e_i)^2)/2) + \delta x_3\big)$

$\qquad + (1 - \varepsilon(6 - 4c - e_i))\big((1 - \delta)(-(e_i + (e_i)^2)/2) + \delta x_4\big)\big),$

(7.12) $y \geq x_3 \geq z$, and $y \geq x_4 \geq z$.

Equivalently, the best and worst symmetric payoffs y and z can be computed by maximizing the difference $y - z$ over all $(y, z, b, c, x_1, x_2, x_3, x_4)$ that satisfy (7.7)–(7.12).

For each x_j, let $q_j = (x_j - z)/(y - z)$. Then the structure of the best and worst publicly correlated equilibria in this repeated game can be implemented using a correlated strategy that can be described in terms of just two possible modes of behavior at each round: a reward mode and a punishment mode. The players begin the best equilibrium at round 1 in reward mode, and they begin the worst equilibrium in punishment mode. At any round when the players are in reward mode, each player chooses some high effort level b. At the end of any round when the players are in reward mode, if there is an accident, then they will switch to punishment mode next round with probability $1 - q_1$ (using some publicly observable random variable, like sunspots); and if there is no accident, then they will switch to punishment mode with probability $1 - q_2$; and otherwise they will stay in reward mode next round. At any round when the players are in punishment mode, each player chooses some relatively low effort level c. At the end of any round when the players are in punishment mode, if there is an accident, then they will switch to reward mode next round with probability q_3; and if there is no accident, then they will switch to reward mode with probability q_4; and otherwise they will stay in punishment mode next round. Other, more complicated equilibria could be devised, in which the players must keep track of the number and timing of all past accidents, but they cannot generate higher or lower expected payoffs.

It is not hard to see that the solution must have $q_2 = 1$ and so $x_2 = y$. That is, to most effectively encourage high efforts in reward mode, the players should plan to switch into punishment mode only if there is an accident. Similarly, to most effectively encourage low efforts in punishment mode, the players should plan to switch back to reward mode only if there is an accident, so a solution must have $q_4 = 0$ and $x_4 = z$. Furthermore, the switch probabilities $1 - q_1$ and q_3 after accidents should be chosen as low as possible, given the other variables, so the first-order optimality conditions for the maximizations in (7.9) and (7.11) will be satisfied even if the maximum occurs at the boundary of the interval [0,1]. That is, we can set 0 equal to the derivative with respect to e_i when e_i equals its maximizing value in (7.9) and (7.11); then solving for q_1 and q_3, we get

$$1 - q_1 = \frac{(1 - \delta)(b - \frac{1}{2})}{\delta(y - z)},$$

$$q_3 = \frac{(1 - \delta)(\frac{1}{2} - c)}{\delta(y - z)}.$$

Substituting back into (7.8) and (7.10) and simplifying, we get

$$y = \varepsilon(4.5b^2 - 4b - 3), \quad z = \varepsilon(4.5c^2 - 4c - 3).$$

Then, when $\delta = .1^\varepsilon$ and $\varepsilon \leq 0.158$, it can be shown that $y - z$ is maximized subject to these constraints by letting $b = 1$ and $c = \frac{4}{9}$, so all players choose their maximum possible effort levels in reward mode and choose effort levels below the stationary equilibrium ($\frac{1}{2}$) in punishment mode. When $\varepsilon = 0.158$, this solution is implemented by switching to punishment mode with probability 1 after an accident in reward mode and by switching to reward mode with probability .111 after an accident in punishment mode. When we break up time into rounds of shorter duration (say, $\varepsilon = 0.001$), the solution remains essentially the same, except that these switch probabilities become slightly smaller ($1 - q_1 = .83$ and $q_3 = .092$).

Now, following Abreu, Milgrom, and Pearce (1988), let us consider a related problem, in which the five players are exerting efforts to increase the probability of making some desirable sales (rather than to reduce the probability of undesirable accidents). To keep the range of possible event probabilities the same (ε to 6ε), let us suppose that the probability of a sale in any round depends on the current efforts $(e_i)_{i \in N}$ according to the formula

$$\varepsilon\left(1 + \sum_{i \in N} e_i\right).$$

As in the previous example, each player's effort e_i must be chosen from the interval [0,1], and costs him $\varepsilon(e_i + (e_i)^2)/2$ in each round. We assume that every player gets an incremental payoff of +1 from each sale and that sales can be observed by everyone, but players cannot observe one anothers' effort levels.

The equations (7.13)–(7.17) that characterize the best (y) and the worst (z) expected discounted average payoffs that can be achieved in symmetric publicly correlated equilibria are, analogous to (7.7)–(7.12),

(7.13) $y \geq x_1 \geq z, \ y \geq x_2 \geq z, \ y \geq x_3 \geq z, \ y \geq x_4 \geq z,$

(7.14) $y = \varepsilon(1 + 5b)\big((1 - \delta)(1 - (b + b^2)/2) + \delta x_1\big)$
$$+ (1 - \varepsilon(1 + 5b))\big((1 - \delta)(-(b + b^2)/2) + \delta x_2\big),$$

(7.15) $b \in \underset{e_i \in [0,1]}{\operatorname{argmax}} \Big(\varepsilon(1 + 4b + e_i)\big((1 - \delta)(1 - (e_i + (e_i)^2)/2) + \delta x_1\big)$
$$+ (1 - \varepsilon(1 + 4b + e_i))\big((1 - \delta)(-(e_i + (e_i)^2)/2) + \delta x_2\big)\Big),$$

(7.16) $z = \varepsilon(1 + 5c)\big((1 - \delta)(1 - (c + c^2)/2) + \delta x_3\big)$
$$+ (1 - \varepsilon(1 + 5c))\big((1 - \delta)(-(c + c^2)/2) + \delta x_4\big),$$

(7.17) $c \in \underset{e_i \in [0,1]}{\operatorname{argmax}} \Big(\varepsilon(1 + 4c + e_i)\big((1 - \delta)(1 - (e_i + (e_i)^2)/2) + \delta x_3\big)$
$$+ (1 - \varepsilon(1 + 4c + e_i))\big((1 - \delta)(-(e_i + (e_i)^2)/2) + \delta x_4\big)\Big).$$

These conditions can be analyzed similarly to (7.7)–(7.12). The main difference in the analysis is that, to maximize $y - z$ here, the probabilities of switching after sales should be set equal to 0, so that $x_1 = y$ and $x_3 = z$. That is, to most effectively encourage effort in reward mode, the players should never switch from reward mode to punishment mode when there is a sale; and to most effectively discourage effort in punishment mode, the players should never switch from punishment mode to reward mode when there is a sale.

When $\delta = .1^{\varepsilon}$ and $\varepsilon \leq \frac{1}{6}$, there is no solution to these equations with $y > z$. The unique solution has $y = z$ and $b = c = \frac{1}{2}$. That is, the only equilibrium is the stationary equilibrium in which all players always choose effort level $\frac{1}{2}$, independently of the past history.

By contrasting these two examples, Abreu, Milgrom, and Pearce (1988) have offered important insights into the design of incentive systems in repeated games with imperfect monitoring of actions. In each example, switches between the best reward mode and the worst punishment mode should occur only when there is bad news: when an accident occurs in the first example, and when no sale occurs in the second example. However, bad news conveys much less information about the unobservable effort levels in the second example than in the first example. For any pair of possible values of an unobservable variable and any observable event, the *likelihood ratio* is defined to be the conditional probability of this event given one of the two possible values of the unobservable variable divided by the conditional probability of the event given the other possible value. Statisticians have shown that the informativeness of the event about the unobservable variable can be measured, in a sense, by the extent to which these likelihood ratios differ from 1. In the second example, for any two possible effort profiles e and $\hat{e}$, the likelihood ratio in the event of bad news (no sale) is

$$\frac{1 - \varepsilon \left(1 + \sum_{i \in N} e_i \right)}{1 - \varepsilon \left(1 + \sum_{i \in N} \hat{e}_i \right)},$$

which is always close to 1 if ε is small. In the first example, however, the likelihood ratio for e and $\hat{e}$ in the event of bad news (an accident) is

$$\frac{\varepsilon \left(6 - \sum_{i \in N} e_i \right)}{\varepsilon \left(6 - \sum_{i \in N} \hat{e}_i \right)},$$

which is independent of ε and different from 1 as long as $\sum_{i \in N} e_i \neq \sum_{i \in N} \hat{e}_i$. Thus, effective collusion is possible in the first example but not in the second, because bad news conveys significant information about the players' unobservable effort levels only in the first (avoiding accidents) example. Good news (making sales) can convey significant information about effort levels in the second example; but this information is of no help to the players because, under an optimal incentive system, they should only switch modes when there is bad news.

7.8 Repeated Games in Large Decentralized Groups

The general feasibility theorem (Theorem 7.2) can be interpreted as a statement about the power of social norms in small groups, such as families, partnerships, and cliques. According to the general feasibility theorem, if the individuals in a group know one another well, can observe one anothers' behaviors, and anticipate a continuing relationship with one another, then social norms can sustain any pattern of group behavior, provided it makes each individual better off than he would be without the group. When we go from small groups into larger social structures, however, the assumption that everyone can observe everyone else may cease to hold, and general feasibility can fail. Thus, the anonymity and privacy of a large society may reduce the set of equilibria and exclude cooperative patterns of behavior.

For an example that illustrates this problem, we consider here a simplified version of a game studied by Milgrom, North, and Weingast (1989). In this repeated game, the set of players is $\{1, 2, \ldots, 2n\}$, for some integer n. At every round, the players are matched into pairs to play the Prisoners' Dilemma. When two players are matched together at round k, their payoffs for this round depend on their moves, generous (g_i) or selfish (f_i), according to Table 7.1. At round 1, each player i in $\{1, 2, \ldots, n\}$ is matched with player $n + i$. At each subsequent round k, player 1 is matched with the player who was matched with player n at round $k - 1$, and each player i in $\{2, \ldots, n\}$ is matched with the player who was matched with player $i - 1$ at round $k - 1$. So two players who are matched at round k will not be matched again until round $k + n$. At any round, each player knows his own past moves and the past moves that other players chose when they were matched with him, but he cannot observe what other players have done when they were not matched with him. The players evaluate their payoff sequences according to a δ-discounted average criterion.

Let us now ask, for what values of n and δ does there exist an equilibrium in which all players will be generous (choose g_i) at all rounds with probability 1 (that is, unless someone deviates from the equilibrium). To deter any player i from being selfish against any player j at round k, player i must anticipate that his selfishness now would cause players matched with him to behave selfishly at some subsequent rounds. Because player j is the only player who actually observes player i's selfishness, one obvious punishment scheme is to suppose that, if i is selfish against j at round k, then i and j will be selfish against each other whenever they are are matched in the future. However, because they are matched at only 1 out of every n rounds, this punishment scheme will deter player i from the initial selfishness only if

$$(6 - 5) + \sum_{l=1}^{\infty} (1 - 5)\delta^{ln} \leq 0.$$

That is, the effect on i's discounted average payoff by increasing his payoff from 5 to 6 at this round must be more than counterbalanced by the decrease in his payoff from 5 to 1 every time he is matched with j in the future. This inequality holds iff

$$\delta^n \geq \tfrac{1}{5}.$$

Other punishment schemes could be designed, but there are limits imposed by the information structure. There is no way that the players who are matched with i during rounds $k + 1$ to $k + n$ could have any information that depends on player i's behavior at round k, so i cannot be punished for a deviation at round k until round $k + n + 1$ at the earliest. However, there is a punishment scheme that would drive player i to his minimax value at every round beginning with round $k + n + 1$. Suppose that each player uses the strategy of being generous at every round until he observes someone being selfish, and thereafter he is selfish at every round (as in the *contagious scenarios* studied by Kandori, 1988). If i deviated from this scenario by being selfish against j at round k, then j would switch to selfishness with all his subsequent partners and so would indirectly induce everyone to behave selfishly by round $k + n + 1$; and so player i would be confronted with selfish behavior at round $k + n + 1$ and every round thereafter. Under this scenario, if player i were selfish at any round, then it would be best for him to be selfish at all rounds thereafter (because he cannot make his punishment any worse); so the initial selfish deviation can be deterred iff

$$\sum_{l=1}^{n} (6 - 5)\delta^{l-1} + \sum_{l=n+1}^{\infty} (1 - 5)\delta^{l-1} \leq 0.$$

This inequality also holds iff $\delta^n \geq \tfrac{1}{5}$; so even the most extreme threat cannot increase the range of parameters in which cooperation can be sustained. Thus, for any given δ, if the number of players $2n$ is bigger than $2\log(5)/\log(1/\delta)$, then even the most extreme punishment scheme (dissolution of all cooperation in the whole society) cannot deter a player from deviating to selfish behavior. This model illustrates how generous or cooperative behavior that can be sustained in small groups may break down in a large decentralized group.

Milgrom, North, and Weingast (1989) show how cooperation may be sustained with arbitrarily large numbers of players if the information structure is changed by adding a central mediator who (acting like a simple judicial system) keeps a list of any players who have violated the cooperative norm. For this example, a player would get onto the mediator's list if he were ever selfish when matched with a player who was not on the list. At each round, the mediator just tells each pair of matched players whether either of them is on the list. If $\delta \geq \frac{1}{5}$, then this game has an equilibrium in which each player is always generous, unless (after some accidental deviation) he is on the list or is matched with someone on the list, in which case he would be selfish.

7.9 Repeated Games with Incomplete Information

In the study of Bayesian games with incomplete information, the players' beliefs about each other's types are assumed to be exogenously given. In many real-life situations, individuals' beliefs about each other are derived from long past experience interacting with each other. In such cases, some information structures may be more likely to arise than others. That is, there may be some things about himself that an individual tends to reveal to others who are involved in long-term relationships with him, and other things that he tends to conceal. Speaking roughly, we might expect that an individual would generally try to let others know about his strengths, but would try to maintain uncertainty about his weaknesses. The study of repeated games with incomplete information provides a framework for formalizing and studying hypotheses such as this.

Aumann and Maschler (1966) introduced the study of repeated games with incomplete information. In most of the literature on these games, the following structures are assumed. There are two or more possible states of nature. The actual state of nature is determined according to some given probability distribution at the beginning of round 1. For each player, there is some function of the state of nature, which we call the player's *type*, that the player learns at the beginning of round 1. Thereafter, the state of nature remains constant throughout the game. After learning his own type at round 1, each player thereafter observes only the moves of the players at the preceding rounds. Payoffs to each player at each round may depend on the state of nature as well as their current moves. To the extent that a player's payoff may depend on the state of nature that is unknown to him, a player will not know the actual payoff that he has gotten at any past round. Each player's objective is to maximize the expected value, given his information, of some measure of his long-term average payoff (say, a limit of average payoffs, or a limit of δ-discounted average payoffs as δ approaches 1, or some other criterion, as discussed in Section 7.2). To date, most work on repeated games with incomplete information has been devoted to the special case of two-person zero-sum games (for a notable exception, see Hart, 1985a).

We consider here a classic example of a repeated two-person zero-sum game with incomplete information, due to Aumann and Maschler (1966). For each player i in $\{1,2\}$, the set of possible moves at each round is $D_i = \{a_i, b_i\}$. The state of nature, a constant throughout all rounds of the repeated game, could be either α or β. The payoffs to the players 1 and 2 depend on the moves and the state of nature as shown in Table 7.7. Let q denote the prior probability that the actual state of nature is α; this parameter q is common knowledge among the players. At the beginning of round 1, one of the two players learns which state is the actual state of nature. Thus, there are two versions of this game that we will consider. In version 1, player 1 knows the state but player 2 does not. In version 2, player 2 knows the state, but player 1 does not. In both versions, the players can observe all of each other's past moves, and the uninformed player can use these observations to try to make inferences about the state of nature.

Let us first consider version 1, in which only player 1 knows the state (and this fact is common knowledge). The analysis of this game is easy. If player 1 knows that the state of nature is α, then he should always

Table 7.7   Payoffs at any round, for all move profiles and states of nature, in a repeated game with incomplete information

State = α

D_1	D_2	
	a_2	b_2
a_1	$-1,1$	$0,0$
b_1	$0,0$	$0,0$

State = β

D_1	D_2	
	a_2	b_2
a_1	$0,0$	$0,0$
b_1	$0,0$	$-1,1$

choose b_1, to guarantee himself his best possible payoff of 0. Similarly, if the state of nature is β, then player 1 should always choose a_1. When player 1 follows this strategy, player 2 will of course be able to infer player 1's type from his actions: α if 1 chooses b_1, β if 1 chooses a_1. Once player 2 has learned what the state is, from player 1's move at round 1, player 2 may well use her weakly undominated move of a_2 if the state is α, b_2 if the state is β, but player 1 will still get his optimum payoff of 0. So it does no harm for player 1 to reveal his information in this game.

Let us next consider version 2, in which only player 2 knows the state. To have some hope of getting payoffs higher than 0, player 2 must try to prevent player 1 from guessing the state. One way for player 2 to do so is to avoid using her information about the state and act as she would have if neither player knew the state. If neither player knew the state of nature, then they would both care only about their expected payoffs, given only that the probability of state α is q. These expected payoffs are shown in Table 7.8.

As a strategic-form game, Table 7.8 has a unique equilibrium $((1 - q)[a_1] + q[b_1], (1 - q)[a_2] + q[b_2])$. The expected payoffs in this equilibrium are $-q(1 - q)$ for player 1, and $q(1 - q)$ for player 2. So, in version 2 of the repeated game, where player 2 knows the state, she can guarantee that her expected payoff (given prior information only) will be $q(1 - q)$ if she uses the strategy of choosing, at each round, a_2 with probability $1 - q$, and b_2 with probability q, independently of her information about the state and all previous moves. Aumann and Maschler (1966) showed that this nonrevealing stationary strategy is essentially the best that player 2 can do in this game.

Thus, in version 1 of this game, the informed player (1) should use his information about the state and reveal it; but in version 2 of this game, the informed player (2) should ignore her information about the

Table 7.8 Expected payoffs for all move profiles

D_1	D_2	
	a_2	b_2
a_1	$-q,q$	$0,0$
b_1	$0,0$	$q - 1, 1 - q$

state, to conceal it. Notice also that player 1's equilibrium payoff in the game shown in Table 7.8 is a convex function of q, whereas player 2's equilibrium payoff in Table 7.8 is a concave function of q. (Let X be a convex subset of some vector space, and let $f:X \to \mathbf{R}$ be any real-valued function on X. Then f is *concave* iff,

$$f(\lambda x + (1 - \lambda)y) \geq \lambda f(x) + (1 - \lambda)f(y),$$

$$\forall x \in X, \quad \forall y \in X, \quad \forall \lambda \in [0,1].$$

On the other hand, f is *convex* iff

$$f(\lambda x + (1 - \lambda)y) \leq \lambda f(x) + (1 - \lambda)f(y),$$

$$\forall x \in X, \quad \forall y \in X, \quad \forall \lambda \in [0,1].)$$

Aumann and Maschler showed that, in general, the convexity or concavity of this equilibrium-payoff function can tell us whether an informed player should reveal or conceal his information in a repeated two-person zero-sum game with incomplete information.

To state Aumann and Maschler's result more precisely, consider the following general model of a repeated two-person zero-sum game with one-sided incomplete information. The set of players is $N = \{1,2\}$. The set of possible states of nature and the sets of possible moves for players 1 and 2 are nonempty finite sets, denoted by Θ, D_1, and D_2, respectively. The payoff to player 1 at each round depends on the state of nature and the moves of both players according to the function $u_1:D_1 \times D_2 \times \Theta \to \mathbf{R}$. The payoff to player 2 is just the opposite of player 1's payoff; that is, $u_2(d_1,d_2,\theta) = -u_1(d_1,d_2,\theta)$. The actual state of nature is constant and determined before round 1 according to an initial probability distribution Q in $\Delta(\Theta)$. At round 1, player 1 learns the actual state of nature, and player 2 gets no informative signal. So player 1 is the *informed* player, and player 2 is *uninformed* (knowing only the distribution Q). At every round after round 1, both players observe the moves chosen at the preceding round.

Let K be some positive integer. Suppose that each player in this repeated two-person zero-sum game with one-sided incomplete information simply applied the criterion of maximizing his or her expected average payoff received during the first K rounds. With this criterion, our infinitely repeated game could be modeled as a finite two-person zero-sum game in extensive form, where each player moves just K times and the terminal payoff is the average of the payoffs in the K rounds.

Let $v_1(K)$ denote the expected payoff to player 1 in an equilibrium of this K-round game.

Now consider the one-round game in which neither player knows the state and their payoffs depend on their moves in $D_1 \times D_2$ according to the formula

$$\hat{u}_i(d_1, d_2) = \sum_{\theta \in \Theta} q(\theta) u_i(d_1, d_2, \theta),$$

where q is some probability distribution in $\Delta(\Theta)$. Let $w_1(q)$ denote the expected payoff to player 1 in any equilibrium, and let $\tau_1(\cdot | q) = (\tau_1(d_1 | q))_{d_1 \in D_1}$ denote an equilibrium strategy in $\Delta(D_1)$ for player 1, in this one-round two-person zero-sum game.

Let $w_1^*:\Delta(\Theta) \to \mathbf{R}$ be the *concave hull* of the function $w_1:\Delta(\Theta) \to \mathbf{R}$. That is, w_1^* is the lowest concave function on $\Delta(\Theta)$ such that $w_1^*(q) \geq w_1(q)$ for every q in $\Delta(\Theta)$. It can be shown that, for any q in $\Delta(\Theta)$, there exists a finite collection of probability distributions $r_1, \ldots, r_J$, each of which is in $\Delta(\Theta)$, and nonnegative numbers $\lambda_1, \ldots, \lambda_J$ such that

(7.18) $$\sum_{j=1}^{J} \lambda_j = 1, \quad \sum_{j=1}^{J} \lambda_j r_j = q, \quad \text{and} \quad \sum_{j=1}^{J} \lambda_j w_1(r_j) = w_1^*(q).$$

In fact, an equivalent way to define the concave hull is to let $w_1^*(q)$ be the highest number that can satisfy (7.18) for some distributions $r_1, \ldots, r_J$ and nonnegative numbers $\lambda_1, \ldots, \lambda_J$.

Let $r_1, \ldots, r_J$ be distributions in $\Delta(\Theta)$ and let $\lambda_1, \ldots, \lambda_J$ be nonnegative numbers that satisfy condition (7.18) when $q = Q$. Now consider the following strategy for player 1 in the repeated game where player 1 knows the actual state of nature and player 2 only knows the distribution Q out of which the state of nature is drawn: at round 1, after learning the actual state, player 1 randomly chooses a number between 1 and J, such that $\lambda_j r_j(\theta)/Q(\theta)$ is the conditional probability of his choosing j given that the actual state is θ; thereafter, given his chosen number j, player 1 chooses his move at each round independently according to the distribution $\tau_1(\cdot | r_j)$ (choosing each move d_1 with probability $\tau_1(d_1 | r_j)$). So the unconditional probability of player 1 choosing number j is $\sum_{\theta \in \Theta} Q(\theta)(\lambda_j r_j(\theta)/Q(\theta)) = \lambda_j$. Notice that, given only that j is the number chosen by player 1, the conditional probability that any θ is the actual state is $Q(\theta)(\lambda_j r_j(\theta)/Q(\theta))/\lambda_j = r_j(\theta)$. So, if we knew only that player 1 was implementing the randomized strategy $\tau_1(\cdot | r_j)$ at each round, then we would infer that the distribution over the unknown state was r_j, and we would assess an expected payoff for player 1 of $w_1(r_j)$ at each round. Thus, before player 1 learns the actual state and chooses his number, his expected payoff under this strategy must be at least $\Sigma_{j=1}^J \lambda_j w_1(r_j) = w_1^*(Q)$. So this strategy guarantees player 1 an expected payoff of $w_1^*(Q)$ at every round in the repeated game.

When player 1 implements this strategy, if $\tau_1(\cdot|r_j) \neq \tau_1(\cdot|r_h)$ whenever $h \neq j$, then player 2 will eventually be able to infer which number j was chosen by player 1. But 2 will never be able to infer anything more about the state of nature by observing 1's moves, because player 1 acts as if he has forgotten the state and remembers only his chosen number. So, after observing player 1 implement this strategy for a long time, player 2's posterior probability distribution for the unknown state of nature should converge to r_j.

When the players only care about the average of their payoffs at the first K rounds, this strategy may not be optimal for player 1, because he might have an incentive to use more of his information about the state toward the end of these K rounds, when the information thereby revealed to player 2 cannot hurt player 1 as much as it might earlier. However, as the horizon K goes to infinity, this strategy is asymptotically optimal. This theorem is the main result of Aumann and Maschler (1968) and can be stated as follows.

THEOREM 7.3.

$$\lim_{K \to \infty} v_1(K) = w_1^*(Q).$$

The full proof can also be found in Sorin (1980). Here, we only sketch the proof. What we need to do is show that player 2 has a strategy that will guarantee that player 1 cannot, in the long run, get an expected average payoff that is better than $w_1^*(Q)$. To show this, we must first present an important theorem of Blackwell (1956).

For any set Z that is a closed convex subset of $\mathbf{R}^\Theta$, and for any point $y = (y_\theta)_{\theta \in \Theta}$ in $\mathbf{R}^\Theta$, there exists a unique point $\xi(y) = (\xi_\theta(y))_{\theta \in \Theta}$ in Z that minimizes the distance to y, that is,

$$\{\xi(y)\} = \operatorname*{argmin}_{z \in Z} \left(\sum_{\theta \in \Theta} (z_\theta - y_\theta)^2 \right)^{1/2}.$$

The *distance* from Z to y is defined to be this minimum distance.

We can extend the function $u_1(\cdot)$ to $\Delta(D_1) \times \Delta(D_2) \times \Theta$ in the usual way, such that

$$u_1(\rho_1,\rho_2,\theta) = \sum_{d_1 \in D_1} \sum_{d_2 \in D_2} \rho_1(d_1)\rho_2(d_2)u_1(d_1,d_2,\theta).$$

A behavioral strategy for player 2 in the repeated game is a sequence $\sigma_2 = (\sigma_2^k)_{k=1}^{\infty}$, where $\sigma_2^1 \in \Delta(D_2)$ and, for each k greater than 1, σ_2^k is a function from $(D)^{k-1}$ into $\Delta(D_2)$. So $\sigma_2^k(d_2|c^1, \ldots, c^{k-1})$ is the probability that, under the strategy σ_2, player 2 would choose move d_2 at round k if $c^1, \ldots, c^{k-1}$ were the history of moves of both players at rounds 1 through $k-1$. (Here each $c^l = (c_1^l, c_2^l)$.) After learning the actual state of nature, player 1 should choose a behavioral strategy that is defined similarly. Once a behavioral strategy for each player is specified, the moves that will be actually carried out by each player at each round become random variables, with well-defined probability distributions that depend on the behavioral strategies. (The formal construction of these probability distributions is described in Section 7.2.) Let $\tilde{d}_i^k$ denote the random variable that is the move that player i will actually carry out at round k, in a realization of these behavioral strategies, and let $\tilde{u}_1^k(\theta)$ $= u_1(\tilde{d}_1^k, \tilde{d}_2^k, \theta)$. Player 2 does not know the actual state, but after playing round k of the game she will know what payoff player 1 would have gotten for each possible state at that round. That is, player 2's knowledge about player 1's realized payoff at round k will be described by the vector

$$\tilde{u}_1^k = (\tilde{u}_1^k(\theta))_{\theta \in \Theta}.$$

We refer to $\tilde{u}_1^k$ as the *realized payoff vector* to player 1 at round k. Let $\tilde{x}^k$ denote the average of the first k realized payoff vectors to player 1. That is, $\tilde{x}^k = (\tilde{x}_\theta^k)_{\theta \in \Theta}$ and

$$\tilde{x}_\theta^k = \frac{\displaystyle\sum_{m=1}^{k} \tilde{u}_1^m(\theta)}{k}.$$

The following important theorem has been proved by Blackwell (1956).

THEOREM 7.4. *Let Z be any subset of $\mathbf{R}^\Theta$ that is closed and convex. Suppose that, for every q in $\mathbf{R}^\Theta$,*

$$\min_{\rho_2 \in \Delta(D_2)} \max_{\rho_1 \in \Delta(D_1)} \sum_{\theta \in \Theta} q(\theta) u_1(\rho_1,\rho_2,\theta) \leq \underset{y \in Z}{\text{supremum}} \sum_{\theta \in \Theta} q(\theta) y_\theta.$$

Then player 2 has some behavioral strategy σ_2 in the repeated game such that, for any strategy that player 1 may use, with probability 1, the distance from Z to $\bar{x}^k$ will converge to 0 as k goes to infinity.

This theorem is known as *Blackwell's approachability theorem*. When the conclusion of the theorem holds, then we say that the set Z is *approachable* by player 2.

Suppose now that the hypothesis of Theorem 7.4 is satisfied. Without giving the full details of the proof, we can easily describe a strategy for player 2 that satisfies the conclusion of Theorem 7.4. At round 1, and at any round k when $\bar{x}^{k-1}$ is in Z, let player 2 choose her move according to any arbitrary distribution over D_2. At any round k when $\bar{x}^{k-1}$ is not in Z, let player 2 choose her move according to any distribution that is in the set

$$\underset{\rho_2 \in \Delta(D_2)}{\text{argmin}} \max_{\rho_1 \in \Delta(D_1)} \sum_{\theta \in \Theta} (\bar{x}_\theta^{k-1} - \xi_\theta(\bar{x}^{k-1})) u_1(\rho_1,\rho_2,\theta),$$

where $\xi(\bar{x}^{k-1})$ is the point in Z that is closest to $\bar{x}^{k-1}$. Blackwell (1956) showed that, when player 2 uses such a strategy, the limit of the distance from Z to the average realized payoff vectors must converge to 0, no matter what behavioral strategy player 1 uses.

To appreciate the full power of Blackwell's approachability theorem, let us now consider instead the case where the hypothesis of the theorem is not satisfied. Then there exists some q such that

$$\sup_{y \in Z} \sum_{\theta \in \Theta} q(\theta) y_\theta < \min_{\rho_2 \in \Delta(D_2)} \max_{\rho_1 \in \Delta(D_1)} \sum_{\theta \in \Theta} q(\theta) u_1(\rho_1,\rho_2,\theta)$$

$$= \max_{\rho_1 \in \Delta(D_1)} \min_{\rho_2 \in \Delta(D_2)} \sum_{\theta \in \Theta} q(\theta) u_1(\rho_1,\rho_2,\theta).$$

That is, there is some q in $\mathbf{R}^\Theta$ and some ρ_1 in $\Delta(D_1)$ such that,

$$\sum_{\theta \in \Theta} q(\theta) u_1(\rho_1,\rho_2,\theta) > \sup_{y \in Z} \sum_{\theta \in \Theta} q(\theta) y_\theta, \quad \forall \rho_2 \in \Delta(D_2).$$

So if player 1 simply uses the stationary strategy of applying the randomized strategy ρ_1 to determine his move independently at every round, then he can guarantee that, with probability 1,

$$\liminf_{k \to \infty} \sum_{\theta \in \Theta} q(\theta) \bar{x}_\theta^k > \sup_{y \in Z} \sum_{\theta \in \Theta} q(\theta) y_\theta.$$

So if the hypothesis of the theorem is not satisfied, then player 1 has a behavioral strategy that guarantees that the limit-infimum of the distance from Z to $\bar{x}^k$ must be strictly positive, no matter what behavioral strategy player 2 uses. In this case, we say that the set Z is *excludable* by player 1. Thus, Blackwell's approachability theorem asserts that any closed convex subset of $\mathbf{R}^\Theta$ is either excludable by player 1 or approachable by player 2. (It is easy to see that a set cannot be both excludable by 1 and approachable by 2!)

Using Blackwell's approachability theorem, we can now sketch the rest of the proof of Theorem 7.3. Because $w_1^*(\cdot)$ is a concave function, there exists some vector z in $\mathbf{R}^\Theta$ such that

$$\sum_{\theta \in \Theta} Q(\theta)z_\theta = w_1^*(Q), \text{ and } \sum_{\theta \in \Theta} q(\theta)z_\theta \geq w_1^*(q), \quad \forall q \in \Delta(\Theta).$$

That is, z is the vector of coefficients of a linear function on $\Delta(\Theta)$ that is tangent to $w_1^*(\cdot)$ at Q. Let $Z = \{y \in \mathbf{R}^\Theta \mid y_\theta \leq z_\theta, \forall \theta \in \Theta\}$.

We now show that this set Z satisfies the hypothesis of Theorem 7.4, so it is approachable by player 2. If q has any negative components, then the supremum in the hypothesis is $+\infty$, so the inequality is trivially satisfied for any ρ_2. If q is the zero vector, then the inequality is also trivially satisfied, with 0 on both sides. For any nonnegative vector q, the inequality can be satisfied iff it can be satisfied for $(1/\sum_{\theta \in \Theta} q_\theta)q$, which is in $\Delta(\Theta)$, because both sides of the inequality are linear in q. But if q is any vector in $\Delta(\Theta)$, then

$$\sup_{y \in Z} \sum_{\theta \in \Theta} q(\theta)y_\theta = \sum_{\theta \in \Theta} q(\theta)z_\theta \geq w_1^*(q) \geq w_1(q)$$

$$= \min_{\rho_2 \in \Delta(D_2)} \max_{\rho_1 \in \Delta(D_1)} \sum_{\theta \in \Theta} q(\theta)u_1(\rho_1, \rho_2, \theta).$$

So this set Z satisfies the hypothesis of Theorem 7.4.

Now, to see why player 1 cannot expect to do substantially better than $w_1^*(Q)$ in the long run, as Theorem 7.3 asserts, suppose that player 2 uses her strategy for approaching this set Z. Then, with probability 1, the limit supremum of each $\bar{x}_\theta^k$, as k goes to infinity, must be less than or equal to z_θ. So (using the fact that payoffs are bounded, so events of infinitesimal probability cannot substantially affect the expected payoff in any state), the limit-supremum, as k goes to infinity, of the expected value of $\sum_{\theta \in \Theta} Q(\theta)\bar{x}_\theta^k$ must be less than or equal to $\sum_{\theta \in \Theta} Q(\theta)z_\theta = w_1^*(Q)$.

7.10 Continuous Time

Although real physical time seems to be a continuous variable, virtually all tractable game-theoretic models have assumed that players can make decisions only at some discrete countable sequence of points in time.

To see, at a basic level, why continuous-time models have been difficult to work with, consider a naive attempt to define a continuous-time version of the repeated Prisoners' Dilemma game. For each nonnegative real number t, each player i must decide at time t whether to be generous (g_i) or selfish (f_i) at time t. Let $e_i(t)$ denote the move chosen by player i at time t. Each player knows all past moves, so player i can implement a strategy in which $e_i(t)$ is some (possibly random) function of the history of past moves $((e_1(s),e_2(s))_{0 \le s < t}$. The objective of each player i is to maximize the expected value of some integral of his payoffs at each point in time, say,

$$\int_0^\infty \delta^t u_i(e_1(t),e_2(t))dt,$$

where $0 < \delta \le 1$.

This model may seem to be a well-defined game, but in fact it is not. For one thing, the integral may not be defined, if the history of moves has too many discontinuities. So some strategies that create too many discontinuities, such as "at each point in time randomize independently between being generous and selfish, each with probability ½," might have to be ruled out.

But there are other deep problems even with strategies that seem to guarantee that the path of payoffs will be integrable. For example, consider the strategy "be generous unless there exists some interval of positive length in the past during which the other player has been selfish, in which case be selfish." A player who implements such a strategy can change his move at most once, and his move will depend continuously on time at all other points in time. But when will he change his move? Suppose that both players decide to implement this strategy. Let K be any arbitrary nonnegative real number, and consider the outcome path

$$(e_1(t),e_2(t)) = (g_1,g_2) \text{ if } t \le K,$$

$$(e_1(t),e_2(t)) = (f_1,f_2) \text{ if } t > K.$$

At any point in time, each player is choosing the move designated by this strategy, given the history of past moves. So, for any K, such an outcome path would seem to be a consistent realization of these strategies. The problem is that, in continuous time, the players can change from generosity to selfishness without there being any first point in time when someone was selfish. To have a well-defined game, either we must somehow restrict the set of possible strategies for each player, to rule out strategies such as this one, or we must find a more sophisticated rule to define the realized outcome path for all pairs of possible strategies. Either way, this seems to be a difficult research problem. (See Anderson, 1983; Fudenberg and Tirole, 1984; Kalai and Stanford, 1985; Simon, 1987; Simon and Stinchcombe, 1989; and Bergin, 1987, for recent work in this direction.)

So the general approach in game theory is to work with discrete-time models. That is, we break up the continuous time line into a countable sequence of intervals, each called a "round" (or "period" or "stage"), and assume that each player can choose a new move only at the beginning of a round. Such a model may reasonably describe a real situation if the length of a round is chosen to be the approximate length of time that it takes each player to respond to new information and change his behavior. To the extent that players can think and react very quickly, however, we are interested in the limit as the length of each round goes to 0.

Simon and Stinchcombe (1989) have probed the question of whether such short-period discrete-time models might nonetheless create artificial phenomena that could not exist in the real continuous-time world. For example, consider the following two-player game. Each player must decide at what time in $\mathbf{R}_+$ he wishes to enter a new market. His payoff is $-\delta^t$ if he is the first to enter and he does so at time t, $-2\delta^t$ if the other player is the first to enter and does so at time t, and $-3\delta^t$ if both players enter simultaneously at time t, where $0 < \delta < 1$. We can describe this game by a discrete-time model in which the length of each round is some small positive number ε, so that the beginning of round k will occur at "real" time $\varepsilon(k-1)$. At the beginning of each round, when he decides whether to enter now or not, each player is assumed to know the decisions of both players (to enter or not) at all past rounds.

Consider the following scenario, which is a subgame-perfect equilibrium of the discrete-time model of this game. Player 1 plans to enter at round 1, while player 2 plans not to enter at round 1. If player 1 did not enter at round 1, then player 2 would enter at round 2 and player 1 would not enter at round 2. Similarly, at every subsequent round, if no one has previously entered, then player 1 would enter and player 2 would not if the round number is odd, but player 2 would enter and player 1 would not if the round number is even. In this equilibrium, player 1 is deterred from delaying his entry at time 0 (round 1) by the fear that, if he did not do so, player 2 would enter at the first subsequent opportunity (round 2, at time ε). Similarly, player 2 would be deterred from delaying her entry after player 1 had failed to enter at time 0 by the fear that, if she failed to enter at her first opportunity after time 0, then player 1 would enter for sure at his first opportunity thereafter. Notice, however, that the logic of this equilibrium depends in an essential way on the artificial assumption that time is divided into discrete rounds. In continuous time, there is no "first opportunity" for someone to enter after another has failed to do so. Thus, we might sometimes want to scrutinize some equilibria of our discrete-time models, to see whether they actually correspond to behavior that would also make sense in continuous time.

The discrete-time model of this game also has symmetric stationary subgame-perfect equilibria in which both players randomize between entering and not at each round until someone enters. To compute these, let q denote the probability that each player would enter at any round when no one has yet entered. At round k, if no one has previously entered and player 2 is expected to use this strategy, then player 1 would be indifferent between entering at round k (time $\varepsilon(k-1)$) or at round $k+1$ (time εk) iff

$$q(-3\delta^{\varepsilon(k-1)}) + (1 - q)(-\delta^{\varepsilon(k-1)}) =$$
$$q(-2\delta^{\varepsilon(k-1)}) + (1 - q)q(-3\delta^{\varepsilon k}) + (1 - q)^2(-\delta^{\varepsilon k}).$$

This quadratic equation has two roots

$$q = \frac{\delta^{\varepsilon} \pm (\delta^{2\varepsilon} + 8\delta^{\varepsilon} - 8)^{1/2}}{4},$$

both of which are in the interval between 0 and 1 when ε is small.

Consider now the equilibrium corresponding to the larger of the two roots. As ε goes to 0, q converges to $\frac{1}{2}$. That is, at each round, each player plans independently to enter with probability close to $\frac{1}{2}$ if no one has entered previously. At each round when no one has yet entered, then the following four events each have probability approximately equal to ¼: player 1 enters alone, player 2 enters alone, both players enter, and neither player enters. For any time t such that $t > 0$, the probability that no one will have entered by time t is approximately $(¼)^{t/\varepsilon}$ (because there are about t/ε rounds before time t), which goes to 0 as $\varepsilon \to 0$ and $q \to ½$. Thus, in real time, the limit of the outcomes of this equilibrium is that someone will enter for sure at (or infinitesimally close to) time 0, and the following three events are equally likely: player 1 enters first alone, player 2 enters first alone, and both players enter first simultaneously. That is, in the limit, each player is entering at time 0 with marginal probability ⅔, and the probability of both entering at time 0 is ⅓. Because ⅓ ≠ (⅔)², it might seem that such behavior would require correlation between the players at time 0; but we have seen that this is in fact the limit of behavior in a discrete-time model in which the players are behaving independently at every point in time, given their shared information about the past. Thus, if we do study continuous-time models, it may be inappropriate to assume that players' behavioral strategies must be stochastically independent at every point in time, even if there is no communication between players.

7.11 Evolutionary Simulation of Repeated Games

Axelrod (1984) used a version of the repeated Prisoners' Dilemma in his work on evolutionary simulation, discussed earlier in Section 3.7. In his results, the strategy of tit-for-tat had more long-run reproductive success than almost any other strategy. This result has a striking qualitative similarity to the results of Kreps, Milgrom, Roberts, and Wilson (1982) on the attractiveness of tit-for-tat in a finitely repeated Prisoners' Dilemma. On the basis of his simulations, Axelrod has argued that cooperative reciprocating behavior should be expected to spontaneously evolve in many social and ecological systems.

Of course, it is hard to know whether other initial distributions might have favored some other strategy in an evolutionary simulation like Axelrod's. The concept of *resistance*, defined in Section 3.7, offers another way to derive and test such conclusions. Consider, for example, the version of the repeated Prisoners' Dilemma given in Table 7.1, with a δ-discounted average payoff criterion. Let us compare the equilibrium in which both players play tit-for-tat with the equilibrium in which both players play selfish-always. In the tit-for-tat equilibrium, both players get δ-discounted average payoffs of 4. In the selfish-always equilibrium, both players get δ-discounted average payoffs of 1. When one player uses tit-for-tat and the other uses selfish-always, their δ-discounted average payoffs are δ and $6 - 5\delta$, respectively. The resistance of the selfish-always equilibrium against tit-for-tat is the highest number λ such that

$$(1 - \lambda)(1) + \lambda(6 - 5\delta) \geq (1 - \lambda)(\delta) + \lambda(4).$$

This number is $(1 - \delta)/(4\delta - 1)$, which goes to 0 as δ approaches 1. That is, in a repeated Prisoners' Dilemma game with a very long-run discounted average criterion, the resistance of the selfish-always equilibrium against the tit-for-tat strategy is almost 0, whereas the resistance of the tit-for-tat equilibrium against the selfish-always strategy is almost 1.

It is not clear to what extent these results generalize to other repeated games, however. From Section 7.4, recall the repeated Chicken game (in Table 7.3) with a δ-discounted average payoff criterion. It can be shown that, as δ goes to 1, the resistance of the (very belligerent) q-positional equilibrium of this repeated game against the tit-for-tat (or getting-even) strategy converges to 4/11, not 0. So it may be harder for tit-for-tat cooperation to evolve in an ecological system where the individuals play repeated Chicken, instead of the repeated Prisoners' Dilemma.

Exercises

Exercise 7.1.   Consider a repeated game with complete state information in which the set of players, the set of possible states of nature, and the set of possible moves for each player are all finite sets. Show by a fixed-point argument that there exists a stationary equilibrium τ and a value vector v that satisfy conditions (7.2) and (7.3).

Exercise 7.2.   Consider a repeated game with standard information, where each player maximizes his expected δ-discounted average payoff, for some δ that is close to but less than 1. Payoffs at each round depend on the players' moves as shown in Table 7.9. Consider the following scenario. At the first round, and at each round where the outcome of the preceding round was (x_1,x_2) or (y_1,y_2), the players independently randomize, each player i choosing y_i with probability q or x_i with prob-

Table 7.9   Payoffs at any round, for all move profiles, in a repeated game

	D_2	
D_1	x_2	y_2
x_1	2,2	0,6
y_1	6,0	1,1

ability $1 - q$. At each round where the outcome of the preceding round was (x_1, y_2), the players choose y_1 and x_2. At each round where the outcome of the preceding round was (y_1, x_2), the players choose x_1 and y_2.

Write a system of equations that determine the probability q that makes this scenario a subgame-perfect equilibrium, once δ is specified, as long as δ is not too small. Show that these equations have a solution where $q = .664183$ when $\delta = .9$. What is the smallest discount factor δ such that the q determined by these equations will be a subgame-perfect equilibrium?

Exercise 7.3.   For the repeated game with standard information described in Table 7.5, with discount factor $\delta = .99$, construct a subgame-perfect equilibrium in which the expected discounted average payoff to each player, at the beginning of the game, is approximately 1.0007009118 or less.

Exercise 7.4.   (A *war of attrition* game.) Two small grocery stores on the same block are feeling the effects of a large supermarket that was recently constructed a half-mile away. As long as both remain in business, each will lose \$1000 per month. On the first day of every month, when the monthly rent for the stores is due, each grocer who is still in business must independently decide whether to stay in business for another month or to quit. If one grocer quits, then the grocer who remains will make \$500 per month profit thereafter. Assume that, once a grocer quits, his or her lease will be taken by some other merchant (not a grocer), so he or she will not be able to reopen a grocery store in this block, even if the other grocer also quits. Each grocer wants to maximize the expected discounted average value of his or her monthly profits, using a discount factor per month of $\delta = .99$.

a. Find an equilibrium of this situation in which both grocers randomize between staying and quitting every month until at least one grocer quits.

b. Discuss another way that the grocers might handle this situation, with reference to other equilibria of this game.

c. Suppose now that grocer 1 has a slightly larger store than grocer 2. As long as both stores remain in business, grocer 1 loses $1200 per month and grocer 2 loses $900 per month. If grocer 1 had the only grocery store on the block, he would earn $700 profit per month. If grocer 2 had the only grocery store on the block, she would earn $400 per month. Find an equilibrium of this situation in which both grocers randomize between staying and quitting every month, until somebody actually quits. In this equilibrium, which grocer is more likely to quit first?

Exercise 7.5. Consider a two-person zero-sum game with incomplete information in which the payoffs depend on the actions and the state as given in Table 7.10. Let q denote the probability of state α.

a. Suppose first that this game is played only once, neither player knows the state, and both agree on q, the probability of α. Draw a graph that shows, as a function of the parameter q, player 1's expected payoff $w_1(q)$ in equilibrium.

Table 7.10 Payoffs at any round, for all move profiles and states of nature, in a repeated game with incomplete information

State = α

D_1	D_2	
	a_2	b_2
a_1	2,−2	0,0
b_1	2,−2	1,−1

State = β

D_1	D_2	
	a_2	b_2
a_1	0,0	1,−1
b_1	0,0	0,0

b. Draw a graph that shows the concave hull w_1^* of the function w_1 that you plotted in part (a). Check that $w_1^*(\frac{1}{2}) = \frac{3}{4} > w_1(\frac{1}{2})$.

c. Now consider an infinitely repeated game in which the state of nature is the same at every round, and each of the two possible states has probability $\frac{1}{2}$ of being the actual state of nature. Suppose that only player 1 observes the state of nature at the start of the game. Player 2 observes player 1's moves but cannot directly observe the state or any payoff that depends on the state. (Suppose player 2's payoffs are going into a blind trust.) Describe a strategy for player 1 that guarantees that the ex ante expected payoff to player 1 at every round will be at least $\frac{3}{4}$, no matter what strategy player 2 may use. (The ex ante expected value is the expected value that would be assessed before the state of nature is determined at the beginning of the game.)

Exercise 7.6. Let $u_2 : D_1 \times D_2 \to \mathbf{R}$ denote the payoff function for player 2 at each round of a two-player repeated game with standard information. Here, for each player i, D_i is the finite set of possible moves at each round for player i. Suppose that $0 \notin D_1$. Let $\Theta = D_1 \cup \{0\}$, and let

$$Z = \{y \in \mathbf{R}^\Theta \mid y_0 \geq \sum_{\theta \in D_1} y_\theta u_2(\theta, c_2), \quad \forall c_2 \in D_2\}.$$

For any (c_1, c_2) in $D_1 \times D_2$, and any θ in Θ, let

$$\begin{aligned}
w_\theta(c_1, c_2) &= 1 && \text{if } \theta = c_1, \\
&= 0 && \text{if } \theta \in D_1 \text{ but } \theta \neq c_1, \\
&= u_2(c_1, c_2) && \text{if } \theta = 0.
\end{aligned}$$

We can write $w(c_1, c_2) = (w_\theta(c_1, c_2))_{\theta \in \Theta}$. When the repeated game is played, let $\tilde{d}_i^k$ denote the move that player i actually chooses at round k, and let

$$\tilde{x}^k = \frac{\displaystyle\sum_{m=1}^{k} w(\tilde{d}_1^m, \tilde{d}_2^m)}{k}.$$

a. Show that, if $\tilde{x}^k \in Z$, then

$$\sum_{m=1}^{k} u_2(\tilde{d}_1^m, \tilde{d}_2^m) \geq \sum_{m=1}^{k} u_2(\tilde{d}_1^m, c_2), \quad \forall c_2 \in D_2.$$

b. Show that, for any ρ_1 in $\Delta(D_1)$, there exists some ρ_2 in $\Delta(D_2)$ such that $w(\rho_1,\rho_2) \in Z$, where

$$w(\rho_1,\rho_2) = \sum_{c_1 \in D_1} \sum_{c_2 \in D_2} \rho_1(c_1)\rho_2(c_2)w(c_1,c_2).$$

c. Show that, for any q in $\mathbf{R}^{\Theta}$,

$$\min_{\rho_2 \in \Delta(D_2)} \max_{\rho_1 \in \Delta(D_1)} \sum_{\theta \in \Theta} q_\theta w_\theta(\rho_1,\rho_2) \leq \sup_{y \in Z} \sum_{\theta \in \Theta} q_\theta y_\theta,$$

and so, by Theorem 7.4, Z is approachable by player 2. Thus, player 2 can guarantee that the limit of her average payoffs will not be below the optimum that she could have gotten if she knew in advance the relative frequency of each of player 1's moves. (HINT: Use (b) and Theorem 3.2. See also Luce and Raiffa, 1957, pages 481–483.)

8

Bargaining and Cooperation in Two-Person Games

8.1 Noncooperative Foundations of Cooperative Game Theory

The concept of cooperation is important in game theory but is somewhat subtle. The term *cooperate* means "to act together, with a common purpose." We might suppose that, for a coalition of two or more individuals to act together with a common purpose, the individuals would have to set aside their separate utility functions and create something completely new—a collective utility function for determining their collective behavior. However, such a conceptual approach is difficult to work with, because all we really have to work with in game theory is our assumption that each player is an intelligent and rational decision-maker, whose behavior is ultimately derived from the goal of maximizing his own expected utility payoff. So we need some model of cooperative behavior that does not abandon the individual decision-theoretic foundations of game theory.

Nash (1951) proposed that cooperation between players can be studied by using the same basic concept of Nash equilibrium that underlies all the areas of noncooperative game theory we have studied in the preceding chapters. He argued that cooperative actions are the result of some process of bargaining among the "cooperating" players, and in this bargaining process each player should be expected to behave according to some bargaining strategy that satisfies the same personal utility-maximization criterion as in any other game situation. That is, in any real situation, if we look carefully at what people can do to reach an agreement on a joint cooperative strategy, in principle we should be able to model it as a game in extensive (or strategic or Bayesian) form and then predict the outcome by analyzing the set of equilibria of this game.

We may define a *cooperative transformation* to be any mapping Ψ such that, if Γ is a game (in extensive, strategic, or Bayesian form), then $\Psi(\Gamma)$ is another game (more complicated, perhaps, but still in extensive, strategic, or Bayesian form) that represents the situation existing when, in addition to the given strategic options specified in Γ, each player has some wide range of options for bargaining with the other players to jointly plan cooperative strategies. This idea was introduced in Chapter 6, where we discussed the concept of transforming a game by adding contract-signing or communication options for the players. In situations where the natural structure of individual options would only admit equilibria that are bad for all players (as in the Prisoners' Dilemma), the players may have a strong incentive to try to transform the structure of the game by adding communication or contract-signing options that give each player some degree of control over the natural decision domain of other players. Such cooperative transformations may change a game with dismal equilibria, like the Prisoners' Dilemma or the game shown in Table 6.1, into a game with equilibria that are better for all of the players, like the games shown in Tables 6.2 and 6.3. *Nash's program* for cooperative game theory is to define cooperative solution concepts such that a cooperative solution for any given game is a Nash equilibrium of some cooperative transformation of the game.

Unfortunately, there is good reason to believe that, if we do a careful job of describing all the things that players can do when they bargain or sign contracts with each other, then we may get a game that has a very large set of equilibria; so Nash's program by itself may fail to determine a unique cooperative solution. Indeed, we saw in Section 6.1 that any individually rational correlated strategy is the equilibrium outcome of a contract-signing game in which all possible contracts (as defined mathematically in Section 6.1) are available to the players. Thus, to get tighter cooperative solution concepts, some theory of *cooperative equilibrium selection* is needed.

Recall from Section 3.5 that the general response to the problem of multiple equilibria in a game is Schelling's (1960) focal-point effect. This theory asserts that, in a game with multiple equilibria, anything that tends to focus the players' attention on one particular equilibrium, in a way that is commonly recognized, tends to make this the equilibrium that the players will expect and thus actually implement. The focal equilibrium could be determined by any of a wide range of possible factors, including environmental factors and cultural traditions (which fall beyond the scope of analysis in mathematical game theory), special mathematical properties of the various equilibria, and preplay statements made by the players or an outside arbitrator. One possible interpretation of the assumption that the players in a game can *cooperate effectively* is that they can use preplay communication to coordinate their expectations on a focal equilibrium that has good welfare properties for some or all of them. Thus, the foundations of cooperative game theory rest, at least in part, on the role of arbitration, negotiation, and welfare properties in determining the focal equilibrium in a game with multiple equilibria.

Recall that a focal arbitrator is an individual who can determine the focal equilibrium in a game by publicly suggesting to the players that they should all implement this equilibrium. For example, in the Battle of the Sexes game (Table 3.5), an arbitrator might announce, "On the basis of many considerations, including a fair coin toss, I have decided to advocate the (s_1, s_2) equilibrium. I urge you to play it." Even though this suggestion may have no binding force, if each player believes that every other player will accept the arbitrator's suggestion, then each player will find it best to do as the arbitrator suggests, provided the arbitrator's suggestion is an equilibrium. To become an effective arbitrator, an individual must be able to communicate with all the players, before the game, in some language that they all understand and that is rich enough to describe any equilibrium. Also, it must be common knowledge among the players that, because of this individual's prestige or authority, all players will attend to and focus on whatever equilibrium he announces. (For example, in the Battle of the Sexes game between a husband and wife, perhaps the wife's father might be well positioned to arbitrate in many situations.)

An impartial arbitrator should try to base his selection on some kind of objective principles. Thus, we might ask what equilibrium would an ideal impartial arbitrator select, in any given game, if his selection should be based on principles that treat the players symmetrically and should depend only on the decision-theoretically relevant structure of the game (so that he would select the same equilibria for two games that are decision-theoretically equivalent in some sense, as discussed in Section 2.3). That is, we can try to develop a normative theory of *impartial arbitration* for games.

For example, recall (from Section 3.5) the *Divide the Dollars* game, in which there are two players who can each make a demand for some amount of money. Each player will get his demand if the two demands sum to $100 or less, but each will get $0 if the demands sum to strictly more than $100. There are infinitely many equilibria of this game, but an impartial arbitrator would probably suggest that the players should implement the (50,50) equilibrium, because it is Pareto efficient and equitable for the two players. But this fact may make (50,50) the focal equilibrium that the players expect and thus implement, even when such an impartial arbitrator is not actually present. That is, because both players know that (50,50) is the efficient and equitable equilibrium that an impartial arbitrator would be most likely to suggest, (50,50) has a strong intrinsic focal property even when there is no arbitrator. In general, whenever the players can identify a unique equilibrium that an impartial arbitrator would select, this equilibrium may become focal just because it has this special property. That is, the welfare properties of *equity* and *efficiency* may determine the focal equilibrium in any game, whether there is an arbitrator or not.

If a focal arbitrator is himself a player in the game, then we call him the *principal* of the game. If one player is the principal in a game with complete information, the focal equilibrium would be an equilibrium that gives the highest possible expected payoff to this principal. For example, if player 1 is the principal in the Battle of the Sexes game, obviously he should simply announce to player 2 that he intends to choose f_1 and expects her to choose f_2.

Another possibility is that a focal equilibrium could be determined by some process of preplay communication between the players in which two or more players have some opportunity to make statements in favor of one equilibrium or another. Any process of preplay communication between players that serves to influence the selection of a focal equilibrium in a game may be called *focal negotiation*.

Developing a formal model of focal negotiation seems to be a very difficult problem. If player 1 has an opportunity to say, "I think that we should play the (f_1, f_2) equilibrium," to player 2 at some time before the Battle of the Sexes game is played, then this situation could be described by an extensive-form game in which his opportunity to say this is represented by an explicit move at some early decision node. But any such game model will have babbling equilibria (like those discussed in Sections 3.5 and 6.2) in which these statements are ignored, and each player i ultimately makes his or her choice in $\{f_i, s_i\}$ according to some equilibrium of the original game (even according to the Pareto-inferior randomized equilibrium) independently of anyone's negotiation statement. So the effectiveness of preplay statements in focal negotiation or arbitration can only be derived from some (cooperative) assumption that goes beyond the basic definition of equilibrium, to rule out these babbling equilibria. Farrell (1988) and Myerson (1989) have proposed that such effectiveness can be built into our analysis by assuming that negotiation and arbitration statements have commonly understood literal meanings and that any statement that passes some formal credibility test should always be interpreted according to its literal meaning. However, such assumptions are technically complicated and can be applied only in models that specify all details about when each player can make a negotiation statement.

One way to lay the foundations for a theory of negotiated outcomes in games, without having to model the details of the negotiation process, is to use the following assumption, which we call the *equity hypothesis*:

> The outcomes of effective negotiations in which the players have equal opportunity to participate should be the same as the recommendations that would be made by an impartial arbitrator who knows the information that is common knowledge among the players during the negotiations.

That is, according to the equity hypothesis, the predictions of a positive theory of cooperative games in which the players have equal negotiating ability should coincide with the prescriptions of a normative theory of impartial arbitration.

We justify the equity hypothesis as follows. On the one hand, an impartial arbitrator should not steer the players to something that could be worse for some players than the outcome they would have negotiated without him, if this outcome is known. A player who did worse in arbitration than he would have in fair negotiations would have justifiable complaint against the arbitrator. On the other hand, if it is obvious what an impartial arbitrator would have suggested, then the most persuasive argument in negotiations may be to settle on this equilibrium. Fisher and Ury (1981) advise negotiators that arguments based on objective and equitable standards are the most effective in real negotiations. Thus, we can expect a close relationship or equivalence between our theories of impartial arbitration and symmetric effective negotiation. In negotiation, just as in arbitration, a focal equilibrium may be determined or selected on the basis of welfare properties like equity and efficiency.

Because we have not yet rigorously defined "symmetric effective negotiation" or "impartial arbitration" for general games, this equity hypothesis cannot now be posed as a formal theorem. It is instead an assertion about the relationship between these two intuitive concepts, both of which we want to formalize. The power of the equity hypothesis comes from the fact that it allows us to carry any restrictions that we can make on one concept over to the other. In the case of the Divide the Dollars game, where we feel that we understand what an impartial arbitrator should do, the equity hypothesis implies that giving the players broad opportunities to negotiate with each other on an equal basis, before playing the game, should tend to focus them on the (50,50) equilibrium. (Thus—to show a testable implication of this theory—we can predict that preplay communication should decrease the observed variance in realized outcomes, relative to the case where the players have no preplay communication, when this game is played by subjects in experimental laboratories.)

8.2 Two-Person Bargaining Problems and the Nash Bargaining Solution

Much of bargaining theory has followed from the remarkable seminal work of Nash (1950, 1953). Nash's formulation of a two-person bargaining problem is based on the following implicit assumption. When two players negotiate or an impartial arbitrator arbitrates, the payoff allocations that the two players ultimately get should depend only on the payoffs they would expect if negotiation or arbitration were to fail to reach a settlement and on the set of payoff allocations that are jointly feasible for the two players in the process of negotiation or arbitration. To justify this assumption, notice that concepts of equity (in the sense that "my gains from our agreement should be commensurate with your gains") generally involve some comparison with what the individuals would get without agreement, and concepts of efficiency generally involve only a comparison with the other feasible allocations.

Thus, we define a *two-person bargaining problem* to consist of a pair (F,v) where F is a closed convex subset of $\mathbf{R}^2$, $v = (v_1,v_2)$ is a vector in $\mathbf{R}^2$, and the set

$$F \cap \{(x_1,x_2) | x_1 \geq v_1 \text{ and } x_2 \geq v_2\}$$

is nonempty and bounded. Here F represents the *set of feasible payoff allocations* or the *feasible set*, and v represents the *disagreement payoff allocation* or the *disagreement point*. The assumption that F is convex can be justified by assuming that players can agree to jointly randomized strategies, so that, if the utility allocations $x = (x_1,x_2)$ and $y = (y_1,y_2)$ are feasible and $0 \leq \theta \leq 1$, then the expected utility allocation $\theta x + (1 - \theta)y$ can be achieved by planning to implement x with probability θ and y otherwise. Closure of F is a natural topological requirement. The nonemptiness and boundedness condition asserts that some feasible allocation is at least as good as disagreement for both players, but unbounded gains over the disagreement point are not possible. We say that a two-person bargaining problem (F,v) is *essential* iff there exists at least one allocation y in F that is strictly better for both players than the disagreement allocation v (i.e., $y_1 > v_1$ and $y_2 > v_2$).

To interpret these structures, we need to be able to specify how they would be derived in the context of some given two-person strategic-form game $\Gamma = (\{1,2\},C_1,C_2,u_1,u_2)$. Two possibilities for F can be suggested, depending on whether the players' strategies can be regulated by binding contracts. If such contracts are possible, then we should let

(8.1) $F = \{(u_1(\mu),u_2(\mu)) | \mu \in \Delta(C)\}$, where $u_i(\mu) = \sum_{c \in C} \mu(c)u_i(c)$.

If there is moral hazard in this game, so strategies cannot be regulated by contracts, then we should let

(8.2) $F = \{(u_1(\mu),u_2(\mu)) | \mu \text{ is a correlated equilibrium of } \Gamma\}$.

To determine the disagreement point v, there are three possible alternatives. One is to let v_i be the minimax value for player i, so

$$v_1 = \min_{\sigma_2 \in \Delta(C_2)} \max_{\sigma_1 \in \Delta(C_1)} u_1(\sigma_1,\sigma_2) \text{ and}$$

$$v_2 = \min_{\sigma_1 \in \Delta(C_1)} \max_{\sigma_2 \in \Delta(C_2)} u_2(\sigma_1,\sigma_2).$$

Another possibility would be to let (σ_1,σ_2) be some focal equilibrium of Γ and let $v_i = u_i(\sigma_1,\sigma_2)$ for each player i. A third possibility is to derive v from some "rational threats," to be defined in Section 8.5, where we also discuss criteria for determining which way for deriving v would make the most sense in any given situation.

A comprehensive theory of negotiation or arbitration would ideally allow us to identify, for any such two-person bargaining problem (F,v), some allocation vector in $\mathbf{R}^2$, which we denote by $\phi(F,v)$, that would be selected as a result of negotiation or arbitration in a situation where F is the set of feasible allocations and v is the disagreement allocation. Thus, our problem of developing a theory of negotiation or arbitration may be considered as a problem of finding some appropriate *solution function* ϕ from the set of all two-person bargaining problems into $\mathbf{R}^2$, the set of payoff allocations.

Nash approached this problem axiomatically. That is, he generated a list of properties that a reasonable bargaining solution function ought to satisfy.

To list Nash's axioms, we let $\phi_i(F,v)$ denote the i-component of $\phi(F,v)$, so

$$\phi(F,v) = (\phi_1(F,v),\phi_2(F,v)).$$

Also, for any two vectors x and y in $\mathbf{R}^2$, we can write

$$x \geq y \text{ iff } x_1 \geq y_1 \text{ and } x_2 \geq y_2, \quad \text{and}$$
$$x > y \text{ iff } x_1 > y_1 \text{ and } x_2 > y_2.$$

That is, we can write an inequality between two vectors iff the inequality is satisfied by each pair of corresponding components in the two vectors. For any two-person bargaining problem (F,v), the axioms for Nash's bargaining solution can now be written as follows.

AXIOM 8.1 (STRONG EFFICIENCY). *$\phi(F,v)$ is an allocation in F, and, for any x in F, if $x \geq \phi(F,v)$, then $x = \phi(F,v)$.*

AXIOM 8.2 (INDIVIDUAL RATIONALITY). *$\phi(F,v) \geq v$.*

AXIOM 8.3 (SCALE COVARIANCE). *For any numbers λ_1, λ_2, γ_1, and γ_2 such that $\lambda_1 > 0$ and $\lambda_2 > 0$, if*

$$G = \{(\lambda_1 x_1 + \gamma_1, \lambda_2 x_2 + \gamma_2) | (x_1,x_2) \in F\}$$
$$and \ w = (\lambda_1 v_1 + \gamma_1, \lambda_2 v_2 + \gamma_2),$$

then

$$\phi(G,w) = (\lambda_1 \phi_1(F,v) + \gamma_1, \lambda_2 \phi_2(F,v) + \gamma_2).$$

AXIOM 8.4 (INDEPENDENCE OF IRRELEVANT ALTERNATIVES). *For any closed convex set G, if $G \subseteq F$ and $\phi(F,v) \in G$, then $\phi(G,v) = \phi(F,v)$.*

AXIOM 8.5 (SYMMETRY). *If $v_1 = v_2$ and $\{(x_2,x_1) | (x_1,x_2) \in F\} = F$, then $\phi_1(F,v) = \phi_2(F,v)$.*

Axiom 8.1 asserts that the solution to any two-person bargaining problem should be feasible and Pareto efficient. That is, there should be no other feasible allocation that is better than the solution for one player and not worse than the solution for the other player. In general, given a feasible set F, we say that a point x in F is *strongly (Pareto) efficient* iff there is no other point y in F such that $y \geq x$ and $y_i > x_i$ for at least one player i. In contrast, we also say that a point x in F is *weakly (Pareto) efficient* iff there is no point y in F such that $y > x$.

Axiom 8.2 asserts that neither player should get less in the bargaining solution than he could get in disagreement. We say that an allocation x in F is *individually rational* iff $x \geq v$.

Axiom 8.3 asserts that if a two-person bargaining problem (G,v) can be derived from a bargaining problem (F,v) by increasing affine utility transformations, which (by Theorem 1.3) will not affect any decision-theoretic properties of the utility functions, then the solution of (G,v) should be derivable from the solution of (F,v) by the same transformation. That is, if we change the way we measure utility when we construct a two-person bargaining problem to represent some real situation but keep our new utility scales decision-theoretically equivalent to the old ones, then the bargaining solution in utility-allocation space should change in the same way, so that it still corresponds to the same real outcome.

Axiom 8.4 asserts that eliminating feasible alternatives (other than the disagreement point) that would not have been chosen should not affect the solution. If an arbitrator were to choose a solution by maximizing some aggregate measure of social gain (so $\{\phi(F,v)\} = \text{argmax}_{x \in F} M(x,v)$, where $M(x,v)$ is his measure of the social gain from choosing allocation x instead of v), then Axiom 8.4 would always be satisfied.

Axiom 8.5 asserts that, if the positions of players 1 and 2 are completely symmetric in the bargaining problem, then the solution should also treat them symmetrically.

Nash's remarkable result is that there is exactly one bargaining solution, called the *Nash bargaining solution*, that satisfies these axioms.

THEOREM 8.1. *There is a unique solution function $\phi(\cdot,\cdot)$ that satisfies Axioms 8.1 through 8.5 above. This solution function satisfies, for every two-person bargaining problem (F,v),*

$$\phi(F,v) \in \underset{x \in F, x \geq v}{\text{argmax}} \ (x_1 - v_1)(x_2 - v_2).$$

Proof. For now, let (F,v) be any essential two-person bargaining problem, so there exists some y in F such that $y_1 > x_1$ and $y_2 > x_2$.

Let x be the unique point in F that achieves the maximum of the function $(x_1 - v_1)(x_2 - v_2)$, called the *Nash product*, over all x in F such that $x \geq v$. This point must satisfy $x > v$, to achieve the strictly positive values of the Nash product that are possible in F. This maximizing point x is unique, given F and v, because maximizing the Nash product is then equivalent to maximizing its logarithm, which is a strictly concave function of x.

Let $\lambda_i = 1/(x_i - v_i)$ and $\gamma_i = -v_i/(x_i - v_i)$, for each i. Define a function $L:\mathbf{R}^2 \to \mathbf{R}^2$ such that

$$L(y) = (\lambda_1 y_1 + \gamma_1, \ \lambda_2 y_2 + \gamma_2),$$

and let $G = \{L(y) | y \in F\}$. For any y in $\mathbf{R}^2$, if $z = L(y)$, then $z_1 z_2 = \lambda_1 \lambda_2 (y_1 - v_1)(y_2 - v_2)$, and $\lambda_1 \lambda_2$ is a positive constant. Thus, because x maximizes the Nash product over F, $L(x)$ must maximize the product $z_1 z_2$ over all z in G. But $L(x) = (1,1)$, and the hyperbola $\{z \in \mathbf{R}^2 | z_1 z_2 = 2\}$ has slope -1 at the point $(1,1)$. Thus, the line $\{z | z_1 + z_2 = 2\}$, which has slope -1 and goes through $(1,1)$, must be above and tangent to the convex set G at $(1,1)$. Let $E = \{z \in \mathbf{R}^2 | z_1 + z_2 \leq 2\}$. Then $G \subseteq E$.

To satisfy Axioms 8.1 and 8.5 (efficiency and symmetry), we must have

$$\phi(E,(0,0)) = (1,1).$$

Then, to satisfy Axiom 8.4 (independence of irrelevant alternatives), we need

$$\phi(G,(0,0)) = (1,1).$$

Then Axiom 8.3 (scale covariance) requires that $L(\phi(F,v)) = \phi(G,(0,0))$, so

$$\phi(F,v) = x.$$

Thus, to satisfy the axioms, ϕ must select the allocation that maximizes the Nash product, among all individually rational allocations in F.

Now, suppose instead that (F,v) is *inessential*, in the sense that there is no point y in F such that $y > v$. By convexity of F, there must then be at least one player i such that, for every y in F, if $y \geq v$, then $y_i = v_i$. (If we could find y and z in F such that $y \geq v$, $z \geq v$, $y_1 > v_1$ and $z_2 > v_2$, then $.5y + .5z$ would be a point in F that was strictly better than v for both players.) Let x be the allocation in F that is best for the player other than i, subject to the constraint that $x_i = v_i$. Then this vector x is the unique point that is strongly Pareto efficient in F and individually rational relative to v. Thus, to satisfy Axioms 8.1 and 8.2, we must have $\phi(F,v) = x$. Obviously x achieves the maximum value of the Nash product, which is 0 for all individually rational allocations in this inessential bargaining problem.

We have thus shown that the five axioms can be satisfied by only one solution function ϕ, which always selects the unique strongly efficient allocation that maximizes the Nash product over all feasible individually rational allocations. It only remains to show that this solution function ϕ does indeed satisfy all of the axioms, a result that is straightforward to verify and is left as an exercise. ■

A weaker version of the efficiency axiom could be written as follows.

AXIOM 8.1′ (WEAK EFFICIENCY). $\phi(F,v) \in F$, *and there does not exist any y in F such that $y > \phi(F,v)$.*

In the above proof, replacing the strong efficiency axiom by the weak efficiency axiom would make no difference in the case of an essential two-person bargaining problem. Furthermore, the individual rationality axiom was also not used in the argument for the essential case. Thus, any solution function that satisfies Axioms 8.1′, 8.3, 8.4, and 8.5 must coincide with the Nash bargaining solution for all essential two-person bargaining problems.

8.3 Interpersonal Comparisons of Weighted Utility

In real bargaining situations, people often make interpersonal comparisons of utility, and they generally do so in two different ways. One way is in applying a principle of *equal gains*, as when a person argues, "You should do this for me because I am doing more for you." For any two-person bargaining problem (F,v), we define the *egalitarian solution* to be the unique point x in F that is weakly efficient in F and satisfies the equal-gains condition

$$x_1 - v_1 = x_2 - v_2.$$

If individuals in negotiation or arbitration are guided by the equal-gains principle, then the outcome should be the egalitarian solution.

Another way of making interpersonal comparisons in bargaining is in applying a principle of *greatest good*, as when a person argues, "You should do this for me because it helps me more than it will hurt you." For any two-person bargaining problem, we define a *utilitarian solution* to be any solution function that selects, for every two-person bargaining problem (F,v), an allocation x such that

$$x_1 + x_2 = \max_{y \in F} (y_1 + y_2).$$

If individuals in negotiation or arbitration are guided by the greatest-good principle, then the outcome should be a utilitarian solution. (Here, "utilitarian" is used in the classical sense of Bentham and Mill.)

The scale-covariance axiom is based on an assumption that only the individual decision-theoretic properties of the utility scales should matter, and interpersonal comparisons of utility have no decision-theoretic significance as long as no player can be asked to decide between being himself or someone else. Thus, it is not surprising that the egalitarian solution and the utilitarian solutions violate the axiom of scale covariance. Given any numbers λ_1, λ_2, γ_1, and γ_2 such that $\lambda_1 > 0$ and $\lambda_2 > 0$, let

$$L(y) = (\lambda_1 y_1 + \gamma_1, \lambda_2 y_2 + \gamma_2), \quad \forall y \in \mathbf{R}^2.$$

Given any two-person bargaining problem (F,v), let

$$L(F) = \{L(y) | y \in F\}.$$

Then the egalitarian solution of $(L(F),L(v))$ is $L(x)$, where x is the unique weakly efficient point in F such that

$$\lambda_1(x_1 - v_1) = \lambda_2(x_2 - v_2),$$

which we refer to as the λ-*egalitarian solution* of (F,v). The egalitarian solution does not satisfy scale covariance because the λ-egalitarian solution is generally different from the simple egalitarian (or (1,1)-egalitarian) solution. Similarly, a utilitarian solution of $(L(F),L(v))$ must be some point $L(z)$, where z is a point in F such that

$$\lambda_1 z_1 + \lambda_2 z_2 = \max_{y \in F} (\lambda_1 y_1 + \lambda_2 y_2).$$

We refer to any such point z as a λ-*utilitarian solution* of (F,v). Egalitarian solutions fail to satisfy scale covariance because a λ-utilitarian solution is generally not a simple utilitarian (or (1,1)-utilitarian) solution.

Thus, to accommodate scale covariance, we must admit that the equal-gains and greatest-good principles each suggest a whole family of bargaining solutions, the λ-egalitarian and the λ-utilitarian solutions. Each of these solutions corresponds to an application of either the equal-gains principle or the greatest-good principle when the payoffs of different players are compared in some λ-*weighted utility scales* (in which each player i's utility is multiplied by a positive number λ_i) that are decision-theoretically equivalent to the originally given utility scales. As λ_1 increases and λ_2 decreases, the λ-egalitarian solutions trace out the individually rational, weakly efficient frontier, moving in the direction of decreasing payoff to player 1. Also, as λ_1 increases and λ_2 decreases, the λ-utilitarian solutions trace out the entire weakly efficient frontier, moving in the direction of increasing payoff to player 1.

Thus, when people want to use interpersonal comparisons of utility in bargaining, there are generally two questions that must be answered: Which of the many decision-theoretically equivalent utility scales for the players should be considered interpersonally comparable? Should the comparisons be used in the equal-gains or greatest-good sense? However, for any essential two-person bargaining problem, there exists an answer to the first question that will make the second question unnecessary. That is, for any essential (F,v), there exists a vector $\lambda = (\lambda_1, \lambda_2)$ such that $\lambda > (0,0)$ and the λ-egalitarian solution of (F,v) is also a λ-utilitarian solution of (F,v). Numbers λ_1 and λ_2 that satisfy this property are called *natural scale factors* for (F,v). The allocation in F that is both

λ-egalitarian and λ-utilitarian in terms of these natural scale factors is the Nash bargaining solution. This result shows that the Nash bargaining solution can be viewed as a natural synthesis of the equal-gains and greatest-good principles.

THEOREM 8.2. *Let (F,v) be an essential two-person bargaining problem, and let x be an allocation vector such that $x \in F$ and $x \geq v$. Then x is the Nash bargaining solution for (F,v) if and only if there exist strictly positive numbers λ_1 and λ_2 such that*

$$\lambda_1 x_1 - \lambda_1 v_1 = \lambda_2 x_2 - \lambda_2 v_2 \text{ and}$$

$$\lambda_1 x_1 + \lambda_2 x_2 = \max_{y \in F} (\lambda_1 y_1 + \lambda_2 y_2).$$

Proof. Let $H(x,v)$ denote the hyperbola

$$H(x,v) = \{y \in \mathbf{R}^2 \,|\, (y_1 - v_1)(y_2 - v_2) = (x_1 - v_1)(x_2 - v_2)\}.$$

The allocation x is the Nash bargaining solution of (F,v) iff the hyperbola $H(x,v)$ is tangent to F at x. But, the slope of the hyperbola $H(x,v)$ at x is $-(x_2 - v_2)/(x_1 - v_1)$, so $H(x,v)$ is tangent at x to the line

$$\{y \in \mathbf{R}^2 \,|\, \lambda_1 y_1 + \lambda_2 y_2 = \lambda_1 x_1 + \lambda_2 x_2\}$$

for any two positive numbers λ_1 and λ_2 such that

(8.3) $$\lambda_1(x_1 - v_1) = \lambda_2(x_2 - v_2).$$

Thus, x is the Nash bargaining solution of (F,v) iff F is tangent at x to a line of the form $\{y \in \mathbf{R}^2 \,|\, \lambda_1 y_1 + \lambda_2 y_2 = \lambda_1 x_1 + \lambda_2 x_2\}$, for some (λ_1,λ_2) satisfying (8.3). ∎

For a simple example, let $v = (0,0)$, and let

$$F = \{(y_1,y_2) \,|\, 0 \leq y_1 \leq 30, \, 0 \leq y_2 \leq (30 - y_1)^{1/2}\}.$$

(F,v) may be interpreted as representing a situation in which the players can divide \$30 in any way on which they agree, or get \$0 each if they cannot agree, and player 1 is risk neutral (has linear utility for money), but player 2 is risk averse, with a utility scale that is proportional to the square root of the monetary value of his gains. To find the Nash bargaining solution, notice that

$$0 = \left(\frac{d}{dy_1}\right)(y_1(30 - y_1)^{1/2}) = (30 - y_1)^{1/2} - \frac{y_1}{2(30 - y_1)^{1/2}}$$

implies that $y_1 = 20$. So the Nash bargaining solution is the utility allocation $(20, 10^{1/2}) = (20, 3.162)$, which corresponds to a monetary allocation of \$20 for player 1 and only \$10 for the risk-averse player 2. Thus, the risk-averse player is under some disadvantage, according to the Nash bargaining solution. (For generalizations of this result, see Roth, 1979.)

Natural scale factors for this problem are $\lambda_1 = 1$ and $\lambda_2 = 40^{1/2} = 6.325$. If we let player 2's utility for a monetary gain of κ dollars be $6.325\kappa^{1/2}$ instead of $\kappa^{1/2}$ (a decision-theoretically irrelevant change), while player 1's utility is still measured in the same units as money, then the representation of this bargaining problem becomes $(G,(0,0))$, where

$$G = \{(y_1, y_2) | 0 \le y_1 \le 30, 0 \le y_2 \le 6.325(30 - y_1)^{1/2}\}.$$

In this representation, the Nash bargaining solution $(20, 20)$, which still corresponds to a monetary allocation of \$20 for player 1 and \$10 for player 2, is both the egalitarian and utilitarian solution.

8.4 Transferable Utility

Transferable utility is an important assumption which can guarantee that the given scale factors in a game will also be the natural scale factors for the Nash bargaining solution.

Given any strategic-form game $\Gamma = (N, (C_i)_{i \in N}, (u_i)_{i \in N})$, to say that Γ is a game *with transferable utility* is to say that, in addition to the strategy options listed in C_i, each player i has the option to give any amount of money to any other player, or even to simply destroy money, and each unit of net monetary outflow or loss decreases i's utility payoff by one unit. That is, saying that a situation can be represented by the game Γ with transferable utility is equivalent to saying that it can be represented by the game

$$\hat{\Gamma} = (N, (\hat{C}_i)_{i \in N}, (\hat{u}_i)_{i \in N}),$$

where, for each i,

$$\hat{C}_i = C_i \times \mathbf{R}_+^N \text{ and}$$

$$\hat{u}_i((c_j,x_j)_{j\in N}) = u_i((c_j)_{j\in N}) + \sum_{j\neq i}(x_j(i) - x_i(j)) - x_i(i).$$

Here $x_j = (x_j(k))_{k\in N}$; for any $k \neq j$, $x_j(k)$ represents the quantity of money given by player j to player k; and $x_j(j)$ denotes the amount of money destroyed by j but not given to any other player. The linear dependence of $\hat{u}_i$ on the transfers x_j expresses an implicit assumption of risk neutrality, which is always assumed when we say that there is transferable utility in a game. (So the example in the preceding section is *not* a game with transferable utility.)

If (F,v) is a two-person bargaining problem derived from a game with transferable utility, then the feasible set F must be of the form

(8.4) $F = \{y \in \mathbf{R}^2 | y_1 + y_2 \leq v_{12}\},$

for some number v_{12} that represents the maximum transferable wealth that the players can jointly achieve. If F is derived under the assumption that the players' strategies can be regulated by binding contracts, then we can let

(8.5) $v_{12} = \max_{\mu\in\Delta(C)} (u_1(\mu) + u_2(\mu)).$

Thus, when there is transferable utility, a two-person bargaining problem can be fully characterized by three numbers: v_1, the disagreement payoff to player 1; v_2, the disagreement payoff to player 2; and v_{12}, the total transferable wealth available to the players if they cooperate. To satisfy the conditions of Theorem 8.2 when (8.4) is satisfied, we must have $\lambda_1 = \lambda_2$, or else $\max_{y\in F}(\lambda_1 y_1 + \lambda_2 y_2)$ would be positive infinity. So the conditions for $\phi(F,v)$ from Theorem 8.2 become

$$\phi_1(F,v) - v_1 = \phi_2(F,v) - v_2 \text{ and } \phi_1(F,v) + \phi_2(F,v) = v_{12}.$$

Solving these equations, we get the following general formulas for the Nash bargaining solution of a game with transferable utility:

(8.6) $\phi_1(F,v) = \dfrac{v_{12} + v_1 - v_2}{2}, \quad \phi_2(F,v) = \dfrac{v_{12} + v_2 - v_1}{2}.$

8.5 Rational Threats

Notice (in equation 8.6, for example) that the payoff to player 1 in the Nash bargaining solution increases as the disagreement payoff to player

2 decreases. That is, a possibility of hurting player 2 in the event of disagreement may actually help player 1 if a cooperative agreement is reached. So the prospect of reaching a cooperative agreement, which will depend on a disagreement point, may give players an incentive to behave more antagonistically before the agreement is determined, as each player tries to create a more favorable disagreement point. This *chilling effect* may be formalized by Nash's (1953) theory of rational threats.

To present this theory, we let $\Gamma = (\{1,2\}, C_1, C_2, u_1, u_2)$ be any finite game in strategic form, and let F be the feasible set derived from Γ either by (8.1) (in the case with binding contracts but without nontransferable utility), or by (8.4) and (8.5) (in the case with binding contracts and transferable utility). Suppose that, before entering into the process of arbitration or negotiation with the other player, each player i must choose a threat τ_i that is a randomized strategy in $\Delta(C_i)$. Suppose also that, if the players failed to reach a cooperative agreement, then each player would be committed independently to carry out the threat that he has chosen. Then, once the threats (τ_1,τ_2) are chosen, the disagreement point in the two-person bargaining problem should be $(u_1(\tau_1,\tau_2), u_2(\tau_1,\tau_2))$. Let $w_i(\tau_1,\tau_2)$ denote the payoff that player i gets in the Nash bargaining solution with this disagreement point, that is,

$$w_i(\tau_1,\tau_2) = \phi_i\big(F, (u_1(\tau_1,\tau_2), u_2(\tau_1,\tau_2))\big).$$

Suppose now that the players expect that they will ultimately reach a cooperative agreement that will depend on the disagreement point according to the Nash bargaining solution. Then the players should not be concerned about carrying out their threats but should instead evaluate their threats only in terms of their impact on the final cooperative agreement. So each player i should want to choose his threat τ_i so as to maximize $w_i(\tau_1,\tau_2)$, given the other player's expected threat. Thus, we say that (τ_1,τ_2) form a pair of *rational threats* iff

$$w_1(\tau_1,\tau_2) \geq w_1(\sigma_1,\tau_2), \quad \forall \sigma_1 \in \Delta(C_1), \quad \text{and}$$
$$w_2(\tau_1,\tau_2) \geq w_2(\tau_1,\sigma_2), \quad \forall \sigma_2 \in \Delta(C_2).$$

That is, rational threats form an equilibrium of the following *threat game*

$$\Gamma^* = (\{1,2\}, \Delta(C_1), \Delta(C_2), w_1, w_2).$$

Existence of rational threats can be proved by an argument using the Kakutani fixed-point theorem.

Figure 8.1 shows how a typical feasible set F can be partitioned by rays or half-lines, such that, for each point z in F, the Nash bargaining solution of (F,z) is the upper endpoint of the ray that goes through z. By Theorem 8.2, the slope of each of these rays is equal in magnitude but opposite in sign to the slope of a tangent line to F at the upper endpoint of the line segment. In the threat game, player 1 wants to choose his threat τ_1 so as to put the disagreement point $(u_1(\tau_1,\tau_2), u_2(\tau_1,\tau_2))$ on the lowest and rightmost possible ray in Figure 8.1, whereas player 2 wants to choose her threat τ_2 so as to put the disagreement point on the highest and leftmost possible ray.

When there is transferable utility, the analysis of the threat game becomes simpler. The efficient frontier of F becomes a line of slope -1, so the rays described above become parallel lines of slope 1. By (8.6), the payoffs in the threat game are

$$w_1(\tau_1,\tau_2) = \frac{v_{12} + u_1(\tau_1,\tau_2) - u_2(\tau_1,\tau_2)}{2},$$

$$w_2(\tau_1,\tau_2) = \frac{v_{12} + u_2(\tau_1,\tau_2) - u_1(\tau_1,\tau_2)}{2},$$

where v_{12} is as defined in equation (8.5). Because v_{12} is a constant, maximizing $w_1(\tau_1,\tau_2)$ is equivalent to maximizing $u_1(\tau_1,\tau_2) - u_2(\tau_1,\tau_2)$, and maximizing $w_2(\tau_1,\tau_2)$ is equivalent to maximizing $u_2(\tau_1,\tau_2) - u_1(\tau_1,\tau_2)$. Thus, when Γ is a game with transferable utility, τ_1 and τ_2 are rational threats for players 1 and 2, respectively, iff (τ_1,τ_2) is an equilibrium of the two-person zero-sum game

$$\Gamma^{**} = (\{1,2\}, \Delta(C_1), \Delta(C_2), u_1 - u_2, u_2 - u_1).$$

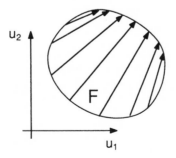

Figure 8.1

We may call Γ^{**} the *difference game* derived from Γ, because the payoff to each player in Γ^{**} is the difference between his own payoff and the other player's payoff in Γ.

In this chapter, we have considered three different ways to determine a disagreement point for any given two-person strategic-form game Γ: by an equilibrium of Γ, by the minimax values, and by rational threats. To compare these in the context of an example, let Γ be the strategic-form game shown in Table 8.1, with transferable utility. The maximum total payoff achievable in this game is $v_{12} = 10$.

This game has a unique noncooperative equilibrium at (b_1, b_2). If we take the payoff at this equilibrium to be the disagreement point, we get $v = (0, 10)$, so the Nash bargaining solution is $\phi(F, v) = (0, 10)$.

Suppose next that we determine the disagreement point by minimax values. The minimax value for player 1 is $v_1 = 0$, which is achieved when player 2 chooses an offensive threat b_2 and player 1 chooses a defensive response of b_1. The minimax value for player 2 is $v_2 = 1$, which is achieved when player 2 chooses b_2 as an optimal defensive strategy while player 1 chooses a_1 as an optimal offensive threat against b_2. Then, with disagreement point $(0, 1)$, the Nash bargaining solution is $\phi(F, v) = (4.5, 5.5)$.

In the threat game Γ^* derived from Γ, the payoffs (w_1, w_2) to all combinations of pure strategies are as shown in Table 8.2. The unique

Table 8.1   A game in strategic form, with transferable utility

C_1	C_2	
	a_2	b_2
a_1	10,0	−5,1
b_1	0,−5	0,10

Table 8.2   The threat game derived from the game in Table 8.1

C_1	C_2	
	a_2	b_2
a_1	10,0	2,8
b_1	7.5,2.5	0,10

equilibrium of this threat game is (a_1, b_2), which leads to the disagreement point $v = (-5, 1)$ (read from Table 8.1), so the Nash bargaining solution with rational threats is $\phi(F, v) = (2, 8)$ (read from Table 8.2).

Each of these three theories of threats and disagreement suggested that player 2 would choose b_2 in any disagreement, because a_2 is dominated by b_2 both with respect to 2's defensive objective of maximizing u_2 and with respect to 2's offensive objective criterion of minimizing u_1. The differences between the three theories arise in their specifications of player 1's disagreement behavior. In the noncooperative-equilibrium theory of disagreement points, it is assumed that player 1's behavior in the event of disagreement would be determined by his purely defensive objective of maximizing u_1; so he would choose b_1 and let player 2 get her maximum possible payoff. In the minimax theory, player 1 is supposed to be able to choose two threats: an offensive threat (a_1) for determining v_2 and a defensive threat (b_1) for determining v_1. In the rational-threats theory, player 1 must choose a single threat that must serve both offensive and defensive purposes simultaneously, so he chooses a_1 because it maximizes an objective $((10 + u_1 - u_2)/2$, or simply $u_1 - u_2)$ that is a synthesis of his offensive and defensive criteria.

Thus, the equilibrium theory of disagreement is appropriate in situations where the players could not commit themselves to any planned strategies in the event of disagreement, until a disagreement actually occurs. The rational-threats theory is appropriate in situations where each player can, before the negotiation or arbitration process (when this process is expected to have virtually no chance of leading to a disagreement), commit himself to a single planned strategy that he would carry out in the event of disagreement, no matter whose final rejection may have caused the disagreement. The minimax-values theory is appropriate in situations where each player can, before the negotiation or arbitration process, commit himself to two planned strategies, one defensive and one offensive, that he would carry out in the event of disagreement, depending on how the disagreement was caused. To interpret these offensive and defensive threats, let us suppose that a player would implement his planned defensive strategy if the final disagreement occurred after he himself rejected the last offer in the negotiation or arbitration process (so the disagreement would be, in a sense, his fault); but he would implement his offensive strategy if final disagreement occurred after the last offer was rejected by the other player.

8.6 Other Bargaining Solutions

The symmetry axiom was used in the proof of Theorem 8.1 only to show that, when $E = \{z \in \mathbf{R}^2 | z_1 + z_2 \leq 2\}$, the solution of the two-person bargaining problem $(E,(0,0))$ must be $(1,1)$. Let us now drop the symmetry axiom and suppose instead that the solution of $(E,(0,0))$ is some efficient point (α,β) such that $\alpha > 0$ and $\beta > 0$. Then an argument similar to the proof of Theorem 8.1 can be used to show that the solution to any essential two-person bargaining problem must be the point that maximizes the following *generalized Nash product* over all individually rational points x in F:

$$(8.7) \qquad (x_1 - v_1)^\alpha (x_2 - v_2)^\beta.$$

(To modify the proof, let $\lambda_1 = \alpha/(x_1 - v_1)$, $\lambda_2 = \beta/(x_2 - v_2)$, let each $\gamma_i = -\lambda_i v_i$, and use the fact that the curve $\{z | (z_1)^\alpha (z_2)^\beta = \alpha^\alpha \beta^\beta\}$ in $\mathbf{R}^2$ has slope -1 at the point (α,β).)

Thus, we say that a bargaining solution function ψ is a *nonsymmetric Nash bargaining solution* iff there exist positive numbers α and β such that, for any two-person bargaining problem (F,v), $\psi(F,v)$ is a point that is individually rational and strongly efficient in F and that maximizes the generalized Nash product (8.7) over all such points. To see how such nonsymmetric Nash bargaining solutions might be used in practice, suppose that player 1 is the head of a family of 4 and player 2 is a single individual, and an arbitrator feels that the appropriate solution to a simple Divide the Dollars game with transferable utility would be to give ⅘ of the available money to player 1 and ⅕ to player 2 (to equalize their per capita income). If this arbitrator wants to satisfy all the Nash axioms other than symmetry, then he should use the nonsymmetric Nash bargaining solution with $\alpha/\beta = 4$ (see Kalai, 1977).

Several other bargaining solution concepts have been proposed, using other systems of axioms (see Roth, 1979). To introduce some of these, we need to restrict our attention to a somewhat narrower class of two-person bargaining problems. Let us say that a two-person bargaining problem (F,v), as defined in Section 8.2, is *regular* iff F is essential (that is, $\{y \in F | y > v\} \neq \emptyset$) and, for any vector y in F,

if $y_1 > v_1$, then $\exists z \in F$ such that $v_1 \leq z_1 < y_1$ and $z_2 > y_2$,

if $y_2 > v_2$, then $\exists \hat{z} \in F$ such that $v_2 \leq \hat{z}_2 < y_2$ and $\hat{z}_1 > y_1$.

That is, in a regular problem, there is a feasible allocation that is strictly better than the disagreement point for each player; and whenever a player is getting strictly more than his disagreement payoff, there is something that he could do that would reduce his own expected payoff and strictly increase the other player's expected payoff. We say that a two-person bargaining problem (F,v) itself is *individually rational* iff the feasible set F only includes individually rational allocations, that is,

$$F \subseteq \{y \in \mathbf{R}^2 \,|\, y \geq v\}.$$

For any regular two-person bargaining problem (F,v), let $m_i(F,v)$ denote the maximum payoff that player i can get in any feasible individually rational allocation, so

$$m_i(F,v) = \max_{y \in F, y \geq v} y_i.$$

For any number z_1 such that $v_1 \leq z_1 \leq m_1(F,v)$, let $h_2(z_1,F)$ denote the highest payoff that player 2 can get in a feasible allocation when player 1 gets z_1. That is,

$$h_2(z_1,F) = \max \{y_2 \,|\, (z_1,y_2) \in F\}.$$

Similarly, let $h_1(z_2,F)$ be the highest payoff that player 1 can get in F when 2 gets z_2, for any z_2 such that $v_2 \leq z_2 \leq m_2(F,v)$.

Kalai and Smorodinsky (1975) suggested a kind of monotonicity axiom, as an alternative to the controversial independence of irrelevant alternatives axiom. They suggested that, if the range of individually rational payoffs that player 1 can get in (F,v) is the same as in (G,v) and, for any feasible individually rational payoff to player 1, the best that player 2 could get is not less in (F,v) than in (G,v), then player 2 should not do worse in (F,v) than in (G,v). Formally, this axiom may be stated as follows.

AXIOM 8.6 (INDIVIDUAL MONOTONICITY). *If* $m_1(G,v) = m_1(F,v)$ *and* $h_2(z_1,G) \leq h_2(z_1,F)$ *for every* z_1 *such that* $v_1 \leq z_1 \leq m_1(F,v)$, *then* $\phi_2(G,v) \leq \phi_2(F,v)$. *Similarly, if* $m_2(G,v) = m_2(F,v)$ *and* $h_1(z_2,G) \leq h_1(z_2,F)$ *for every* z_2 *such that* $v_2 \leq z_2 \leq m_2(F,v)$, *then* $\phi_1(G,v) \leq \phi_1(F,v)$.

Kalai and Smorodinsky showed that there is a unique bargaining solution for regular two-person bargaining problems that satisfies Axioms 8.1, 8.3, 8.5, and 8.6 (strong efficiency, scale covariance, symmetry, and individual monotonicity). For any regular bargaining problem (F,v), this Kalai-Smorodinsky solution is the unique efficient point x in F such that

$$\frac{x_2 - v_2}{x_1 - v_1} = \frac{m_2(F,v) - v_2}{m_1(F,v) - v_1}.$$

Given any sequence of two-person bargaining problems $(F(k),v(k))_{k=1}^{\infty}$, we can write $\lim_{k \to \infty} (F(k),v(k)) = (\overline{F},\overline{v})$ iff $\overline{v} = \lim_{k \to \infty} v(k)$ in $\mathbf{R}^2$ and $\overline{F}$ is the set of limits in $\mathbf{R}^2$ of all convergent sequences $(y(k))_{k=1}^{\infty}$ such that $y(k) \in F(k)$ for every k. (This topology for closed sets is called *Hausdorff convergence*.) We say that a bargaining solution function ϕ is *continuous* iff, for any sequence of two-person bargaining problems $(F(k),v(k))_{k=1}^{\infty}$,

if $\lim_{k \to \infty} (F(k),v(k)) = (\overline{F},\overline{v})$,

then $\lim_{k \to \infty} \phi(F(k),v(k)) = \phi(\overline{F},\overline{v})$.

For any two bargaining problems (F,v) and (G,w) and any number α between 0 and 1, we write

$$\alpha F + (1 - \alpha)G = \{\alpha y + (1 - \alpha)z \,|\, y \in F,\, z \in G\}, \quad \text{and}$$
$$\alpha(F,v) + (1 - \alpha)(G,w) = (\alpha F + (1 - \alpha)G,\, \alpha y + (1 - \alpha)z).$$

To interpret this two-person bargaining problem, suppose that the bargaining problem that players 1 and 2 will face tomorrow depends on whether it will rain tonight or not. If it rains, which will occur with probability α, then tomorrow's bargaining problem will be (F,v), otherwise it will be (G,w). If players 1 and 2 bargain today over conditional plans, they can achieve any expected payoff of the form $\alpha y + (1 - \alpha)z$, where $y \in F$ and $z \in G$, by agreeing to the plan "implement y if it rains and z if it does not rain." Thus, if the players bargain over plans today, their problem is $\alpha(F,v) + (1 - \alpha)(G,w)$. In such circumstances, they will both want to bargain today over plans, rather than wait and bargain tomorrow, if their bargaining solution function ϕ is *concave*, in the sense that

$$\phi(\alpha(F,v) + (1 - \alpha)(G,w)) \geq \alpha\phi(F,v) + (1 - \alpha)\phi(G,w)$$

for any (F,v) and (G,w) and any α in $[0,1]$.

Concave solution functions include λ-egalitarian and λ-utilitarian solutions discussed in the preceding section (see Myerson, 1981b). However, Perles and Maschler (1981) have proved that there is a solution function for regular individually rational two-person bargaining problems that is concave, continuous, and satisfies the axioms of strong efficiency, scale covariance, and symmetry. If $h_1(\cdot,F)$ and $h_2(\cdot,F)$ are differentiable functions, then this Perles-Maschler solution selects the unique efficient point x in F such that

$$\int_{v_1}^{x_1} (-h_2'(z_1,F))^{1/2}dz_1 = \int_{v_2}^{x_2} (-h_1'(z_2,F))^{1/2}dz_2.$$

Raiffa (see Luce and Raiffa, 1957, pages 136–137) discussed bargaining solution concepts that are closely related to the Kalai-Smorodinsky solution. Given any regular two-person bargaining problem (F,v), we inductively define a sequence of allocation vectors $(w(k))_{k=1}^{\infty}$, where each $w(k) = (w_1(k),w_2(k)) \in \mathbf{R}^2$, such that $w(1) = v$ and, for every positive integer k,

$$w(k + 1) = \tfrac{1}{2}\big(h_1(w_2(k),F), w_2(k)\big) + \tfrac{1}{2}\big(w_1(k), h_2(w_1(k),F)\big).$$

It can be shown that this sequence is nondecreasing componentwise and converges to some w^* that is an efficient allocation vector in F. This limit w^* is Raiffa's sequential bargaining solution of (F,v).

Raiffa's sequential bargaining solution may be justified along the lines of Nash's program, by analyzing a specific model of the bargaining process in which the players can make offers in up to k rounds, where k is some (large) number. At each round, a player is selected at random to make an offer, which can be any point in F. After getting this offer, the other player can either accept or reject it. The first offer that is accepted is the payoff allocation that the players get in the game. If no offer is accepted at any of the k rounds, then the players get the disagreement payoff allocation v. At each round, each player has probability $\tfrac{1}{2}$ of being the one to make an offer, independently of the past history of the game.

This game has a unique subgame-perfect equilibrium in which the first-round offer is accepted, and the expected payoff to each player i is $w_i(k + 1)$. In this equilibrium, at the lth round from the end, for any l in $\{1,2,\ldots,k\}$, each player i would accept any offer that gives him at least $w_i(l)$ (because that would be his expected payoff in the subgame consisting of the last $l - 1$ rounds if no offer were accepted before), and so player 1 would make the offer $(h_1(w_2(l),F),\ w_2(l))$, and player 2 would make the offer $(w_1(l),\ h_2(w_1(l),F))$. Thus, Raiffa's sequential bargaining solution is the limit of the expected outcome of this game in equilibrium, as the number of possible offers goes to infinity.

8.7 An Alternating-Offer Bargaining Game

Stahl (1972) and Rubinstein (1982) considered a somewhat different model of bargaining, in which the players alternate making offers until one is accepted. Rubinstein considered a general model in which there is no finite bound on the number of offers that may be made but in which each player has some cost of time, so his payoff depends on both the offer accepted and the number of the round in which this acceptance took place. Rubinstein elegantly characterized the subgame-perfect equilibria of these alternating-offer bargaining games for a wide class of cost-of-time formulas and showed that the subgame-perfect equilibrium is unique when each player's cost of time is given by some discount factor δ. Binmore (see Binmore and Dasgupta, 1987, chap. 4) showed that, with the δ-discounting cost-of-time formula, if the per-round discount factor δ is close to 1, then the outcome of the unique subgame-perfect equilibrium is close to the Nash bargaining solution.

In this section, we present a special case of these results. However, instead of assuming discounting or making some other assumptions about players' trade-offs between prizes received at different points in time, we assume here that the cost of delay in bargaining is derived from an exogenous positive probability, after each round, that the bargaining process may permanently terminate in disagreement if no offer has yet been accepted.

Let (F,v) be any regular two-person bargaining problem, and let p_1 and p_2 be numbers such that $0 < p_1 < 1$ and $0 < p_2 < 1$. In our alternating-offer bargaining game, player 1 makes an offer in every odd-numbered round, and player 2 makes an offer in every even-numbered round, beginning in round 1 and continuing until some offer is accepted or the game ends in disagreement. At each round, after one player makes the offer, the other player can either accept or reject. At any round when player 1 makes the offer and player 2 rejects it, there is a probability p_1 that the bargaining will then end in a state of disagreement in which player 2 will get payoff v_2 and player 1 will get some payoff w_1 such that $w_1 \leq m_1(F)$. Similarly, in any round when player 2 makes the offer and player 1 rejects it, there is a probability p_2 that the bargaining will then end in a disagreement in which player 1 will get payoff v_1, and player 2 will get some payoff w_2 such that $w_2 \leq m_2(F)$. An offer can be any payoff allocation in the feasible set F. When an offer is accepted, the game ends in agreement and the players get the payoffs specified in the accepted offer. (Notice that the players get payoffs at only one round in this game, when either an offer is accepted or bargaining ends in disagreement.)

THEOREM 8.3. *The above alternating-offer game has a unique subgame-perfect equilibrium in which player 1 plans to always offer some allocation $\bar{x}$, player 2 plans to always offer some allocation $\bar{y}$, player 1 would accept any offer that gives him at least $\bar{y}_1$, and player 2 would accept any offer that would give her at least $\bar{x}_2$. Thus, in equilibrium, the game will end in an agreement on $\bar{x}$ at round 1. These vectors $\bar{x}$ and $\bar{y}$ are strongly efficient allocations in F and satisfy*

$$\bar{y}_1 = (1 - p_2)(\bar{x}_1 - v_1) + v_1 \ \text{ and } \ \bar{x}_2 = (1 - p_1)(\bar{y}_2 - v_2) + v_2.$$

Proof. Notice first that all subgames that begin at an odd round (player 1 offers first) must have the same set of subgame-perfect equilibria as the original game. Similarly, all subgames that begin at an even round (player 2 offers first) must have the same set of subgame-perfect equilibria. Let x_1 be the supremum and let $\hat{x}_1$ be the infimum of the set of all possible expected payoffs that player 1 could get in a subgame-perfect equilibrium of any subgame in which player 1 offers first. Similarly, let y_2 be the supremum and let $\hat{y}_2$ be the infimum of the set of all possible expected payoffs that player 2 could get in a subgame-perfect equilibrium of any subgame in which player 2 offers first.

In any subgame-perfect equilibrium of this game, player 2 would always accept any offer giving her more than $(1 - p_1)y_2 + p_1 v_2$ at an odd-numbered round, because this payoff is the best that she could possibly expect when she rejects an offer from player 1. Thus, player 1 cannot do worse in any subgame-perfect equilibrium where he offers first than $h_1((1 - p_1)y_2 + p_1 v_2, F)$. Furthermore, for any positive number ε, there is a subgame-perfect equilibrium in which player 2 would expect to get within ε of her best payoff y_2 in any subgame beginning at round 2; so she could guarantee herself at least $(1 - p_1)(y_2 - \varepsilon) + p_1 v_2$ at round 1. Player 1 would expect not more than $h_1((1 - p_1)(y_2 - \varepsilon) + p_1 v_2, F)$ in this equilibrium. (Because $w_i \leq m_i(F)$ for each i, the expected utility allocation in any equilibrium cannot be above the efficient frontier of the convex set F.) Thus, the infimum of player 1's expected payoffs over all subgame-perfect equilibria when he offers first is

(8.8) $\hat{x}_1 = h_1((1 - p_1)y_2 + p_1v_2, F)$.

A similar argument, using the fact that player 1 would surely accept any offer giving him more than $(1 - p_2)x_1 + p_2v_1$ at any even-numbered round, can be used to show that

(8.9) $\hat{y}_2 = h_2((1 - p_2)x_1 + p_2v_1, F)$.

On the other hand, player 2 would never accept any offer giving her less than $(1 - p_1)\hat{y}_2 + p_1v_2$ at an odd-numbered round, because this payoff is the least she could expect to get by rejecting the current offer in a subgame-perfect equilibrium; so player 1 could never do better than $h_1((1 - p_1)\hat{y}_2 + p_1v_2, F)$. Furthermore, for any positive ε, there exists a subgame-perfect equilibrium in which player 2 would expect to get within ε of $\hat{y}_2$ in any subgame beginning at round 2, so she would accept any offer giving her $(1 - p_1)(\hat{y}_2 + \varepsilon) + p_1v_2$ at round 1; and player 1 would get at least $h_1((1 - p_1)(\hat{y}_2 + \varepsilon) + p_1v_2, F)$ in this equilibrium. So the supremum of all payoffs that player 1 could get in an equilibrium when he moves first is

(8.10) $x_1 = h_1((1 - p_1)\hat{y}_2 + p_1v_2, F)$.

A similar argument, using the fact that player 1 would never accept any offer giving him less than $(1 - p_2)\hat{x}_1 + p_2v_1$ at any even-numbered round, can be used to show that

(8.11) $y_2 = h_2((1 - p_2)\hat{x}_1 + p_2v_1, F)$.

Let us complete the definition of vectors x, $\hat{x}$, y, and $\hat{y}$ in $\mathbf{R}^2$ by letting

$$x_2 = h_2(x_1, F), \quad \hat{x}_2 = h_2(\hat{x}_1, F), \quad y_1 = h_1(y_2, F), \quad \text{and} \quad \hat{y}_1 = h_1(\hat{y}_2, F),$$

so these vectors are all on the efficient frontier of F. Regularity and (8.8)–(8.11) then imply that

(8.12) $\hat{x}_2 = (1 - p_1)y_2 + p_1v_2$,

(8.13) $\hat{y}_1 = (1 - p_2)x_1 + p_2v_1$,

(8.14) $x_2 = (1 - p_1)\hat{y}_2 + p_1v_2$,

(8.15) $y_1 = (1 - p_2)\hat{x}_1 + p_2v_1$.

To verify these equations, notice that regularity implies that any weakly efficient individually rational vector in F is also strongly efficient. This fact in turn implies that, for any allocation z that is individually rational and efficient in F, if $\hat{z}_1 \geq v_1$ and $z_2 = h_2(\hat{z}_1, F)$, then $\hat{z}_1 = z_1$; and if $\hat{z}_2 \geq v_2$ and $z_1 = h_1(\hat{z}_2, F)$, then $\hat{z}_2 = z_2$.

There is a unique pair of vectors $\bar{x}$ and $\bar{y}$ such that $\bar{x}$ and $\bar{y}$ are individually rational and efficient in F, and

$$\bar{y}_1 = (1 - p_2)(\bar{x}_1 - v_1) + v_1 \text{ and } \bar{x}_2 = (1 - p_1)(\bar{y}_2 - v_2) + v_2.$$

To show this, notice that these conditions are equivalent to

(8.16) $\bar{x}_1 - \bar{y}_1 = p_2(\bar{x}_1 - v_1),$

(8.17) $\bar{y}_2 - \bar{x}_2 = \dfrac{p_1(\bar{x}_2 - v_2)}{1 - p_1} .$

As we increase $\bar{x}_1$ from v_1 to $m_1(F, v)$, $\bar{x}_2$ decreases from $m_2(F, v)$ to v_2, so the right-hand side of (8.17) decreases monotonically to 0. On the other hand, as we increase $\bar{x}_1$ from v_1 to $m_1(F, v)$, the right-hand side of (8.16) increases monotonically from 0. Furthermore, given that $\bar{x}$ and $\bar{y}$ are on the efficient frontier of the convex set F, when we increase both $\bar{x}_1$ and $\bar{x}_1 - \bar{y}_1$, the difference $\bar{y}_2 - \bar{x}_2$ must increase monotonically as well and must be 0 when $\bar{x}_1 - \bar{y}_1$ is 0. So when we require that $\bar{x}$ and $\bar{y}$ be efficient in F and satisfy (8.16), the left-hand side of (8.17) increases monotonically from 0 as $\bar{x}_1$ increases from v_1 to $m_1(F, v)$. Thus, using continuity of these conditions, we can guarantee the existence and uniqueness of the solutions $\bar{x}$ and $\bar{y}$ on the individually rational efficient frontier of F.

Thus, conditions (8.12) and (8.15) imply that $\hat{x} = \bar{x}$ and $y = \bar{y}$. Conditions (8.13) and (8.14) similarly imply that $x = \bar{x}$ and $\hat{y} = \bar{y}$. Thus, in every subgame-perfect equilibrium, player 1 must get expected payoff $\bar{x}_1$ in all subgames that begin with player 1 making the offer, and player 2 must get expected payoff $\bar{y}_2$ in all subgames that begin with player 2 making the offer. By the way that $\bar{x}$ and $\bar{y}$ are constructed, it is straightforward to check that the strategies described in Theorem 8.3 form the only subgame-perfect equilibrium with this property. (CHECK: In any equilibrium where 2 expected to get $\bar{y}_2$ in the subgame starting in the second round, if 1 offered 2 more than $\bar{x}_2$, then she would accept but he would get less than $\bar{x}_1$ himself, whereas if he offered her less than $\bar{x}_2$, then she would reject him and he would expect at best $(1 - p_1)\bar{y}_1 + p_1 v_1$, which is strictly less than $\bar{x}_1$.) ∎

In this game, the parameter p_i is a measure of player i's *power of commitment*, because it is the probability that any offer he makes may, if rejected, be final. The analysis of this alternating-offer bargaining game suggests that bargainers can be expected to reach an agreement in which their relative shares depend crucially on their relative powers of commitment p_1/p_2, even if the absolute level of each player's power of commitment p_i is quite low. To see how this conclusion would follow from Theorem 8.3, let α, β, and ε be positive numbers, and suppose that

$$1 - p_1 = (1 - \varepsilon)^\alpha \text{ and } 1 - p_2 = (1 - \varepsilon)^\beta.$$

So if ε is small, then $p_1 \approx \varepsilon\alpha$ and $p_2 \approx \varepsilon\beta$. The conditions in Theorem 8.3 then imply that

$$(\bar{x}_1 - v_1)^\alpha (\bar{x}_2 - v_2)^\beta = (\bar{x}_1 - v_1)^\alpha (\bar{y}_2 - v_2)^\beta (1 - \varepsilon)^{\alpha\beta}$$
$$= (\bar{y}_1 - v_1)^\alpha (\bar{y}_2 - v_2)^\beta.$$

Thus, $\bar{x}$ and $\bar{y}$ are two intersections of the efficient frontier of F with one isoquant of the generalized Nash product (8.7), so the nonsymmetric Nash bargaining solution with parameters α and β must be between $\bar{x}$ and $\bar{y}$. Furthermore, if we let ε go to 0, holding α and β fixed, then p_1 and p_2 go to 0, so (by 8.16 and 8.17) the differences $\bar{x}_1 - \bar{y}_1$ and $\bar{y}_2 - \bar{x}_2$ must go to 0. Thus, as ε goes to 0, holding α and β fixed, the outcome of the unique subgame-perfect equilibrium of this alternating-offer bargaining game converges to the nonsymmetric Nash bargaining solution with parameters α and β, where $\alpha/\beta = \lim_{\varepsilon\to 0} p_1/p_2$.

For example, in a two-person bargaining problem with transferable utility, if $p_1 = .004$ and $p_2 = .001$, then the unique subgame-perfect equilibrium of this alternating-offer bargaining game will lead to an outcome in which player 1 will get approximately $\frac{4}{5}$ of the transferable gains above the disagreement outcome (more accurately, 80.064% when 1 offers first, or 79.984% in a subgame when 2 offers first), and player 2 will only get about $\frac{1}{5}$.

This analysis suggests that a good bargainer should try to create the perception that there is a relatively high probability that bargaining may terminate in disagreement whenever one of his offers is rejected, whereas there is a relatively low probability that bargaining may terminate in disagreement when he rejects someone else's offer. Notice, however, that at no time in this subgame-perfect equilibrium does either player actually want to terminate the bargaining process in disagreement after his offer is rejected; he is always better off waiting to receive the new offer from the other player. Thus, in this model, increasing p_i means encouraging the belief that player i might irrationally walk out and terminate the bargaining process when his latest offer is rejected. In practical terms, a good bargainer may try to raise his own apparent power of commitment by making offers firmly, with an appearance of finality, while he may try to lower the other player's power of commitment by making rejections politely and amicably.

8.8 An Alternating-Offer Game with Incomplete Information

In the preceding section, it was assumed to be common knowledge that the players are rational utility maximizers with disagreement payoffs v_1 and v_2. This assumption implies that player 1 could never convince anyone that there are some offers that would be impossible for him to accept even though they would be better for him than v. However, in real bargaining, people often try to convince each other that they could never accept any offer outside of some range. To understand the importance of such tactics, we must drop the assumption of complete information.

Given a regular two-person bargaining problem (F,v), let us again consider the alternating-offer bargaining game that we analyzed in the preceding section, but let us introduce a second possible type for player 1. Let r be any efficient allocation vector in F such that $r_1 > v_1$ and $r_2 > v_2$, and suppose that there is some reason to suspect that player 1 might have some kind of irrational commitment that compels him to insist on this allocation r. To be specific, suppose that there is a small but strictly positive probability q that player 1 might have been replaced (before the game began) by a robot or agent who mechanically implements the following *r-insistent strategy* for player i: at odd rounds he always offers r, and at even rounds he accepts an offer y if and only if $y_1 \geq r_1$. Also, to simplify our analysis, suppose that disagreement payoffs are 0 in the alternating-offer game, so

$$0 = v_1 = v_2 = w_1 = w_2.$$

THEOREM 8.4. *In any equilibrium of the alternating-offer game with incomplete information, as described above, there exists some number $J(F,v,r,q)$,* *(which does not depend on p_1 or p_2) such that, if player 1 follows the r-insistent strategy, then player 2 is sure to offer r or accept r within the first $J(F,v,r,q)$ rounds. Thus, the expected payoff to player 1 in equilibrium cannot be less than $r_1(1 - \max\{p_1,p_2\})^{J(F,v,r,q)}$. If p_1 and p_2 are small, then this lower bound on 1's expected equilibrium payoff is close to r_1.*

Proof. Let $\delta = (1 - p_1)(1 - p_2)$. Because $0 < r_2 < m_2(F,v)$ and p_1, p_2, and q are all positive, we can choose positive integers K and I large enough so that

$$\delta^K < \frac{0.5r_2}{m_2(F,v)} \quad \text{and}$$

$$\left(1 - \frac{0.5r_2}{m_2(F,v)}\right)^I < q.$$

Now consider player 2's position at any round where player 1 has offered r and has always previously behaved according to the r-insistent strategy, and player 2 is following some strategy c_2 that has positive probability in an equilibrium of the alternating-offer game. Notice that this event has positive probability, because $q > 0$, so c_2 must be sequentially rational for player 2 given the information available to her at this round. (Recall Theorem 4.4.) Suppose that, under c_2, player 2 plans to withstand the r-insistent strategy, by neither accepting r nor making an offer in which player 1 gets at least r_1, during the next $2K$ rounds. Let π denote the probability that player 1 (by rational choice according to his equilibrium strategy, or as a robot) will not observably deviate from the r-insistent strategy during the next $2K$ rounds if the bargaining game does not end in disagreement during this period. Player 2 can get payoff r_2 by accepting r now, but if she rejects and avoids r for the next $2K$ rounds then her expected payoff cannot be more than

$$m_2(F,v)(1 - \pi(1 - \delta^K)),$$

because there is a probability of at least $\pi(1 - \delta^K)$ that bargaining will end in disagreement during the next $2K$ rounds. Thus, sequential rationality of 2's strategy at this round implies that

$$r_2 \leq m_2(F,v)\left(1 - \pi\left(1 - \frac{0.5r_2}{m_2(F,v)}\right)\right),$$

so

$$\pi \le 1 - \frac{0.5r_2}{m_2(F,v)} .$$

By repeating this argument I times, assuming that player 2 would continue to withstand 1's r-insistent strategy throughout the first $2KI$ rounds, the probability that 1 will not deviate from the r-insistent strategy during the first $2KI$ rounds, if the bargaining does not end in disagreement during this period, could not be greater than $(1 - 0.5r_2/m_2(F,v))^I$. However, this probability is less than q, by definition of I, so there cannot be any strategy that player 2 uses with positive probability in equilibrium under which she would withstand 1's r-insistent strategy for more than $2KI$ rounds.

The above argument does not prove the theorem, because K and I depend on p_1 and p_2, but it does prove that, in any equilibrium, there is some finite number L such that player 2 would not withstand the r-insistent strategy for more than L rounds. That is, as long as player 1 is always observed to follow the r-insistent strategy, then player 2 is sure to accept r or offer something that is at least as good as r for player 1 by round L. If L were an even number, then player 2 would be willing to accept r at round $L - 1$, so we can assume without loss of generality that L is odd and player 1 offers at round L.

Now let us work backward from round L. At any even-numbered round $L + 1 - 2k$, where $k \ge 1$, if player 1 has not deviated from the r-insistent strategy, then he knows that he can get at least $r_1(1 - p_1)^{k-1}(1 - p_2)^k$ by following the r-insistent strategy, so he will not accept any offer from player 2 that gives him less. Similarly, player 1 will not make his first deviation from the r-insistent strategy at any odd-numbered round $L - 2k$ unless his expected payoff from doing so is at least $r_1(1 - p_1)^k(1 - p_2)^k$. Thus, if player 1 makes his first deviation from the r-insistent strategy at round $L + 1 - 2k$ or $L - 2k$, then player 2's expected payoff when this deviation occurs cannot be more than $h_2(r_1\delta^k, F)$.

By regularity, there exists some line that has negative slope and is tangent to F at r. So let A and B be positive numbers such that $A - Br_1 = r_2$ and $A - By_1 \ge y_2$ for every y in F. Let M be a positive integer such that $M > 2A/(A - Br_1) = 2A/r_2$.

For any positive integer k, consider player 2's position at round $L - 2Mk$ when player 1 has never deviated from the r-insistent strategy and has just made an offer of r to player 2. Suppose that player 2 is following a pure strategy c_2 that has positive probability in equilibrium and under which she would withstand the r-insistent strategy, by neither accepting r nor making an offer in which player 1 gets at least r_1, in all rounds before round L. Let ξ denote the probability that player 1 (either by rational choice or as a robot) will not deviate from the r-insistent strategy against c_2 during the next $2(M - 1)k$ rounds at least (that is, at least until round $L - 2k$) if the bargaining game does not end in disagreement during this period. Then player 2's conditionally expected payoff cannot be greater than

$$(1 - \xi)(A - Br_1\delta^{Mk}) + \xi\delta^{(M-1)k}(A - Br_1\delta^k).$$

(In the $(1 - \xi)$-probability event that player 1 does deviate from the r-insistent strategy before round $L - 2k$, he would not do so for an expected payoff of less than $r_1\delta^{Mk}$, even if he deviated in the next round $(L + 1 - 2Mk)$; so player 2 could not expect to do better than $A - Br_1\delta^{Mk}$. In the ξ-probability event that player 1 will not deviate from the r-insistent strategy before round $L - 2k$, there is a probability of only $\delta^{(M-1)k}$ that the bargaining game will not terminate in disagreement during this period; and thereafter player 1 would not deviate for less than $r_1\delta^k$, giving player 2 not more than $A - Br_1\delta^k$.) Thus, for continuing with c_2 (instead of accepting r at round $L - 2Mk$) to be optimal for player 2 in this situation (as it must be, because this situation could occur in the equilibrium with positive probability), we must have

$$A - Br_1 \leq (1 - \xi)(A - Br_1\delta^{Mk}) + \xi\delta^{(M-1)k}(A - Br_1\delta^k);$$

so

$$\xi \leq \left(\frac{Br_1}{A}\right)\left(\frac{1 - \delta^{Mk}}{1 - \delta^{(M-1)k}}\right).$$

Using some basic calculus (including l'Hospital's rule), it can be shown that, whenever $0 < \delta < 1$, the right-hand side of this inequality is never more than $Br_1M/((M - 1)A)$; and this quantity is less than $1 - 1/(M - 1)$, because M is greater than $2A/(A - Br_1)$. So $\xi \leq 1 - 1/(M - 1)$.

Applying this argument H times, with $k = 1, M, M^2, \ldots, M^{H-1}$, we can see that the probability that player 1 will not deviate from the r-insistent strategy for more than $2M^H$ rounds, if player 2 neither accepts nor offers r and the game does not end in disagreement during this period, cannot be greater than $(1 - 1/(M - 1))^H$. Now let H be a positive integer large enough so that

$$\left(1 - \frac{1}{M-1}\right)^{H} < q,$$

and let $J(F,v,r,q) = 2M^{H}$. Given that the probability that player 1 is an r-insistent robot is q, there cannot be any equilibrium in which player 2 would withstand the r-insistent strategy for more than $J(F,v,r,q)$ rounds. Notice that our construction of M, H, and J did not depend on the value of δ or p_1 or p_2. ■

When p_1 and p_2 are both small, Theorem 8.4 gives insights into bargaining that contrast sharply with the insights generated by Theorem 8.3. Theorem 8.3 suggests that the ratio of the players' powers of commitment p_1/p_2 may be critical to determining the relative shares of players 1 and 2 in a cooperative agreement. However, Theorem 8.4 suggests that, if player 1 can create some doubt (in player 2's mind) about whether he might have some irrational commitment to an allocation r, then he can expect to get an agreement that is close to r or better than r for himself, no matter what the ratio p_1/p_2 may be.

More generally, a comparison of Theorems 8.3 and 8.4 should lead us to expect that other small structural perturbations (e.g., assuming that both players have small positive probabilities of independently being replaced by robots or agents who would irrationally insist on some allocation) could make other large changes in the set of equilibria of our alternating-offer game. In the next section, we consider one such perturbation in which, when p_1 and p_2 are small, almost any individually rational allocation can be achieved in equilibrium.

8.9 A Discrete Alternating-Offer Game

Let F_0 be a finite subset of the feasible set F, in a regular two-person bargaining problem (F,v). We can think of F_0 as being a very large subset. Given any positive number ε, we can say that F_0 is ε-*dense* in F iff, for each y in F there exists some $\hat{y}$ in F_0 such that $|y_i - \hat{y}_i| \leq \varepsilon$ for both $i = 1,2$.

Suppose now that the alternating-offer bargaining game from Section 8.7 (where there was no incomplete information) is changed only by requiring that all offers must be in the set F_0. For example, if utility is measured in dollars, but allocations that split pennies cannot be offered, then the players might be constrained to make their offers in the set

F_0, when F_0 is the set of all vectors $y = (y_1, y_2)$ in F such that y_1 and y_2 are both integer multiples of $1/100$.

In addition to this assumption, let us again make the restriction that disagreement payoffs are 0, so $0 = v_1 = v_2 = w_1 = w_2$.

For each player i, let $t(i) = (t_1(i), t_2(i))$ be the individually rational allocation that is optimal for player i in F_0. If there is more than one such optimal allocation for player i, then let $t(i)$ be the allocation that is best for the player other than i among these allocations that are optimal for i. That is, among the allocation vectors that are strongly efficient within the set F_0 (when F_0 is considered to be the feasible set), $t(1)$ is the best allocation for player 1, and $t(2)$ is the best allocation for player 2. Thus,

$$t(i) \in \operatorname*{argmax}_{y \in F_0, y \geq (0,0)} y_i,$$

and, when $j \neq i$,

if $z \in F_0$ and $z_j > t_j(i)$, then $z_i < t_i(i)$.

When F_0 is a large set, then $t_1(2)$ and $t_2(1)$ should be small numbers, close to the disagreement payoff of 0.

Suppose that p_1 and p_2 are small enough to allow the following two conditions to hold.

(8.18) $\forall z \in F_0$, if $z_2 > t_2(1)$, then $z_1 < (1 - p_2)t_1(1)$.

(8.19) $\forall z \in F_0$, if $z_1 > t_1(2)$, then $z_2 < (1 - p_1)t_2(2)$.

Given our assumption that F_0 is a finite set, there must exist strictly positive values of p_1 and p_2 that do satisfy these conditions.

THEOREM 8.5. *Under the above assumptions, for any allocation x in F_0 such that $x_1 \geq t_1(2)$ and $x_2 \geq t_2(1)$, there exists a subgame-perfect equilibrium of the alternating-offer game (with offers required to be in F_0) in which player 1 offers x and player 2 accepts it at round 1.*

Proof. (See also van Damme, Selten, and Winter, 1990.) Consider first the special case of $x = t(1)$. There is a subgame-perfect equilibrium in which player 1 would always offer $t(1)$ and accept only an offer that gives him $t_1(1)$ or more, and player 2 would also offer $t(1)$ and accept any offer that gives her $t_2(1)$ or more. We call this the *2-submissive equilibrium*.

To verify that the 2-submissive equilibrium is a subgame-perfect equilibrium, notice that, at any round where player 2 makes the offer, player 1 expects to be able to offer $t(1)$ and get it accepted if the play continues until the next round. Thus, player 1 should be willing to accept an offer from player 2 only if it gives him $(1 - p_2)t_1(1)$ or more. By (8.18), there is no such allocation that would be better for player 2 than $t(1)$. Thus, player 2 cannot expect to do better than to offer $t(1)$ at any even-numbered round and to accept $t(1)$ at any odd-numbered round. When player 1 makes his offer, $t(1)$ is clearly optimal for him, because it is the best individually rational offer that can be made and player 2 is expected to accept it, in this equilibrium.

Similarly, there is a subgame-perfect equilibrium, called the *1-submissive equilibrium*, in which both players would always offer $t(2)$, which each would always accept, but neither would accept any offer that gives him or her less than $t(2)$.

Now let x be any allocation in F_0 such that $x_1 \geq t_1(2)$ and $x_2 \geq t_2(1)$. Consider the following strategy for a player: as long as no one has ever made an offer different from x, offer x and accept x when offered; but if anyone has ever made an offer different from x, play according to the i-submissive equilibrium, where i is the first player to have made an offer different from x. When both players follow this strategy, the outcome will be an agreement on x at the first round, and this pair of strategies forms a subgame-perfect equilibrium. For example, suppose player 1 made the first offer different from x, by offering some other vector y in F_0. In the subgame that begins after his deviant offer, player 1 would get an expected payoff of $t_1(2)(1 - p_1)$ if $y_2 < t_2(2)$, and he would get an expected payoff of y_1 if $y_2 \geq t_2(2)$. Notice, however, that $y_2 \geq t_2(2)$ implies that $y_1 \leq t_1(2)$, by definition of $t(2)$. ∎

The allocations that can occur in equilibrium according to this theorem include all of the allocations that are strongly efficient within the finite feasible set F_0 (in the sense that there is no other allocation in F_0 that is better for one player and not worse for the other). However, very inefficient allocations can also be equilibrium outcomes, including any allocations in F_0 as low as $(t_1(2), t_2(1))$, which may be close to the disagreement allocation $(0,0)$.

Even if F_0 only included efficient allocations, there generally exist randomized equilibria of this game in which both players get low expected payoffs. Let us add the weak assumption that $t_1(1) > t_1(2)$ and $t_2(2) > t_1(1)$. Then conditions (8.18) and (8.19) imply that

$$(1 - p_2)t_1(1) > t_1(2) \text{ and } (1 - p_1)t_2(2) > t_2(1).$$

So there exist numbers s_1 and s_2, each between 0 and 1, such that

$$t_2(1) = (1 - p_1)(s_1 t_2(2) + (1 - s_1)(1 - p_2)t_2(1)),$$
$$t_1(2) = (1 - p_2)(s_2 t_1(1) + (1 - s_2)(1 - p_1)t_1(2)).$$

Now consider the following equilibrium. If player 1 has always offered $t(1)$ and player 2 has always offered $t(2)$, then each player i always offers $t(i)$ at the rounds when he makes offers, and, at each round where he receives the other player j's offer $t(j)$ he randomizes, accepting it with probability s_i and rejecting it with probability $1 - s_i$. If any player i has ever deviated from offering $t(i)$, then the players behave as in the i-submissive equilibrium, where player i is the first player to have made such a deviant offer.

It can be verified that these strategies form a subgame-perfect equilibrium. At the first round, where player 1 makes the first offer, player 1's expected payoff under this equilibrium is $t_1(2)/(1 - p_2)$ and player 2's expected payoff is $t_2(1)$, so the expected payoff allocation is very close to the worst equilibrium allocation for both players, $(t_1(2), t_2(1))$. This equilibrium may be called a *standoff equilibrium*, because each player is always making his most selfish offer, which has, at any round, only a low probability of being accepted. Neither player i could gain by making an offer that is less selfish than $t(i)$, because the other player j would then reject this offer in the expectation that i would soon concede even more and accept $t(j)$ in the next round. In this logic, this standoff equilibrium is in some ways similar to the positional equilibrium of the repeated game discussed in Section 7.4.

Under the assumptions of Theorem 8.5, virtually any individually rational offer could be accepted in some equilibrium. This result suggests that the focal-point effect may be the crucial determinant of the outcome of the bargaining process. In his original work on the focal-point effect, Schelling (1960) actually included a discussion of equilibria that are very similar to those that we constructed in our proof of Theorem 8.5. Schelling argued that, once the players focus on any one particular allocation (whether for reasons of historical tradition, qualitative prominence, or apparent equity and efficiency under some criterion), each player in the bargaining game may be afraid to offer anything that makes the other player better off than in this allocation, because the other player might interpret such a concession as an invitation to take everything. Schelling argued, for example, that the back-and-forth fighting in the Korean War had to end with a truce along the relatively narrow waist of the country, because it was the one line that either side could withdraw to without giving the other side the impression that it was willing to be pushed out of Korea entirely. International borders, once established, are rarely renegotiated for the same reason. If we ceded a small piece of our country, then we might give our neighbors the impression that they could take as much as they want. So we must hold the line at the existing border, lest we be unable to convincingly hold a line anywhere.

Theorem 8.5 has implications about good tactics in negotiations and bargaining that complement and contrast with the implications of Theorems 8.3 and 8.4. Theorem 8.3 emphasizes the importance of developing a perceived power of commitment or potential finality of one's offers in general. Theorem 8.4 emphasizes the importance of encouraging the belief that one may be irrationally committed to a specific allocation (that is favorable to oneself). Theorem 8.5 emphasizes the importance of the focal-point effect in negotiating the equilibrium that will be played. For example, the incomplete-information model of Theorem 8.4 suggests that a player might try to achieve a particular allocation by announcing, in preliminary negotiations, that he will have to insist on this allocation, and by trying to seem fanatical or unreasonable when making this announcement. In contrast, the discrete model of Theorem 8.5 suggests that the player might instead achieve a particular allocation by first describing some objective criterion according to which this allocation is in some sense the most natural or reasonable one, and then arguing that neither bargainer could offer concessions beyond this obviously reasonable allocation without giving the impression that he would give everything away.

The uniqueness of the equilibria in the models of Theorems 8.3 and 8.4 contrasts strikingly with the ubiquity of the equilibria in Theorem 8.5. However, given any small positive number ε, if F_0 is ε-dense in F, then all of the equilibria in Theorem 8.5 correspond to ε-equilibria of the game in Theorem 8.3. Thus, in comparison with Theorem 8.3,

Theorem 8.5 shows that we should beware of placing too much significance on the uniqueness of an equilibrium in a game that has a large set of ε-equilibria, all of which may be equilibria of games that differ only slightly from the given game and that might be equally good as models of any given real situation.

On the other hand, from the point of view of Theorem 8.5, the models of Theorem 8.3 and 8.4 might be interpreted as criteria that can determine the focal equilibrium that gets played. In general, when players are involved in some given game that has a very large set of equilibria, if there is some obvious or salient perturbation of the game that has only one equilibrium, then the players might naturally focus on playing the equilibrium of the given game that is closest to the equilibrium of the salient perturbation. Thus, for example, suppose that two risk-neutral players are bargaining over how to divide $30.00 in an alternating-offer game with $p_1 = p_2 = .0001$, but they can only make offers in which both payoffs are integer multiples of $.01. Then conditions (8.18) and (8.19) are satisfied, so any individually rational outcome is an equilibrium of this discrete alternating-offer game. However, the players might naturally compare their situation to the corresponding continuous alternating-offer game, in which the restriction of offers to integer multiples of $.01 is dropped. Among the possible offers in the discrete alternating-offer game, (15.00,15.00) is closest to the unique equilibrium outcome of this continuous alternating-offer game; and this fact may make focal an equilibrium in which this allocation is the outcome in the discrete game that is actually being played.

(The preceding three sections are only an introduction to the analysis of sequential-offer bargaining games, on which there is a large and growing literature. See also Sobel and Takahashi, 1983; Fudenberg, Levine, and Tirole, 1985; Binmore, Rubinstein, and Wolinsky, 1986; Sutton, 1986; Rubinstein, 1987; and Ausubel and Deneckere, 1989.)

8.10 Renegotiation

The analysis in the preceding section illustrates that it may be difficult or impossible to derive, from the noncooperative analysis of a realistic bargaining game, the simple conclusion that players will achieve an efficient utility allocation. However, there are many situations in which it may be reasonable to assume that players will, through negotiation statements like "Let's play this equilibrium, which is better for both of us," be able to coordinate their focal expectations on a Pareto-efficient equilibrium. So if Pareto efficiency cannot be derived from noncooperative analysis, then it may be reasonable to make efficiency a fundamental assumption in the analysis of any game model of a situation where the players have ample opportunities to negotiate effectively in a rich common language at the beginning of the game. That is, if players can negotiate effectively at the beginning of an extensive-form game, but only at the beginning, then we can assume that the equilibrium that is actually played will be one that is at least weakly efficient within the sequential equilibrium set, in the sense that there is no sequential equilibrium that gives higher expected payoffs for all players.

However, the impact of negotiations may be somewhat different if players can negotiate effectively at more than one point in time during a game. We should not expect that intelligent players at the beginning of the game could persuade themselves, through some negotiation process, to focus on playing one equilibrium if, at some later stage in the game, a similar negotiation process would lead them to focus on some different equilibrium. The players should not be able to negotiate to an equilibrium that would be renegotiated later. A variety of definitions of *renegotiation-proof equilibria* and related concepts have been offered by Bernheim, Peleg, and Whinston (1987); Bernheim and Ray (1989); Farrell and Maskin (1987); Pearce (1987); and Benoit and Krishna (1989). As an introduction to this subject, we consider here a simple example studied by Benoit and Krishna (Table 8.3).

This is a two-person game played as follows in K rounds, where K is some given positive integer. At each round, knowing all past moves of both players, each player i must choose a move in the set $\{a_i, b_i, c_i, d_i\}$. The incremental contributions to the payoffs of players 1 and 2, re-

Table 8.3   Payoffs at any round, for all move profiles, in a finitely repeated game

1's moves	2's moves			
	a_2	b_2	c_2	d_2
a_1	0,0	2,4	0,0	6,0
b_1	4,2	0,0	0,0	0,0
c_1	0,0	0,0	3,3	0,0
d_1	0,6	0,0	0,0	5,5

spectively, depend on their moves according to Table 8.3. The payoff to each player in the game is the sum of the incremental contributions to his payoff in all K rounds.

There exist sequential equilibria of this game in which each player's expected total payoff is $5K - 2$ (which corresponds to an average of $5 - 2/K$ per round). In one such equilibrium, the players are supposed to do (d_1, d_2) at every round until the last round and (c_1, c_2) at the last round, unless someone deviates from his d_i move before the last round; if anyone does deviate from (d_1, d_2) before the last round, then the players are supposed thereafter to do (a_1, b_2) if player 1 deviated first, or (b_1, a_2) if player 2 deviated first, or (c_1, c_2) if both deviated first simultaneously. We call this the *grim equilibrium*. This equilibrium is subgame-perfect and sequential, because no player could ever gain by unilaterally deviating from his equilibrium strategy, as long as the other player is expected to follow his equilibrium strategy.

Suppose now that the players can effectively renegotiate the equilibrium at the beginning of each round. Consider the situation that would arise at the next round after some player made an initial deviation from the (d_1, d_2) moves prescribed by the equilibrium. Let J denote the number of rounds remaining in the game, and suppose that $J \geq 2$ and the deviator was player 1. In the subgame that remains to be played, according to the grim equilibrium for the original K-round game, player 1's total payoff will be $2J$ and player 2's payoff will be $4J$. But both players could do better and get $5J - 2$ in this subgame if they renegotiated and played instead according to the restarted grim equilibrium for the J-round game that remains, ignoring the past $K - J$ rounds! That is, either player would have an incentive to make the following negotiation speech, if it would be heeded.

> A mistake has just been made. We have been expecting to go into a punishment mode after such a mistake, and such punishment might even be justified, but it will not do either of us any good now. Let us forgive the errors of the past and go back to doing d_1 and d_2. But let us also consider ourselves fully warned now that future deviations will not be so equally forgiven. Whoever makes the next deviation in the future should expect that he really will get only 2 at every round thereafter.

If the players can renegotiate the equilibrium in this way, then the grim equilibrium may not be credible, because each player would expect to be able to avoid punishment of any deviation before the second-to-last round by such renegotiation. That is, the grim equilibrium is not Pareto efficient in some subgames, even though it is Pareto efficient in the game overall, and as a result it is not renegotiation proof.

Renegotiation-proof equilibria for such two-person finitely repeated games (with complete state information and perfectly observable moves) may be defined by induction as follows. In any one-round game, a renegotiation-proof equilibrium is an equilibrium such that no other equilibrium would be, in this one-round game, as good or better for both players and strictly better for at least one player. In any J-round game, a renegotiation-proof equilibrium is a subgame-perfect equilibrium such that it is renegotiation proof in all $(J - 1)$-round subgames (beginning at the second round of the J-round game), and there is no other subgame-perfect equilibrium that is renegotiation proof in all $(J - 1)$-round subgames and that is, in the J-round game, as good or better for both players and strictly better for at least one player.

If the failure of the grim equilibrium to be renegotiation proof implies any kind of irrationality, it may be a *cooperative* or *collective* irrationality, but not an *individual* irrationality. In the context of the grim equilibrium (or in any other sequential equilibrium), neither player could ever expect to increase his expected payoff by unilaterally changing his own strategy in any subgame. For intelligent players, the act of renegotiating an equilibrium must be an essentially cooperative act, in which both players simultaneously change the strategies that they plan to use.

In this example, for any even number K, there is an essentially unique renegotiation-proof equilibrium, with the following structure. At every even-numbered round, the players are supposed to choose (d_1, d_2). At every odd-numbered round, the players are supposed to choose (c_1, c_2) if the moves were (d_1, d_2) at the preceding round, (b_1, a_2) if the moves were (d_1, a_2) at the preceding round, (a_1, b_2) if the moves were (a_1, d_2) at the preceding round, and some pair in $\{(c_1, c_2), (b_1, a_2), (a_1, b_2)\}$ following any other possible history of past moves. In this equilibrium, with probability 1, the incremental payoffs will therefore be $(5,5)$ at every even round and $(3,3)$ at every odd round, for a total payoff of $4K$ for each player in the overall game (an average of 4 per round).

Benoit and Krishna (1989) proved this result by induction. In a one-round game, (c_1, c_2), (b_1, a_2), and (a_1, b_2) are the only equilibria that are Pareto efficient within the set of all equilibria. In a two-round game, there is no equilibrium that is better for either player than the equilibria sketched above, in which each player gets an expected payoff of 5 +

$3 = 8$. Thus, (8,8) is the unique payoff allocation that can be generated by a renegotiation-proof equilibrium of a two-round subgame. Now assume inductively that, given any even number K such that $K \geq 4$, $(4(K - 2), 4(K - 2))$ is the unique payoff allocation that can be generated in a renegotiation-proof equilibrium of a $(K - 2)$-round game. Thus, in a renegotiation-proof equilibrium of a K-round game, the players cannot expect their moves at the first two rounds to have any impact on their payoffs in the last $K - 2$ rounds. Therefore, the analysis of the first two rounds of a K-round game must be the same as the analysis of a two-round game. Subject to the constraint that the expected payoff for each player must be $4(K - 2)$ in any $(K - 2)$-round subgame, no equilibrium is better for either player in the K-round game than an equilibrium in which they play (d_1, d_2) in the first round and (c_1, c_2) in the second round, unless someone profitably deviated in the first round, in which case he should be punished by the one-round equilibrium in which he gets 2 in the second round. In such an equilibrium, both players get an expected total payoff of $4K$.

Notice that, for this example, there exist subgame-perfect equilibria (such as the grim equilibrium described earlier) that are strictly better for both players than any renegotiation-proof equilibrium. So this example shows that the opportunity to renegotiate at later rounds of a game may reduce the set of equilibria that can be negotiated at earlier rounds, in a way that may hurt both players. That is, both players may be hurt in a game by a constraint that their behavior should be collectively rational in all subgames.

Exercises

Exercise 8.1.   John and Mary are discussing whether to go to the ballet, to the boxing match, or to stay home tonight. They have a random-number generator, and they can agree to let their decision depend on its output in any way, so as to create a lottery with any probability distribution over the three possible outcomes (go to the ballet, go to boxing, or stay home). If they cannot agree, they will stay home. John wants to go to the boxing match, and he is indifferent between going to the ballet and staying home. For Mary, going to the ballet would be best, staying home would be worst, and she would be indifferent between going to the boxing match or taking a lottery with probability ¼ of going to the ballet and probability ¾ of staying home. They both satisfy the axioms of von Neumann-Morgenstern utility theory.

a. Describe this situation by a two-person bargaining problem (F,v) and compute its Nash bargaining solution. How should they implement the Nash bargaining solution?

b. The above description of this situation does not specify a specific two-person bargaining problem, as defined in Section 8.2. Which axiomatic properties of the Nash bargaining solution permitted you to answer part (a) anyway?

c. Find natural utility scales for John and Mary in which the egalitarian and utilitarian solutions both coincide with the Nash bargaining solution.

d. If the television set were broken, the prospect of staying home would become much worse for John. To be specific, John would be indifferent between going to the ballet and a lottery with probability ⅔ of going to boxing and probability ⅓ of staying home with a broken television set. A broken television set would not affect Mary's preferences. If Mary breaks the television set, her brother (who is a television repairman, and who is coming for breakfast tomorrow) will fix it for free. How would breaking the television set change the probability of going to the ballet in the Nash bargaining solution?

Exercise 8.2. Suppose that ϕ is a solution concept for two-person bargaining problems with bounded feasible sets, and ϕ does not actually depend on the disagreement point. That is, $\phi(F,v) = \phi(F,w)$ for any feasible set F that is a closed, convex, and bounded subset of $\mathbf{R}^2$, and for any two disagreement points v and w. Suppose also that ϕ satisfies Axioms 8.1, 8.3, and 8.4 (strong efficiency, scale covariance, and independence of irrelevant alternatives) from Section 8.2. Show that either ϕ must always select the best strongly efficient allocation for player 1, or ϕ must always select the best strongly efficient allocation for player 2. (This result is related to many general results in social choice theory; see Arrow, 1951; Sen, 1970.)

Exercise 8.3. Suppose that $x \in F$, $y \in F$, $x_1 = v_1$, $y_2 = v_2$, and $.5x + .5y$ is a strongly efficient allocation in F. Show that $.5x + .5y$ is the Nash bargaining solution of (F,v).

Exercise 8.4. Suppose that player 1 has made a demand for an allocation x in F, and player 2 has made a demand for an allocation y in F, where $x_1 > y_1$ and $y_2 > x_2 > v_2$. For each player i, let q_i be the highest probability such that player i would be willing to accept a lottery with probability q_i of giving the disagreement point v and probability $1 - q_i$ of giving the allocation vector that i demanded, when the alternative would be to accept the allocation vector that the other player demanded. Show that, if player 1's demand is the Nash bargaining solution of (F,v) and player 2's demand is any other vector in F, then $q_1 > q_2$. (See Harsanyi, 1956.)

Exercise 8.5. Consider the strategic-form game shown in Table 8.4 as a cooperative game with transferable utility.

a. Find the unique noncooperative equilibrium outcome. If this outcome is the disagreement point, then what is the Nash bargaining solution?

b. Calculate the minimax value for each player. If the minimax values are the disagreement payoffs for each player, then what is the Nash bargaining solution?

c. Calculate the rational threats for players 1 and 2, as defined in Section 8.5. Show the resulting disagreement point in utility allocation space and compute the resulting Nash bargaining solution.

d. In the rational-threats model that we considered in Section 8.5, there was no chance of threats actually being carried out. Let us consider now a more elaborate model of a three-stage bargaining process, in which threats may be carried out.

At the first stage of the bargaining process, each player i independently chooses a probability q_i (or an *agreeability index*) in the interval $[0,1]$. At the second stage, after these probabilities q_1 and q_2 have been revealed publicly to both players, each player i independently chooses

Table 8.4 A game in strategic form, with transferable utility

C_1	C_2		
	x_2	y_2	z_2
x_1	10,0	7,1	−4,−4
y_1	0,10	4,0	−5,−5

a threat strategy c_i in C_i. At the third stage, either agreement or disagreement occurs, and the probability of agreement is q_1q_2. If disagreement occurs, then each player i must carry out his chosen threat strategy c_i, and they get the resulting payoffs (as specified above in the original strategic-form game). If agreement occurs, then the players get the Nash bargaining solution of this game, where the disagreement point is the payoff allocation that they would get if both carried out their chosen threat strategies.

Find a subgame-perfect equilibrium of this three-stage bargaining game in which the probability of agreement is positive.

Exercise 8.6. Suppose that the alternating-offer game of Section 8.7 is played with feasible set $F = \{(x_1,x_2) | x_1 + x_2 \leq v_{12}\}$, disagreement payoffs (v_1,v_2), and termination probabilities (p_1,p_2), where $v_{12} > v_1 + v_2$. As a function of these parameters $(v_{12}, v_1, v_2, p_1, p_2)$, what is the general formula for the offer that player 1 would always make and the offer that player 2 would always make in the subgame-perfect equilibrium?

Exercise 8.7. (Analysis of arbitration, following Crawford, 1979, 1981.) Players 1 and 2 have taken their bargaining problem to an arbitrator. The set of feasible utility allocations for players 1 and 2

$$F = \{(x_1,x_2) | 2x_1 + x_2 \leq 150 \text{ and } x_1 + 2x_2 \leq 150\}.$$

Thus, for example, (75,0), (50,50), and (0,75) are strongly efficient allocations in F. In arbitration, each player must independently specify an offer, which is an allocation in F, and the arbitrator will then select one of these two offers. Unfortunately, the arbitrator has strange criteria and will select the offer for which the quantity $(x_1)^2 + (x_2)^2$ is *lower*. If this quantity is the same for both players' offers, then the arbitrator will randomize, choosing either offer with probability $\frac{1}{2}$.

a. Suppose first that no further bargaining is allowed after the arbitrator's selection, so the offer selected by the arbitrator will be the realized payoff allocation for the players. (This is called *final-offer arbitration.*) Find an equilibrium of this game.

b. Suppose now that, after the arbitrator's selection is announced, the player whose offer has been selected by the arbitrator can propose one subsequent offer in F. Then the other player (whose offer was not selected) can choose between the offer selected by the arbitrator and the subsequent offer, and the offer that he then chooses will be the realized payoff allocation for the players. Find a subgame-perfect equilibrium of this game. In this equilibrium, what is the expected payoff for each player?

c. Now suppose that, after the arbitrator's selection is announced, the player whose offer has *not* been selected by the arbitrator can propose one subsequent offer in *F*. Then the player whose offer was selected by the arbitrator can choose between the offer selected by the arbitrator and the subsequent offer, and the offer that he then chooses will be the realized payoff allocation for the players. Find a subgame-perfect equilibrium of this game. In this equilibrium, what is the expected payoff for each player?

9

Coalitions in Cooperative Games

9.1 Introduction to Coalitional Analysis

In the preceding chapter, we considered bargaining and cooperation only in two-player games. It may at first seem easy to generalize the definition of a bargaining problem and the Nash bargaining solution to games with more than two players. Let $N = \{1, 2, \ldots, n\}$ be the set of players, and let F be a closed and convex subset of $\mathbf{R}^N$, representing the set of feasible payoff allocations that the players can get if they all work together. Let $(v_1, \ldots, v_n)$ be the disagreement payoff allocation that the players would expect if they did not cooperate, and suppose that $\{y \in F \mid y_i \geq v_i, \forall i \in N\}$ is a nonempty bounded set. Then the pair $(F, (v_1, \ldots, v_n))$ may be called an *n-person bargaining problem*.

Given any such feasible set F that is a subset on $\mathbf{R}^N$, we say that x is *weakly (Pareto) efficient* in F iff $x \in F$ and there is no other vector y in F such that $y_i > x_i$ for every i in N. We say that an allocation x is *strongly (Pareto) efficient* in F iff $x \in F$ and there is no other vector y in F such that $y_i \geq x_i$ for every i in N and $y_j > x_j$ for at least one j in N.

Suppose also that $(F, (v_1, \ldots, v_n))$ is essential, in the sense that there exists some y in F such that $y_i > v_i$ for every i. Then the Nash bargaining solution for such an n-person bargaining problem (F, v) can be defined to be the unique strongly efficient vector x that maximizes the Nash product

$$\prod_{i \in N} (x_i - v_i)$$

over all vectors x in F such that $x_i \geq v_i$ for all i. It is not hard to show that this bargaining solution can be derived from generalized versions of Nash's axioms.

However, this n-person Nash bargaining solution is not widely used for the analysis of cooperative games when n is greater than 2 because it completely ignores the possibility of cooperation among subsets of the players. Any nonempty subset of the set of players may be called a *coalition*. A general theory of cooperative games with more than two players must include a theory of coalitional analysis that goes beyond this simple model of an n-person bargaining problem.

For an introduction to the problems of coalitional analysis in cooperative games with more than two players, consider the following two examples, given as games in strategic form. In each game, the set of players is $N = \{1,2,3\}$ and the set of strategy options for each player i is

$$C_i = \{(x_1,x_2,x_3) \in \mathbf{R}^3 \,|\, x_1 + x_2 + x_3 \leq 300 \text{ and}, \; \forall j, \; x_j \geq 0\}.$$

That is, in both games, each player can propose a payoff allocation for the three players such that no player's payoff is negative and the sum of their payoffs does not exceed 300. In *Example 9.1*, the players get 0 unless all three players propose the same allocation, in which case they get this allocation. That is, we let

$$u_i^1(c_1,c_2,c_3) = x_i \quad \text{if } c_1 = c_2 = c_3 = (x_1,x_2,x_3),$$
$$u_i^1(c_1,c_2,c_3) = 0 \quad \text{if } c_j \neq c_k \text{ for some } j \text{ and } k.$$

In *Example 9.2*, the players get 0 unless player 1 and 2 propose the same allocation, in which case they get this allocation. That is, we let

$$u_i^2(c_1,c_2,c_3) = x_i \quad \text{if } c_1 = c_2 = (x_1,x_2,x_3),$$
$$u_i^2(c_1,c_2,c_3) = 0 \quad \text{if } c_1 \neq c_2.$$

In both of these games, the players can jointly achieve any allocation in which their payoffs are nonnegative and sum to 300 or less, and the minimax value that each player can guarantee himself is 0. That is, we could describe this game as a three-person bargaining problem by letting

$$F = \{(x_1,x_2,x_3) \in \mathbf{R}^3 \,|\, x_1 + x_2 + x_3 \leq 300 \text{ and}, \; \forall j, \; x_j \geq 0\} \text{ and}$$
$$(v_1,v_2,v_3) = (0,0,0).$$

With this feasible set and disagreement allocation, the 3-person Nash bargaining solution discussed above would select the allocation (100,100,100) for both of these games, and indeed this might be a reasonable outcome for Example 9.1. However, the allocation (100,100,100) should seem quite unreasonable for Example 9.2. In Example 9.2, players 1 and 2 can together determine the payoff allocation, without any help from player 3. Thus, we might expect players 1 and 2 simply to divide the available payoff between themselves. If 1 and 2 ignored 3 and acted as if they were in a two-player cooperative game, then the Nash bargaining solution would specify that they should divide equally between themselves the maximum total payoff that they can get, giving nothing to 3; so the outcome would be the payoff allocation (150,150,0).

Before we completely dismiss (100,100,100) as an unreasonable prediction for Example 9.2, let us carefully examine the assumptions implicit in this rejection. Given that the players must choose their proposals simultaneously, it would be an equilibrium for both players 1 and 2 to propose (100,100,100), just as it would be an equilibrium for them both to propose (150,150,0). Even if we add the possibility of costless nonbinding preplay communication, there is still an equilibrium in which players 1 and 2 both expect each other to ignore anything that might be said (including statements such as, "Let us cut out player 3 and propose (150,150,0)," because each interprets the other's statements as meaningless babble rather than English), and both choose the final proposal (100,100,100). If player 3 has any influence in such matters, he would certainly want to promote such mutual misunderstanding between players 1 and 2, so that they might focus on such an equilibrium. Thus, the key assumption that we need to dismiss (100,100,100) as an unreasonable outcome for Example 9.2 is that players 1 and 2 can negotiate effectively, to focus themselves on an equilibrium that they prefer, before choosing their strategies in C_1 and C_2.

In general, when we say that the members of a coalition of players can *negotiate effectively*, we mean that, if there were a feasible change in the strategies of the members of the coalition that would benefit them all, then they would agree to actually make such a change, unless it contradicted agreements that some members of the coalition might have made with other players outside this coalition, in the context of some other equally effective coalition. The key assumption that distinguishes cooperative game theory from noncooperative game theory is this assumption that players can negotiate effectively. The n-person Nash bargaining solution might be relevant if the only coalition that can negotiate effectively is the *grand coalition* that includes all the players in N together. Instead, however, we shall assume in most of this chapter (except in Section 9.5) that all coalitions can negotiate effectively.

Thus, in cooperative games involving three or more players, we must also take into account the possibility that some subset of the players might form a cooperative coalition without the other players. In Example 9.1, no coalition that is smaller than $\{1,2,3\}$ can guarantee more than 0 to its members, but in Example 9.2 the coalition $\{1,2\}$ could guarantee its members any payoff allocation that they could get in $\{1,2,3\}$. The n-person bargaining problem is an inadequate representation of a cooperative game with more than two players because it suppresses all information about the power of multiplayer coalitions other than the grand coalition N.

Cooperative game theory is greatly complicated by the possibility of competition between overlapping coalitions. For example, consider now *Example 9.3*, which differs from the previous examples in that the players get 0 unless there is some pair of players ($\{1,2\}$, $\{2,3\}$, or $\{1,3\}$) who propose the same allocation, in which case they get this allocation. That is, we let

$$u_i^3(c_1,c_2,c_3) = x_i \quad \text{if} \quad c_j = c_k = (x_1,x_2,x_3) \text{ for some } j \neq k,$$
$$u_i^3(c_1,c_2,c_3) = 0 \quad \text{if} \quad c_1 \neq c_2 \neq c_3 \neq c_1.$$

When we assume that every coalition can negotiate effectively, the analysis of this game can become rather complicated. Because the three players have symmetric roles in this game and can get up to a total payoff of 300 by acting cooperatively, it may be quite reasonable to predict that the outcome of this game should be the allocation (100,100,100). However, players 1 and 2 could do better if, negotiating effectively in the $\{1,2\}$ coalition, they agreed instead to both propose the allocation (150,150,0). If this were the expected outcome of the game, however, player 3 would be very eager to persuade one other player to form instead an effective coalition with him. Rather than see (150,150,0) be the outcome of the game, player 3 would be willing to negotiate an agreement with player 2 to both propose (0,225,75), for example. But if (0,225,75) were the expected outcome in the absence of further negotiations, then player 1 would be willing to negotiate an agreement with player 3 to both propose an allocation that is better for both of them, such as (113,0,187). Indeed, at any equilibrium of this game, there is always at least one pair of players who could both do strictly better by jointly agreeing to change their two strategies. That is, there is always an incentive for two players who are getting less than everything to cut out the third player and divide his payoff between themselves. At some point, the process of coalitional negotiation must stop.

One way to explain how such coalitional negotiations might stop is to assume that, after a player has negotiated an agreement as a part of a coalition, he cannot later negotiate a different agreement with another coalition that does not contain all the members of the first coalition. Under this assumption, if the grand coalition {1,2,3} negotiated the agreement (100,100,100) before any two-player coalition could negotiate separately, then no two-player coalition could thereafter overturn this agreement by negotiating against the third player. On the other hand, if the coalition {1,2} were the first coalition to negotiate and they agreed to (150,150,0), then player 3 would be unable to increase his payoff by negotiating with player 2 or player 1 separately. Thus, under this assumption, the order in which coalitions can meet and negotiate may crucially determine the outcome of the game, and the advantage is to coalitions that negotiate earlier.

The alternative is to assume that negotiated agreements are only tentative and nonbinding; so a player who negotiates sequentially in various coalitions can nullify his earlier agreements and reach a different agreement with a coalition that negotiates later. This assumption also implies that the outcome of the game can depend on the order in which the various coalitions may negotiate, or on the rules by which this order may be determined. However, this assumption gives an advantage to coalitions that get to negotiate later, because no coalition can expect to implement an agreement that would be renegotiated by another coalition later. For example, if the players were expected to meet in negotiations in the exogenously given order {1,2} first, {2,3} second, {1,3} third, and {1,2,3} last, then the outcome of the game would have to give a payoff of 0 to player 2. Any agreement by an earlier coalition to give players 1 and 3 a total payoff less than 300 would be overturned by players 1 and 3 when they negotiate, and player 2 would not be able to get them to deviate from this agreement in the {1,2,3} coalition. For another example, suppose player 1 believed that, if he ever negotiated an agreement with player 2 in the coalition {1,2}, then player 3 would thereafter get an opportunity to negotiate with player 2 in the coalition {2,3}. In this case, player 1 might rationally refuse to negotiate with player 2 for the allocation (150,150,0) against an earlier agreement on (100,100,100), because such negotiations would in turn stimulate further negotiations between 2 and 3 that could result in player 1 ultimately getting a payoff of 0.

In most real situations where individuals can negotiate in various coalitions, the potential structure of these coalitional negotiations is more complicated than any rule we could hope to analyze. That is, any tractable rule for specifying which set of players can meet to negotiate at each point in time (possibly as a function of the earlier history of the negotiation process) is bound to be unrealistic in practice. Thus, there is a need for theories of cooperative games that can give us some sense of what to expect from the balance of power among all the possible coalitions without specifying such an ordering of the $2^n - 1$ different coalitions. On the other hand, the fact that such an ordering may be quite important tells us to be prepared to find some difficulties in the interpretation or application of any such order-independent theory.

9.2 Characteristic Functions with Transferable Utility

Because interactions among $2^n - 1$ different coalitions in an n-player game can be so complex, the simplifying assumption of *transferable utility* is often used in cooperative game theory. That is, as discussed in Section 8.4, there is often assumed to be a commodity—called money—that players can freely transfer among themselves, such that any player's utility payoff increases one unit for every unit of money that he gets.

With transferable utility, the cooperative possibilities of a game can be described by a *characteristic function* v that assigns a number $v(S)$ to every coalition S. Here $v(S)$ is called the *worth* of coalition S, and it represents the total amount of transferable utility that the members of S could earn without any help from the players outside of S. In any characteristic function, we always let

$$v(\emptyset) = 0,$$

where $\emptyset$ denotes the empty set. A characteristic function can also be called a *game in coalitional form* or a *coalitional game*.

In Example 9.1, the characteristic function is

$$v(\{1,2,3\}) = 300, \quad v(\{1,2\}) = v(\{1,3\}) = v(\{2,3\}) = 0, \quad \text{and}$$
$$v(\{1\}) = v(\{2\}) = v(\{3\}) = 0.$$

In Example 9.2, the characteristic function is

$$v(\{1,2,3\}) = 300 = v(\{1,2\}), \quad v(\{1,3\}) = v(\{2,3\}) = 0, \quad \text{and}$$
$$v(\{1\}) = v(\{2\}) = v(\{3\}) = 0.$$

In Example 9.3, the characteristic function is

$$v(\{1,2,3\}) = 300 = v(\{1,2\}) = v(\{1,3\}) = v(\{2,3\}) \quad \text{and}$$
$$v(\{1\}) = v(\{2\}) = v(\{3\}) = 0.$$

Notice that, under the assumption of transferable utility, specifying a single number for each coalition is sufficient to describe what allocations of utility can be achieved by the members of the coalition. Thus, inspecting the characteristic function for Example 9.3, we can see that, in this game, the three players together or any pair of players who cooperate can divide 300 units of payoff (say, "dollars") among themselves in any way that might be mutually agreeable, but each player alone can guarantee himself only 0.

In Chapter 8 (Sections 8.2 and 8.5), we considered three different ways of deriving a disagreement point from a two-person strategic-form game: by minimax values, noncooperative equilibria, and rational threats. Each of these ideas can be generalized to provide a way of deriving a characteristic function from a n-person strategic-form game.

Given a game in strategic form $\Gamma = (N, (C_i)_{i \in N}, (u_i)_{i \in N})$ with transferable utility, von Neumann and Morgenstern (1944) suggested that the characteristic function should be defined by

$$(9.1) \qquad v(S) = \min_{\sigma_{N \setminus S} \in \Delta(C_{N \setminus S})} \max_{\sigma_S \in \Delta(C_S)} \sum_{i \in S} u_i(\sigma_S, \sigma_{N \setminus S}).$$

Here, $N \setminus S$ denotes the set of all players in N who are not in the coalition S. For any coalition T, $C_T = \times_{j \in T} C_j$, so $\Delta(C_T)$ is the set of correlated strategies available to coalition T. We let $u_i(\sigma_S, \sigma_{N \setminus S})$ denote player i's expected payoff, before transfers of money, when the correlated strategies σ_S and $\sigma_{N \setminus S}$ are independently implemented; that is,

$$u_i(\sigma_S, \sigma_{N \setminus S}) = \sum_{c_S \in C_S} \sum_{c_{N \setminus S} \in C_{N \setminus S}} \sigma_S(c_S) \sigma_{N \setminus S}(c_{N \setminus S}) u_i(c_S, c_{N \setminus S}).$$

Thus, (9.1) asserts that $v(S)$ is the maximum sum of utility payoffs that the members of coalition S can guarantee themselves against the best offensive threat by the complementary coalition $N \backslash S$. If (9.1) is satisfied, then we say that v is the *minimax representation* in coalitional form of the strategic-form game Γ with transferable utility.

Characteristic functions can also be derived by assuming that complementary coalitions would play an essentially defensive pair of equilibrium strategies against each other. That is, we say that v is a *defensive-equilibrium representation* in coalitional form of the strategic-form game Γ with transferable utility iff, for every pair of complementary coalitions S and $N \backslash S$, there exist strategies $\bar{\sigma}_S$ and $\bar{\sigma}_{N \backslash S}$ such that

$$\bar{\sigma}_S \in \underset{\sigma_S \in \Delta(C_S)}{\text{argmax}} \sum_{i \in S} u_i(\sigma_S, \bar{\sigma}_{N \backslash S}),$$

$$\bar{\sigma}_{N \backslash S} \in \underset{\sigma_{N \backslash S} \in \Delta(C_{N \backslash S})}{\text{argmax}} \sum_{j \in N \backslash S} u_j(\bar{\sigma}_S, \sigma_{N \backslash S}),$$

$$v(S) = \sum_{i \in S} u_i(\bar{\sigma}_S, \bar{\sigma}_{N \backslash S}), \text{ and } v(N \backslash S) = \sum_{j \in N \backslash S} u_j(\bar{\sigma}_S, \bar{\sigma}_{N \backslash S}).$$

Harsanyi (1963) proposed that characteristic functions can be derived by a generalization of Nash's rational-threats criterion. We say that v is a *rational-threats representation* in coalitional form of the strategic-form game Γ with transferable utility iff, for every pair of complementary coalitions S and $N \backslash S$, there exist strategies $\bar{\sigma}_S$ and $\bar{\sigma}_{N \backslash S}$ such that

$$\bar{\sigma}_S \in \underset{\sigma_S \in \Delta(C_S)}{\text{argmax}} \left(\sum_{i \in S} u_i(\sigma_S, \bar{\sigma}_{N \backslash S}) - \sum_{j \in N \backslash S} u_j(\sigma_S, \bar{\sigma}_{N \backslash S}) \right),$$

$$\bar{\sigma}_{N \backslash S} \in \underset{\sigma_{N \backslash S} \in \Delta(C_{N \backslash S})}{\text{argmax}} \left(\sum_{j \in N \backslash S} u_j(\bar{\sigma}_S, \sigma_{N \backslash S}) - \sum_{i \in S} u_i(\bar{\sigma}_S, \sigma_{N \backslash S}) \right),$$

$$v(S) = \sum_{i \in S} u_i(\bar{\sigma}_S, \bar{\sigma}_{N \backslash S}), \text{ and } v(N \backslash S) = \sum_{j \in N \backslash S} u_j(\bar{\sigma}_S, \bar{\sigma}_{N \backslash S}).$$

As discussed in Section 8.5, the distinctions between these three ways of representing a strategic-form game in coalitional form can be interpreted in terms of alternative assumptions about the ability of coalitions to commit themselves to offensive and defensive threats. For example, the minimax representation in coalitional form implicitly assumes that a coalition S should be concerned that $N \backslash S$ would attack S offensively if the members of S decided to cooperate with each other but without the players in $N\backslash S$. However, offensively minimizing the sum of payoffs to the players in S would generally not be in the best interests of the players in $N\backslash S$, who really want to maximize their own payoffs. To justify an assumption that the members of $N\backslash S$ might act offensively against S, we need to assume that, at a time when it is expected that all players will ultimately cooperate together as a part of the grand coalition N and the players are negotiating over the possible divisions of the worth $v(N)$, the players in $N\backslash S$ can jointly commit themselves to an offensive threat that would be carried out only in the unexpected event that the players in S break off negotiations with the players in $N\backslash S$. During these negotiations, the members of $N\backslash S$ should want the members of S to fear that failing to reach an agreement would stimulate an offensive attack of $N\backslash S$ against S, because such fear would make the members of S more eager to avoid a disagreement with the members of $N\backslash S$ and thus more willing to concede a larger share of $v(N)$ to the members of $N\backslash S$.

Notice that, in all three representations discussed above, the worth of the grand coalition is the same, and can be written

$$v(N) = \max_{c_N \in C_N} \sum_{i \in N} u_i(c_N).$$

For example, consider a three-player game with transferable utility in which $C_i = \{a_i, b_i\}$ for each i in $\{1,2,3\}$, and the payoffs (u_1, u_2, u_3) depend on the three players' strategies as shown in Table 9.1. Here a_i can be interpreted as a "generous" strategy and b_i as a "selfish" strategy. In the minimax representation, each coalition S gets the most that it could guarantee itself if the players in the complementary coalition were selfish. Thus, the minimax representation in coalitional form is

$$v(\{1,2,3\}) = 12, \quad v(\{1,2\}) = v(\{1,3\}) = v(\{2,3\}) = 4, \quad \text{and}$$
$$v(\{1\}) = v(\{2\}) = v(\{3\}) = 1.$$

Table 9.1   A three-player game in strategic form, with transferable utility

C_1	$C_2 \times C_3$			
	a_2,a_3	b_2,a_3	a_2,b_3	b_2,b_3
a_1	4,4,4	2,5,2	2,2,5	0,3,3
b_1	5,2,2	3,3,0	3,0,3	1,1,1

The members of a two-player coalition could actually increase the sum of their payoffs by both being generous, so the defensive-equilibrium representation can be calculated by supposing that, if the players acted in two complementary coalitions, a single player alone would be selfish, but two players in a coalition together would be generous. Thus, the defensive-equilibrium representation of this game in coalitional form is

$$v(\{1,2,3\}) = 12, \quad v(\{1,2\}) = v(\{1,3\}) = v(\{2,3\}) = 4, \quad \text{and}$$
$$v(\{1\}) = v(\{2\}) = v(\{3\}) = 5.$$

Notice that this representation imputes an advantage to a player who acts selfishly alone against a generous two-player coalition. When both offensive and defensive considerations are taken into account, the rational-threats criterion would have all coalitions smaller than N choose selfishness in this game. Thus, the rational-threats representation of this game in coalitional form is

$$v(\{1,2,3\}) = 12, \quad v(\{1,2\}) = v(\{1,3\}) = v(\{2,3\}) = 2, \quad \text{and}$$
$$v(\{1\}) = v(\{2\}) = v(\{3\}) = 1.$$

If all three of these representations in coalitional form coincide, then we can say that the game has *orthogonal coalitions*. Games with orthogonal coalitions often arise in economic situations, where the worst thing that one coalition can do against another is refuse to trade with it. For example, consider the following class of *pure trading* games. Let M denote a finite set of commodities. For each player i and each commodity k, let $\omega_k(i)$ denote the amount of commodity k that i initially has to trade. Let $U_i(\chi(i))$ denote the (transferable) utility that player i would derive from consuming a bundle $\chi(i) = (\chi_k(i))_{k \in M}$. A strategy for player i is any specification of how much of each commodity he will deliver, out of his initial endowment, to each other player. For such a game, under any of the above-defined representations, the worth of any coalition S is

$$v(S) = \max \left\{ \sum_{i \in S} U_i(\chi(i)) \,\middle|\, \sum_{i \in S} \chi(i) = \sum_{i \in S} \omega(i) \right\}.$$

We say that a characteristic function v is *superadditive* iff, for every pair of coalitions S and T,

$$\text{if } S \cap T = \varnothing, \text{ then } v(S \cup T) \geq v(S) + v(T).$$

It can be shown that, if v is the minimax representation of a strategic-form game, then v must be superadditive. The defensive-equilibrium and rational-threats representations are not necessarily superadditive.

For any game v in coalitional form, we can define the *superadditive cover* of v to be the superadditive game w in coalitional form with lowest possible worths for all coalitions such that $w(S) \geq v(S)$ for every coalition S. A *partition* of any set S is a collection of nonempty sets $\{T_1, \ldots, T_k\}$ such that

$$T_1 \cup T_2 \cup \ldots \cup T_k = S \text{ and, } \forall j, \ \forall l \neq j, \ T_j \cap T_l = \varnothing.$$

Let $\mathscr{P}(S)$ denote the set of all partitions of S. Then the superadditive cover w of the game v in coalitional form satisfies the equation

$$w(S) = \max \left\{ \sum_{j=1}^{k} v(T_j) \, | \, \{T_1, \ldots, T_k\} \in \mathscr{P}(S) \right\}, \quad \forall S \subseteq N.$$

That is, the worth of a coalition in the superadditive cover is the maximum worth that the coalition could achieve by breaking up into a set of smaller disjoint coalitions. The concept of superadditive cover gives us a way to define a superadditive game corresponding to any game in coalitional form.

Once a representation in coalitional form has been specified, we can try to predict the outcome of bargaining among the players. Such an analysis is usually based on the assumption that the players will form the grand coalition and divide the worth $v(N)$ among themselves after some bargaining process, but that the allocation resulting from a focal equilibrium of this bargaining process will depend on the power structure rather than on the details of how bargaining proceeds. A player's *power* is his ability to help or hurt any set of players by agreeing to cooperate with them or refusing to do so. Thus, a characteristic function is a summary description of the power structure of a game.

9.3 The Core

Let $v = (v(S))_{S \subseteq N}$ be any game in coalitional form with transferable utility, where the set of players is $N = \{1, 2, \ldots, n\}$. A *payoff allocation* is any vector $x = (x_i)_{i \in N}$ in $\mathbf{R}^N$, where each component x_i is interpreted as the utility payoff to player i.

We say that an allocation y is *feasible for a coalition S* iff

$$\sum_{i \in S} y_i \le v(S),$$

and so the players in S can achieve their components of this allocation by dividing among themselves the worth $v(S)$ that they can get by cooperating together. When we say that an allocation is *feasible* without reference to any particular coalition, we mean that it is feasible for the grand coalition N.

We say that a coalition S can *improve on* an allocation x iff $v(S) > \sum_{i \in S} x_i$. That is, S can improve on x iff there exists some allocation y such that y is feasible for S and the players in S all get strictly higher payoff in y than in x.

An allocation x is in the *core* of v iff x is feasible and no coalition can improve on x. That is, x is in the core iff

$$\sum_{i \in N} x_i = v(N) \quad \text{and} \quad \sum_{i \in S} x_i \ge v(S), \quad \forall S \subseteq N.$$

Thus, if a feasible allocation x is not in the core, then there is some coalition S such that the players in S could all do strictly better than in x by cooperating together and dividing the worth $v(S)$ among themselves.

Given a strategic-form game Γ with transferable utility, if we plan to apply the core as our solution concept, then the minimax representation in coalitional form is probably the appropriate way to derive a characteristic function from Γ. When v is the minimax representation, the inequality $\sum_{i \in S} x_i < v(S)$ is necessary and sufficient to imply that the players in S can guarantee themselves payoffs that are strictly better than in x, no matter what the other players in $N \backslash S$ do.

The core is a very appealing solution concept, in view of our assumption that any coalition can negotiate effectively. For Example 9.1, in which the players can get 300 only if they all cooperate, the core is

$$\{x \in \mathbf{R}^3 | x_1 + x_2 + x_3 = 300, \, x_1 \ge 0, \, x_2 \ge 0, \, x_3 \ge 0\}.$$

That is, the core treats the three players symmetrically in this game and includes all individually rational Pareto-efficient allocations.

For Example 9.2, in which players 1 and 2 can get 300 together without player 3, the core is

$$\{x \in \mathbf{R}^3 | x_1 + x_2 = 300, \, x_1 \ge 0, \, x_2 \ge 0, \, x_3 = 0\}.$$

So the weakness of player 3 in this game is reflected in the core, where he always gets 0.

Unfortunately, the core may be empty, as it is for Example 9.3, which may be called the *three-person majority game*. In this game, the worth available to the grand coalition can also be achieved by any two players. Thus, when any player i gets a positive payoff in a feasible allocation, the other two players must get less than the worth (300) that they could get by themselves. The emptiness of the core may help to explain why the dynamics of coalitional negotiations seemed so important and yet so complicated for this example. No matter what allocation may ultimately occur, there is always a coalition that could gain if it could get one more final opportunity to negotiate effectively against this allocation.

There are also some games in which the core is nonempty but may seem rather extreme. For example, consider a game in which there are 2,000,001 players, among whom 1,000,000 players can each supply one left glove, and the other 1,000,001 players can each supply one right glove. Let N_L denote the set of left-glove suppliers and let N_R denote the set of right-glove suppliers. The worth of each coalition is defined to be the number of matched pairs (one left glove and one right glove) that it can make. That is,

$$v(S) = \min\{|S \cap N_L|, |S \cap N_R|\}.$$

(Here, for any coalition T, $|T|$ denotes the number of players in T.) The core of this game consists of the unique allocation x such that

$$x_i = 1 \text{ if } i \in N_L, \quad x_j = 0 \text{ if } j \in N_R.$$

To see why this is the only allocation in the core, suppose that some right-glove supplier has positive payoff in a feasible allocation. Then the total payoff to the other 2,000,000 players must be less than 1,000,000, which they could improve on by making 1,000,000 matched pairs among themselves without this one right-glove supplier. To interpret the unique core allocation economically, we might say that, because right gloves are in excess supply, they have a market price of 0. However, the sensitivity of this result is quite dramatic. If we added just two more left-glove suppliers, the unique core allocation would switch to the allocation in which every left-glove supplier gets payoff 0 and every right-glove supplier gets payoff 1.

This instability of the core for large games can be mitigated by considering ε-cores. For any number ε, an allocation x is in the ε-*core*, or the ε-*approximate core*, of the coalitional game v iff

$$\sum_{i \in N} x_i = v(N) \text{ and } \sum_{i \in S} x_i \geq v(S) - \varepsilon|S|, \quad \forall S \subseteq N.$$

That is, if x is in the ε-core then no coalition S would be able to guarantee all its members more than ε above what they get in x. For this left glove–right glove game, an allocation x is in the ε-core iff

$$\min\{x_i | i \in N_L\} + \min\{x_j | j \in N_R\} \geq 1 - 2\varepsilon.$$

That is, the worst-off left-glove supplier and the worst-off right-glove supplier together must get at least $1 - 2\varepsilon$. So with 1,000,000 left-glove suppliers and 1,000,001 right-glove suppliers, if $\varepsilon > 0.0000005$ then, for any number α such that $0 \leq \alpha \leq 1$, the allocation x such that

$$x_i = 1 - \alpha, \quad x_j = \alpha/1.000001, \quad \forall i \in N_L, \quad \forall j \in N_R,$$

is in the ε-core.

Although the logical appeal of the core is self-evident, its possible emptiness forces us to critically question its significance. In fact, given a game v and a feasible allocation x that is not in the core of v, we cannot conclude that the players could not cooperatively agree to the allocation x. If a coalition can make binding agreements that its members cannot subsequently break without the consent of all others in the coalition, then the existence of a coalition that can improve on x may be irrelevant, because some players in S could be prevented from acting as members of this coalition by their previous binding agreements with other coalitions. On the other hand, if coalitions cannot make binding agreements, then the members of a coalition S that can improve on x might nonetheless abstain from negotiating against this allocation, because they may fear that whatever improvement they might agree on might itself be overturned by subsequent negotiations by other coalitions, with the ultimate result of making some members of S worse off than in x.

Furthermore, even if an allocation x is in the core of v and coalitional agreements are binding, there may still be some forms of coalitional bargaining under which an attempt to achieve x might be blocked by some coalition. For example, suppose that $N = \{1,2,3\}$ and

$$v(\{1,2,3\}) = 9, \quad v(\{1,2\}) = v(\{1,3\}) = v(\{2,3\}) = 6, \quad \text{and}$$
$$v(\{1\}) = v(\{2\}) = v(\{3\}) = 0.$$

Suppose that a lawyer offers to help the players by promising to give them the core allocation (3,3,3) if they will all sign a contract to give him power of attorney, to act on their behalf in all coalitional bargaining. Suppose that a second lawyer announces that, if the three players all sign power of attorney over to him, then they will get the noncore allocation (4,4,1). Suppose furthermore that each lawyer announces that, if only two players sign with him, then they will each get a payoff of 3 whereas the other player will get 0. Given these two lawyers' offers, there is an equilibrium of the implicit contract-signing game in which all players sign with the first attorney and get the core allocation (3,3,3). However, there is also an equilibrium in which the three players all sign with the second lawyer and get (4,4,1). The problem is that in the noncooperative contract-signing game each player's decision is not between the options "get allocation (3,3,3)" and "sign with second lawyer" but between "sign with first lawyer (who might give (3,3,3) but might give 0)" and "sign with second lawyer." Because the first lawyer cannot guarantee (3,3,3) unless everyone signs with him, the second lawyer can block the (3,3,3) allocation in equilibrium.

Thus, the logical appeal of the core is based implicitly on assumptions that, when a coalition S contemplates blocking an allocation x by negotiating for some alternative allocation y that is feasible for them, (1) they are not prevented from doing so by prior agreements, (2) their agreement on such an alternative allocation y would be final, and (3) if they do not agree as a coalition on this alternative y, then they really will get the allocation x.

Although such assumptions may seem questionable in the context of cooperative games with small numbers of players, they are rather close in spirit to the kinds of assumptions made in the analysis of large competitive markets. In economic theory, it is generally assumed that any small set of individuals perceive that the general market equilibrium offers them terms of trade that are fixed independently of their own trading decisions, although they may be free to negotiate other terms of trade among themselves. Building on this insight, Debreu and Scarf (1963) and Aumann (1964) proved a classical conjecture of Edgeworth (1881) that, for large market games, the core is essentially equivalent to the set of competitive (Walrasian) equilibria. Furthermore, Kaneko and

Wooders (1982), Wooders (1983), and Wooders and Zame (1984) have proved that, in a certain sense, all large games have nonempty approximate cores.

To understand these core existence theorems, we begin with a general characterization of games in which the core is nonempty; this characterization is due to Bondereva (1963) and Shapley (1967) and is based on duality theory of linear programming. Consider the problem of finding the least amount of transferable utility that is necessary for an allocation that no coalition can improve on. This problem is a linear programming problem and can be formulated as follows:

$$(9.2) \qquad \underset{x \in \mathbf{R}^N}{\text{minimize}} \sum_{i \in N} x_i \text{ subject to } \sum_{i \in S} x_i \geq v(S), \quad \forall S \subseteq N.$$

(The optimal solutions to this linear programming problem are called the *balanced aspirations* for the players in v, and they coincide with the core when it is nonempty; see Cross, 1967, Albers, 1979, and Bennett, 1983.) We let $L(N)$ denote the set of all coalitions among the players in N, so

$$L(N) = \{S \,|\, S \subseteq N \text{ and } S \neq \varnothing\}.$$

Then the dual of this linear programming problem, as defined in Section 3.8, can be written:

$$(9.3) \qquad \underset{\mu \in \mathbf{R}_+^{L(N)}}{\text{maximize}} \sum_{S \subseteq N} \mu(S)v(S) \text{ subject to } \sum_{S \supseteq \{i\}} \mu(S) = 1, \quad \forall i \in N.$$

It is not difficult to show that the constraints of these two linear programming problems can be satisfied by some appropriate vectors x and μ. Thus, by the duality theorem of linear programming, there exist optimal solutions to both problems and the optimal values of their objective functions are equal. That is, for any given game v in coalitional form, there exists an allocation x in $\mathbf{R}^N$ and a vector μ in $\mathbf{R}^{L(N)}$ such that

$$(9.4) \qquad \mu(S) \geq 0 \text{ and } \sum_{i \in S} x_i \geq v(S), \quad \forall S \subseteq N,$$

$$(9.5) \qquad \sum_{S \supseteq \{i\}} \mu(S) = 1, \quad \forall i \in N,$$

$$(9.6) \qquad \sum_{S \subseteq N} \mu(S)v(S) = \sum_{i \in N} x_i.$$

For example, if v is as given in Example 9.3 (the three-player majority game with maximal worth 300), then these conditions are satisfied when

$$x = (150,150,150), \quad \mu(\{1,2\}) = \mu(\{1,3\}) = \mu(\{2,3\}) = \tfrac{1}{2},$$

and all other $\mu(S) = 0$.

A coalitional game v on the set of players N is *balanced* iff, for any μ in $\mathbf{R}_+^{L(N)}$,

$$\text{if } \mu \text{ satisfies (9.5) then } \sum_{S \subseteq N} \mu(S)v(S) \le v(N).$$

The fact that the optimal solutions to the linear programming problems (9.2) and (9.3) have the same optimal values of the objective functions implies that a game v has a nonempty core if and only if it is balanced.

To interpret this result, we need to use some additional facts about linear programming. If a linear programming problem has an optimal solution, then it must have an optimal solution that is an extreme point or corner of the set of vectors that satisfy the constraints. There are finitely many of these extreme points. If the coefficients in the constraints are rational numbers, then the components of each extreme point are all rational numbers (see Chvatal, 1983).

Notice that the constraints of the dual problem (9.3) depend only on the set of players N, not on the game v. Thus, given N, there is a finite set of rational vectors—the corners of the feasible set—such that, for any coalitional game v on N, one of these vectors is an optimal solution of the dual problem (9.3). Let K be the lowest common denominator of the components of these vectors. Then, for any coalitional game v on the set of players N, there exists some μ such that μ is an optimal solution of the dual problem (9.3), and, for every coalition S, $K\mu(S)$ is a nonnegative integer. Let $\eta(S) = K\mu(S)$, for each S. Then (9.5) and (9.6) can be rewritten:

$$(9.7) \qquad \sum_{S \supseteq \{i\}} \eta(S) = K, \quad \forall i \in N,$$

$$(9.8) \qquad \sum_{S \subseteq N} \eta(S)v(S) = K \sum_{i \in N} x_i.$$

Of course, K itself is a positive integer. So suppose that we replicate the original game K times. That is, consider a large game in which, for every player in the original game v, there are K different individuals who are identical to this individual in their ability to form coalitions. We can formally define the *K-fold replicated game* v' that is generated by v as follows. Let the new set of players be

$$N' = N \times \{1, 2, \ldots, K\}.$$

We say that a subset S' of N' is a *replicated version* of a subset S of N iff, for every i in N, the set $\{l \mid (i, l) \in S'\}$ has one member if $i \in S$, and has no members if $i \notin S$. For any subset S' of N', if S' is a replicated version of some subset S of N, then let $w(S') = v(S)$, and otherwise let $w(S') = \min_{i \in N} v(\{i\})$. Then let the K-fold replicated game v' be the superadditive cover of this coalitional game w on the set of players N'.

Condition (9.7) asserts that we can make a partition of N' in which, for each subset S of N, there are $\eta(S)$ replicated versions of the coalition S. (Recall that $\eta(S)$ is a nonnegative integer.) Such a partition of N' would use exactly the K copies $((i, l) \in N'$, for $l = 1, \ldots, K)$ of each player i in N. Condition (9.8) asserts that such a partition of the players in N' will generate a total transferable worth that is large enough to give each player (i, l) in N' the payoff x_i. Finally, (9.4) asserts that this payoff allocation cannot be improved on by any coalition S' in the coalitional game v' on N', because x cannot be improved on by any coalition S in the coalitional game v on N. That is, the core of the K-fold replicated game v' on the set of players N' must be nonempty. This argument proves the following theorem, which is due to Kaneko and Wooders (1982).

THEOREM 9.1. *Given the finite set N, there exists a positive integer K such that, for any game v in coalitional form with transferable utility on the player set N, the core of the K-fold replicated game that is generated by v is nonempty.*

For example, let v be Example 9.3, the three-person majority game where any pair of players has worth 300. Then the K-fold replicated game has a nonempty core whenever K is even. With an even total number of players, every player can be matched into a two-person coalition with another player and get payoff 150, which no coalition can improve on. On the other hand, suppose that K is odd. Then the K-fold replicated game has an empty core, because if we try to partition the $3K$ players into pairs, then we must have one unpaired player left over. However, with $3K$ players, it is possible to form $(3K - 1)/2$ disjoint pairs, each worth 300, and then transfer utility to achieve the allocation in which each player gets $(300(3K - 1)/2)/3K = 150 - 50/K$. For any positive number ε, this allocation is in the ε-core when $K \geq 50/\varepsilon$. So the ε-core of the K-fold replicated game is nonempty whenever K is sufficiently large.

The following theorem, which extends this result to general games in coalitional form, has been proved by Kaneko and Wooders (1982).

THEOREM 9.2. *For any coalitional game v with transferable utility on the finite set N and for any positive number ε, there exists an integer K^* such that, for every k that is greater than K^*, the ε-core of the k-fold replicated game that is generated by v is nonempty.*

(I give here only a sketch of the proof. The idea is to let K^* be a large multiple of the number K from Theorem 9.1. Then all but a small fraction of the players can be grouped into disjoint coalitions of size K, where they can get core allocations, which become ε-core allocations after each player pays a small tax of ε. This small tax on everyone can then be used to finance an ε-core allocation for the remaining small fraction of the players.)

In a large replicated game, any player has a large number of twins who are perfect substitutes for himself in any coalition. However, this property is not sufficient to guarantee nonemptiness of approximate cores. For a counterexample, consider a game in which there are $3K$ players, and the worth of any coalition S is

$$v(S) = 2K \text{ if } |S| \geq 2K,$$
$$v(S) = 0 \text{ if } |S| < 2K.$$

For any K, no matter how large, the core of this game is empty.

The *maximum individual contribution* in a coalitional game v is the maximum amount that a single player can increase the worth of any coalition, that is,

$$\max\{v(S) - v(S - i) | S \subseteq N, i \in S\}.$$

The maximum individual contribution of the K-fold replicated game that is generated by v is the same as the maximum individual contribution of the game v itself, so it stays constant as K increases. However, in the above game with an empty core, the maximum individual contribution is $2K$, which diverges to infinity as K gets large. Wooders and

Zame (1984) have shown that, in a general class of games with bounded maximum individual contributions, approximate cores of large games are nonempty.

The duality conditions (9.4), (9.5), and (9.6) can also be directly interpreted in terms of a *dynamic matching process* that players continually enter and leave over time. In this process, N is reinterpreted as the set of *classes* of players, and each player who enters the process belongs to exactly one class in N (e.g., the class of left-glove suppliers). Suppose that, for each i in N, one new player of class i enters the matching process during each unit of time (say, each day). Suppose also that every player must be matched or assigned to a coalition that contains at most one player of each class, after which he leaves the matching process. A player may wait to be matched, so the players in a coalition may have entered the matching process at different times. Whenever an S *coalition* (that is, a coalition containing exactly one player from each class in the set S) is formed, it generates $v(S)$ units of transferable utility that can be divided among its members in any way.

For each S such that $S \subseteq N$, let $\mu(S)$ denote the average number of S coalitions that are formed per unit time in this matching process. For every i in N, to match all class-i players in the long run, μ must satisfy $\Sigma_{S \supseteq \{i\}} \mu(S) = 1$. Thus, the existence of a solution to conditions (9.4)–(9.6) implies that there exists a way to match all the players in this process such that the expected allocation of payoffs to the various classes of players cannot be improved on by any coalition.

9.4 The Shapley Value

The core of a coalitional game may be empty or quite large, a situation that makes the core difficult to apply as a predictive theory. The best that we could hope for would be to derive a theory that predicts, for each game in coalitional form, a unique expected payoff allocation for the players. That is, we would like to identify some mapping $\phi : \mathbf{R}^{L(N)} \to \mathbf{R}^N$ such that, when the players in N play any coalitional game v, the expected payoff to each player i would be $\phi_i(v)$. Here we use the notation $\phi(v) = (\phi_i(v))_{i \in N}$.

Shapley (1953) approached this problem axiomatically. That is, he asked what kinds of properties we might expect such a solution concept to satisfy, and he characterized the mappings ϕ that satisfy these properties.

A *permutation* of the set of players N is any function $\pi{:}N \to N$ such that, for every j in N, there exists exactly one i in N such that $\pi(i) = j$ (that is, $\pi{:}N \to N$ is one-to-one and onto). Given any such permutation π and any coalitional game v, let πv be the coalitional game such that

$$\pi v(\{\pi(i) | i \in S\}) = v(S), \quad \forall S \subseteq N.$$

That is, the role of any player i in v is essentially the same as the role of the player $\pi(i)$ in πv. Shapley's first axiom asserts that only the role of a player in the game should matter, not his specific names or label in the set N.

AXIOM 9.1 (SYMMETRY). *For any v in $\mathbf{R}^{L(N)}$, any permutation $\pi{:}N \to N$, and any player i in N, $\phi_{\pi(i)}(\pi v) = \phi_i(v)$.*

We say that a coalition R is a *carrier* of a coalitional game v iff

(9.9) $$v(S \cap R) = v(S), \quad \forall S \subseteq N.$$

If R is a carrier of v, then all the players who are not in R are called *dummies* in v, because their entry into any coalition cannot change its worth. Notice that (9.9) and the convention $v(\varnothing) = 0$ imply that, if R is a carrier and $i \notin R$, then $v(\{i\}) = 0$. Shapley's second axiom asserts that the players in a carrier set should divide their joint worth (which equals the worth of the grand coalition) among themselves, allocating nothing to the dummies.

AXIOM 9.2 (CARRIER). *For any v in $\mathbf{R}^{L(N)}$ and any coalition R, if R is a carrier of v, then $\Sigma_{i \in R} \phi_i(v) = v(R)$.*

In place of Shapley's third axiom—additivity—we consider here a closely related linearity axiom. For any two coalitional games v and w on the player set N and any number p that is between 0 and 1, let $pv + (1 - p)w$ denote the coalitional game such that, for each coalition S,

$$(pv + (1 - p)w)(S) = pv(S) + (1 - p)w(S).$$

To interpret this game, suppose that the players in N will play tomorrow either the coalitional game v or the game w, depending on some random event that will be observable tomorrow (say, v if it rains tomorrow and w if it does not). Suppose that p is the probability that tomorrow's game will be v, and $1 - p$ is the probability that tomorrow's game will be w.

If the players bargain and negotiate today, planning their strategies for tomorrow's moves in advance, then the situation that they face today can be represented by the coalitional game $pv + (1 - p)w$, because planned cooperative behavior by any coalition S would earn an expected worth of $pv(S) + (1 - p)w(S)$. According to the function ϕ, if the players bargain today, then the expected payoff to player i should be $\phi_i(pv + (1 - p)w)$; but if the players bargain tomorrow, then i's expected payoff should be $p\phi_i(v) + (1 - p)\phi_i(w)$. The linearity axiom asserts that the expected payoff to each player should not depend on whether the players bargain about their coalitional strategies today (before the resolution of uncertainty) or tomorrow (after the resolution of uncertainty).

AXIOM 9.3 (LINEARITY).   *For any two coalitional games v and w in $\mathbf{R}^{L(N)}$, any number p such that $0 \le p \le 1$, and any player i in N,*

$$\phi_i(pv + (1 - p)w) = p\phi_i(v) + (1 - p)\phi_i(w).$$

Remarkably, Shapley showed that there is a unique mapping ϕ—called the *Shapley value*—that satisfies these three axioms.

THEOREM 9.3.   *There is exactly one function $\phi : \mathbf{R}^{L(N)} \to \mathbf{R}^N$ that satisfies Axioms 9.1, 9.2, and 9.3 above. This function satisfies the following equation, for every i in N and every v in $\mathbf{R}^L$:*

$$\phi_i(v) = \sum_{S \subseteq N - i} \frac{|S|!(|N| - |S| - 1)!}{|N|!} (v(S \cup \{i\}) - v(S)).$$

(Here, for any positive integer k, $k! = 1 \times 2 \times \ldots \times k$, and $0! = 1$.)

The formula in Theorem 9.3 can be interpreted as follows. Suppose that we plan to assemble the grand coalition in a room, but the door to the room is only large enough for one player to enter at a time, so the players randomly line up in a queue at the door. There are $|N|!$ different ways that the players might be ordered in this queue. For any set S that does not contain player i, there are $|S|!(|N| - |S| - 1)!$ different ways to order the players so that S is the set of players who are ahead of player i in the queue. Thus, if the various orderings are equally likely, $|S|!(|N| - |S| - 1)!/|N|!$ is the probability that, when player i enters the room, he will find the coalition S there ahead of him. If i finds S ahead of him when he enters, then his marginal contribution to the worth of the coalition in the room when he enters is $v(S \cup \{i\}) -$

$v(S)$. Thus, under this story of randomly ordered entry, the Shapley value of any player is his expected marginal contribution when he enters.

If v is superadditive, then the Shapley value must be individually rational, in the sense that

$$\varphi_i(v) \geq v(\{i\}), \quad \forall i \in N.$$

To verify this fact, notice that superadditivity implies that

$$v(S \cup \{i\}) \geq v(S) + v(\{i\}), \quad \forall S \subseteq N - i,$$

so i's marginal contribution to any coalition is never less than $v(\{i\})$.

Proof of Theorem 9.3. Given the random order interpretation, it is straightforward to verify that the formula in the theorem satisfies the three axioms. The formula is linear in v and treats the various players symmetrically. (In the formula for φ_i, all that matters about a coalition is whether it contains i and the number of players that it contains.) Furthermore, for any ordering, the scheme of paying of marginal contributions guarantees that the total payoff $v(N)$ will be divided among the players and dummies will get 0; so any carrier R must get $v(N)$, which equals $v(R)$.

We now show that there can exist at most one function φ that satisfies the three axioms. The carrier axiom implies that $\varphi_i(\mathbf{0}) = 0$ for every player i, where $\mathbf{0}$ is the coalitional game that assigns worth 0 to every coalition. This fact, together with Axiom 9.3, implies that φ must be a linear mapping from $\mathbf{R}^{L(N)}$ to $\mathbf{R}^N$. (Axiom 9.3 by itself actually only asserts that φ must be an affine mapping.) Notice that $\mathbf{R}^{L(N)}$, the set of coalitional games on N, is a $(2^{|N|} - 1)$-dimensional vector space.

For any coalition R, let w_R be the coalitional game that is defined so that

(9.10) $w_R(S) = 1$ if $R \subseteq S$, $w_R(S) = 0$ otherwise.

That is, a coalition has worth 1 in w_R if it contains all the players in R, but it has worth 0 in w_R if it does not contain all the players in R. (We call w_R the *simple R-carrier game*.) By the carrier axiom, we must have

$$\sum_{i \in R} \varphi_i(w_R) = 1, \quad \text{and} \quad \varphi_j(w_R) = 0, \quad \forall j \notin R,$$

because R and $R \cup \{j\}$ are both carriers of w_R. By the symmetry axiom, all players in R must get the same payoff. So

(9.11) $\phi_i(w_R) = 1/|R|$, $\forall i \in R$;   and $\phi_j(w_R) = 0$, $\forall j \notin R$.

There are $2^{|N|} - 1$ such games w_R, one for each coalition R. Furthermore, these games are linearly independent in the vector space $\mathbf{R}^{L(N)}$. To prove linear independence, we must show that the equation $\Sigma_{R \in L(N)} \alpha_R w_R = \mathbf{0}$ implies that all α_R equal 0. If not, let S be any coalition of minimal size such that $\alpha_S \neq 0$; then we get

$$0 = \sum_{R \in L(N)} \alpha_R w_R(S) = \sum_{R \subseteq S, R \neq \varnothing} \alpha_R = \alpha_S,$$

which is a contradiction. So $\{w_R | R \in L(N)\}$ is a basis for $\mathbf{R}^{L(N)}$, because it is linearly independent and contains as many vectors as the dimension of the space. But a linear mapping is completely determined by what it does on a basis of the domain. Thus, there is a unique linear mapping $\phi:\mathbf{R}^{L(N)} \to \mathbf{R}^N$ that satisfies (9.11) for every game w_R. ∎

The Shapley value is a powerful tool for evaluating the power structure in a coalitional game. For Example 9.1, where all three players must agree to get $300, the Shapley value is (100,100,100). For Example 9.2, where only players 1 and 2 need to agree to get $300, the Shapley value is (150,150,0). For Example 9.3, where any pair of players can divide $300, the Shapley value is (100,100,100), by symmetry.

In the left glove–right glove game with 2,000,001 players, including one extra right-glove supplier, the Shapley value of any right-glove supplier is the probability that, in the randomly ordered entry process, the players who enter before him will include more left-glove suppliers than right-glove suppliers. This probability is very close to .5. Aumann (1987b) has calculated that the Shapley value of this game gives .499557 to each right-glove supplier and .500443 to each left-glove supplier. So the Shapley value shows just a slight disadvantage for the right-glove suppliers in this game. This conclusion may seem quite reasonable, because the excess supply of right gloves is so small in relation to the whole market.

For another interesting game, consider the *apex game* (due to Maschler). In this game, $N = \{1,2,3,4,5\}$,

$v(S) = 1$ if $1 \in S$ and $|S| \geq 2$,

$v(S) = 1$ if $|S| \geq 4$,

and $v(S) = 0$ otherwise. In this game, we call player 1 the big player, and the other four, small players. The big player with any one or more of the small players can earn a worth of 1, and the four small players can together also earn a worth of 1. The core of this game is empty. In the randomly ordered entry process, the big player will have a marginal contribution of 1 unless he enters first or last, which happens with probability $\frac{2}{5}$, so $\phi_i(v) = \frac{3}{5}$. By symmetry, the four small players must get the same values. By the carrier axiom, the values of all five players must sum to 1. Thus, the Shapley value of this game is

$$\phi(v) = (\frac{3}{5}, \frac{1}{10}, \frac{1}{10}, \frac{1}{10}, \frac{1}{10}).$$

The formula for the Shapley value can be equivalently written

$$(9.12) \qquad \phi_i(v) = \sum_{R \subseteq N-i} \frac{|R|!(|N| - |R| - 1)!}{|N|!} (v(N \backslash R) - v(R)).$$

So the value of each player depends only on the differences between the worths of complementary coalitions. For each pair of complementary coalitions, S and $N \backslash S$, the values of the players in $N \backslash S$ increase as $v(N \backslash S) - v(S)$ increases, and the values of the players in S increase as $v(S) - v(N \backslash S)$ increases.

This result has implications for the analysis of cooperative games with transferable utility in strategic form. Given a strategic-form game Γ as above, suppose that each coalition S must choose a threat σ_S in $\Delta(C_S)$, and these threats will determine a characteristic function by the formula

$$v(S) = \sum_{i \in S} u_i(\sigma_S, \sigma_{N \backslash S}), \quad \text{and} \quad v(N \backslash S) = \sum_{j \in N \backslash S} u_j(\sigma_S, \sigma_{N \backslash S}).$$

Suppose that the Shapley value $\phi(v)$ is the expected payoff allocation that the players will ultimately get, as a result of some bargaining process in which all of these coalitional threats are taken into account. Then each member of coalition S should want to choose σ_S to maximize

$$v(S) - v(N \backslash S) = \sum_{i \in S} u_i(\sigma_S, \sigma_{N \backslash S}) - \sum_{j \in N \backslash S} u_j(\sigma_S, \sigma_{N \backslash S})$$

while each member of $N \backslash S$ should want to choose $\sigma_{N \backslash S}$ to minimize this difference. Thus, as Harsanyi (1963) has argued, the rational-threats representation (discussed in Section 9.2) may be the appropriate way to derive a game in coalitional form from a game in strategic form when the Shapley value is being used to determine the outcome. (For a systematic analysis of more general systems of threats and values, see Myerson, 1978b.) With this representation, the ultimate payoff to each player i in the cooperative game will be

$$(9.13) \quad \sum_{R \subseteq N-i} \frac{|R|!(|N| - |R| - 1)!}{|N|!} \delta(N\backslash R, R),$$

where $\delta(N\backslash R, R)$ equals

$$\max_{\sigma_{N\backslash R} \in \Delta(C_{N\backslash R})} \min_{\sigma_R \in \Delta(C_R)} \left(\sum_{j \in N\backslash R} u_j(\sigma_{N\backslash R}, \sigma_R) - \sum_{k \in R} u_k(\sigma_{N\backslash R}, \sigma_R) \right).$$

Aumann and Shapley (1974) extended the theory of the Shapley value to games with infinitely many players. In this work, they used a variety of approaches, defining values both axiomatically and by taking the limit of large finite games. Their results are both mathematically elegant and rich in economic intuition. To avoid the measure-theoretic prerequisites of their work, we can only offer here a sketch of their results for a small subset of the range of games that they considered. (For a study of noncooperative equilibria of games with infinitely many players, see Schmeidler, 1973.)

Let H be a finite set, representing the set of classes of players, and suppose that there are infinitely many players of each class. Let us describe coalitions here by the fraction of the players of each class who are in the coalition. So we can say that $r = (r_h)_{h \in H}$ is a *fractional coalition* iff r is in the set $[0,1]^H$, where each component r_h is interpreted as the fraction of all class-h players who are in the coalition. When we interpret our infinite-player model as an approximation of a very large finite-player game, then r_h can also be interpreted as the number of class-h players in the coalition divided by the total number of class-h players in the game. Then an infinite game in coalitional form can be represented by a function $f:[0,1]^H \to \mathbf{R}$ where, for each r in $[0,1]^H$, $f(r)$ denotes the worth of a fractional coalition r in the game. We require $f(\mathbf{0}) = 0$, where $\mathbf{0}$ is the vector in which all components are 0. We also assume here that the function $f(\cdot)$ is continuous on $[0,1]^H$ and, for each h in H, the partial derivative of $f(r)$ with respect to r_h, denoted by $f'_h(r)$ or $(\partial f/\partial r_h)(r)$, is nonnegative and continuous at every r such that $r_h > 0$. (So f satisfies the hypotheses of Proposition 10.17 of Aumann and Shapley, 1974.)

Let **1** denote the vector in $[0,1]^H$ in which all components are 1. For any number α, $\alpha\mathbf{1}$ then denotes the vector in which all components are α. The set $\{\alpha\mathbf{1} \mid 0 < \alpha < 1\}$ is called the *main diagonal* of the set $[0,1]^H$.

In a large finite game, when the players form coalitions according to the model of randomly ordered entry described above, the set of players who precede any specific player should be, with high probability, a statistically large sample randomly drawn from the overall population. (Only if the player is one of the first few to enter would the sample fail to be "statistically large." The key idea from basic statistics that we are applying here is that the statistical properties of a sample depend on the size of the sample but not on the size of the population from which it is drawn.) Thus, by the law of large numbers, the relative numbers of the various classes in the sample that precedes the player should, with very high probability, be very close to the same as the relative numbers of these classes in the set of all players in the game. That is, the fractional coalition that precedes any given player in the coalition-formation process is very likely to be close to the main diagonal, when the game is large.

When a class-h player enters a fractional coalition r, his marginal contribution is proportional to $f'_h(r)$, the partial derivative of $f(r)$ with respect to r_h. Because the coalition that precedes any player is almost surely on (or almost on) the main diagonal in large games but might be drawn from anywhere on this diagonal according to the uniform distribution, the value of the class-h players in the game f should be

$$(9.14) \qquad \int_0^1 f'_h(\alpha\mathbf{1})d\alpha.$$

Actually, (9.14) should be interpreted not as the value of a single class-h player but as the total value or sum of the values of all class-h players in the game f. By symmetry, each class-h player must get an equal share of the value in (9.14). That is, if the infinite game f is an approximation to some large finite game in which there are K players of each class, then the value of each class-h player in the large finite game should be approximately equal to the quantity in (9.14) divided by K.

We say that f is *1-homogeneous* iff, for every r in $[0,1]^H$ and every α in $[0,1]$, $f(\alpha r) = \alpha f(r)$, so the worth of a coalition is proportional to its size (when the ratios between the numbers of different classes of players are kept constant). If f is 1-homogeneous, then its partial derivatives are constant along the main diagonal. Thus, if f is 1-homogeneous, then the value of the class-h players is just $f'_h(\mathbf{1})$, their marginal contribution in the grand coalition. If we add the assumption that f is also concave, then it can be shown that this value allocation is also the unique point in the core. (A vector x in $\mathbf{R}^H$ represents an allocation in the core of f iff $\Sigma_{h \in H} \, r_h x_h \geq f(r)$ for every r in $[0,1]^H$.) Thus, there is a wide and important class of games with infinitely many players in which the core and the value coincide.

9.5 Values with Cooperation Structures

In the Shapley value, the payoff for any player i depends on the worth of every possible coalition. For any coalitional game v with transferable utility, where N is the set of players, all $2^{|N|} - 1$ coalitional worths enter into the formula for $\phi_i(v)$, and the coefficient of any $v(S)$ depends only on the number of players in S and on whether i is in S. There are some situations, however, in which such symmetric treatment of all coalitions may be unrealistic. In general, we use the term *cooperation structure* to refer to any mathematical structure that describes which coalitions (within the set of all $2^{|N|} - 1$ possible coalitions) can negotiate or coordinate effectively in a coalitional game.

There may be exogenous factors that make some coalitions intrinsically more important than others. Geographical, sociological, or even linguistic barriers may make some coalitions easier to form than others. For example, if a firm is viewed as a game played by the various managers and workers, the coalition of all managers and workers whose first names begin with "R" is unlikely to be able to cooperate or negotiate effectively, whereas the coalition of all workers who work together in a particular building is likely to be able to cooperate.

The relatively greater effectiveness of some coalitions may also arise as an endogenous consequence of players' decisions. For example, in a three-player majority game like Example 9.3 (where any two players can achieve the same worth as all three players together), two players may have strong incentive to form an effective coalition together and to refuse to negotiate separately with the third player. On the other hand, in a three-player unanimity game like Example 9.1 (where only the grand coalition can earn a positive worth), there would probably be much less incentive for players to form such exclusive negotiation agreements.

Thus, we need a theory to offer answers to two questions. First, if some coalitions can be more effective than others, how should the outcome of any given coalitional game depend on the cooperation structure? Second, if the cooperation structure can be influenced by exclusive negotiation agreements among the players, what kinds of cooperation structures should we expect to observe, in any given coalitional game? Clearly, the second question of predicting endogenous cooperation structures can be understood only after we have some answer to the first question of how expected payoffs might depend on the cooperation structure. Thus, most of this section is devoted to the first question. I describe several ways that the Shapley value has been extended to depend on the cooperation structure.

Cooperation structures can be most simply described if we assume that each player will belong to one active or effective coalition. Under this assumption, the cooperation structure in a game can be described by a partition of the set of all players N. Then, given any coalitional game v on the set of players N, for each player i and each coalition S that contains i, we need to specify the payoff that player i would get if S were the active coalition to which he belonged. Harsanyi (1963) and Aumann and Dreze (1974) suggested that this payoff should be the Shapley value of player i in the coalitional game v restricted to subsets of S, which we denote $\phi_i(v,S)$. That is, we let

$$(9.15) \quad \phi_i(v,S) = \sum_{T \subseteq S-i} \frac{|T|!(|S| - |T| - 1)!}{|S|!} (v(T \cup \{i\}) - v(T)).$$

Of course, $\phi_i(v,N) = \phi_i(v)$.

Myerson (1980) has shown that this coalitional value ϕ is the unique solution to the following system of equations:

$$(9.16) \quad \sum_{i \in S} \phi_i(v,S) = v(S), \quad \forall S \subseteq N;$$

$$(9.17) \quad \phi_i(v,S) - \phi_i(v,S-j) = \phi_j(v,S) - \phi_j(S-i), \quad \forall S \subseteq N, \quad \forall i \in S, \forall j \in S.$$

(When S is any set, then we let $S - i$ denote the set differing from S by the removal of i, so $S - i = S \backslash \{i\}$.) Condition (9.16) asserts that the total payoff to the members of an active coalition S should be the worth of that coalition. Condition (9.17) asserts that, for any two members of an active coalition S, the amounts that each player would gain or lose by the other's withdrawal from the coalition should be equal. We can view

(9.16) as an efficiency condition and (9.17) as an equity or *balanced-contributions* condition.

If the contributions that are equated in (9.17) are all nonnegative, then we can suppose that players will all welcome one another into coalitions and that restricted-negotiation agreements should therefore be unlikely. For instance, in Example 9.1 we have

$$\phi_i(v,S) = 100 \text{ if } |S| = 3, \quad \phi_i(v,S) = 0 \text{ if } |S| \leq 2.$$

On the other hand, if some contributions $\phi_i(v,S) - \phi_i(v,S - j)$ are negative, then there may be some tendency for players to try to exclude one another and form active coalitions that are smaller than N. For instance, in Example 9.2,

$$\phi_i(v,S) = 100 \text{ if } |S| = 3,$$
$$\phi_i(v,S) = 150 \text{ if } |S| = 2,$$
$$\phi_i(v,S) = 0 \quad \text{ if } |S| = 1;$$

so $\phi_i(v,N) - \phi_i(v,N - j) = -50$ for any two players i and j.

Myerson (1977) suggested that cooperation structures can be described by graphs in which the nodes correspond to the players and the branches or links represent communication links or friendship relationships between pairs of players. So let $L_2(N)$ denote the set of all pairs of players; that is,

$$L_2(N) = \{\{i,j\} | i \in N, j \in N, i \neq j\}.$$

Then any subset of $L_2(N)$ can be called a *graph* on the set of players or a *graphical cooperation structure*. For any such graph G, any coalition S, and any players i and j in S, we can say that i and j are *connected by G within S* iff $i = j$ or there exists some $\{h_1,h_2, . . . ,h_k\}$ such that $\{h_1,h_2, . . . ,h_k\} \subseteq S$, $h_1 = i$, $h_k = j$, and $\{h_l,h_{l+1}\} \in G$ for every l. Let S/G denote the partition of S into the sets of players who are connected by G within S, so

$$S/G = \{\{i | i \text{ and } j \text{ are connected by } G \text{ within } S\} | j \in S\}.$$

We say that S is *internally connected by G* iff $S/G = \{S\}$. That is, S is internally connected iff its members could communicate with one another through a telephone system that only had lines between players who are paired in G and are both in S. Given any graphical cooperation structure, we can think of the internally connected coalitions as the set of coalitions that can negotiate effectively.

Given any coalitional game v on the set of players N, an allocation rule for v should then specify a payoff $\psi_i(v,G)$ for every player i and every graphical cooperation structure G. A *fair allocation rule* is defined to be any such function that satisfies the following equations:

(9.18) $\displaystyle\sum_{i \in S} \psi_i(v,G) = v(S), \quad \forall G \subseteq L_2(N), \quad \forall S \in N/G,$

(9.19) $\psi_i(v,G) - \psi_i(v,G - \{i,j\}) = \psi_j(v,G) - \psi_j(v,G - \{i,j\}),$

$\qquad \forall G \subseteq L_2(N), \ \forall\{i,j\} \in G.$

Condition (9.18) is an efficiency condition, asserting that each maximal internally connected coalition will divide its worth among its members. Condition (9.19) is an equity condition, asserting that any two players should gain or lose equally by creating a link between themselves.

Myerson (1977) proved that there exists a unique fair allocation rule for any coalitional game v. Furthermore, if we let $L_2(S)$ denote the graph consisting of all pairs of players in S, we have

$$\psi_i(v,L_2(S)) = \phi_i(v,S), \forall S \subseteq N, \quad \forall i \in S.$$

Thus, for the complete graph $L_2(N)$ that includes all pairs of players, the fair allocation rule coincides with the Shapley value

$$\psi_i(v,L_2(N)) = \phi_i(v).$$

More generally, if we let v/G denote the coalitional game such that

$$(v/G)(S) = \sum_{R \in S/G} v(R), \quad \forall S \subseteq N,$$

then the fair allocation rule for v satisfies the equation

(9.20) $\psi_i(v,G) = \phi_i(v/G).$

If v is superadditive, then both sides of (9.19) are nonnegative; that is,

(9.21) $\psi_i(v,G) - \psi_i(v,G - \{i,j\}) \geq 0, \quad \forall G \subseteq L_2(N), \ \forall\{i,j\} \in G.$

So in superadditive games, a player can never lose by forming links with other players. For instance, in Example 9.3, the fair allocation rule is shown in Table 9.2. (To illustrate how these allocations are computed,

Table 9.2   The fair allocation rule for Example 9.3

G	$(\psi_1(v,G),\ \psi_2(v,G),\ \psi_3(v,G))$
$\varnothing$	$(0, 0, 0)$
$\{\{1,2\}\}$	$(150, 150, 0)$
$\{\{1,3\}\}$	$(150, 0, 150)$
$\{\{2,3\}\}$	$(0, 150, 150)$
$\{\{1,2\}, \{1,3\}\}$	$(200, 50, 50)$
$\{\{1,2\}, \{2,3\}\}$	$(50, 200, 50)$
$\{\{1,3\}, \{2,3\}\}$	$(50, 50, 200)$
$\{\{1,2\}, \{1,3\}, \{2,3\}\}$	$(100, 100, 100)$

consider the fair allocation for $\{\{1,2\},\{1,3\}\}$. Given the other allocations above, it must satisfy

$$\psi_1(v, \{\{1,2\},\{1,3\}\}) - 150 = \psi_2(v, \{\{1,2\},\{1,3\}\}) - 0,$$

$$\psi_1(v, \{\{1,2\},\{1,3\}\}) - 150 = \psi_3(v, \{\{1,2\},\{1,3\}\}) - 0, \ \text{and}$$

$$\psi_1(v, \{\{1,2\},\{1,3\}\}) + \psi_2(v, \{\{1,2\},\{1,3\}\})$$
$$+ \ \psi_3(v, \{\{1,2\},\{1,3\}\}) = 300.$$

Thus, $3\psi_1(v, \{\{1,2\},\{1,3\}\}) - 300 = 300$, so player 1 gets 200 when he is the only player linked to both other players.) Now consider a link-formation process in which each player independently writes down a list of the players with whom he wants to form a link, and the payoff allocation is the fair allocation above for the graph that contains a link for every pair of players who have named each other. In this game, the unique perfect equilibrium is for every player to name all other players, so that the outcome will be the Shapley value.

Aumann and Myerson (1988) considered a different model of endogenous link formation. In their model, pairs of players sequentially decide whether to form links, but all such links must be formed publicly, and the process of link formation does not stop until every pair of unlinked players has had another opportunity to form a link after the last link was formed. For Example 9.3, the graphical cooperation structure $\{\{1,2\}\}$ could be the result of a subgame-perfect equilibrium in such a link-formation game. In this equilibrium, after the $\{1,2\}$ link has been formed, player 1 would understand that, if he formed a link with player 3, then player 2 would subsequently form a link with player 3 as well, so 1's initial gains (200 − 150) by forming a link with player 3 would be followed by greater losses (100 − 200).

Consider the apex game discussed in Section 9.4 above. For this game,

$$\phi_1(v,\{1,2\}) = \phi_2(v,\{1,2\}) = \tfrac{1}{2} = \psi_1(v,\{\{1,2\}\}) = \psi_2(v,\{\{1,2\}\}).$$

That is, under either of the extended values defined above, when the big player forms a coalition with one small player, they divide their worth equally. This result may seem unreasonable, because the big player has more outside options than the small player. Essentially, the problem is that the fair allocation rule ignores the impact of any coalition that is not internally connected by the graphical cooperation structure. Owen (1977) proposed a different value concept for games with cooperation structures that does take some unconnected coalitions into account (see also Hart and Kurz, 1983).

We say that $\mathcal{F}$ is a *nested coalition structure* iff $\mathcal{F}$ is a subset of $L(N)$ and

$$\forall S \in \mathcal{F}, \ \forall T \in \mathcal{F}, \ \text{if} \ S \cap T \neq \emptyset, \ \text{then} \ S \subseteq T \ \text{or} \ T \subseteq S.$$

In Owen's model, a cooperation structure is described by such a nested set of active coalitions, which Owen calls unions. That is, each player can belong to several active coalitions, but the active coalitions to which he belongs must be ranked by inclusion.

A nested coalition structure $\mathcal{F}$ can also be viewed as a nested family of partitions. For any coalition S and any player i in S, let $\mathcal{F}_i(S)$ denote the largest coalition R such that

$$R \in \mathcal{F}, \ i \in R, \ \text{and} \ R \subset S \ \text{(that is,} \ R \subseteq S \ \text{and} \ R \neq S),$$

if any such coalition exists; otherwise, let $\mathcal{F}_i(S) = \{i\}$. Let $\mathcal{F}_*(S)$ denote the partition of S into such sets; that is,

$$\mathcal{F}_*(S) = \{\mathcal{F}_i(S) \,|\, i \in S\} \cdot$$

Then $\mathcal{F}$ divides the set of players N into the partition $\mathcal{F}_*(N)$, and each coalition S in this partition is subdivided by $\mathcal{F}$ into the partition $\mathcal{F}_*(S)$, and so on.

For any nested coalition structure $\mathcal{F}$, any coalitional game v, and any player i, the Owen value assigns a payoff denoted by $\theta_i(v,\mathcal{F})$. This value is linear in the game, so $\theta_i(0,\mathcal{F}) = 0$ and

$$\theta_i(pv + (1 - p)w, \ \mathcal{F}) = p\theta_i(v,\mathcal{F}) + (1 - p)\theta_i(w,\mathcal{F})$$

for any coalitional games v and w and any number p between 0 and 1. To complete the specification of the Owen value for all games, then, it suffices to define the Owen value for the simple carrier games (which form a linear basis of $\mathbf{R}^{L(N)}$).

So let R be any coalition, and let w_R be the simple R-carrier game defined by (9.10). The Owen value of w_R allocates the whole worth of the grand coalition to the members of the carrier R and gives 0 to every dummy player outside of R; that is,

$$(9.22) \qquad \sum_{i \in R} \theta_i(w_R, \mathscr{F}) = 1 \text{ and } \theta_j(w_R, \mathscr{F}) = 0, \quad \forall j \in N \backslash R.$$

The allocation within the carrier R is defined by the following condition

$$(9.23) \qquad \forall Q \in \mathscr{F} \cup \{N\}, \ \forall S \in \mathscr{F}_*(Q), \ \forall T \in \mathscr{F}_*(Q),$$

$$\text{if } S \cap R \neq \varnothing \text{ and } T \cap R \neq \varnothing,$$

$$\text{then } \sum_{i \in S} \theta_i(w_R, \mathscr{F}) = \sum_{j \in T} \theta_j(w_R, \mathscr{F}).$$

Conditions (9.22) and (9.23) determine a unique value $\theta(w_R, \mathscr{F}) = (\theta_i(w_R, \mathscr{F}))_{i \in N}$, for every coalition R and every nested coalition structure $\mathscr{F}$.

The Owen value is based on an assumption that each union in the nested coalition structure $\mathscr{F}$ appoints an agent to conduct all negotiations between members of the union and players who are not in the union. Thus, from the perspective of players not in the union, the union will behave like a single player. (Any one-player coalition $\{i\}$ would be represented by player i as the agent for himself.) We may suppose that, after these agents are appointed, the agents who represent coalitions in $\mathscr{F}_*(N)$ will meet and bargain to divide among themselves the worth of the grand coalition N. Then, for each union Q that was represented in the previous round of bargaining, the agents for the coalitions in $\mathscr{F}_*(Q)$ will meet and bargain to divide among themselves the value obtained by the agent for Q in the previous stage of bargaining. This process continues until all unions in $\mathscr{F}$ have allocated their payoffs down to the level of individuals (one-player coalitions). Condition (9.23) asserts that, at any stage in this process, when agents meet to divide the transferable utility that is available to them collectively (either because they together form N or some union in $\mathcal{F}$ that has completed its external bargaining), any two agents who represent players in the carrier R should get the same shares of the available utility. Such agents should get equal shares because each such agent has an equal ability to reduce the worth of any larger coalition to 0, by withdrawing the players that he represents. An implicit assumption used here is that agents for different unions have equal bargaining ability, independent of the size of the union that they represent.

The Owen value can be equivalently described by a story of randomly ordered entry of players into the coalition-formation process. Given a nested coalition structure $\mathcal{F}$, let us define an $\mathcal{F}$-*permissible ordering* of the players to be any strict ordering of N such that, for any union S in $\mathcal{F}$ and any two players i and j who are both in S, all players who come between i and j in the ordering must also be in S. Now suppose that we randomly select one of the $\mathcal{F}$-permissible orderings, such that every $\mathcal{F}$-permissible ordering has an equal chance of being selected. Let $\tilde{S}(i)$ denote the set of all players who will precede player i in this ordering. Then the Owen value for player i is the expected value of $v(\tilde{S}(i) \cup \{i\}) - v(\tilde{S}(i))$.

For the apex game with the nested coalition structure $\mathcal{F} = \{\{1,2\}\}$, the Owen value gives 0 to players 3, 4, and 5, value $\frac{1}{4}$ to player 2, and value $\frac{3}{4}$ to player 1. (To check this, notice that, among the $\mathcal{F}$-permissible orderings, player 2 has a marginal contribution of 1 in only two orderings. One is the ordering where the union $\{1,2\}$ comes first among the coalitions in $\mathcal{F}_*(N)$, which occurs with probability $1/|\mathcal{F}_*(N)| = \frac{1}{4}$, and player 2 comes second in $\{1,2\}$, which has an independent probability $\frac{1}{2}$. The second is the ordering where the union $\{1,2\}$ comes last among the coalitions in $\mathcal{F}_*(N)$, which occurs with probability $\frac{1}{4}$, and player 2 comes first in $\{1,2\}$, which has an independent probability $\frac{1}{2}$. Thus, player 2 has a marginal contribution of 1 in two orderings that have a combined probability of $\frac{1}{4}$; otherwise player 2 has a marginal contribution of 0.) Thus, unlike the fair allocation rule of Myerson (1977), the Owen value allows player 1 to get some benefit from his stronger power to form winning coalitions with players 3, 4, and 5, even when he has formed a union with player 2. However, if the nested coalition structure were changed to $\{\{1,2\},\{3,4,5\}\}$, then the Owen value would give $\frac{1}{2}$ to each of players 1 and 2, because they would have the same marginal contribution to the union of the other players.

9.6 Other Solution Concepts

The core and the Shapley value are only two of many solution concepts for coalitional games with transferable utility that have been studied by game theorists. To describe some of the other solution concepts that have been studied, we should first consider some basic definitions. Throughout this section, let v be a coalitional game with transferable utility and let N be the set of all players.

For any two allocations x and y in $\mathbf{R}^N$ and any coalition S, we can write

$$x >_S y \text{ iff } x_i > y_i, \quad \forall i \in S.$$

That is, $x >_S y$ iff x is strictly preferred to y by all the members of coalition S. Similarly, we can write

$$x \geq_S y \text{ iff } x_i \geq y_i, \quad \forall i \in S.$$

For any partition Q of the set of players, let $I(Q)$ denote the set of all payoff allocations in which every player gets at least what he could get by himself and every coalition in the partition is dividing its worth among its members. That is,

$$I(Q) = \{x \in \mathbf{R}^N | x_i \geq v(\{i\}), \quad \forall i \in N, \quad \text{and}$$

$$\sum_{j \in R} x_j = v(R), \quad \forall R \in Q\}.$$

Any allocation in $I(\{N\})$ is called an *imputation*.

For any coalition S and any allocation x, let

$$e(S,x) = v(S) - \sum_{i \in S} x_i.$$

This quantity $e(S,x)$ is called the *excess* of S at x and is the net transferable worth that S would have left after paying x_i to each member i. Notice, $e(S,x) \geq 0$ if and only if coalition S could by itself achieve its share of the allocation x.

The first solutions studied by von Neumann and Morgenstern (1944) are now known as *stable sets*. A stable set is any set of imputations Z such that the following two properties hold:

(9.24) $\forall y \in I(\{N\})$, if $y \notin Z$, then $\exists x \in Z$ and $\exists S \in L(N)$
    such that $e(S,x) \geq 0$ and $x >_S y$.

(9.25) $\forall y \in Z, \ \forall x \in Z, \ \forall S \in L(N),$ if $x >_S y$, then $e(S,x) < 0$.

The idea behind these conditions is as follows. We are supposed to think of a stable set as the set of allocations that the players consider to be possible outcomes of the game, without knowing which one will ultimately be chosen. Condition (9.24) asserts that, if any other imputation y is proposed, then there is at least one coalition S that could block the adoption of y by insisting on getting their share of some possible outcome in Z that is feasible for them. Condition (9.25) asserts that nothing in Z itself can be blocked in this way by another possible outcome in Z.

For Example 9.3 (a version of the three-person majority game), there are many stable sets. The set $\{(150,150,0),(150,0,150),(0,150,150)\}$ is a stable set. Also, for any player i and any number α such that $0 \le \alpha < 150$, the set $\{x \in I(\{N\}) \mid x_i = \alpha\}$ is a stable set. Thus, although the core of this game is empty, every imputation is in at least one stable set. However, Lucas (1969) showed that some games have no stable sets.

Aumann and Maschler (1964) introduced the concept of the bargaining set, offering several alternative definitions, of which we discuss here one version (known as $M_1^{(i)}$). The idea behind the bargaining set is that a player might not make an objection to a proposed payoff allocation if he feared that his objection might prompt a counterobjection by another player.

An *objection* by a player i against another player j and a payoff allocation x is a pair (y,S) such that

$$y \in \mathbf{R}^N, S \subseteq N, i \in S, j \notin S, e(S,y) = 0, \text{ and } y >_S x.$$

That is, the players in S can jointly achieve their share of y, which is strictly better for them all, including player i, than the allocation x. A *counterobjection* to i's objection (y,S) against j and x is any pair (z,T) such that

$$z \in \mathbf{R}^N, T \subseteq N, j \in T, i \notin T,$$
$$T \cap S \ne \varnothing, e(T,z) = 0, z \ge_T x, \text{ and } z \ge_{T \cap S} y.$$

That is, in the counterobjection, player j can form a coalition T that takes away some of i's partners in the objection (but not i himself) and makes them at least as well off as in the objection; thus, j can restore himself and the other members of T to payoffs at least as good as they had in x.

The bargaining set is defined relative to a partition of the players. So let Q denote any partition of N. An allocation x is in the *bargaining set* of v, relative to the partition Q, iff $x \in I(Q)$ and, for any coalition R in Q and any two players i and j who are in R, there exists a counterobjection to any objection by i against j and x. It is straightforward to check that the core is a (possibly empty) subset of the bargaining set of v relative to $\{N\}$. Peleg (1963) proved that, for any partition Q, if $I(Q)$ is nonempty (as it must be if v is superadditive) then the bargaining set of v relative to Q is nonempty.

For example, the bargaining set of the apex game with the partition $\{\{1,2,3,4,5\}\}$ (all players together in the grand coalition) is $\{(1 - 4\alpha,\alpha,\alpha,\alpha,\alpha)\,|\,\frac{1}{13} \le \alpha \le \frac{1}{7}\}$. If any small player i got less than another small player j, then i would have an objection with the big player 1 that j could not counter. If the small players all got some amount more than $\frac{1}{7}$, then player 1 would have an objection $((\frac{3}{7},\frac{4}{7},0,0,0),\{1,2\})$ that player 3 could not counter. If the small players all got some amount less than $\frac{1}{13}$, so that player 1 got more than $\frac{9}{13}$, then player 2 would have an objection $((0,\frac{1}{13},\frac{4}{13},\frac{4}{13},\frac{4}{13}),\{2,3,4,5\})$ that player 1 could not counter.

With the partition $\{\{1,2\},\{3\},\{4\},\{5\}\}$, on the other hand, the bargaining set of the apex game becomes $\{(1-\alpha,\alpha,0,0,0)\,|\,\frac{1}{4} \le \alpha \le \frac{1}{2}\}$.

Davis and Maschler (1965) defined the *kernel* of v, relative to the partition Q, to be the set of all allocations x such that $x \in I(Q)$ and, for any coalition T in Q and any two players i and j in T,

$$\max_{S \subseteq N-j, i \in S} e(S,x) = \max_{T \subseteq N-i, j \in T} e(T,x).$$

That is, if players i and j are in the same coalition in the partition Q, then the highest excess that i can make in a coalition without j is equal to the highest excess that j can make in a coalition without i. The kernel is always a nonempty subset of the bargaining set. For example, the kernel of the apex game with the partition $\{\{1,2,3,4,5\}\} = \{N\}$ is $\{(\frac{3}{7},\frac{1}{7},\frac{1}{7},\frac{1}{7},\frac{1}{7})\}$. The kernel of the apex game with the partition $\{\{1,2\},\{3\},\{4\},\{5\}\}$ is $\{(\frac{1}{2},\frac{1}{2},0,0,0)\}$.

For any allocation x, let $\zeta_k(x)$ be the kth largest excess generated by any coalition at x. That is,

$$|\{S \in L(N)\,|\,e(S,x) \ge \zeta_k(x)\}| \ge k,$$
$$|\{S \in L(N)\,|\,e(S,x) > \zeta_k(x)\}| < k.$$

Thus, x is in the core iff $\zeta_1(x) \leq 0$.

Let $A(1)$ be the set of all imputations that minimize ζ_1. That is,

$$A(1) = \underset{x \in I(\{N\})}{\operatorname{argmin}} \ \zeta_1(x).$$

If the core is nonempty, then $A(1)$ must be a subset of the core. Now, inductively define $A(k)$, for all $k = 2,3,\ldots,2^{|N|} - 1$, by the equation

$$A(k) = \underset{x \in A(k-1)}{\operatorname{argmin}} \ \zeta_k(x).$$

Schmeidler (1969) showed that $A(2^{|N|} - 1)$ must consist of a single point, which he called the *nucleolus* of v. The nucleolus is in the kernel and the bargaining set of v with the partition $\{N\}$, and it is in the core of v if the core is nonempty. For example, the nucleolus of the apex game is $(3/7, 1/7, 1/7, 1/7, 1/7)$.

It is helpful to categorize cooperative solution concepts by the outcomes that they designate for the simple two-player game where

$$N = \{1,2\}, \quad v(\{1\}) = v(\{2\}) = 0, \quad v(\{1,2\}) = 1.$$

If a solution concept is derived only from the criterion of averting coalitional objections, then the "solution" for this game should be the set of all imputations, $\{(\alpha, 1-\alpha) \mid 0 \leq \alpha \leq 1\}$. Such *unobjectionable* solution concepts include the core, the bargaining set (relative to $\{N\}$), and the stable sets. On the other hand, if a solution concept is also based on some consideration of equity between players (as well as efficiency), then the "solution" for this game should be the allocation $(1/2, 1/2)$. Such *equitable* solution concepts include the Shapley value, the nucleolus, and the kernel (relative to $\{N\}$).

In Chapter 8 we discussed Nash's argument that cooperative games should be analyzed by computing equilibria of a fully specified model of the bargaining process. As a criticism of the existing literature in cooperative game theory, this argument may be more relevant to the unobjectionable solution concepts than to the equitable solution concepts. The unobjectionable solutions are supposed to include all the payoff allocations that the players would accept without forming coalitions to demand reallocation, so it does seem reasonable to ask for a full description of the strategic process by which players form coalitions and make demands. (Whatever this process is, it has some Nash equilibria, by the general existence theorem, so the core cannot be generally identified with the set of equilibria of a bargaining process.)

On the other hand, the equitable solutions can be defended against Nash's argument. Like the Nash bargaining solution, the Shapley value and other equitable solutions can be interpreted as arbitration guidelines, or as determinants of focal equilibria in bargaining. We only need to assume that the unspecified bargaining process has a sufficiently large set of equilibria and that the focal equilibrium will be determined by its properties of equity and efficiency. Here we use "equity" to mean that each player's gains from cooperating with others should be commensurate (in some sense) with what his cooperation contributes to other players. So equitable solutions should depend on the power structure that is summarized by the representation of the game in coalitional form.

9.7 Coalitional Games with Nontransferable Utility

So far in this chapter, we have considered only cooperative games with transferable utility, or *TU games*. Let us now consider games without transferable utility, which may be called *NTU* (or *nontransferable utility*) games for short. A generalized concept of coalitional form for NTU games was developed by Aumann and Peleg (1960).

An *NTU coalitional game* (or a game in *NTU coalitional form*) on the set of players N is any mapping $V(\cdot)$ on the domain $L(N)$ such that, for any coalition S,

(9.26) $V(S)$ is a nonempty closed convex subset of $\mathbf{R}^S$,

and

(9.27) $\{x | x \in V(S)$ and $x_i \geq v_i, \quad \forall i \in S\}$ is a bounded subset of $\mathbf{R}^S$,

where

(9.28) $v_i = \max\{y_i | y \in V(\{i\})\} < \infty, \quad \forall i \in N$.

Here, $V(S)$ represents the set of expected payoff allocations that the members of coalition S could guarantee for themselves (in some sense) if they acted cooperatively. Closedness and convexity (9.26) of $V(S)$ are natural technical conditions. In particular, convexity may follow from an assumption that members of a cooperative coalition can use jointly randomized strategies. Condition (9.28) asserts that the maximum payoff that any player can guarantee himself alone is finite. Condition (9.27) asserts that a coalition cannot offer unbounded payoffs to any player, unless it gives less to some other player in the coalition than he could get alone.

We say that V is *compactly generated* iff there exists some set Ω that is a closed and bounded (i.e., compact) subset of $\mathbf{R}^N$ such that, for each coalition S and each allocation x in $V(S)$, there exists some allocation y in Ω such that $(y_i)_{i \in S}$ is a vector in $V(S)$ and $x_i \leq y_i$ for every player i in S. We say that V is *comprehensive* iff, for each coalition S and each vector z in $\mathbf{R}^S$, if there exists some vector x such that $x \in V(S)$ and $z_i \leq x_i$ for every i in S, then $z \in V(S)$.

An NTU coalitional game V is *superadditive* iff, for any two coalitions S and T,

if $S \cap T = \varnothing$, then
$$V(S \cup T) \supset \{x \in \mathbf{R}^{S \cup T} | (x_i)_{i \in S} \in V(S), (x_j)_{j \in T} \in V(T)\}.$$

So superadditivity means that $S \cup T$ can give its members any allocations that they could get in the disjoint coalitions S and T separately.

The NTU coalitional form is a generalization of the coalitional form with transferable utility. Any TU coalitional game v in $\mathbf{R}^{L(N)}$ is equivalent to the NTU coalitional game such that

$$(9.29) \quad V(S) = \left\{ x \in \mathbf{R}^S | \sum_{i \in S} x_i \leq v(S) \right\}.$$

When condition (9.29) holds, V is a superadditive NTU coalitional game iff v is a superadditive coalitional game with transferable utility. However, V cannot be compactly generated if it satisfies (9.29) for some v in $\mathbf{R}^{L(N)}$.

Two-person bargaining problems may also be viewed as a special class of NTU coalitional games. That is, when $N = \{1,2\}$, an NTU coalitional game V may be identified with a two-person bargaining problem $(F,(v_1,v_2))$ where v_1 and v_2 are as specified in (9.28) and $F = V(\{1,2\})$.

Let $\Gamma = (N, (C_i)_{i \in N}, (u_i)_{i \in N})$ be any finite game in strategic form, without transferable utility. Aumann and Peleg (1960) suggested two ways that such a game might be represented in NTU coalitional form.

An allocation vector x in $\mathbf{R}^S$ is *assurable* in Γ for coalition S iff there exists some joint correlated strategy σ_S in $\Delta(C_S)$ such that, for every strategy $\sigma_{N \setminus S}$ in $\Delta(C_{N \setminus S})$,

(9.30) $u_i(\sigma_S,\sigma_{N\backslash S}) \geq x_i, \quad \forall i \in S.$

That is, x is assurable for S iff the players in S can guarantee that they all get at least as much as in x, when they must announce their joint correlated strategy before the joint correlated strategy of the players in $N\backslash S$ is chosen. The *assurable representation* of Γ is the NTU coalitional game V^α such that

$$V^\alpha(S) = \{x \in \mathbf{R}^S | x \text{ is assurable in } \Gamma \text{ for } S\}, \quad \forall S \in L(N).$$

An allocation vector x in $\mathbf{R}^S$ is *unpreventable* in Γ for coalition S iff, for each strategy $\sigma_{N\backslash S}$ in $\Delta(C_{N\backslash S})$ that the complementary coalition could use, there exists some strategy σ_S in $\Delta(C_S)$ that satisfies condition (9.30). That is, x is unpreventable for S iff the players in S could guarantee that they all get at least as much as in x, when they can choose their joint correlated strategy after the joint correlated strategy of the players in $N\backslash S$ is announced. The *unpreventable representation* of Γ is the NTU coalitional game V^β such that

$$V^\beta(S) = \{x \in \mathbf{R}^S | x \text{ is unpreventable in } \Gamma \text{ for } S\}, \quad \forall S \in L(N).$$

It is straightforward to verify that V^α and V^β both satisfy the conditions of closedness, convexity, and boundedness that are required of an NTU coalitional game. Aumann (1961) showed that V^α and V^β are both superadditive, and they are both compactly generated (with, for example, Ω equal to the set of all x such that $|x_i| \leq \max_{c\in C} |u_i(c)|$ for every i) and comprehensive. Aumann (1959, 1967) and Mertens (1980) have shown that the unpreventable representation V^β has some conceptual advantages over V^α in the study of repeated games.

Clearly, $V^\alpha(S) \subseteq V^\beta(S)$ for any coalition S. For a simple example where the assurable and unpreventable representations differ, consider the coalition $\{2,3\}$ in the three-player game shown in Table 9.3, where the pure strategy set for each player i is $C_i = \{a_i,b_i\}$.

Table 9.3 A three-player game in strategic form

C_1	$C_2 \times C_3$			
	a_2,a_3	b_2,a_3	a_2,b_3	b_2,b_3
a_1	1,2,2	0,5,0	2,3,0	1,3,1
b_1	1,1,3	0,0,5	2,0,3	1,2,2

If player 1 chose a_1, then the coalition $\{2,3\}$ could not guarantee more than 2 utility units to player 3, even if it could move after player 1. Similarly, if player 1 chose b_1, then the coalition $\{2,3\}$ could not guarantee more than 2 to player 2. After player 1 announced any randomized strategy σ_1, the coalition $\{2,3\}$ could guarantee at least 2 to each of its members by letting $\sigma_{\{2,3\}}(a_2,a_3) = \sigma_1(a_1)$ and $\sigma_{\{2,3\}}(b_2,b_3) = \sigma_1(b_1)$. Thus, the unpreventable set for coalition $\{2,3\}$ is

$$V^\beta(\{2,3\}) = \{(x_2,x_3) \,|\, x_2 \le 2,\, x_3 \le 2\}.$$

Now consider the case where coalition $\{2,3\}$ must announce $\sigma_{\{2,3\}}$ first, before player 1 chooses σ_1. No matter what $\sigma_{\{2,3\}}$ might be, the worst outcome for player 2 would be if player 1 chose b_1, and the worst outcome for player 3 would be if player 1 chose a_1. So choosing $\sigma_{\{2,3\}}$ can only guarantee an allocation x for $\{2,3\}$ if

$$x_2 \le 1\sigma_{\{2,3\}}(a_2,a_3) + 0\sigma_{\{2,3\}}(b_2,a_3) + 0\sigma_{\{2,3\}}(a_2,b_3) + 2\sigma_{\{2,3\}}(b_2,b_3),$$

$$x_3 \le 2\sigma_{\{2,3\}}(a_2,a_3) + 0\sigma_{\{2,3\}}(b_2,a_3) + 0\sigma_{\{2,3\}}(a_2,b_3) + 1\sigma_{\{2,3\}}(b_2,b_3).$$

(The coefficients in the first inequality are 2's payoffs when 1 chooses b_1, and the coefficients in the second inequality are 3's payoffs when 1 chooses a_1.) Thus, the assurable set for coalition $\{2,3\}$ is

$$V^\alpha(\{2,3\}) = \{(x_2,x_3) \,|\, x_2 \le 2,\, x_3 \le 2,\, x_2 + x_3 \le 3\},$$

which has extreme points at $(2,1)$ and $(1,2)$.

Since we have invested so much in the study of TU games in coalitional form, it will be analytically helpful to have ways to generate TU coalitional games that might correspond (in some sense) to a given NTU coalitional game. One obvious way to generate such a game is to let the worth of each coalition be the maximum sum of payoffs that it can achieve for its members, that is,

$$v(S) = \max_{x \in V(S)} \sum_{i \in S} x_i.$$

However, this TU coalitional game is not an adequate representation of the NTU game V when utility is not really transferable, because it contains no information about the distribution of payoffs among the players. For example, consider the two-player NTU game where

$$V(\{i\}) = \{x_i \,|\, x_i \le 0\}, \quad \text{for } i = 1,2,$$

$$V(\{1,2\}) = \{(x_1,x_2) \,|\, 2x_1 + x_2 \le 0,\, x_1 \le 0,\, x_2 \le 2\}.$$

Because $(-1,2) \in V(\{1,2\})$, we would get $v(\{1,2\}) = 1$, but the $\{1,2\}$ coalition could not really do anything with this worth because it could be achieved only by forcing player 1 to accept less than he could get on his own. If utility is not transferable, then there is no way for player 2 to make such a sacrifice worthwhile for player 1.

There is a way, however, in which an NTU coalitional game can be adequately represented by a family of TU coalitional games, which we now consider. Here, the symbol $\mathbf{R}^N_+$ denotes the set of all vectors in $\mathbf{R}^N$ in which all components are nonnegative, and $\mathbf{R}^N_{++}$ denotes the set of all vectors in $\mathbf{R}^N$ in which all components are strictly positive.

Recall from Chapters 1 and 2 that no decision-theoretic properties of a game are transformed if we change to measuring a player's payoffs in some scale that is derived by multiplying his given utility scale by a positive constant or weighting factor. Thus, a natural generalization of the assumption of transferable utility is the assumption of *transferable λ-weighted utility*, where $\lambda = (\lambda_i)_{i \in N}$ is any vector in $\mathbf{R}^N_{++}$. We say that λ-weighted utility is transferable iff there exists some freely transferable (and disposable) currency such that, for each player i, the utility payoff to i would be increased by λ_i for every additional unit of this currency that i gets. If we alter the situation represented by an NTU coalitional game V by allowing transferable λ-weighted utility, then the feasible set of each coalition S would become

$$\left\{ y \in \mathbf{R}^S \,\middle|\, \sum_{i \in S} \lambda_i y_i \leq \max_{x \in V(S)} \sum_{i \in S} \lambda_i x_i \right\}.$$

So given any positive vector λ in $\mathbf{R}^N_{++}$, and any compactly generated NTU coalitional game V, we define a TU coalitional game v^λ by

(9.31) $v^\lambda(S) = \max_{x \in V(S)} \sum_{i \in S} \lambda_i x_i, \quad \forall S \in L(N).$

That is, $v^\lambda(S)$ is the highest sum of λ-weighted payoffs that coalition S could achieve by any allocation that is feasible for it. As a coalitional game, v^λ characterizes the situation that would exist if λ-weighted payoffs were transferable. This coalitional game v^λ is called the λ-*transfer game* generated by V.

The following theorem shows that the family of all λ-transfer games generated by V together contain essentially all the information in the structure of V itself.

THEOREM 9.4. *Let V be any compactly generated and comprehensive NTU coalitional game. For any coalition S and any vector y in $\mathbf{R}^S$, $y \in V(S)$ if and only if, for every λ in $\mathbf{R}^N_{++}$,*

$$\sum_{i \in S} \lambda_i y_i \le v^\lambda(S),$$

where v^λ denotes the λ-transfer game generated by V.

Proof. If y is in $V(S)$, then the maximization in (9.31) includes the case of $x = y$, so the inequality in the theorem follows.

The converse step requires the *separating hyperplane theorem* (see, for example, Rockafellar, 1970). The separating hyperplane theorem asserts that, for any closed convex subset Z of any finite dimensional vector space $\mathbf{R}^S$, and for any vector y in $\mathbf{R}^S$, $y \notin Z$ if and only if there exists some vector γ in $\mathbf{R}^S$ such that

$$\sup_{z \in Z} \sum_{i \in S} \gamma_i z_i < \sum_{i \in S} \gamma_i y_i.$$

So suppose now that y is not in $V(S)$. The separating hyperplane theorem immediately implies that there exists some γ in $\mathbf{R}^S$ such that

$$\sup_{x \in V(S)} \sum_{i \in S} \gamma_i x_i < \sum_{i \in S} \gamma_i y_i.$$

By comprehensiveness, if γ had any negative component γ_i, then the supremum here would be $+\infty$, because the component x_i can be made arbitrarily far below 0, so the inequality could not be satisfied. Thus, γ must be in $\mathbf{R}^N_+$. This conclusion implies that the term "sup(remum)" here may be replaced by "max(imum)," because V is compactly generated (so the maximum exists in the set).

For any positive number ε, let $\lambda_i(\varepsilon) = \gamma_i + \varepsilon$ for every i, and notice that $\lambda(\varepsilon) = (\lambda_i(\varepsilon))_{i \in S} \in \mathbf{R}^S_{++}$. We now show that, for some sufficiently small ε,

$$\max_{x \in V(S)} \sum_{i \in S} \lambda_i(\varepsilon) x_i < \sum_{i \in S} \lambda_i(\varepsilon) y_i,$$

so the inequality in the theorem cannot be satisfied for all λ in $\mathbf{R}^S_{++}$ when $y \notin V(S)$. If not, then for every ε there exists some vector $z(\varepsilon)$ in the closed and bounded set Ω (that compactly generates V) such that the vector $(z_i(\varepsilon))_{i \in S}$ is in $V(S)$ and

$$\sum_{i \in S} \lambda_i(\varepsilon) z_i(\varepsilon) \geq \sum_{i \in S} \lambda_i(\varepsilon) y_i.$$

By compactness of Ω, there exists a sequence of possible values for ε, converging to 0, and there exists a vector $\bar{z}$ such that $\lim_{\varepsilon \to 0} z(\varepsilon) = \bar{z}$. But then $(\bar{z}_i)_{i \in S}$ is a vector in $V(S)$ and

$$\sum_{i \in S} \gamma_i \bar{z}_i \geq \sum_{i \in S} \gamma_i y_i,$$

which contradicts the way that γ was constructed. ∎

9.8 Cores without Transferable Utility

Given an NTU coalitional game V on the set of players N, the *core* of V may be defined as the set of all allocation vectors x in $\mathbf{R}^N$ such that $x \in V(N)$ and, for any coalition S and any allocation y in $\mathbf{R}^S$,

if $y_i > x_i$, $\forall i \in S$, then $y \notin V(S)$.

That is, x is in the core iff it is feasible for the grand coalition N, and no coalition has a feasible vector that is strictly better for all its members (see Scarf, 1967).

By taking into account the possibility of randomization among coalitions, we can generate a second concept of a core for NTU games. This solution concept is known as the *inner core* and has great theoretical importance (see Qin, 1989).

In Section 9.3, we argued that the appeal of the core relied on an implicit assumption that, when players in a set S are offered a chance to form a blocking coalition against an allocation x, the players expect that they will really get the allocation x if any player in S refuses to join the blocking coalition. That is, the proposed blocking coalition S is assumed to be final, and no other coalition will subsequently block x if S does not. Maintaining this assumption, we now broaden our scope to consider blocking coalitions that are organized by a mediator who uses a randomized strategy.

To describe the plan of such a mediator, let a *randomized blocking plan* be any pair (η, Y) such that η is a probability distribution on the set of coalitions, so $\eta \in \Delta(L(N))$, and Y is a function on $L(N)$ that satisfies the feasibility condition

(9.32) $Y(S) \in V(S)$, $\forall S \in L(N)$.

Here $\eta(S)$ represents the probability that the mediator will make S a blocking coalition S, and $Y(S) = (Y_i(S))_{i \in S}$ represents the allocation that the members of S would take for themselves if they formed a blocking coalition together. Given such a randomized blocking plan (η, Y), suppose that the mediator first chooses, randomly according to η, the blocking coalition that he will attempt to form, and then he invites each member of this coalition separately to participate in the blocking coalition. Suppose that each player i knows only the plan (η, Y), but not the actual blocking coalition, when the mediator invites him to participate. Then, conditional on being invited to participate in the blocking coalition, the expected payoff to player i if everyone accepts the mediator's invitation is

$$\frac{\displaystyle\sum_{S \supseteq \{i\}} \eta(S) Y_i(S)}{\displaystyle\sum_{S \supseteq \{i\}} \eta(S)}.$$

Suppose also that, if any member of his blocking coalition refuses to participate in it, then the coalition cannot act and no further blocking against x will be attempted. Thus, there is an equilibrium in which every player would be willing to accept the mediator's invitation iff, for any player i who has a positive probability of being invited by the mediator, i's conditionally expected payoff when he is invited to participate is at least x_i. This condition holds iff

$$(9.33) \qquad \sum_{S \supseteq \{i\}} \eta(S) Y_i(S) \geq \sum_{S \supseteq \{i\}} \eta(S) x_i, \quad \forall i \in N.$$

We say that a randomized blocking plan (η, Y) is *viable* against x iff it satisfies (9.33). An allocation x in $\mathbf{R}^N$ is *strongly inhibitive* for the game V iff there does not exist any viable randomized blocking plan against it. That is, an allocation is strongly inhibitive iff it could not be blocked by any random one-stage coalition-formation plan.

The following theorem asserts that any strongly inhibitive allocation for the NTU coalitional game V corresponds to a λ-weighted payoff allocation that cannot be blocked in the λ-transfer game generated by V, for some λ.

THEOREM 9.5.   *Suppose V is a compactly generated NTU coalitional game. An allocation x is strongly inhibitive for V if and only if there exists some vector* λ *in* $\mathbf{R}^N_{++}$ *such that*

$$\sum_{i \in S} \lambda_i x_i > v^\lambda(S), \quad \forall S \in L(N),$$

where v^λ *denotes the* λ*-transfer game generated by V.*

Proof. Let $Z(x)$ be the subset of $\mathbf{R}^N$ such that $z \in Z(x)$ iff there exists some randomized blocking plan (η, Y) such that

(9.34) $z_i \leq \sum_{S \supseteq \{i\}} \eta(S)(Y_i(S) - x_i), \quad \forall i \in N.$

It is straightforward to show that $Z(x)$ is closed, when V is compactly generated. Furthermore, $Z(x)$ is convex. To show convexity, let z and $\hat{z}$ be any two vectors in $Z(x)$, and let p be a number between 0 and 1. Suppose that (η, Y) verifies (9.34) for z, and $(\hat{\eta}, \hat{Y})$ verifies the same condition for $\hat{z}$. To verify that $pz + (1 - p)\hat{z}$ is in $Z(x)$, it suffices to consider the randomized blocking plan $(\overline{\eta}, \overline{Y})$ such that, for any coalition S

$$\overline{\eta}(S) = p\eta(S) + (1 - p)\hat{\eta}(S),$$
$$\overline{\eta}(S)\overline{Y}(S) = p\eta(S)Y(S) + (1 - p)\hat{\eta}(S)\hat{Y}(S).$$

Such a plan $(\overline{\eta}, \overline{Y})$ exists, because each $V(S)$ is convex.

The allocation x is strongly inhibitive iff the zero vector in $\mathbf{R}^N$ is not in $Z(x)$. Thus, by the separating hyperplane theorem, x is strongly inhibitive iff there exists some vector λ in $\mathbf{R}^N$ such that,

$$\sup_{z \in Z(x)} \sum_{i \in N} \lambda_i z_i < 0.$$

We know that λ_i cannot be negative, for any i, because reducing the component z_i toward $-\infty$ would never take the vector z out of the set $Z(x)$ but would increase $\sum_{i \in N} \lambda_i z_i$ to $+\infty$ if λ had any negative components. Thus, the vector λ satisfies this condition iff, for every randomized blocking plan (η, Y),

$$\sum_{i \in N} \lambda_i \sum_{S \supseteq \{i\}} \eta(S)(Y_i(S) - x_i) < 0.$$

This inequality can be rewritten

$$\sum_{S \in L(N)} \eta(S) \sum_{i \in S} \lambda_i(Y_i(S) - x_i) < 0.$$

But this condition can be satisfied for every randomized blocking plan (η, Y) iff

$$\max_{y \in V(S)} \sum_{i \in S} \lambda_i(y_i - x_i) < 0, \quad \forall S \in L(N).$$

So x is strongly inhibitive iff there exists some λ in $\mathbf{R}_+^N$ such that

$$v^\lambda(S) - \sum_{i \in S} \lambda_i x_i < 0, \quad \forall S \in L(N).$$

This condition could not be satisfied when $S = \{i\}$ if λ_i were 0, so λ must be in $\mathbf{R}_{++}^N$. ∎

We say that x is *inhibitive* iff there exists some sequence of strongly inhibitive allocations that converge to x. That is, an allocation is inhibitive iff, by perturbing the allocation an arbitrarily small amount, we could get an allocation against which there is no viable randomized blocking plan. Thus, if an inhibitive allocation can be blocked, it can only be a knife-edge phenomenon, with at least one player willing to veto the blocking coalition and insist on x.

The *inner core* of V is the set of all inhibitive allocations in $V(N)$. It is straightforward to verify that the inner core is a subset of the core. By Theorem 9.5, the inner core has a natural interpretation in terms of the λ-transfer games in coalitional form generated by V. In particular, x is in the inner core of V iff $x \in V(N)$ and, for each positive number ε, there exists some vector λ in $\mathbf{R}_{++}^N$ such that the allocation vector $(\lambda_i(x_i + \varepsilon))_{i \in N}$ is in the core of the TU coalitional game $v^{\lambda, \varepsilon}$, that is defined by the equations

$$v^{\lambda, \varepsilon}(N) = v^\lambda(N) + \varepsilon \left(\sum_{i \in N} \lambda_i \right), \text{ and } v^{\lambda, \varepsilon}(S) = v^\lambda(S), \quad \forall S \neq N,$$

where v^λ is the λ-transfer game generated by V.

Consider the three-player game V such that

$$V(\{1,2,3\}) = \left\{x \,\middle|\, \sum_{i \in S} x_i \le 9, \; \forall S \in L(N)\right\},$$

$$V(\{1,2\}) = \{(x_1,x_2) \,|\, x_1 + 9x_2 \le 9, \; x_1 \le 9, \; x_2 \le 1\},$$

$$V(\{2,3\}) = \{(x_2,x_3) \,|\, x_2 + 9x_3 \le 9, \; x_2 \le 9, \; x_3 \le 1\},$$

$$V(\{1,3\}) = \{(x_1,x_3) \,|\, x_3 + 9x_1 \le 9, \; x_3 \le 9, \; x_1 \le 1\},$$

$$V(\{i\}) = \{(x_i) \,|\, x_i \le 0\}, \quad \text{for } i = 1,2,3.$$

This game is compactly generated and superadditive. Furthermore, its core is nonempty, including the allocation $(3,3,3)$, for example. However, its inner core is empty. To see why $(3,3,3)$ is not in the inner core, consider the randomized blocking plan (η, Y) such that

$$\eta(\{1,2\}) = \eta(\{2,3\}) = \eta(\{1,3\}) = \tfrac{1}{3},$$

$$Y(\{1,2\}) = (9,0), \quad Y(\{2,3\}) = (9,0), \quad Y(\{1,3\}) = (0,9).$$

Conditional on his being invited to participate in the randomized blocking coalition, a player would have an expected payoff of 4.5, which is strictly better than what he gets in $(3,3,3)$.

As we saw in Section 9.3, there is a sense in which large games tend to have nonempty cores if only small coalitions matter. To formalize this proposition for the inner core, let us temporarily reinterpret N as a set of classes of players, rather than as the set of players itself. As at the end of Section 9.3, consider a *dynamic matching process* in which, for each i in N, new individuals of class i enter the matching process at some positive birth rate ρ_i. Every individual who enters the matching process is to be assigned to some coalition with at most one player of each class, and he leaves the matching process in this coalition. Suppose that $V(S)$ describes the set of payoff allocations that players in a coalition can get if S is the set of classes of the players in the coalition.

To characterize the output of such a dynamic matching process, a *matching plan* is any pair (μ, Y) such that $Y(\cdot)$ satisfies the feasibility condition (9.32), μ is in $\mathbf{R}_+^{L(N)}$, and μ satisfies the following *balance* condition:

$$(9.35) \qquad \sum_{S \supseteq \{i\}} \mu(S) = \rho_i, \quad \forall i \in N.$$

For any S in $L(N)$ and any i in S, $Y_i(S)$ denotes the expected payoff that a class-i player would get in this matching process if he were assigned to a coalition where the set of classes represented is S, and $\mu(S)$ denotes the rate at which such S coalitions are being formed in the matching process. The balance condition (9.35) asserts that the rate at which class-i players are assigned into coalitions to leave the matching process must equal the rate at which class-i players enter the matching process. The expected payoff to a class-i player in this matching plan is then

$$\sum_{S \supseteq \{i\}} \frac{\mu(S)Y_i(S)}{\rho_i} \, .$$

The following theorem asserts that inhibitive allocations can always be achieved when the game V is infinitely replicated and turned into such a dynamic matching process.

THEOREM 9.6. *Suppose that V is compactly generated and $\rho \in \mathbf{R}^N_{++}$. Then there exists some matching plan (μ, Y), satisfying feasibility (9.32) and balance (9.35), such that the vector*

$$\left(\sum_{S \supseteq \{i\}} \frac{\mu(S)Y_i(S)}{\rho_i} \right)_{i \in N}$$

is an inhibitive allocation for V.

Proof. Let Ω be the closed and bounded set within which V can be compactly generated. Let b be a number big enough so that

$$|x_i| + \rho_i + 1 < b, \quad \forall x \in \Omega, \forall i \in N.$$

If x is an allocation vector that is not strongly inhibitive, then define the set $f(x)$ such that $y \in f(x)$ iff there exists some randomized blocking plan (η, Y) such that (η, Y) is viable against x and

$$y_i = x_i + \sum_{S \supseteq \{i\}} \eta(S), \quad \forall i \in N.$$

Now define a correspondence $g : [-b,b]^N \twoheadrightarrow [-b,b]^N$ as follows. If x is strongly inhibitive, then let $g(x) = \{x - \rho\}$. If x is not inhibitive, then let $g(x) = f(x)$. Finally, if y is inhibitive but not strongly inhibitive, then let $f(x)$ be the smallest convex set that contains both $\{x - \rho\}$ and $f(x)$ as subsets. It can be verified that the range of g on $[-b,b]^N$ is contained in $[-b,b]^N$. If $x_i < -b + \rho_i$ for any i, then x is not inhibitive, because $v_i \geq -b + \rho_i$, so $\eta(\{i\}) = 1$ and $Y_i(\{i\}) = v_i$ would characterize a viable blocking plan (recall equation 9.28); and so any y in $g(x)$ must be in $f(x)$ and must satisfy $x_i \leq y_i \leq x_i + 1$. On the other hand, if $x_i > b - 1$ for some i, then there cannot be any viable blocking plan (η, Y) against x such that

$\Sigma_{S \supseteq \{i\}} \eta(S)$ is positive, so any y in $g(x)$ must have $y_i \leq x_i$. Furthermore, it can be checked that $g(x)$ is always a nonempty convex set and is upper-hemicontinuous.

Thus, by the Kakutani fixed-point theorem, there exists some x such that $x \in g(x)$. Such an x must be inhibitive but not strongly inhibitive, and there must exist some viable blocking plan (η, Y) against x such that the vector

$$\left(\sum_{S \supseteq \{i\}} \eta(S) \right)_{i \in N}$$

is proportional to ρ. That is, there must exist some positive number q such that $\Sigma_{S \supseteq \{i\}} \eta(S) = q\rho_i$ for every player i. So let $\mu(S) = \eta(S)/q$ for every S in $L(N)$. Then (μ, Y) is a matching plan that satisfies the balance condition (9.35), and, by the viability condition (9.33),

$$\sum_{S \supseteq \{i\}} \frac{\mu(S)Y_i(S)}{\rho_i} \geq x_i, \quad \forall i \in N.$$

But if x is inhibitive, then any vector that is higher in all components is also inhibitive. So the vector of expected payoffs $(\Sigma_{S \supseteq \{i\}}\mu(S)Y_i(S)/\rho_i)_{i \in N}$ is inhibitive. ■

9.9 Values without Transferable Utility

Just as the inner core of an NTU coalitional game V can be defined in terms of the cores of (slight perturbations of) the λ-transfer games generated by V, so NTU versions of the Shapley value and other solution concepts can be defined in terms of these λ-transfer games (see Shapley, 1969).

The Shapley NTU value can be defined for NTU coalitional games as follows. Let V be an NTU coalitional game, and suppose that V is compactly generated. For any λ in $\mathbf{R}^N_{++}$, let v^λ be the λ-transfer game generated by V. The Shapley value $\phi(v^\lambda)$ must be interpreted as an allocation of λ-weighted utility, because the coalitional worths in v^λ are sums of λ-weighted utilities for the players. Thus, we must divide each component of $\phi(v^\lambda)$ by the corresponding component of λ, to compute the allocation of payoffs in the original utility scales that corresponds to this allocation of λ-weighted utility. Let $\Phi(V,\lambda) = (\Phi_i(V,\lambda))_{i \in N}$ be the allocation vector computed in this way, so

$$\Phi_i(V,\lambda) = \frac{\phi_i(v^\lambda)}{\lambda_i}, \quad \forall i \in N, \ \forall \lambda \in \mathbf{R}^N_{++}.$$

It can be shown that, if V is compactly generated, then v^λ and $\Phi(V,\lambda)$ depend continuously on λ in $\mathbf{R}^N_{++}$. Furthermore, for any λ,

$$(9.36) \qquad \sum_{i \in N} \lambda_i \Phi_i(V,\lambda) = \sup_{x \in V(N)} \sum_{i \in N} \lambda_i x_i,$$

and if V is superadditive, then

$$(9.37) \qquad \Phi_i(V,\lambda) \geq v_i, \quad \forall i \in N.$$

Let $\Phi^*(V)$ denote the subset of $\mathbf{R}^N$ that is the closure of $\{\Phi(V,\lambda) | \lambda \in \mathbf{R}^N_{++}\}$. That is, $x \in \Phi^*(V)$ iff there exists some sequence $(\lambda(k))^\infty_{k=1}$ such that $x = \lim_{k \to \infty} \Phi(V,\lambda(k))$. (Taking this closure will be essential to proving the existence theorem below.)

If the Shapley value of a TU coalitional game is considered to be a fair and reasonable solution, then any feasible allocation x in $V(N)$ that equals $\Phi(V,\lambda)$ for some λ in $\mathbf{R}^N_{++}$ might similarly be considered as a fair solution to V, by a kind of "independence of irrelevant alternatives" argument. That is, an impartial arbitrator might reason as follows: "The players would have bargained to x if they believed that transfers of λ-weighted utility were possible; so the players should be willing to stay with x (which is feasible without transfers) even when such transfers are not possible." By extension, we can apply this argument to any feasible allocation that is in $\Phi^*(V)$ and so is arbitrarily close to allocations that would be fair by such λ-transfer standards. Finally, we might argue that, if an allocation x would be a reasonable solution if it were feasible, then any feasible allocation that is better than x for all players should be considered to be a reasonable solution. Thus, we define a *Shapley NTU value* of V to be any allocation y such that $y \in V(N)$ and there exists some vector x such that $x \in \Phi^*(V)$ and $x_i \leq y_i$ for every i in N. That is, the set of Shapley NTU values is

$$V(N) \cap \{x + z | x \in \Phi^*(V), z \in \mathbf{R}^N_+\}.$$

Under this definition, a sufficient condition for x to be a Shapley NTU value of V is that there exists some λ in $\mathbf{R}^N_{++}$ such that

$$x = \left(\frac{\phi_i(v^\lambda)}{\lambda_i} \right)_{i \in N} \quad \text{and} \quad x \in V(N).$$

When this condition is satisfied, we refer to λ as the vector of *natural scale factors* (or *natural utility weights*) supporting the Shapley NTU value x.

Remarkably, the Shapley NTU value generalizes both the Shapley value and the Nash bargaining solution. If V is equivalent to a TU coalitional game v, in the sense of equation (9.29), then the supremum in (9.36) will be finite only when λ is a positive scalar multiple of the vector of all ones $(1,1,\ldots,1)$; so the original Shapley value $\phi(v)$ will be the only vector in $\Phi^*(V)$ and the unique Shapley NTU value of V. On the other hand, if V is equivalent to an essential two-person bargaining problem $(F,(v_1,v_2))$, in the sense that $N = \{1,2\}$, v_1 and v_2 are as defined in (9.28), and $F = V(\{1,2\})$, then by Theorem 8.2 the Nash bargaining solution of $(F,(v_1,v_2))$ will be the unique Shapley NTU value of V.

An NTU coalitional game may have more than one Shapley NTU value, but a general existence theorem can be proved.

THEOREM 9.7. *Suppose that V is compactly generated and superadditive. Then there exists at least one Shapley NTU value of V.*

Proof. Actually, we only need to consider utility-weight vectors λ where the components sum to 1. Let ε be any small positive number less than $1/|N|$, and let

$$\Delta^\varepsilon(N) = \left\{\lambda \in \mathbf{R}^N \Big| \sum_{i\in N} \lambda_i = 1, \lambda_j \geq \varepsilon, \forall j \in N\right\}.$$

Let Ω be a closed, convex, and bounded subset of $\mathbf{R}^N$ such that, for any x in $V(N)$ there exists some y in $\Omega \cap V(N)$ such that $y_i \geq x_i$ for all i.

For any pair (λ,y), define the set $f_\varepsilon(\lambda,y)$ such that $(\mu,x) \in f_\varepsilon(\lambda,y)$ iff

$$x \in \underset{z\in\Omega\cap V(N)}{\text{argmax}} \sum_{i\in N} \lambda_i z_i,$$

$$\mu \in \Delta^\varepsilon(N),$$

and, for each player i,

$$\text{if } \Phi_i(V,\lambda) - y_i < \max_{j\in N} (\Phi_j(V,\lambda) - y_j), \text{ then } \mu_i = \varepsilon.$$

It can be verified that the correspondence $f_\varepsilon : \Delta^\varepsilon(N) \times (\Omega \cap V(N)) \longrightarrow\!\!\!\rightarrow \Delta^\varepsilon(N) \times (\Omega \cap V(N))$ is upper-hemicontinuous and nonempty convex-valued, so it satisfies all of the conditions of the Kakutani fixed-point theorem. Thus, there exists some $(\lambda(\varepsilon), y(\varepsilon))$ such that

$$(\lambda(\varepsilon), y(\varepsilon)) \in f_\varepsilon(\lambda(\varepsilon), y(\varepsilon)).$$

Let $S = \text{argmax}_{j \in N} \left(\Phi_j(V, \lambda(\varepsilon)) - y_j(\varepsilon) \right)$. Using (9.36) yields

$$
\begin{aligned}
0 &= \sum_{i \in N} \lambda_i(\varepsilon)(\Phi_i(V, \lambda) - y_i(\varepsilon)) \\
&= (1 - |N \backslash S|\varepsilon) \max_{j \in N} \left(\Phi_j(V, \lambda(\varepsilon)) - y_j(\varepsilon) \right) \\
&\quad + \sum_{i \in N \backslash S} \varepsilon \left(\Phi_i(V, \lambda(\varepsilon)) - y_i(\varepsilon) \right).
\end{aligned}
$$

So, using (9.37) yields

$$
\max_{j \in N} \left(\Phi_j(V, \lambda(\varepsilon)) - y_j(\varepsilon) \right) \leq \sum_{i \in N} \frac{\varepsilon \, \max\{0, (y_i(\varepsilon) - v_i)\}}{1 - |N|\varepsilon}.
$$

A sequence $(\varepsilon_k)_{k=1}^\infty$ can be found such that $\lim_{k \to \infty} \varepsilon_k = 0$ and $(y(\varepsilon_k))_{k=1}^\infty$ is a convergent sequence with some limit $\bar{y}$ in the closed and bounded set $\Omega \cap V(N)$. Along this sequence, the right-hand side of the above inequality converges to 0. Thus, a subsequence can be found along which, for each player j, the numbers $\Phi_j(V, \lambda(\varepsilon_k))$ also converge to some limit $\bar{x}_j$ such that $v_j \leq \bar{x}_j \leq \bar{y}_j$. So the allocation $\bar{x}$ is in $\Phi^*(V)$, and the feasible allocation $\bar{y}$ is a Shapley NTU value of V. ∎

Aumann (1985) derived the Shapley NTU value from a set of axioms similar to the axioms for the Shapley value (in the TU case) and for the Nash bargaining solution. We present here a modified version of his derivation.

Let $\partial_+ V(N)$ be the upper boundary of the closed convex set $V(N)$, that is, the set of all weakly Pareto-efficient allocations in $V(N)$. Because $V(N)$ is convex, it can be shown that, for any allocation x in $\partial_+ V(N)$, there exists at least one vector λ in $\Delta(N)$ such that

$$(9.38) \qquad x \in \text{argmax}_{y \in V(N)} \sum_{i \in N} \lambda_i y_i.$$

Such a vector λ is called a *supporting vector* for $V(N)$ at x. (The general existence of nonzero supporting vectors for a convex set at any point on its boundary is known as the *supporting hyperplane theorem*. This result is a straightforward corollary of the separating hyperplane theorem.) Geometrically, the supporting vector λ that satisfies (9.38) is orthogonal to a tangent plane to $V(N)$ at x. If there is a corner of $V(N)$ at x, then there may be many such supporting vectors in $\Delta(N)$ for $V(N)$ at x. Aumann's smoothness condition rules out the possibility of such corners. In addition, he uses a condition that all weakly Pareto-efficient allocations are also strongly Pareto efficient. We can subsume these two assumptions by saying that $V(N)$ is *positively smooth* iff, for every x in $\partial_+V(N)$, there exists a unique supporting vector λ in $\Delta^0(N)$ that satisfies condition (9.38). (Recall from Section 1.6 that $\Delta^0(N) = \Delta(N) \cap \mathbf{R}^N_{++}$.)

For any two NTU coalitional games V and W, and any number p between 0 and 1, let $pV + (1 - p)W$ be defined so that, for any coalition S, $(pV + (1 - p)W)(S)$ is the closure in $\mathbf{R}^S$ of the set

$$\{py + (1 - p)z | y \in V(S), z \in W(S)\}.$$

For any vector λ in $\mathbf{R}^N_{++}$, let λV denote the NTU coalitional game such that

$$(\lambda V)(S) = \{(\lambda_i y_i)_{i \in S} | y \in V(S)\}, \quad \forall S \in L(N).$$

Let $\Psi(\cdot)$ be a mapping that determines a set of "solutions" $\Psi(V)$ for any NTU coalitional game V. We can then consider the following axioms.

AXIOM 9.4 (EFFICIENCY). *For any NTU coalitional game V, $\Psi(V) \subseteq \partial_+V(N)$.*

AXIOM 9.5 (SCALE COVARIANCE). *For any λ in $\mathbf{R}^N_{++}$ and any NTU coalitional game V, $\Psi(\lambda V) = \{(\lambda_i x_i)_{i \in N} | x \in \Psi(V)\}$.*

AXIOM 9.6 (EXTENSION OF SHAPLEY VALUE). *If V is equivalent to a TU coalitional game v, in the sense of (9.29), then $\Psi(V) = \{\phi(v)\}$.*

AXIOM 9.7 (CONDITIONAL LINEARITY). *Suppose that V and W are positively smooth NTU coalitional games, p is a number between 0 and 1, $pV + (1 - p)W$ satisfies the conditions (9.26)–(9.28) of a NTU coalitional game, $y \in \Psi(V)$, $z \in \Psi(W)$, and $py + (1 - p)z \in \partial_+(pV + (1 - p)W)(N)$. Then $py + (1 - p)z \in \Psi(pV + (1 - p)W)$.*

Conditional linearity is a weaker version of a natural generalization of the linearity axiom for the Shapley value in the TU case. (Actually, Aumann used a conditional additivity axiom that is slightly simpler. Here we use linearity instead of additivity to be more consistent with Section 9.4.) Axiom 9.7 asserts that, if randomizing between the solutions of V and W is efficient in $pV + (1 - p)W$, then it is a solution of $pV + (1 - p)W$. It can be shown that these four axioms are satisfied when $\Psi(V)$ denotes the set of Shapley NTU values of V. (Without the positive smoothness condition, however, the Shapley NTU value would not satisfy conditional linearity.)

THEOREM 9.8. *Suppose that $\Psi(\cdot)$ satisfies the above four axioms. Let V be any positively smooth NTU coalitional game, and let x be any allocation in $\Psi(V)$. Then x is a Shapley NTU value of V.*

Proof. Let λ be the unique supporting vector in $\Delta^0(N)$ for $V(N)$ at x. By scale covariance, $(\lambda_i x_i)_{i \in N}$ is in $\Psi(\lambda V)$. For each coalition S, let

$$W(S) = \left\{ z \in \mathbf{R}^S \,\middle|\, \sum_{i \in S} z_i \leq v^\lambda(S) \right\},$$

where v^λ is the λ-transfer game generated by V. By extension of the Shapley value, $\Psi(W) = \{\phi(v^\lambda)\}$. It can be shown that

$$.5(\lambda V) + .5W = W.$$

Furthermore, $.5(\lambda_i x_i)_{i \in N} + .5\phi(v^\lambda) \in \partial_+ W(N)$, because $\Sigma_{i \in N} \lambda_i x_i = v^\lambda(N)$. So conditional additivity implies that

$$.5(\lambda_i x_i)_{i \in N} + .5\phi(v^\lambda) = \phi(v^\lambda);$$

and so $x_i = \phi_i(v^\lambda)/\lambda_i$ for every player i. Thus, x is a Shapley NTU value. ∎

Other definitions of NTU values have been suggested by Harsanyi (1963) and Owen (1972). The Harsanyi NTU value may be described as follows. Suppose that V is a comprehensive NTU coalitional game. For any S, let

$$\partial_+ V(S) = \{x \in V(S) \,|\, \forall y \in \mathbf{R}^S, \text{ if } y_i > x_i \,\forall i \in S, \text{ then } y \notin V(S)\}.$$

An allocation x is a Harsanyi NTU value if there exists some vector λ in $\mathbf{R}^N_{++}$ and some function $X(\cdot)$ such that

$$X(S) = (X_i(S))_{i \in S} \in \partial_+ V(S), \quad \forall S \in L(N),$$

$$\lambda_i X_i(S) - \lambda_i X_i(S-j) = \lambda_j X_j(S) - \lambda_j X_j(S-i), \quad \forall S \in L(N),$$

$$\forall i \in S, \quad \forall j \in S,$$

$$x = X(N) \in \underset{y \in V(N)}{\operatorname{argmax}} \sum_{i \in N} \lambda_i y_i.$$

(The original definition by Harsanyi, 1963, was somewhat more complicated, but this formulation is equivalent; see Myerson, 1980.) The Harsanyi NTU value also generalizes both the TU Shapley value and the two-player Nash bargaining solution, and it has been axiomatically derived by Hart (1985b).

For example, consider the *Banker Game* from Owen (1972), where

$$V(\{i\}) = \{(x_i) | x_i \leq 0\}, \text{ for } i = 1,2,3,$$

$$V(\{1,2\}) = \{(x_1,x_2) | x_1 + 4x_2 \leq 100, x_1 \leq 100\},$$

$$V(\{1,3\}) = \{(x_1,x_3) | x_1 \leq 0, x_3 \leq 0\},$$

$$V(\{2,3\}) = \{(x_2,x_3) | x_2 \leq 0, x_3 \leq 0\},$$

$$V(\{1,2,3\}) = \{(x_1,x_2,x_3) | x_1 + x_2 + x_3 \leq 100\}.$$

The idea is that player 1 can get \$100 with the help of player 2. To reward player 2, player 1 can send him money; but without player 3, there is a 75% chance of losing the money that is sent. Player 3 is a banker who can prevent such loss in transactions. How much should player 1 pay to player 2 for his help and to player 3 for his banking services?

The unique Shapley NTU value for this game is $(50,50,0)$, supported by the natural scale factors $\lambda = (1,1,1)$. With these weights,

$$v^\lambda(\{1,2\}) = 100 = v^\lambda(\{1,2,3\}),$$

because $(100,0)$ is feasible for the coalition $\{1,2\}$, and every other coalition S gets $v^\lambda(S) = 0$; so $\phi(v^\lambda) = (50,50,0) \in V(N)$.

The unique Harsanyi NTU value is $(40,40,20)$, supported by $\lambda = (1,1,1)$ and

$$X_i(\{i\}) = 0, \text{ for } i = 1,2,3,$$

$$X_1(\{1,2\}) = X_2(\{1,2\}) = 20,$$

$$X_1(\{1,3\}) = X_2(\{1,3\}) = 0,$$

$$X_2(\{2,3\}) = X_3(\{2,3\}) = 0,$$

$$X_1(\{1,2,3\}) = 40 = X_2(\{1,2,3\}), \quad X_3(\{1,2,3\}) = 20.$$

The Harsanyi NTU value gives less to players 1 and 2, because it takes account of the fact that the {1,2} coalition could achieve the maximal sum of payoffs only by choosing an allocation that would be rather unfair to player 2. If the {1,2} coalition were constrained to choose a feasible allocation that gave them equal gains in λ-weighted utility over what each could get alone, then they would have to settle for a sum of payoffs of at most 40. The Owen NTU value also gives the banker a positive payoff and gives more to player 1 than to player 2.

There is a possible rationale for the Shapley NTU value giving 0 to the banker. Getting 0, player 3 is indifferent between accepting the outcome specified by the Shapley NTU value or not, so it is not unreasonable to assume that he will probably accept it. (Think of his payoff in the Shapley NTU value as positive but infinitesimal, whereas his cost of providing banking services is 0.) So suppose that there is only some small probability q that player 3 will refuse to accept his NTU-value allocation and will break up the grand coalition. As long as $q \leq \frac{1}{2}$, players 1 and 2 can accommodate this possibility with no loss of expected utility to either of them and no reduction in player 3's payoff when he cooperates. They can simply plan to choose $(100,0)$ if player 3 rejects the grand coalition, and $(100 - 50/(1 - q), 50/(1 - q), 0)$ if player 3 agrees to cooperate.

Now let i equal 1 or 2, and suppose instead that there is a small probability q that player i would reject the Shapley NTU value (even though it is better for him than what he could get alone) and break up the grand coalition. In this case, the expected payoffs to the other two players could not sum to more than $50(1 - q)$ without reducing player i's payoff in the case of agreement. That is, a low-probability threat by either player 1 or player 2 would cause real losses in the expected payoffs of the other players, and in a symmetric manner; but such a threat by player 3 would have no effect on expected payoffs if it were anticipated correctly. In this sense, players 1 and 2 have equal bargaining power and player 3 has none, so $(50,50,0)$ may be a reasonable bargaining solution.

In general, let x be an efficient payoff allocation for the grand coalition, suppose that λ in $\mathbf{R}_{++}^N$ is a supporting vector for $V(N)$ at x, and suppose that V is positively smooth. Then, to a first-order approximation, small transfers of λ-weighted utility are feasible near x for the players in the grand coalition. That is, for any sufficiently small number ε, if player i allowed his payoff to be reduced from x_i to $x_i - \varepsilon/\lambda_i$, then the payoff of any other player j could be increased from x_j to $x_j + \varepsilon/\lambda_j$, minus some "transactions cost" that is very small in proportion to ε, without affecting any other player's payoff or leaving the feasible set $V(N)$.

Now suppose that the players are expected to accept the allocation x almost surely, except that, with some small probability, a smaller coalition S might have to choose something feasible for themselves. Suppose also that the plans for what S would do in this case can be agreed on by the members of S before they learn whether the grand coalition will cooperate or not. In such a situation, the promise of a small transfer of λ-weighted utility in the likely event that N cooperates would be worth a much larger transfer of λ-weighted utility in the unlikely event that S has to act alone. Thus, when the members of coalition S decide what to do if they must act alone, λ-weighted utility is effectively transferable among themselves, where the medium of exchange is a promise to make a small feasible reallocation away from x (without affecting the payoffs to players in $N \backslash S$) in the much more likely event that x is accepted. By this argument, it may be appropriate to analyze this bargaining game as if λ-weighted utility were transferable for any coalition S. If such analysis confirms the initial assumption that x would be a reasonable outcome to the bargaining process when the grand coalition N cooperates (i.e., if $\lambda_i x_i = \phi_i(v^\lambda)$ for every i), then it is reasonable to argue that x should be a cooperative solution for the NTU game V. In this sense, the Shapley NTU values are reasonable cooperative solutions for V.

Roth (1980) and Shafer (1980) have studied other games where Shapley NTU values appear counterintuitive. For example, consider the following NTU coalitional game, due to Roth (1980).

$$V(\{i\}) = \{(x_i)|x_i \leq 0\}, \text{ for } i = 1,2,3,$$
$$V(\{1,2\}) = \{(x_1,x_2)|x_1 \leq \tfrac{1}{2}, x_2 \leq \tfrac{1}{2}\},$$
$$V(\{1,3\}) = \{(x_1,x_3)|x_1 \leq \tfrac{1}{4}, x_3 \leq \tfrac{3}{4}\},$$
$$V(\{2,3\}) = \{(x_2,x_3)|x_2 \leq \tfrac{1}{4}, x_3 \leq \tfrac{3}{4}\},$$
$$V(\{1,2,3\}) = \{(x_1,x_2,x_3)|x_1 + x_2 + x_3 \leq 1, x_1 \leq \tfrac{1}{2}, x_2 \leq \tfrac{1}{2}\}.$$

For this game, the unique Shapley NTU value is $(\tfrac{1}{3},\tfrac{1}{3},\tfrac{1}{3})$, supported by the natural scale factors $\lambda = (1,1,1)$.

Roth argued that the only reasonable cooperative outcome for this game is $(\tfrac{1}{2},\tfrac{1}{2},0)$, because $\tfrac{1}{2}$ is the best payoff that player 1 or player 2 could get in any coalition, and they can both get $\tfrac{1}{2}$ without any help from player 3. Thus, there seems to be no reason for any disagreement between players 1 and 2, and no reason for either to bargain with player 3. Notice that this argument depends on the fact that there is a corner in $V(\{1,2,3\})$ at $(\frac{1}{2},\frac{1}{2},0)$, so this argument applies because this game violates the smoothness assumption in Theorem 9.8.

One possible response to the difficulty with the Shapley NTU value in this example is to study NTU values with cooperation structures and theories about how such cooperation structures might be endogenously determined. For this game, a reasonable theory of endogenous cooperation structure should specify that players 1 and 2 would cooperate together without player 3.

There are alternative ways to define Shapley NTU values for strategic-form games directly, without working through the NTU coalitional form. Let $\Gamma = (N, (C_i)_{i \in N}, (u_i)_{i \in N})$ be any finite game in strategic form. For any vector λ in $\mathbf{R}_{++}^N$, let the λ-rescaled version of Γ be

$$\lambda * \Gamma = (N, (C_i)_{i \in N}, (\lambda_i u_i)_{i \in N}).$$

That is, $\lambda * \Gamma$ differs from Γ only in that the utility function of each player i is multiplied by λ_i. Let w^λ denote the representation in coalitional form that we would compute for the strategic-form game $\lambda * \Gamma$ with transferable utility, according to any one of the definitions given in Section 9.2. For example, using the minimax representation gives us

$$w^\lambda(S) = \min_{\sigma_{N \backslash S} \in \Delta(C_{N \backslash S})} \max_{\sigma_S \in \Delta(C_S)} \sum_{i \in S} \lambda_i u_i(\sigma_S, \sigma_{N \backslash S}).$$

Then let

$$\Phi_i(\Gamma, \lambda) = \frac{\phi_i(w^\lambda)}{\lambda_i}, \quad \forall i \in N.$$

Then, as before, let $\Phi^*(\Gamma)$ denote the closure in $\mathbf{R}^N$ of the set

$$\{\Phi(\Gamma, \lambda) \,|\, \lambda \in \mathbf{R}_{++}^N\}.$$

Then we can say that x is a Shapley NTU value of Γ iff there exists some joint strategy σ_N in $\Delta(C_N)$ and there exists some allocation y in $\Phi^*(\Gamma)$ such that

$$x_i = u_i(\sigma_N) \geq y_i, \quad \forall i \in N.$$

The general existence of Shapley NTU values for finite strategic-form games can be proved, using conditions similar to (9.36) and (9.37), as in the proof of Theorem 9.7.

Exercises

Exercise 9.1. Consider the following four-person game in coalitional form, with transferable utility.

$$v(\{i\}) = 0, \quad \forall i \in \{1,2,3,4\},$$

$$v(\{1,2\}) = v(\{1,3\}) = v(\{2,4\}) = v(\{3,4\}) = 1,$$

$$v(\{1,4\}) = v(\{2,3\}) = 0,$$

$$v(\{2,3,4\}) = v(\{1,3,4\}) = v(\{1,2,4\}) = 1,$$

$$v(\{1,2,3\}) = 2 = v(\{1,2,3,4\}).$$

(Notice that the worth of each coalition except $\{1,2,3\}$ is equal to the number of disjoint pairs that consist of one player from $\{1,4\}$ and one player from $\{2,3\}$ that can be formed among the coalition's members.)

a. Show that the core of this game consists of a single allocation vector.

b. Compute the Shapley value of this game. (Notice the symmetry between players 2 and 3.)

c. Suppose that the worth of $\{1,2,3\}$ were changed to $v(\{1,2,3\}) = 1$. Characterize the core of the new game, and show that all of the new allocations in the core are strictly better for player 1 than the single allocation in the core of the original game.

d. Compute the Shapley value of the new game from (c).

Exercise 9.2. Prove that, if v is the minimax representation of a strategic-form game, then v must be superadditive.

Exercise 9.3. Suppose that there are three firms producing the same product. Firm 1 is a small firm and can produce either zero units or one unit per day. Firms 2 and 3 are large firms, and each can produce either two or three units per day. The market price P depends on the total daily output, according to the formula

$$P = 8 - (c_1 + c_2 + c_3),$$

where c_i is the daily output of firm i. The transferable payoff to firm i is its daily revenue Pc_i.

a. Model this game in strategic form.

b. Calculate the minimax representation in coalitional form of this game with transferable utility. Calculate the Shapley value of this coalitional game and check whether it is in the core.

c. Calculate the rational-threats representation in coalitional form of this game with transferable utility. Is this coalitional game superadditive? Calculate the Shapley value of this coalitional game. Is this Shapley value individually rational, in the sense that each player i gets at least the worth of his one-person coalition $v(\{i\})$?

d. What are the defensive-equilibrium representations in coalitional form of this game with transferable utility?

Exercise 9.4. A *simple game* is a game in coalitional form, with transferable utility, such that $v(N) = 1$, where N is the grand coalition of all players, $v(\{i\}) = 0$ for every player i in N, and, for each other coalition S, $v(S)$ equals either 0 or 1. Player i is a *veto player* for a simple game v iff

$$v(S) = 0, \quad \forall S \subseteq N - i.$$

Show that the core of any simple game can be characterized in terms of its set of veto players. (HINT: Your characterization should imply that the core is empty if there are no veto players.)

Exercise 9.5. The Executive Council of a major international organization has five members. Countries 1 and 2 are big countries; and countries 3, 4, and 5 are small countries. The Executive Council can approve an action on any issue relating to international security provided both big countries and at least one small country vote for the proposed action. The power structure on this Council may be described mathematically by a coalitional game in which any coalition that can approve an action (by the votes of its members alone) has worth 1, and any coalition that cannot approve an action has worth 0.

a. What is the core of this game?

b. What is the Shapley value of this game?

c. Suppose that the three small countries form closer relations among themselves in the Executive Council, so there is an effective nested coalition structure $\{\{1\},\{2\},\{3,4,5\}\}$. What is the Owen value of this game with this coalition structure? Does the union help the small countries, according to the Owen value?

d. Compute the Owen value of this game with the nested coalition structure $\{\{1,2\},\{3,4,5\}\}$. If the small countries form a union, do the big countries gain by also forming a union, according to the Owen value?

Exercise 9.6.   Consider a three-person coalitional game with transferable utility in which $v(\{1\}) = v(\{2\}) = v(\{3\}) = 0$, $v(\{1,2\}) = 8$, $v(\{1,3\}) = v(\{2,3\}) = 5$, $v(\{1,2,3\}) = 9$.

a. Find the Shapley value and the core of this game.

b. Compute the fair allocation rule ψ for this game, and verify that it satisfies condition (9.21).

Exercise 9.7.   Three cities are building a district water system to utilize an existing reservoir. All cities must be connected to the reservoir by a network of pipes. The costs of laying a pipeline (in millions of dollars) between each pair of cities, and between each city and the reservoir, are as follows.

Site 1	City 1	City 2	City 3
Reservoir	18	21	27
City 1		15	12
City 2			24

Based on these costs, the cheapest way to build the system is to build a pipeline from the reservoir to City 1, and then to build pipelines from City 1 to City 2 and from City 1 to City 3, for a total cost of $18 + 15 + 12 = 45$. The game-theoretic problem is to determine how this cost should be divided among the cities, viewed as players in a game.

a. Model this problem as a three-person game in coalitional form with transferable utility, where the players are the cities, and the worth of each coalition S is minus the cost of the cheapest pipeline system that would connect only the cities in S to the reservoir. For example, $v(\{1,2,3\}) = -45$.

b. It has been proposed that each city should pay for the pipeline flowing into it, regardless of whether other cities also use this pipeline. This would lead to the allocation $(-18, -15, -12)$. (All payoffs are negative, because they represent costs that must be paid.) Show that this allocation is in the core.

c. Compute the Shapley value of this game. Is it in the core?

d. Express this game as a linear combination of the simple carrier games, defined by equation (9.10).

Exercise 9.8.   Given a finite strategic-form game $\Gamma = (N, (C_i)_{i \in N}, (u_i)_{i \in N})$, let the NTU coalitional games V^α and V^β be respectively the assurable representation of Γ and the unpreventable representation of Γ.

a. Show that, for any coalition S, $V^\alpha(S)$ and $V^\beta(S)$ are convex sets.

b. Prove that V^α is superadditive.

c. Prove that V^β is superadditive. (HINT: You may use the Kakutani fixed-point theorem.)

Exercise 9.9. Suppose that, in the conditional linearity axiom discussed in Section 9.9, we dropped the requirement that V and W be positively smooth. Construct an example to show that the Shapley NTU value would not satisfy this stronger axiom.

Exercise 9.10. Consider the strategic-form game in Table 9.3, without transferable utility.

a. For this game, complete the construction of the assurable representation V^α and the unpreventable representation V^β in NTU coalition form.

b. Show that the allocation $(1,2,2)$ is in the core of V^α and V^β but is not in the inner core of V^α or V^β. To prove that it is not in the inner core, find a randomized blocking plan such that each player's conditionally expected payoff, when invited to join a blocking coalition, is strictly greater than what he gets in $(1,2,2)$. (HINT: It can be done by assigning positive probabilities to the coalitions $\{1,2,3\}$, $\{1,2\}$, and $\{1,3\}$.)

c. Compute Shapley NTU values for V^α and V^β.

d. As suggested at the end of Section 9.9, let w^λ denote the minimax representation of the λ-rescaled version of the strategic-form game in Table 9.3. Letting $\lambda_1 = \lambda_2 = \lambda_3 = 1$, compute the TU coalitional game w^λ, and show that its Shapley value $\phi(w^\lambda)$ is feasible for the players in the original game without transferable utility.

Exercise 9.11. Characterize the core and the inner core of the Banker Game, from Section 9.9. Find a randomized blocking plan that would be viable against the allocation $(41,41,21)$, and thus show that the Harsanyi NTU value $(40,40,20)$ is not inhibitive for this game.

Bibliographic Note

Owen (1982) and Lucas (1972) survey a wide range of solution concepts for games in characteristic function or coalitional form. In the past decade or so, most interest in this area has been focused on the Shapley value and closely related solution concepts: see Kalai and Samet (1985,

1987), Hart and Mas-Colell (1989), Young (1985), and the survey volume of Roth (1988). Bennett (1983) has studied solution concepts that are more closely related to the core.

Lucas and Thrall (1963) defined a generalized coalitional form, called *partition function form*, in which the worth of a coalition may depend on how the other players are assigned to coalitions (see also Myerson, 1978b).

For applications of cooperative game theory to the study of voting, see Shapley and Shubik (1954), Peleg (1984), and Moulin (1988). Values of nonatomic games have been applied to cost allocation problems by Billera, Heath, and Ranaan (1978) and others; see Tauman (1988).

10

Cooperation under Uncertainty

10.1 Introduction

In this chapter we consider the elements of a theory of cooperative games with incomplete information. Adding incomplete information raises many new conceptual questions for cooperative game theory. In particular, to understand cooperative games with incomplete information, we will need to consider not only questions of how different players should compromise with each other but also questions of how a single player should compromise between the goals of his true type and the goals of his other possible types, to maintain an inscrutable facade in negotiations.

In this chapter, to simplify the analysis, we consider Bayesian collective-choice problems and Bayesian bargaining problems, but we do not attempt a general treatment of Bayesian games with general incentive constraints (as formulated in Section 6.3). Recall from Section 6.4 that a Bayesian collective-choice problem may be written in the form

(10.1) $(N, C, (T_i)_{i \in N}, (p_i)_{i \in N}, (u_i)_{i \in N})$,

where N is the set of players, C is the set of possible *outcomes* or options that the players can jointly choose among, p_i is the probability function that represents player i's beliefs about other players' types as a function of his own type, and u_i is player i's utility payoff function.

To simplify formulas, we assume in most of this chapter (except in Section 10.6) that the players' beliefs are consistent with some prior probability distribution in $\Delta(T)$ under which the players' types are independent random variables. (Recall $T = \times_{i \in N} T_i$.) That is, we assume that for every player i there exists a probability distribution $\bar{p}_i$ in $\Delta(T_i)$, such that $\bar{p}_i(t_i)$ is the prior marginal probability that player i's type will be t_i and

(10.2) $p_i(t_{-i}|t_i) = \prod_{j\in N-i} \bar{p}_j(t_j), \quad \forall i \in N, \; \forall t_{-i} \in T_{-i}, \; \forall t_i \in T_i.$

(As usual, we let $T_{-i} = \times_{j\in N-i} T_j$, $t_{-i} = (t_j)_{j\in N-i}$, and $p_i(t_{-i}|t_i)$ is the probability that player i would assign to the event that t was the profile of all players' types when he knows only that his own type is t_i.) In the prior probability distribution itself, the probability that some t in T will be the true combination of types for the players is

$$p(t) = \prod_{i\in N} \bar{p}_i(t_i).$$

We assume here that all types have positive probability, so

$$\bar{p}_i(t_i) > 0, \quad \forall i \in N, \; \forall t_i \in T_i.$$

In Section 2.8, it was shown that any finite Bayesian game is equivalent to a Bayesian game that satisfies this condition of consistency with an independent prior distribution.

A *mediation plan* or *mechanism* for the Bayesian collective-choice problem (10.1) is any function $\mu : T \to \Delta(C)$. We assume that players' types are not verifiable by a mediator, so an *incentive-compatible* mechanism must satisfy the following informational incentive constraints:

(10.3) $U_i(\mu|t_i) \geq U_i^*(\mu, s_i|t_i), \quad \forall i \in N, \; \forall t_i \in T_i, \; \forall s_i \in T_i,$

where

(10.4) $U_i(\mu|t_i) = \sum_{t_{-i}\in T_{-i}} \sum_{c\in C} \left(\prod_{j\in N-i} \bar{p}_j(t_j) \right) \mu(c|t) u_i(c,t),$

(10.5) $U_i^*(\mu, s_i|t_i) = \sum_{t_{-i}\in T_{-i}} \sum_{c\in C} \left(\prod_{j\in N-i} \bar{p}_j(t_j) \right) \mu(c|t_{-i}, s_i) u_i(c,t).$

As defined in Section 6.4, a *Bayesian bargaining problem* Γ^β is a Bayesian collective-choice problem together with a specification of the disagreement outcome d^* that will occur if the players fail to agree on a mechanism. With the independent prior assumption, we can write

(10.6) $\Gamma^\beta = (N, C, d^*, (T_i)_{i\in N}, (\bar{p}_i)_{i\in N}, (u_i)_{i\in N}).$

For most applications, it is convenient to define utility scales so that the disagreement outcome always gives zero payoffs

(10.7) $u_i(d^*,t) = 0, \quad \forall i \in N, \ \forall t \in T,$

and we shall apply this convention throughout this chapter. Thus, a mechanism μ is said to be *individually rational* iff it satisfies the participation (or individual-rationality) constraints

(10.8) $U_i(\mu|t_i) \geq 0, \quad \forall i \in N, \ \forall t_i \in T_i,$

so no type of any player would expect to do worse under μ than in the disagreement outcome.

For a Bayesian bargaining problem, we say that a mechanism is *incentive feasible* iff it is both incentive-compatible and individually rational, in the sense of conditions (10.3) and (10.8). For a Bayesian collective-choice problem, where there are no participation constraints, we use "incentive feasibility" and "incentive compatibility" synonymously, so a mechanism is *incentive feasible* for a Bayesian collective-choice problem iff it satisfies condition (10.3).

10.2 Concepts of Efficiency

The concept of Pareto efficiency is central in cooperative game theory. To develop a theory of cooperation under uncertainty, the first step must be to extend the definition of Pareto efficiency to games with incomplete information.

A game theorist or a mediator who analyzes the Pareto efficiency of behavior in a game with incomplete information must use the perspective of an outsider, so he cannot base his analysis on the players' private information. An outsider may be able to say how the outcome will depend on the players' types, but he cannot generally predict the actual outcome without knowing the players' actual types. That is, he can know the mechanism but not its outcome. Thus, Holmstrom and Myerson (1983) argued that the concept of efficiency for games with incomplete information should be applied to mechanisms, rather than to outcomes, and the criteria for determining whether a particular mechanism μ is efficient should depend only on the commonly known structure of the game, not on the privately known types of the individual players.

Thus, a definition of Pareto efficiency in a Bayesian collective-choice problem or a Bayesian bargaining problem must look something like this:

A mechanism is efficient iff no other feasible mechanism can be found that might make some other individuals better off and would certainly not make other individuals worse off.

However, this definition is ambiguous in several ways. In particular, we must specify what information is to be considered when determining whether an individual is "better off" or "worse off."

One possibility is to say that an individual is made worse off by a change that decreases his expected utility payoff as would be computed before his own type or any other individuals' types are specified. This standard is called the *ex ante welfare criterion*. Thus, we say that a mechanism v is *ex ante Pareto superior* to another mechanism μ iff

$$\sum_{t \in T} \sum_{c \in C} p(t)v(c|t)u_i(c,t) \geq \sum_{t \in T} \sum_{c \in C} p(t)\mu(c|t)u_i(c,t), \quad \forall i \in N,$$

and this inequality is strict for at least one player in N. Notice that

$$\sum_{t \in T} \sum_{c \in C} p(t)\mu(c|t)u_i(c,t) = \sum_{t_i \in T_i} \bar{p}_i(t_i)U_i(\mu|t_i).$$

Another possibility is to say that an individual is made worse off by a change that decreases his conditionally expected utility, given his own type (but not given the type of any other individuals). An outside observer, who does not know any individual's type, would then say that a player i "would certainly not be made worse off (by some change of mechanism)" in this sense if this conditionally expected utility will not be decreased (by the change) for any possible type of player i. This standard is called the *interim welfare criterion*, because it evaluates each player's welfare after he learns his own type but before he learns any other player's type. Thus, we say that a mechanism v is *interim Pareto superior* to another mechanism μ iff

$$U_i(v|t_i) \geq U_i(\mu|t_i), \quad \forall i \in N, \ \forall t_i \in T_i,$$

and this inequality is strict for at least one type of one player in N.

Yet another possibility is to say that an individual is made worse off by a change that decreases his conditionally expected utility, given the types of all individuals. An outside observer would then say that a player "would certainly not be made worse off" in this sense if his conditionally expected utility would not be decreased for any possible combination of types for all the players. This standard is called the *ex post welfare criterion*, because it uses the information that would be available after all individuals have revealed their types. Thus, we say that a mechanism ν is *ex post Pareto superior* to another mechanism μ iff

$$\sum_{c \in C} \nu(c|t)u_i(c,t) \geq \sum_{c \in C} \mu(c|t)u_i(c,t), \quad \forall i \in N, \ \forall t \in T,$$

and this inequality is strict for at least one player in N and at least one possible combination of types t in T such that $p(t) > 0$.

Another ambiguity in the above verbal definition of efficiency is in the term "feasible." One sense of this term, which can be called *classical feasibility*, is to consider that any function from T to $\Delta(C)$, selecting a randomized joint strategy for each possible combination of the players' types, is feasible. However, the revelation principle implies that a mechanism cannot be implemented, by any equilibrium of a communication game induced by any communication system, unless the mechanism is incentive compatible (and, where relevant, individually rational). Thus, it is appropriate to recognize the unavoidability of incentive constraints by defining the "feasible" mechanisms to be the incentive-feasible mechanisms (that is, the incentive-compatible mechanisms for a Bayesian collective-choice problem, or the incentive-compatible individually rational mechanisms for a Bayesian bargaining problem).

Given any concept of feasibility, the three welfare criteria (ex ante, interim, ex post) give rise to three different concepts of efficiency. For any set of mechanisms F (to be interpreted as the set of "feasible" mechanisms in some sense), we say that a mechanism μ is *ex ante efficient* in the set F iff μ is in F and there exists no other mechanism ν that is in F and is ex ante Pareto superior to μ. Similarly, μ is *interim efficient* in F iff μ is in F and there exists no other mechanism ν that is in F and is interim Pareto superior to μ; and μ is *ex post efficient* in F iff μ is in F and there exists no other mechanism ν that is in F and is ex post Pareto superior to μ.

If ν is interim Pareto superior to μ, then ν is also ex ante Pareto superior to μ. Similarly, if ν is ex post Pareto superior to μ, then ν is also interim Pareto superior to μ. Thus, for any given set F, the set of ex ante efficient mechanisms in F is a subset of the set of interim efficient mechanisms in F, which is in turn a subset of the set of ex post efficient mechanisms in F.

The distinction between these different concepts of efficiency can be represented by restrictions on a class of social welfare functions that one might consider. Suppose that C and T are finite sets, and suppose that F is a set of mechanisms defined by a finite number of linear constraints (so F is geometrically a convex polyhedron in a finite-dimensional vector space). Then (by the supporting hyperplane theorem) a mechanism is ex post efficient in F iff it is an optimal solution to an optimization problem of the form

$$(10.9) \qquad \underset{\mu \in F}{\text{maximize}} \sum_{t \in T} p(t) \sum_{c \in C} \mu(c \,|\, t) \sum_{i \in N} \lambda_i(t) u_i(c, t),$$

where the *utility weight* $\lambda_i(t)$ is a positive number for each player i and each combination of types t. If we require that the utility weights for each player depend only on the player's own type, then we get the interim efficient mechanisms. That is, an interim efficient mechanism in F is any mechanism that is an optimal solution to an optimization problem of the form

$$(10.10) \qquad \underset{\mu \in F}{\text{maximize}} \sum_{t \in T} p(t) \sum_{c \in C} \mu(c \,|\, t) \sum_{i \in N} \lambda_i(t_i) u_i(c, t),$$

where $\lambda_i(t_i)$ is a positive number for each player i and each type t_i. Finally, if we require that the utility weight for each player is independent of his type, then we get the ex ante efficient mechanisms. That is, an ex ante efficient mechanism in F is any mechanism that is any optimal solution to an optimization problem of the form

$$(10.11) \qquad \underset{\mu \in F}{\text{maximize}} \sum_{t \in T} p(t) \sum_{c \in C} \mu(c \,|\, t) \sum_{i \in N} \lambda_i u_i(c, t),$$

where λ_i is a positive number for each player i.

When we say that there is incomplete information in a game, we mean that each player knows his own type but does not know anyone else's type at the beginning of the game, when plans and strategies are first chosen. Thus, each player is actually concerned with his own conditionally expected payoff, given his own true type, at this initial decision-making stage. Holmstrom and Myerson (1983) have argued that, to take proper account of such players' concerns, as well as the restrictions implied by the revelation principle, the most appropriate concept of efficiency for Bayesian games with incomplete information is interim efficiency in the set of incentive-feasible mechanisms. Because of the importance of this concept, it has been given the shorter alternative name of *incentive efficiency*. That is, an incentive-efficient mechanism is an incentive-feasible mechanism μ such that no other incentive-feasible mechanism is interim Pareto superior to μ. If a mechanism μ is incentive efficient in this sense, then a mediator who does not know any player's actual type could not propose any other incentive-compatible mechanism that every player is sure to prefer.

The ex post welfare criterion evaluates the expected payoffs that would be attributed to the players by someone who knew all of their types. If a mediator knew everyone's type, then he would not need to be concerned with informational incentive constraints, so the ex post welfare criterion seems logically connected to the classical feasibility concept. Thus, when we say that a mechanism is *ex post efficient*, without mentioning any specific feasible set, we mean that the mechanism is ex post efficient in the set of all functions from T to $\Delta(C)$.

10.3 An Example

To illustrate these ideas, consider a trading problem discussed in Section 6.4. In this example, which we call *Example 10.1*, player 1 is the only seller and player 2 is the only buyer of some divisible commodity. Player 1 has one unit available, and he knows whether it is of good quality (in which case his type is "1.a") or of bad quality (in which case his type is "1.b"). If it is good quality, then it is worth $40 per unit to the seller and $50 per unit to the buyer. If it is bad quality, then it is worth $20 per unit to the seller and $30 per unit to the buyer. The buyer has no private information, so her unique type may be called "2.0"; and she is known to believe that the probability of good quality is .2. We suppose that the buyer cannot verify any claims the seller might make about the quality, that it is not possible for the seller to offer any quality-contingent warranties, and that these two players do not expect to make any further transactions in the future. To formulate this as a Bayesian game, we let $C = C_1 \times C_2 = [0,1] \times \mathbf{R}_+$, where $C_1 = [0,1]$ represents the set of possible quantities of the commodity that player 1 can deliver to player 2 and $C_2 = \mathbf{R}_+$ represents the set of possible quantities of money that player 2 can pay to player 1. Payoffs to each player are defined to be his net dollar gains from trade, as in Section 6.4. The disagreement outcome in this Bayesian bargaining problem is $d* = (0,0)$, where player 2 gets none of the commodity and pays nothing. That is, each player could guarantee himself a payoff of 0 by refusing to trade.

Because of the linearity of the utility functions and the convexity of C, we can restrict our attention to deterministic trading mechanisms, that is, functions from T to C, instead of functions from T to $\Delta(C)$. So, as in Section 6.4, we can represent a mechanism by a pair $(Q(\cdot),Y(\cdot))$, where, for each t_1 in T_1, if 1's reported type is t_1, then $Q(t_1)$ is the quantity of the commodity that player 1 delivers to 2 and $Y(t_1)$ is the quantity of money that player 2 pays to 1, under the terms of the mechanism. In this notation, the expected utilities for each type of each player are then

$$U_1(Q,Y|1.a) = Y(1.a) - 40Q(1.a),$$
$$U_1(Q,Y|1.b) = Y(1.b) - 20Q(1.b),$$
$$U_2(Q,Y|2.0) = .2(50Q(1.a) - Y(1.a)) + .8(30Q(1.b) - Y(1.b)).$$

In this Bayesian bargaining problem, a mechanism is incentive feasible iff it is incentive compatible and individually rational, because player 1 can lie about his type and either player can refuse to trade. The informational incentive constraints are

(10.12) $U_1(Q,Y|1.a) \geq U_1^*(Q,Y,1.b|1.a) = Y(1.b) - 40Q(1.b),$

(10.13) $U_1(Q,Y|1.b) \geq U_1^*(Q,Y,1.a|1.b) = Y(1.a) - 20Q(1.a),$

and the individual-rationality (or participation) constraints are

(10.14) $U_1(Q,Y|1.a) \geq 0, U_1(Q,Y|1.b) \geq 0, U_2(Q,Y|2.0) \geq 0.$

We showed in Section 6.4 that, if (Q,Y) is any incentive-compatible mechanism for this game, then

(10.15) $.3U_1(Q,Y|1.a) + .7U_1(Q,Y|1.b) + U_2(Q,Y|2.0) \leq 8.$

Thus, any mechanism that satisfies

(10.16) $.3U_1(Q,Y|1.a) + .7U_1(Q,Y|1.b) + U_2(Q,Y|2.0) = 8$

must be incentive efficient.

For example, consider the following two mechanisms, which we call (Q^1,Y^1) and (Q^2,Y^2).

$$Q^1(1.a) = .2, Y^1(1.a) = 9, Q^1(1.b) = 1, Y^1(1.b) = 25,$$

$$Q^2(1.a) = .25, \quad Y^2(1.a) = 10.20, \quad Q^2(1.b) = 1, \quad Y^2(1.b) = 25.20.$$

In mechanism (Q^1, Y^1), player 1 sells .2 units of the commodity to player 2 at a price $9/.2 = \$45$ per unit if the quality is good, and he sells one unit to the buyer at a price of \$25 if the quality is bad. In mechanism (Q^2, Y^2), player 1 sells .25 units of the commodity to player 2 at a price $10.20/.25 = \$40.80$ per unit if the quality is good, and he sells one unit to the buyer at a price of \$25.20 if the quality is bad. It has been shown in Section 6.4 that mechanism (Q^1, Y^1) is incentive compatible. To show that mechanism (Q^2, Y^2) is incentive compatible, notice that, when the quality is bad, the seller gets a profit of $\$25.20 - \$20 = \$5.20$ if he is honest; and he would get an expected profit of $.25 \times (\$40.80 - \$20) = \$5.20$ if he lied. So he has no incentive to lie about the quality of his commodity.

The type-contingent expected payoffs for these mechanisms are

$$U_1(Q^1, Y^1 | 1.a) = .2 \times (45 - 40) = 1,$$
$$U_1(Q^1, Y^1 | 1.b) = 25 - 20 = 5,$$
$$U_2(Q^1, Y^1 | 2.0) = .2 \times (.2 \times (50 - 45)) + .8 \times (30 - 25) = 4.2,$$
$$U_1(Q^2, Y^2 | 1.a) = .25 \times (40.8 - 40) = 0.2,$$
$$U_1(Q^2, Y^2 | 1.b) = 25.2 - 20 = 5.2,$$
$$U_2(Q^2, Y^2 | 2.0) = .2 \times (.25 \times (50 - 40.8)) + .8 \times (30 - 25.2)$$
$$= 4.3.$$

The ex ante expected payoff for player 1 is the same in (Q^1, Y^1) and (Q^2, Y^2), because

$$.2 \times 1 + .8 \times 5 = 4.2 = .2 \times 0.2 + .8 \times 5.2.$$

Thus, because $U_2(Q^2, Y^2 | 2.0) > U_2(Q^1, Y^1 | 2.0)$, (Q^2, Y^2) is ex ante Pareto superior to (Q^1, Y^1). However, because $U_1(Q^2, Y^2 | 1.a) < U_1(Q^1, Y^1 | 1.a)$, neither of these mechanisms is interim Pareto superior to the other. In fact, (Q^1, Y^1) and (Q^2, Y^2) both satisfy equation (10.16), so both are incentive efficient.

To understand why the interim welfare criterion is so important, suppose that the two players expect to implement the mechanism (Q^1, Y^1) unless they both agree to change to some other mechanism. Imagine now that some mediator offers to help them implement the ex ante superior mechanism (Q^2, Y^2) instead. Should they both agree to make this change? Obviously player 1 would not agree to the change if his type were 1.a. So, to make matters more interesting, suppose that, unknown to player 2 and the mediator, player 1's type is actually 1.b. Then both players actually prefer (Q^2, Y^2) over (Q^1, Y^1), because

$$U_1(Q^2, Y^2 | 1.b) > U_1(Q^1, Y^1 | 1.b) \quad \text{and}$$

$$U_2(Q^2, Y^2 | 2.0) > U_2(Q^1, Y^1 | 2.0).$$

However, this agreement of privately known preferences cannot be translated into an actual agreement to change from (Q^1, Y^1) to (Q^2, Y^2). Player 2 should recognize that player 1 would agree to change from (Q^1, Y^1) to (Q^2, Y^2) only if his type were 1.b. When player 1's type is 1.b, player 2 would get payoff $30 - 25 = 5$ in (Q^1, Y^1), which is better than the payoff $30 - 25.20 = 4.8$ that she would get in (Q^2, Y^2). Thus, player 2 should recognize that player 1 would agree to the change only if the change would be disadvantageous to player 2. (This effect is an example of the *winner's curse* in bidding and bargaining.) So player 2 should reject the change from (Q^1, Y^1) to (Q^2, Y^2), even though (Q^2, Y^2) is ex ante Pareto superior and appears to be strictly better than (Q^1, Y^1) for player 2 given her current information.

An agreement to change from (Q^1, Y^1) to (Q^2, Y^2) could be reached if player 2 believed that player 1 agreed to the change before learning his own type (at the ex ante stage). But if this game is to be played only once and player 2 knows that player 1 knows his type before he bargains over the mechanism, then player 2 should reject any proposal to change from (Q^1, Y^1) to (Q^2, Y^2) that requires player 1's approval to be effective, because she should recognize that player 1 will not use the ex ante welfare criterion in his own decision-making.

A similar argument against change can be made if the players expected to implement the incentive-efficient mechanism (Q^2, Y^2) unless they unanimously agreed to some other incentive-compatible mechanism. In this case, player 2 should reject any proposal to change to (Q^1, Y^1), because she should recognize that player 1 will also accept the proposed change only if his type is 1.a, but then (Q^2, Y^2) would be better than (Q^1, Y^1) for player 2.

It can be shown that, if a mechanism (Q, Y) is ex ante efficient in the set of all incentive-feasible mechanisms for this Bayesian bargaining problem, then player 1's expected payoff when the quality is good

$(U_1(Q,Y|1.a))$ must equal 0. To see why, recall condition (10.15), which implies that (Q,Y) must satisfy

$$.3U_1(Q,Y|1.a) + .7U_1(Q,Y|1.b) \le 8 - U_2(Q,Y|2.0).$$

The ex ante expected payoff to player 1 is

$$.2U_1(Q,Y|1.a) + .8U_1(Q,Y|1.b);$$

and this quantity could be increased, without changing $U_2(Q,Y|2.0)$ or violating the above constraint, by decreasing $U_1(Q,Y|1.a)$ to 0 and increasing $U_1(Q,Y|1.b)$ by $^3/_7$ times the decrease in $U_1(Q,Y|1.a)$.

In fact, a mechanism is ex ante efficient in the set of all incentive-feasible mechanisms iff it is of the following form:

$$Q(1.a) = z/20 - 1, \quad Y(1.a) = 40(z/20 - 1),$$
$$Q(1.b) = 1, \quad Y(1.b) = z,$$

where $20 \le z \le 31^3/_7$. In all of these mechanisms, however, the seller with a good quality commodity (type 1.a) can only sell the commodity at a price of $40 per unit, so he cannot get a positive profit. If in fact he knows that his quality is good, it is hard to see why he should participate in any such mechanism. He would have nothing to lose if instead he made an offer to sell some quantity for a price that is strictly higher than $40 per unit, which would at least give him some chance of making a positive profit.

There are no incentive-feasible mechanisms for this example that are ex post efficient (in the set of all mechanisms). To satisfy ex post efficiency, player 1 must sell his entire supply of the commodity to player 2, no matter what his type is, because the commodity is always worth more to 2 than to 1. But then, for incentive compatibility, the expected price cannot depend on his type. Individual rationality for type 1.b would then require that price to be at least $40; but such a price would require that individual rationality be violated for player 2, because she would expect to lose by paying more than $34 = 0.2 \times 50 + 0.8 \times 30$.

10.4 Ex Post Inefficiency and Subsequent Offers

The impossibility of ex post efficiency in this example implies that, no matter what mechanism is being implemented, there is a positive probability that player 1 will end up holding a positive amount of the commodity, even though it is worth more to player 2. This result may be somewhat disturbing, if we suppose that the players can always make another offer to buy or sell more of the commodity. For example, suppose that the players implement mechanism (Q^1, Y^1) and player 1 sells .2 units for $45 \times .2 = \$9$. After this outcome, by Bayes's rule, player 2 should believe that player 1's type is 1.a with probability 1. So if player 1 followed the outcome of (Q^1, Y^1) by a subsequent offer to sell his remaining .8 units of the commodity for $45 \times .8 = \$36$, it might seem that player 2 should be willing to accept, because she is convinced that the quality is good.

Of course, if player 1 believed at the beginning of the mechanism that such a subsequent offer would be accepted, then he would want to pretend that his type was 1.a in the mechanism even if it were 1.b, because selling one unit for $45 (with a slight delay for the sale of the last .8 units) would surely be better than selling one unit for $25! To deter such behavior, it is necessary that, when player 1 reports his type to the mediator implementing the mechanism (Q^1, Y^1), player 1 must believe that no such additional offer would be accepted. However, player 2's actual behavior after the mechanism has no way of changing player 1's beliefs at the beginning of the mechanism, so player 2 should not be concerned about destroying the incentive compatibility of the preceding mechanism when she considers accepting a subsequent offer.

To understand how opportunities for subsequent offers can affect the analysis of this example, we need to consider an explicit model of trading over time, including a specification of the cost of waiting. To be specific, suppose that there are infinitely many points in time, or rounds, numbered $1, 2, 3, \ldots$, at which the players may trade. Suppose also that a profit of x dollars at round k would be worth $x\delta^{k-1}$ at round 1, where δ is a *discount factor* such that $0 < \delta < 1$. Given any equilibrium of a sequential-offer bargaining game, let $\bar{q}_k(t_1)$ be the expected quantity of the commodity that will be traded at round k if player 1's type is t_1, and let $\bar{y}_k(t_1)$ be the expected payment from player 2 to player 1 at round k if 1's type is t_1. Then let

$$Q(t_1) = \sum_{k=1}^{\infty} \bar{q}_k(t_1)\delta^{k-1}, \quad Y(t_1) = \sum_{k=1}^{\infty} \bar{y}_k(t_1)\delta^{k-1}.$$

It can be shown that expected payoffs for the players in this multiround game depend on Q and Y, with this interpretation, exactly as in the original one-round game considered in Section 10.3. For example, if player 1's type is 1.b and he uses his correct equilibrium strategy for this type, then his expected discounted payoff in equilibrium is

$$\sum_{k=1}^{\infty} (\bar{y}_k(1.b) - 20\bar{q}_k(1.b))\delta^{k-1} = Y(1.b) - 20Q(1.b);$$

but if he followed the equilibrium trading strategy of type 1.a when his type is really 1.b, then his expected discounted payoff would be

$$\sum_{k=1}^{\infty} (\bar{y}_k(1.a) - 20\bar{q}_k(1.a)) \delta^{k-1} = Y(1.a) - 20Q(1.a).$$

So an equilibrium of the sequential-offer bargaining game must satisfy

$$Y(1.b) - 20Q(1.b) \geq Y(1.a) - 20Q(1.a),$$

as well as the other informational incentive constraint and individual rationality constraints that we got in the original one-round version of the game. Thus, if we take waiting costs into account along with costs of failing to trade, for any equilibrium of a discounted sequential-offer game, there must exist an incentive-feasible mechanism (Q,Y) for the original one-round game that gives the same expected payoff to each type of each player.

This conclusion leaves us with the question, Might the set of incentive-feasible mechanisms actually be reduced by the assumption that players 1 and 2 have an inalienable right to keep offering to trade any amount of the commodity that player 1 might still hold? In effect, the players' right to keep bargaining could create a moral-hazard problem that constrains the set of incentive-compatible mechanisms, because of the need to give each player an incentive to never trade more than the quantity stipulated in the mechanism. However, there are reasonable models of sequential-offer bargaining games in which these moral-hazard constraints do not significantly reduce the set of feasible mechanisms; so any mechanism (Q,Y) that satisfies (10.12)–(10.14) can in fact be implemented by an equilibrium of a sequential-offer bargaining game with mediated communication, an unbounded time horizon, and discounting (see Ausubel and Deneckere, 1989).

One way to derive such a general feasibility result is to assume, as we did in Section 8.9, that offers must be in some (large) finite set and the discount factor is close to 1. For example, suppose that price offers must be in integer multiples of $0.01 and that the discount factor is $\delta > .999$. (This discount factor may seem quite large, but remember that a "round" is supposed to represent the length of time necessary for a player to formulate and present a new offer after the most recent one has been rejected. A large discount factor corresponds to the fact that waiting such a short time interval is not very costly.) Then, after the mediator has implemented the mechanism (Q,Y) and all trades stipulated by the mechanism have been completed at round 1, trading at future rounds can be prevented if the players follow a *standoff equilibrium* in their bargaining at all subsequent rounds. In this standoff equilibrium, player 1 always offers to sell his remaining supply at a price of $50 per unit, and player 2 always offers to buy the remaining supply at a price of $20 per unit. If player 2 ever deviates from this standoff behavior by offering any price higher than $20 at any round, then player 1 will expect that player 2 will submissively accept the $50 offer at the next round, so 1 will reject 2's offer (because even 49.99 − 40 is less than $(50 − 40)\delta$). If player 1 ever deviates from this standoff behavior by offering any price lower than $50, then player 2 will infer from this surprise that player 1's type is 1.b (such an inference is Bayes consistent even if she assigned zero probability to type 1.b before the deviation, because the deviation was a zero-probability event), and she will expect that player 1 will submissively accept the $20 offer at the next round; so 2 will reject 1's offer.

With two-sided incomplete information, such standoff equilibria can be sustained even when offers that split pennies arbitrarily are allowed and the players have inalienable opportunities to alternate offers. For example, suppose that we perturb Example 10.1 slightly by supposing that there is a very small positive probability that player 2 might be type "2.1," in which case the value of the commodity to her would be $100 per unit. Suppose that a mediator implements at round 1 an incentive-feasible mechanism in which player 1 only retains some of his supply if he is type 1.a and player 2 is type 2.0. In the event that player 1 still holds some of his supply at round 2, after the mechanism has been implemented, there is a standoff equilibrium in which player 1 always offers to sell his remaining supply at a price of $70 per unit and player 2 always offers to buy the remaining supply at a price of $25, and neither ever accepts (because 70 > 50 and 25 < 40). If player 1 ever offered any price less than $70, then player 2 would infer from this zero-probability event that player 1 was actually type 1.b and would play thereafter as in the equilibrium of the alternating-offer game described in Section 8.7 for the complete-information bargaining game where the commodity is worth \$20 per unit to 1 and \$30 per unit to 2. In this equilibrium, player 2 always offers a price close to \$25 and expects 1 to accept it. On the other hand, if player 2 ever offered any price higher than \$25, then player 1 would infer from this zero-probability event that player 2 was actually type 2.1 and would play thereafter as in the equilibrium for the complete-information bargaining game where the commodity is worth \$40 per unit to 1 and \$100 per unit to 2; in this equilibrium, player 1 always offers a price close to \$70 and expects 2 to accept it. So neither player could gain from deviating from the standoff offers of \$70 and \$25 when their types are actually 1.a and 2.0, and trade after the mediation at round 1 would never occur.

10.5 Computing Incentive-Efficient Mechanisms

Consider now a Bayesian collective-choice problem or a Bayesian bargaining problem (like that in Section 10.1) in which the type sets and outcome sets are all finite. Under the finiteness assumption, the set of incentive-compatible mechanisms is a convex polyhedron. Thus, by the supporting hyperplane theorem, an incentive-feasible mechanism $\overline{\mu}$ is incentive-efficient iff there exist some positive numbers $\lambda_i(t_i)$ for each type t_i of each player i such that $\overline{\mu}$ is an optimal solution to the optimization problem

$$\underset{\mu:T\rightarrow\Delta(C)}{\text{maximize}} \sum_{i\in N} \sum_{t_i\in T_i} \lambda_i(t_i)U_i(\mu|t_i)$$

subject to $U_i(\mu|t_i) \geq U_i^*(\mu,s_i|t_i)$, $\forall i \in N$, $\forall t_i \in T_i$, $\forall s_i \in T_i$.

This optimization problem is equivalent to (10.10), when F is the set of incentive-compatible mechanisms. Furthermore, this optimization problem is a linear programming problem, because the objective and constraints are all linear in μ, including the probability constraints

$$\sum_{c\in C} \mu(c|t) = 1 \text{ and } \mu(d|t) \geq 0, \quad \forall d \in C, \forall t \in T,$$

which are implicit in the restriction that μ is a function from T to $\Delta(C)$.

For such optimization problems, a *Lagrangean function* can be formed by multiplying constraint functions by variables called *Lagrange multipliers* and adding them into the objective function. (Essentially the same construction is used in the general definition of a dual problem in Section 3.8.) Let $\alpha_i(s_i|t_i)$ denote the Lagrange multiplier for the constraint that asserts that player i should not expect to gain by reporting type s_i when his actual type is t_i. Then the Lagrangean function can be written

$$\sum_{i \in N} \sum_{t_i \in T_i} \lambda_i(t_i) U_i(\mu \mid t_i) + \sum_{i \in N} \sum_{t_i \in T_i} \alpha_i(s_i \mid t_i)(U_i(\mu \mid t_i) - U_i^*(\mu, s_i \mid t_i)).$$

To simplify this expression, let

$$(10.17) \quad v_i(c,t,\lambda,\alpha) = \Big(\big(\lambda_i(t_i) + \sum_{s_i \in T_i} \alpha_i(s_i \mid t_i) \big) u_i(c,t)$$

$$- \sum_{s_i \in T_i} \alpha_i \big((t_i \mid s_i) u_i(c,(t_{-i}, s_i)) \big) \Big) / \bar{p}_i(t_i).$$

This quantity $v_i(c,t,\lambda,\alpha)$ is called the *virtual utility* payoff to player i from outcome c, when the type profile is t, with respect to the utility weights λ and the Lagrange multipliers α. Then the above Lagrangean function is equal to

$$\sum_{t \in T} p(t) \sum_{c \in C} \mu(c \mid t) \sum_{i \in N} v_i(c,t,\lambda,\alpha).$$

Standard results in optimization theory (using the duality theorem of linear programming) then imply the following theorem.

THEOREM 10.1. *Suppose that* μ *is an incentive-feasible mechanism. Then* μ *is incentive efficient if and only if there exist vectors*

$$\lambda = (\lambda_i(t_i))_{i \in N, t_i \in T_i} \quad and \quad \alpha = (\alpha_i(s_i \mid t_i))_{i \in N, s_i \in T_i, t_i \in T_i}$$

such that

$$\lambda_i(t_i) > 0, \quad \forall i \in N, \ \forall t_i \in T_i,$$

$$\alpha_i(s_i \mid t_i) \geq 0, \quad \forall i \in N, \ \forall s_i \in T_i, \ \forall t_i \in T_i,$$

$$\alpha_i(s_i \mid t_i)(U_i(\mu \mid t_i) - U_i^*(\mu, s_i \mid t_i)) = 0, \quad \forall i \in N, \ \forall s_i \in T_i, \ \forall t_i \in T_i,$$

and

$$\sum_{c \in C} \mu(c \mid t) \sum_{i \in N} v_i(c,t,\lambda,\alpha) = \max_{c \in C} \sum_{i \in N} v_i(c,t,\lambda,\alpha), \quad \forall t \in T.$$

To interpret these conditions, we need to develop some terminology. If λ and α satisfy the conditions in Theorem 10.1 for μ and $\alpha_i(s_i \mid t_i) > 0$, then we say that type t_i *jeopardizes* type s_i. The condition

$$\alpha_i(s_i | t_i)(U_i(\mu | t_i) - U_i^*(\mu, s_i | t_i)) = 0$$

in Theorem 10.1 asserts that t_i can only jeopardize s_i in the mechanism μ if the constraint that player i should not be tempted to pretend that his type is s_i when it really is t_i is a binding constraint for the mechanism μ.

Equation (10.17) defines the virtual utility for any type t_i of player i to be a positive multiple of his actual utility minus a weighted sum of the utility payoffs for the types that jeopardize t_i. Thus, the virtual utility of t_i differs qualitatively from the actual utility of t_i in that it exaggerates the difference from types that jeopardize t_i.

The condition

$$\sum_{c \in C} \mu(c | t) \sum_{i \in N} v_i(c, t, \lambda, \alpha) = \max_{c \in C} \sum_{i \in N} v_i(c, t, \lambda, \alpha)$$

in Theorem 10.1 asserts that the incentive-efficient mechanism μ must put positive probability only on outcomes that maximize the sum of the players' virtual utilities, for every joint state of the players' information. That is, if the players got transferable payoffs in the virtual utility scales with respect to λ and α, then the mechanism μ would be ex post efficient.

Consider the example from Section 10.3. Referring to condition (10.15), we may try the parameters

$$\lambda_1(1.a) = 0.3, \quad \lambda_1(1.b) = 0.7, \quad \lambda_2(2.0) = 1,$$
$$\alpha_1(1.a | 1.b) = 0.1, \quad \alpha_1(1.b | 1.a) = 0.$$

As in Section 6.4, let q denote the amount of the commodity that the seller delivers to the buyer and let y denote the amount of money that the buyer pays to the seller. The actual utility functions are

(10.18) $u_1((q,y),1.a) = y - 40q,$

(10.19) $u_2((q,y),1.a) = 50q - y,$

(10.20) $u_1((q,y),1.b) = y - 20q,$

(10.21) $u_2((q,y),1.b) = 30q - y.$

(We suppress the dependence on player 2's constant type 2.0.) The virtual utility payoffs for player 1 are then

$$v_1((q,y),1.a,\lambda,\alpha) = \frac{0.3(y - 40q) - 0.1(y - 20q)}{.2} = y - 50q,$$

$$v_1((q,y),1.b,\lambda,\alpha) = \frac{(0.7 + 0.1)(y - 20q)}{.8} = y - 20q,$$

$$v_2((q,y),t_1,\lambda,\alpha) = u_2((q,y),t_1,\lambda,\alpha), \quad \forall t_1 \in T_1.$$

Thus, the only difference between virtual utility and actual utility with these parameters is that the seller has a virtual valuation of $50 for each unit of the commodity, instead of an actual valuation of $40. That is, because the bad type 1.b jeopardizes the good type 1.a, the good type's virtual valuation ($50) differs from its actual valuation ($40) in a way that exaggerates the difference from the bad type's actual valuation ($20) for the commodity.

By Theorem 10.1, μ is incentive efficient if it always maximizes the sum of the players' virtual utilities and has a binding incentive constraint whenever one type jeopardizes another in α. When 1's type is 1.a, the sum of the virtual utilities is

$$v_1((q,y),1.a,\lambda,\alpha) + v_2((q,y),1.a,\lambda,\alpha) = (y - 50q) + (50q - y) = 0,$$

so any outcome would maximize this sum. When 1's type is 1.b, the sum of virtual utilities is

$$v_1((q,y),1.b,\lambda,\alpha) + v_2((q,y),1.b,\lambda,\alpha) = (y - 20q) + (30q - y) = 10q,$$

so virtual ex post efficiency requires that the quantity sold should be the maximum possible, that is, $q = 1$. Because $\alpha_1(1.a|1.b) > 0$ but $\alpha_1(1.b|1.a) = 0$, the only binding incentive constraint that we require is

$$U_1(Q,Y|1.b) = U_1^*(Q,Y,1.a|1.b).$$

Thus, an incentive-feasible mechanism for this example is incentive efficient if it satisfies the following two properties: the seller delivers all of his supply of the commodity to the buyer when the seller's type is 1.b (bad-quality supplier), and the seller with type 1.b would be indifferent between being honest about his type and dishonestly reporting type 1.a.

Notice that both mechanisms (Q^1,Y^1) and (Q^2,Y^2) in Section 10.3 satisfy these two properties. In fact, it can be shown that all of the incentive-efficient mechanisms satisfy these two properties, because the same λ and α satisfy Theorem 10.1 for all incentive-efficient mechanisms in this simple example.

In general, the virtual utility formula (10.17) can give insights into the problem of finding good signals in any situation where people have difficulty trusting each other's claims about unverifiable private information. For example, consider a labor market (as presented in Spence, 1973) where there are high-ability and low-ability workers, and each worker knows what his ability type is, but the ability of any given worker is not directly observable by the prospective employers. We can expect that a worker's low-ability type would jeopardize his high-ability type in negotiations with an employer; that is, an employer may have difficulty preventing a low-ability worker from claiming to have high ability. So the virtual utility of a high-ability worker should exaggerate the difference from a low-ability worker. Suppose that there is some useless educational process, which contributes nothing to a worker's productive skills and would be mildly unpleasant for a high-ability worker but would be extremely painful for a low-ability worker. Then a high-ability worker may get positive virtual utility from this education, because the large negative payoff that this education would generate for a low-ability worker has a negative coefficient in the linear formula for the high-ability type's virtual utility. Thus, as in Spence's (1973) labor-market equilibria, an incentive-efficient mechanism may force a high-ability worker to go through this costly and unproductive educational process (from which he gains only virtual utility), as a signal before he can be hired at a high wage.

On the other hand, it seems unlikely that a high-ability worker would be tempted to claim that he had low ability in such a labor market, so we can expect that a low-ability type would not be jeopardized. Therefore, a low-ability worker's virtual utility would be just some positive multiple of his real utility. Thus, an incentive-efficient mechanism should be ex post efficient (in terms of real utility, as well as virtual utility) when it places a low-ability worker, and the low-ability worker should not be made to suffer any such unproductive educational experiences.

In general, a good signal for a type t_i of player i is any observable activity that would be more costly for the other types s_i that jeopardize t_i than for type t_i itself and so can increase the virtual utility of type t_i. (Here the coefficients $\lambda_i(t_i)$ and $\alpha_i(t_i|s_i)$ determine the weighted utility scales for such intertype comparisons.) Of course, an ideal signal for type t_i would be some activity that would be costless for type t_i but would be impossible (that is, infinitely costly) for any other type s_i of player i. However, if such an ideal signal existed, then the coefficient $\alpha_i(t_i|s_i)$ would be equal to 0 and the constraint that "type s_i should not be tempted to claim to be t_i" could be effectively dropped from the list of informational incentive constraints, because player i would have a cost-less way to prove that his type was t_i, not s_i.

10.6 Inscrutability and Durability

Having established by the revelation principle that a mediator or social planner can restrict his attention to incentive-compatible mechanisms, which form a mathematically simple set, we can naturally suppose that intelligent players in a cooperative game with incomplete information would themselves bargain over the set of incentive-compatible mechanisms. By the definition of a game with incomplete information, each player privately knows his type already, at the time when fundamental economic plans and decisions are made. Thus, a theory of cooperative games with incomplete information should be a theory of mechanism selection by individuals who have private information.

During any process of bargaining over mechanisms, if the players reach a unanimous agreement to change from some mechanism μ to another mechanism ν, then, at the moment when the agreement is reached, it must be common knowledge among the players that they all prefer to change from μ to ν. Such unanimity for change can be common knowledge when the players know only their own types if and only if μ is not incentive efficient. We formally state this result below as Theorem 10.2 (due to Holmstrom and Myerson, 1983).

To state Theorem 10.2 nontrivially, we must drop for now the assumption that players' types are independent random variables. Instead, let us suppose only that their type-conditional beliefs $(p_i)_{i \in N}$ are consistent with some common prior probability distribution p on T, so

$$p_i(t_{-i}|t_i) = \frac{p(t)}{\bar{p}_i(t_i)}, \quad \forall t \in T, \ \forall i \in N,$$

where we suppose that

$$\bar{p}_i(t_i) = \sum_{s_{-i} \in T_{-i}} p(s_{-i}, t_i) > 0, \quad \forall i \in N, \ \forall t_i \in T_i.$$

For any set R that is a subset of T, we can let R_i denote the set of all types in T_i that could occur when the combination of players' types is in R; that is,

$$R_i = \{r_i | r \in R\}.$$

We say that R is a *common-knowledge event* iff R is a nonempty subset of T and, for every player i, every r_i in R_i, and every t_{-i} in T_{-i},

if $(t_{-i}, r_i) \notin R$, then $p_i(t_{-i}|r_i) = 0$.

That is, whenever the profile of players' types is in the common-knowledge event R, every player assigns zero probability to all profiles of types outside of R. (This condition implies that, for any t in T, if there is at least one player i for whom $t_i \in R_i$ and at least one player j for whom $t_j \notin R_j$, then $p(t) = p_i(t_{-i}|t_i) \, \bar{p}_i(t_i) = 0$; so $p_j(t_{-j}|t_j) = p(t)/\bar{p}_j(t_j) = 0$.)

We say that v is *interim superior to* μ *within* R iff

$$U_i(v|r_i) \geq U_i(\mu|r_i), \quad \forall i \in N, \ \forall r_i \in R_i,$$

with a strict inequality for at least one player i and type r_i in R_i.

THEOREM 10.2. *A incentive-feasible mechanism μ is incentive efficient if and only if there does not exist any other incentive-compatible mechanism v that is interim superior to μ within some common-knowledge event R.*

Proof. If μ is not incentive efficient, then some other incentive-compatible mechanism v is interim superior to μ within T, which is itself a common-knowledge event.

Conversely, suppose that there exists some incentive-compatible mechanism v that is interim superior to μ within a common-knowledge event R. Let η be the mechanism defined such that

$$\eta(c|t) = v(c|t) \text{ if } t \in \underset{i \in N}{\times} R_i,$$

$$\eta(c|t) = \mu(c|t) \text{ if } t \notin \underset{i \in N}{\times} R_i.$$

Then η is interim Pareto superior to μ (in all of T) because

$$U_i(\eta|t_i) = U_i(\mu|t_i) \text{ if } t_i \notin R_i,$$
$$U_i(\eta|r_i) = U_i(v|r_i) \geq U_i(\mu|r_i) \text{ if } r_i \in R_i.$$

Furthermore, it is straightforward to show that η is incentive feasible. If $r_i \in R_i$ and $t_i \in R_i$, then

$$U_i^*(\eta, t_i|r_i) = U_i^*(v, t_i|r_i) \leq U_i(v|r_i) = U_i(\eta|r_i).$$

If $r_i \in R_i$ and $t_i \notin R_i$, then

$$U_i^*(\eta,t_i|r_i) = U_i^*(\mu,t_i|r_i) \le U_i(\mu|r_i) \le U_i(\eta|r_i).$$

If $s_i \notin R_i$, then

$$U_i^*(\eta,t_i|s_i) = U_i^*(\mu,t_i|s_i) \le U_i(\mu|s_i) = U_i(\eta|s_i),$$

because type s_i would expect every other player j to report a type not in R_j. So μ is not incentive efficient. ∎

Theorem 10.2 asserts that, if μ is incentive efficient, then the players cannot reach a unanimous common-knowledge agreement to change from μ to some other incentive-compatible mechanism without changing some player's beliefs about other players' types. That is, a unanimous agreement to change away from an incentive-efficient mechanism may be possible, but only after some communication process in which some player has revealed substantive information about his type to some other player.

When we consider a mechanism-selection game, in which individuals can bargain over mechanisms, there should be no loss of generality in restricting our attention to equilibria in which there is one incentive-feasible mechanism that is selected with probability 1, independently of anyone's type. This proposition, called the *inscrutability principle*, can be justified by viewing the mechanism-selection process itself as a communication game induced from the original Bayesian game or collective-choice problem and by applying the revelation principle. For example, suppose that there is an equilibrium of the mechanism-selection game in which some mechanism ν would be selected if the profile of players' types were in some set A (where $A \subseteq T$), and some other mechanism μ would be selected otherwise. Then there should exist an equivalent equilibrium of the mechanism-selection game in which the players always select a mechanism η, which coincides with mechanism ν when the players report (to the mediator who implements η) a profile of types in A and which coincides with μ when they report a profile of types that is not in A; that is,

$$\eta(\cdot|t) = \nu(\cdot|t) \text{ if } t \in A,$$
$$\eta(\cdot|t) = \mu(\cdot|t) \text{ if } t \notin A.$$

In effect, any information that the players could have revealed during the mechanism-selection process can be revealed instead to the mediator after the mechanism has been selected, and the mediator can use this information in the same way that it would have been used in the mechanism-selection process. Thus, when the players bargain over mechanisms, they can agree to share information during the implementation of the mechanism without sharing any information during the mechanism-selection process.

However, Theorem 10.2 and the inscrutability principle do not imply that the possibility of revealing information during a mechanism-selection process is irrelevant. There may be some incentive-feasible mechanisms that we should expect not to be selected by the players in such a process, precisely because some players would choose to reveal information about their types rather than let these mechanisms be selected. For example, consider the following Bayesian collective-choice problem, due to Holmstrom and Myerson (1983). There are two players in $N = \{1,2\}$, $T_1 = \{1.a,1.b\}$, $T_2 = \{2.a,2.b\}$, and

$$\frac{1}{4} = p(1.a,2.a) = p(1.a,2.b) = p(1.b,2.a) = p(1.b,2.b).$$

The set of possible outcomes or social choices is $C = \{x,y,z\}$, and each player's utility for these options depends on his type according to Table 10.1. The incentive-efficient mechanism that maximizes the ex ante expected sum of the two players' payoffs is μ such that

$$\mu(x|1.a,2.a) = 1, \quad \mu(y|1.a,2.b) = 1,$$
$$\mu(z|1.b,2.a) = 1, \quad \mu(y|1.b,2.b) = 1.$$

That is, a mediator implementing this mechanism would choose x if the reported types are 1.a and 2.a, z if the reported types are 1.b and 2.a, and y if player 2 reports type 2.b. (Choosing y when the reported types are 1.a and 2.b serves here to deter player 2 from reporting 2.b when her type is 2.a.) However, Holmstrom and Myerson argued that such a mechanism would not be chosen in a mechanism-selection game that is played when player 1 already knows his type. When player 1's type is

Table 10.1 Payoffs to each type of each player, for all social choice options

Option	1.a	1.b	2.a	2.b
x	2	0	2	2
y	1	4	1	1
z	0	9	0	−8

1.a, he could do better by proposing the mechanism that always chooses x, and 2 would always want to accept this proposal. That is, because 1 would have no incentive to conceal his type from 2 in a mechanism-selection game if his type were 1.a (when his interests would have no conflict with 2's), we should not expect the individuals in a mechanism-selection game to inscrutably agree to an incentive-efficient mechanism like the one above that implicitly puts so much weight on the payoff for type 1.b.

Some economists, following Coase (1960), have argued that we should expect to observe efficient allocations in any economic situation where there is complete information and bargaining costs are small. According to this argument, often called the *Coase theorem*, an inefficient allocation of resources could not endure, because someone could suggest an alternative allocation that would make everyone better off and everyone would agree to this alternative. (NOTE: Informational incentive constraints may be a source of some of the "bargaining costs" that are discussed in this literature.) As we have seen in Sections 8.9 and 10.4, this argument may not be valid when there is a range of Pareto-superior alternatives over which individuals can bargain, because agreement about which alternative to choose may be stalled in a standoff equilibrium, in which each individual insists on an alternative that is most favorable to himself. However, if the players are considering only one alternative to the status quo and if that alternative is Pareto superior to the status quo, then a unanimous agreement to change from the status quo to the alternative should be assured, at least in the case of complete information.

Before trying to extend the Coase theorem to games with incomplete information, let us formalize this result for the complete information case. Suppose that there is a given subset of $\mathbf{R}^N$ that represents the set of feasible payoff allocations for the players in N. Consider the following voting game between a given status quo and a given alternative, each of which is a feasible payoff allocation. In this voting game, the players vote simultaneously for either the status quo or the alternative. After the vote, the alternative will be implemented if all players vote for it, but the status quo will be implemented if one or more players votes for it. That is, we suppose that a change from the status quo to the alternative will be implemented if and only if there is unanimous agreement for it.

In such a voting game, there is always an equilibrium in which everyone votes for the status quo, regardless of his preferences, because each player anticipates that his vote cannot change the outcome. Such a trivial equilibrium can be eliminated by several natural restrictions or refinements of the equilibrium set. The weakest such restriction is that every player who would strictly prefer the alternative to the status quo (if everyone else were voting for the alternative) should himself vote for the alternative; let us say that an equilibrium of this voting game is *active* iff it satisfies this property. Then a feasible allocation w is Pareto efficient (in the weak sense) if and only if, when w is the status quo, for each alternative in the feasible set, this voting game has an active equilibrium in which at least one player is sure to vote for the status quo w.

Things become more complicated when we try to extend this positive characterization of efficiency to the case of incomplete information. With incomplete information, the status quo and alternative are generally mechanisms, rather than simple allocations. The voting game must be defined to include, not only the voting decision of each player, but also the decision that each player must make about his report to the mediator in the subsequent implementation of the mechanism that wins the vote. The result of the vote may convey information to the players about one anothers' types, so we cannot generally assume that honest reporting would necessarily be equilibrium behavior when the mechanism is implemented after the vote, even if this mechanism would be incentive compatible in terms of the information available to the players before the vote. (For the example in Table 10.1, if the above mechanism μ is implemented as the status quo, after a vote against the alternative of choosing x for sure, then player 2 will believe player 1's type must be 1.b, because only 1.b would have voted for the status quo. Then player 2 will want to report type 2.b even if her type is 2.a.) So a sequential equilibrium of the voting game must include a specification of what each player would believe about the other players' types if he voted for the alternative and it won unanimous endorsement in the vote. These beliefs must be consistent with the equilibrium voting strategies. Also, the players' reporting strategies must be sequentially rational given these beliefs. (We assume here that, after the vote, the players learn which of the two mechanisms is to be implemented, but they learn nothing else about each other's votes.)

Holmstrom and Myerson (1983) say that an incentive-compatible mechanism is *durable* iff, when it is the status quo, for each alternative mechanism, there exists an active sequential equilibrium of this voting game in which, for every possible profile of types, at least one player is sure to vote for the status quo, which will then be implemented with all players reporting honestly. Holmstrom and Myerson have shown that durable incentive-efficient mechanisms always exist, for any finite Bayesian collective-choice problem. Furthermore, when only one player has private information, all incentive-efficient mechanisms are durable.

For example, recall the two incentive-efficient mechanisms (Q^1, Y^1) and (Q^2, Y^2) for Example 10.1, in which only player 1 has private information. It was shown in Section 10.3 that, if either one of these mechanisms is the status quo and the other is the alternative, then the information that player 1 would vote for a change from status quo to alternative should make player 2 expect to lose from the change. Thus, player 2 would always vote for whichever of these two incentive-efficient mechanisms is the status quo.

However, when two or more players have private information, there may also exist incentive-efficient mechanisms that are not durable. For the example in Table 10.1, the mechanism μ discussed above is incentive efficient but is not durable. Furthermore, some durable mechanisms may not be incentive efficient. For example, suppose that there are two players in $N = \{1,2\}$, $T_1 = \{1.a, 1.b\}$, $T_2 = \{2.a, 2.b\}$, $C = \{x, y\}$, $\frac{1}{4} = p(1.a, 2.a) = p(1.a, 2.b) = p(1.b, 2.a) = p(1.b, 2.b)$, and, for each player i,

$$u_i(x,t) = 2, \quad \forall t \in T,$$
$$u_i(y,t) = 3 \text{ if } t = (1.a, 2.a) \text{ or } t = (1.b, 2.b),$$
$$u_i(y,t) = 0 \text{ if } t = (1.a, 2.b) \text{ or } t = (1.b, 2.a).$$

Let η be the mechanism such that $\eta(x|t) = 1$ for all t. Then η is not incentive efficient but is durable. The players would both gain by choosing y when their types are the same; but in the voting game with any alternative mechanism, there is always an active sequential equilibrium in which the players all vote for η, independently of their types, and they would report (babble) independently of their true types if the alternative were approved. (There is no constant mechanism in which the outcome does not depend on the players' types that the players would prefer to η.)

This discussion of durability suggests that the positive interpretation of efficiency, which is expressed by the Coase theorem, may be less applicable to games with incomplete information. However, the concept of durability is limited by the assumption that only one alternative to the status quo will be considered, with no story about how this unique alternative is to be specified. (See also Crawford, 1985.) Furthermore, the above example with a durable but inefficient mechanism also has a unique incentive-efficient mechanism that is durable and that is likely to be the focal equilibrium outcome in a situation where the players can negotiate effectively. In general, a positive interpretation of efficiency may be simply derived from an assumption that incentive efficiency is one determinant of the focal equilibrium of the mechanism-selection process. That is, in a process of bargaining over mechanisms, it may be a focal equilibrium for everyone to inscrutably endorse some specific mechanism on the basis of its incentive efficiency and its equity (in some sense) for all types of all players. Such equity considerations should disqualify mechanisms such as the nondurable mechanism μ for the example in Table 10.1, which gives too little to type 1.a and too much to type 1.b. We consider some efforts to develop such a theory of equity in the next two sections.

10.7 Mechanism Selection by an Informed Principal

To develop a theory of bargaining over mechanisms in a game with incomplete information, it is best to start with the simplest case in which one player, whom we call the *principal*, has all of the negotiating ability. After all, if we cannot understand what mechanism a player would choose when it is wholly his own choice, then we cannot hope to understand what mechanism a player should negotiate for in a mechanism-selection process where he must bargain and compromise with others. (Recall the discussion in Section 6.7.)

Let us consider *Example 10.2*, which differs from Example 10.1 in Section 10.3 only in that the probability of the good type 1.a is changed to .9 (instead of .2). Thus, in this Bayesian bargaining problem, the expected payoff to player 2 from a mechanism (Q,Y) is

$$U_2(Q,Y \mid 2.0) = .9(50 \ Q(1.a) - Y(1.a)) + .1(30 \ Q(1.b) - Y(1.b)).$$

Except for this change, the basic formulas and constraints that characterize the incentive-feasible mechanisms are the same as for Example 10.1.

Suppose that player 1 is the principal and that he knows his type when he selects the mechanism. Among all the feasible mechanisms, the following maximizes both $U_1(Q,Y \mid 1.a)$ and $U_1(Q,Y \mid 1.b)$:

$$Q^3(1.a) = 1, \quad Y^3(1.a) = 48, \quad Q^3(1.b) = 1, \quad Y^3(1.b) = 48.$$

That is, the best incentive-feasible mechanism for player 1, no matter what his type may be, is to simply offer to sell his unit of the commodity to player 2 for \$48, which is the expected value of the commodity to her (.9 × 50 + .1 × 30 = 48). This mechanism gives expected payoffs

$$U_1(Q^3, Y^3 | 1.a) = 8, \quad U_1(Q^3, Y^3 | 1.b) = 28, \quad U_2(Q^3, Y^3 | 2.0) = 0.$$

Thus, it seems clear that player 1 should select this mechanism (Q^3, Y^3) in this modified example if he is the principal.

Before we move on from this conclusion, we should examine its logical foundations more closely, however, by carefully analyzing the mechanism-selection process as a noncooperative game. In this game, player 1, knowing his type, selects a mechanism to be implemented by a mediator. Players 1 and 2 then decide whether to participate in this mechanism, by signing a contract to accept the outcome that will be specified by the mediator as a function of player 1's type report, or to take the no-trade option and get payoff 0. If both agree to participate, then player 1 reports his type to the mediator (with the possibility of lying) and the mediator implements the trade specified by the mechanism for this report. There is a sequential equilibrium of this mechanism-selection game in which player 1 always selects (Q^3, Y^3) and the players subsequently participate honestly in this mechanism.

However, there are other sequential equilibria of this game. For example, consider the following mechanism

$$Q^4(1.a) = \frac{1}{3}, \quad Y^4(1.a) = 50/3, \quad Q^4(1.b) = 1, \quad Y^4(1.b) = 30,$$

which gives

$$U_1(Q^4, Y^4 | 1.a) = 3\frac{1}{3}, \quad U_1(Q^4, Y^4 | 1.b) = 10, \quad U_2(Q^4, Y^4 | 2.0) = 0.$$

In this mechanism, player 1 has the options either to sell his entire supply of one unit for a price of \$30 or to sell $\frac{1}{3}$ unit at a price of \$50 per unit; and he chooses the former when his type is 1.b $(30 - 20 \geq \frac{1}{3}(50 - 20))$ and the latter when his type is 1.a. Among the set of all incentive-feasible mechanisms that have the property that player 2 never pays a price higher than her actual value for the commodity, this mechanism is the best mechanism for both types of player 1. There is another sequential equilibrium of the mechanism-selection game in which player 1 always selects (Q^4, Y^4) and the players subsequently participate honestly in this mechanism. If player 1 proposed any other mechanism in which player 2 would expect to pay a price higher than her true value against at least one type of player 1, then (in this sequential equilibrium) player 2 would infer from this surprising proposal that player 1's type was the type against which she would pay more than her true value; so she would refuse to participate. For example, if player 1 proposed (Q^3, Y^3) in this sequential equilibrium, then player 2 would believe that his type was 1.b; so she would refuse to participate (because $30 - 48 < 0$). Notice, however, that (Q^4, Y^4) is not incentive-efficient, because (Q^3, Y^3) is interim Pareto superior to it.

So to get incentive efficiency as a result of a mechanism-selection process, we need some further assumptions, like those of cooperative game theory. In particular, we need to assume that a player who selects an incentive-feasible mechanism also has some ability to negotiate for both the inscrutable equilibrium of the mechanism-selection game and the honest participation equilibrium in the implementation of the mechanism itself. That is, we need to assume that the principal can supplement his selection of a mechanism with a negotiation statement of the following form:

> You should not infer anything about my type from the fact that I am selecting this mechanism, because I would have proposed it no matter what my type was. Notice that, given that no one is making any such inferences, it is an equilibrium for everyone to participate honestly in this mechanism. So let us all plan to participate honestly in this mechanism.

We then need to assume that, if this statement passes some kind of credibility test, then all players will believe it and participate honestly in the mechanism. (See Farrell, 1988; Grossman and Perry, 1986; and Myerson, 1989; for general discussions of such credibility tests.)

Now let us consider again the original Example 10.1, in which the probability of the good type 1.a is .2. Among all incentive-feasible mechanisms for this example, the one that maximizes player 2's expected payoff is

$$Q^5(1.a) = 0, \quad Y^5(1.a) = 0, \quad Q^5(1.b) = 1, \quad Y^5(1.b) = 20,$$

which gives

$$U_1(Q^5, Y^5 | 1.a) = 0, \quad U_1(Q^5, Y^5 | 1.b) = 0, \quad U_2(Q^5, Y^5 | 2.0) = 8.$$

That is, in player 2's best mechanism, player 1 sells his whole supply for $20 if the quality is bad (type 1.b), and there is no trade if the quality is good (type 1.a). Notice that this mechanism always gives payoff

0 to player 1 and is incentive efficient. Even though there is a probability .2 that player 2 will not buy a commodity that is always worth more to her than to the seller, this is the best incentive-feasible mechanism for player 2. Thus, if player 2 has all of the negotiating ability, then we should expect her to negotiate for this mechanism, in which she essentially makes a take-it-or-leave-it offer of $20 for the commodity.

Things become more complicated when we assume that player 1 is the principal in Example 10.1, because the feasible mechanism that is best for player 1 (in the sense of maximizing his expected payoff given his true type) depends on what his type is. For the good type (1.a), the best feasible mechanism is

$$Q^6(1.a) = 0, \quad Y^6(1.a) = 8, \quad Q^6(1.b) = 1, \quad Y^6(1.b) = 28,$$

which gives

$$U_1(Q^6, Y^6 | 1.a) = 8, \quad U_1(Q^6, Y^6 | 1.b) = 8, \quad U_2(Q^6, Y^6 | 2.0) = 0.$$

In this mechanism, player 2 pays a nonrefundable fee of $8 to player 1 in exchange for the right to then make a take-it-or-leave-it offer to buy one unit of the commodity for an additional payment of $20 (if 1 accepts the offer). On the other hand, the best incentive-feasible mechanism for type 1.b of player 1 is

$$Q^7(1.a) = 4/7, \quad Y^7(1.a) = 40 \times 4/7 = 22.86,$$
$$Q^7(1.b) = 1, \quad Y^7(1.b) = 31.43,$$

which gives

$$U_1(Q^7, Y^7 | 1.a) = 0, \quad U_1(Q^7, Y^7 | 1.b) = 11.43, \quad U_1(Q^7, Y^7 | 2.0) = 0.$$

In this mechanism player 2 buys 4/7 units of the commodity at the price of $40 per unit if player 1's type is 1.a, and player 2 buys one unit of the commodity for $31.43 if player 1's type is 1.b. In this mechanism, the losses that player 2 would get if 1's type is 1.b just cancel in expected value the gains that player 2 would get if 1's type is 1.a.

Naive analysis of mechanism selection by player 1 might suggest that player 1 would select the mechanism (Q^6, Y^6) if he is type 1.a and (Q^7, Y^7) if he is type 1.b. But then player 2, being intelligent, would know player 1's type as soon as he chooses the mechanism. So if player 1 chose (Q^6, Y^6), then player 2 would infer that 1 was type 1.a, so she would refuse to participate in the mechanism (anticipating that 1 would pocket the \$8 fee and reject the offer to sell for \$20). The mechanism (Q^6, Y^6) satisfied the participation constraint only when it was assumed that both types of player 1 would choose it.

Similarly, participating in (Q^7, Y^7) ceases to be rational for player 2 if she understands the fact that this mechanism would be selected only if player 1's type is 1.b. Another way to see this difficulty is to apply the inscrutability principle and observe that this naive theory is equivalent to the theory that player 1 would inscrutably select the mechanism that coincides with (Q^6, Y^6) if he is type 1.a and (Q^7, Y^7) if he is type 1.b. This inscrutable equivalent mechanism is

$$Q^8(1.a) = 0, \quad Y^8(1.a) = 8, \quad Q^8(1.b) = 1, \quad Y^8(1.b) = 31.43,$$

which gives

$$U_1(Q^8, Y^8 | 1.a) = 8, \quad U_1(Q^8, Y^8 | 1.b) = 11.43,$$
$$U_2(Q^8, Y^8 | 2.0) = -2.74,$$

so it is not individually rational for player 2.

So no matter what the type of player 1 may be, he cannot implement the feasible mechanism that is best for him unless player 2 believes that both types would have selected the same mechanism. Because these are different mechanisms for the different types, and because player 2 is assumed to be intelligent, player 1 must sometimes select a mechanism other than the feasible mechanism that he actually most prefers. That is, player 1 must make some compromise between what he really wants and what he might have wanted if his type had been different, in order to select the mechanism inscrutably.

In general, to understand mechanism selection by a principal with private information, we need a theory of how he should make such an *inscrutable intertype compromise*. This is a difficult problem, but there is a class of situations where a strong solution to the informed principal's problem can be defined.

Given any Bayesian collective-choice problem or Bayesian bargaining problem where one player i is specified as the principal, we say that a mechanism is a *safe* proposal for the principal iff it is incentive feasible and it would remain incentive feasible even if all the players knew the principal's true type, no matter what that type may be. We say that a mechanism μ is *interim inferior for the principal* to some other mechanism ν iff there is at least one type of the principal who would expect a higher payoff in v than in μ (in the sense that $U_i(v|t_i) > U_i(\mu|t_i)$) and there does not exist any type of the principal who would expect a lower payoff in v than in μ. A mechanism is a *strong solution* for the principal iff it is safe for the principal and is not interim inferior for the principal to any other incentive-feasible mechanism.

Many Bayesian games have no strong solution. For the modified Example 10.2 discussed at the beginning of this section, the best safe mechanism for player 1 is (Q^4, Y^4), which is interim inferior to (Q^3, Y^3) for him. So there is no strong solution for player 1 in Example 10.2 (even though there is a mechanism that is best for both types of player 1).

In Example 10.1 (where (Q^3, Y^3) is not incentive feasible), player 1's best safe mechanism (Q^4, Y^4) is not interim inferior to any other incentive-feasible mechanism for player 1. (Notice that $.3 \times 3\frac{1}{3} + .7 \times 10 = 8$, so (Q^4, Y^4) satisfies the incentive-efficiency condition (10.16) with $U_2(Q^4, Y^4|2.0) = 0$.) So (Q^4, Y^4), in which player 1 has the option to either sell his entire supply of one unit for a price of \$30 or sell $\frac{1}{3}$ unit at a price of \$50 per unit, is a strong solution for player 1 in Example 10.1.

Myerson (1983) argued that, when such a strong solution exists, it is the most reasonable solution to the principal's mechanism-selection problem. No matter what the other players might infer about the principal's type when he selects his strong solution, they would still be willing to participate honestly in the μ mechanism, because it is safe. On the other hand, suppose that μ is a strong solution for the principal, but suppose that he actually selected some other mechanism v that would (if implemented with honest participation by all players) be better than μ for at least one of his types. Let $S_i(v,\mu)$ denote the set of the principal's types that would prefer v to μ, that is

$$S_i(v,\mu) = \{t_i \in T_i | U_i(v|t_i) > U_i(\mu|t_i)\},$$

where i is the principal. If the principal had been expected to select μ, then the other players might naturally infer from his selection of v that his type was in $S_i(v,\mu)$. However, if they calculate their new beliefs by Bayes's rule on the basis of this inference, then in the game with these new beliefs the mechanism v must violate at least one informational incentive constraint or participation constraint. If v were incentive feasible for the game with these new beliefs, then the mechanism η defined such that, for each c in C,

$$\eta(c|t) = \nu(c|t) \text{ if } t_i \in S_i(\nu,\mu),$$
$$\eta(c|t) = \mu(c|t) \text{ if } t_i \notin S_i(\nu,\mu),$$

would be incentive feasible for the game with the original beliefs, and μ would be interim inferior to η for the principal; but this result would contradict the fact that μ is not interim inferior for the principal to any other incentive-feasible mechanism. Thus, any other mechanism ν would be rendered infeasible (either not incentive compatible or not individually rational) as soon as the other players infer that the principal prefers ν to the strong solution.

Furthermore, all strong solutions must be essentially equivalent for the principal, in the sense that, if i is the principal and μ and μ' are both strong solutions for him, then $U_i(\mu|t_i)$ must equal $U_i(\mu'|t_i)$ for every type t_i in T_i. If not, then by the preceding argument μ' would not be incentive feasible if the other players inferred that player i's type was in the nonempty set $\{t_i \in T_i | U_i(\mu'|t_i) > U_i(\mu|t_i)\}$, which would contradict the assumption that μ' is safe for player i.

Myerson (1983) axiomatically derived a *neutral optimum* solution concept for mechanism selection by an informed principal. An incentive-efficient neutral optimum always exists, for any given principal player. If a strong solution exists, then it is a neutral optimum. Furthermore, a neutral optimum is durable and is not interim inferior to any other incentive-compatible mechanism for the principal. In Example 10.1, (Q^4, Y^4) is the unique strong solution and neutral optimum for player 1. In the modified Example 10.2, where player 1 has no strong solution, (Q^3, Y^3) is the unique neutral optimum for player 1. We omit here the formal definition of a neutral optimum (see Myerson, 1983), and instead we develop next a closely related solution concept for mechanism selection by two players who have equal negotiating ability.

10.8 Neutral Bargaining Solutions

Harsanyi and Selten (1972) first studied the question of how to define a generalization of the Nash bargaining solution that can be applied to games with incomplete information. Myerson (1984a,b) developed generalizations of both the Nash bargaining solution and the Shapley NTU value for games with incomplete information. In this section, we consider this generalization of the Nash bargaining solution, called the neutral bargaining solution, for two-player Bayesian bargaining problems of the form

$$\Gamma^\beta = (\{1,2\}, C, d^*, T_1, T_2, \bar{p}_1, \bar{p}_2, u_1, u_2),$$

as defined in Section 10.1.

The neutral bargaining solution can be derived from two axioms. One of these axioms asserts something about what the solution should look like for a class of problems that are fairly straightforward to understand (like Nash's symmetry axiom), and the other axiom asserts something about how the solutions to different games should be related (like Nash's independence of irrelevant alternatives axiom).

AXIOM 10.1 (RANDOM DICTATORSHIP). *Suppose that there exist two outcomes b^1 and b^2 in C such that*

$$u_2(b^1,t) = 0 \quad \text{and} \quad u_1(b^2,t) = 0, \quad \forall t \in T,$$

and the mechanisms $[b^1]$, $[b^2]$, and $.5[b^1] + .5[b^2]$ are all incentive efficient. (Here $[b^i]$ is the mechanism that always chooses b^i, and $.5[b^1] + .5[b^2]$ is the mechanism that always gives probability .5 to each of b^1 and b^2, no matter what types are reported.) Then $5[b^1] + .5[b^2]$ is a bargaining solution for Γ.

AXIOM 10.2 (EXTENSIONS). *Suppose that μ is an incentive-efficient mechanism for a Bayesian bargaining problem Γ^β, where*

$$\Gamma^\beta = (\{1,2\}, C, d^*, T_1, T_2, \bar{p}_1, \bar{p}_2, u_1, u_2).$$

Suppose also that, for each positive number ε, there exists a Bayesian bargaining problem $\hat{\Gamma}^\beta(\varepsilon)$ and a bargaining solution η for $\hat{\Gamma}^\beta(\varepsilon)$ such that

$$\hat{\Gamma}^\beta(\varepsilon) = (\{1,2\}, \hat{C}, d^*, T_1, T_2, \bar{p}_1, \bar{p}_2, \hat{u}_1, \hat{u}_2),$$
$$\hat{C} \supseteq C,$$
$$\hat{u}_i(c,t) = u_i(c,t), \quad \forall i \in \{1,2\}, \quad \forall c \in C, \forall t \in T, \text{ and}$$
$$U_i(\mu|t_i) \geq \hat{U}_i(\eta|t_i) - \varepsilon, \quad \forall i \in \{1,2\}, \forall t_i \in T_i.$$

Then μ must be a bargaining solution for Γ. (Here $\hat{U}_i(\eta|t_i)$ is the expected payoff to type t_i of player i from the mechanism η in the game $\hat{\Gamma}^\beta(\varepsilon)$.)

Suppose, as stated in the hypotheses of Axiom 10.1, that $[b^1]$ is an incentive-efficient mechanism that always gives player 2 the worst payoff that she can guarantee herself. Then $[b^1]$ cannot be interim inferior to any other incentive-feasible mechanism for player 1. Furthermore, there obviously can never be any difficulty satisfying informational incentive constraints for a mechanism that always selects the same outcome, so $[b^1]$ must be a safe mechanism for player 1. Thus, the hypotheses of Axiom 10.1 imply that $[b^1]$ is a strong solution for player 1. Similarly, $[b^2]$ is a strong solution for player 2. So Γ^β is an example in which we can confidently predict what mechanism each player would select if he were the principal.

Assuming the hypotheses of Axiom 10.1, suppose that we gave each player a probability .5 of acting as the principal and selecting the mechanism dictatorially. Then the overall result of this random dictatorship process would be to implement $.5[b^1] + .5[b^2]$, which is also given to be incentive efficient. Because the random dictatorship is clearly equitable to the two players and it is efficient, it has a fair claim to being a bargaining solution.

Of course, the hypotheses of Axiom 10.1 are quite restrictive and are likely to be satisfied by relatively few games, but that just means that this axiom is very weak. Only when it is clear what a random dictatorship scheme would lead to and only when it is also efficient, do we then argue that the random-dictatorship mechanism should be a bargaining solution.

To appreciate how restrictive these hypotheses are, consider a two-person bargaining problem $(F,(0,0))$ (with complete information) as defined in Chapter 8. In such a complete-information context, the hypotheses of Axiom 10.1 imply that the allocation $.5(h_1(0,F),0) + .5(0,h_2(0,F))$ is Pareto efficient in F; but that can happen only if the individually rational Pareto frontier of F is a straight line and its midpoint $.5(h_1(0,F),0) + .5(0,h_2(0,F))$ is then the Nash bargaining solution. (Here h_i is as defined in Section 8.6.)

Axiom 10.2 is related to Nash's independence of irrelevant alternatives axiom, together with a kind of upper-hemicontinuity condition. If $\hat{\Gamma}^\beta(\varepsilon)$ satisfies the hypotheses of Axiom 10.2, then $\hat{\Gamma}^\beta(\varepsilon)$ differs from Γ^β only in that $\hat{\Gamma}^\beta(\varepsilon)$ contains some possible outcomes that are not in Γ^β; so we say that $\hat{\Gamma}^\beta(\varepsilon)$ is an *extension* of Γ^β. However, in spite of the fact that the mechanism μ does not use any of these additional outcomes (it is feasible for the more restricted game Γ^β), it is almost interim Pareto superior to a bargaining solution for the extension $\hat{\Gamma}^\beta(\varepsilon)$. Axiom 10.2 asserts that if there are ways to extend the Bayesian bargaining problem Γ^β such that the incentive-efficient mechanism μ is interim Pareto superior to a bargaining solution of the extension, or comes arbitrarily close to being interim Pareto superior to a bargaining solution of the extension, then μ must be a bargaining solution itself.

There are many bargaining solution concepts that satisfy these two axioms. For example, both axioms would be satisfied by letting the set of "bargaining solutions" be the set of all incentive-efficient mechanisms. A *neutral bargaining solution* is defined to be any mechanism that is a solution for *every* bargaining solution concept that satisfies these two axioms. Myerson (1984a) proved that the neutral bargaining solutions themselves satisfy the two axioms, and the set of neutral bargaining solutions is nonempty for any finite two-player Bayesian bargaining problem. The following characterization theorem is also proved by Myerson (1984a).

THEOREM 10.3. *A mechanism μ is a neutral bargaining solution for a finite two-player Bayesian bargaining problem Γ^β if and only if, for each positive number ε, there exist vectors λ, α, and w (which may depend on ε) such that*

$$(10.22) \quad \left((\lambda_i(t_i) + \sum_{s_i \in T_i} \alpha_i(s_i|t_i)) \, w_i(t_i) - \sum_{s_i \in T_i} \alpha_i(t_i|s_i) \, w_i(s_i) \right) / \bar{p}_i(t_i)$$

$$= \sum_{t_{-i} \in T_{-i}} \bar{p}_{-i}(t_{-i}) \max_{c \in C} \sum_{j \in \{1,2\}} v_j(c,t,\lambda,\alpha)/2, \quad \forall i \in N, \; \forall t_i \in T_i;$$

$$\lambda_i(t_i) > 0 \; \text{ and } \; \alpha_i(s_i|t_i) \geq 0, \quad \forall i \in N, \; \forall s_i \in T_i, \; \forall t_i \in T_i;$$

$$\text{and } U_i(\mu|t_i) \geq w_i(t_i) - \varepsilon, \quad \forall i \in N, \; \forall t_i \in T_i.$$

(In this two-player context, we let $-i$ denote the player other than i; and $v_j(\cdot)$ is the virtual utility function defined in Section 10.5.)

Proof. Given any vector λ of positive utility weights and any vector α of nonnegative Lagrange multipliers, consider an extension $\hat{\Gamma}^\beta$ in which

$$\hat{C} = C \cup \{b^1, b^2\},$$
$$\hat{u}_1(b^2,t) = 0 = \hat{u}_2(b^1,t), \quad \forall t \in T,$$

$$\left((\lambda_1(t_1) + \sum_{s_1 \in T_1} \alpha_1(s_1|t_1)) \, \hat{u}_1(b^1,t) - \sum_{s_1 \in T_1} \alpha_1(t_1|s_1) \, \hat{u}_1(b^1,(s_1,t_2)))\right)/\bar{p}_1(t_1)$$

$$= \max_{c \in C} \sum_{j \in \{1,2\}} v_j(c,t,\lambda,\alpha), \quad \forall t \in T,$$

$$\left((\lambda_2(t_2) + \sum_{s_2 \in T_2} \alpha_2(s_2|t_2)) \, \hat{u}_2(b^2,t) - \sum_{s_2 \in T_2} \alpha_2(t_2|s_2) \, \hat{u}_2(b^2,(t_1,s_2)))\right)/\bar{p}_2(t_2)$$

$$= \max_{c \in C} \sum_{j \in \{1,2\}} v_j(c,t,\lambda,\alpha), \quad \forall t \in T.$$

Then the mechanisms $[b^1]$, $[b^2]$, and $.5[b^1] + .5[b^2]$ all are incentive-efficient for $\hat{\Gamma}^\beta$, because they satisfy the conditions of Theorem 10.1 with λ and α. Thus, to satisfy the random-dictatorship axiom, $.5[b^1] + .5[b^2]$ must be a bargaining solution for $\hat{\Gamma}^\beta$. Furthermore, the system of equations (10.22) in the theorem can then be satisfied by letting

$$w_i(t_i) = \hat{U}_i(.5[b^1] + .5[b^2]|t_i) = \sum_{t_{-i} \in T_{-i}} \frac{\bar{p}_{-i}(t_{-i})\hat{u}_i(b^i,t)}{2} .$$

It can be shown (see Lemma 1 of Myerson, 1983) that, for any λ and α that satisfy the positivity and nonnegativity conditions in Theorem 10.3, the system of equations (10.22) has a unique solution $w = (w_i(t_i))_{t_i \in T_i, i \in N}$. Thus, any such solution must correspond to the allocation of interim expected payoffs to the players in a bargaining solution of an extension of Γ^β. So the extensions axiom implies that any incentive-efficient mechanism μ that satisfies the conditions of Theorem 10.3 must be a bargaining solution of Γ.

To complete the proof, it remains to show that the set of all incentive-efficient mechanisms μ for which the conditions in Theorem 10.3 can be satisfied is a solution set that itself satisfies the two axioms. The technical details of this argument can be found in Myerson (1984a). ∎

Notice that the vectors λ, α, and w may depend on the positive number ε in Theorem 10.3. However, if we can find λ, α, and w that satisfy the conditions of this theorem for the case of $\varepsilon = 0$ (as is possible in many examples) then the same λ, α, and w will satisfy Theorem 10.3 for every positive ε. Furthermore, it can be shown that such λ and α will also satisfy the incentive-efficiency conditions in Theorem 10.1 for the mechanism μ.

To interpret the conditions in Theorem 10.3, consider the (λ,α) *virtual bargaining problem*, a fictitious game that differs from Γ^β in the following three ways. First, the players' types are verifiable (as if, for example, each player has an identity card that says what his type is, which he can be asked to show to prove that he is not lying about his type), so there are no informational incentive constraints. Second, each player's payoffs are in the virtual utility scales $v_i(\cdot,\cdot,\lambda,\alpha)$ with respect to λ and α, instead of $u_i(\cdot,\cdot)$. Third, these virtual utility payoffs are transferable among the players.

For the (λ,α) virtual bargaining problem, it is easy to identify a mechanism that would be both equitable and efficient in almost any reasonable sense of these terms: for any combination of types, the outcome in C should be one that maximizes the sum of the players' transferable virtual-utility payoffs (for efficiency), and transfers should then be used to allocate this transferable payoff equally among the players (for equity). (Notice that each player's virtual-utility payoff in the disagreement outcome is always 0, by (10.7) and (10.17).) Because types are verifiable, there would be no incentive constraints to prevent an arbitrator or mediator from implementing such an equitable and efficient mechanism for the (λ,α) virtual bargaining problem.

Thus, considerations of equity and efficiency in the (λ,α) virtual bargaining problem lead to a mechanism in which the expected payoff to player i when his type is t_i is

$$\sum_{t_{-i}\in T_{-i}} \bar{p}_{-i}(t_{-i})\max_{c\in C} \sum_{j\in\{1,2\}} \frac{v_j(c,t,\lambda,\alpha)}{2} \,.$$

By definition of virtual utility (10.17), the system of equations (10.22) asserts that, for each player i, the vector $(w_i(t_i))_{t_i\in T_i}$ is the allocation of expected utility payoffs for the possible types of player i that corresponds to these equitable and efficient expected virtual-utility payoffs in the (λ,α) virtual bargaining problem. Thus, any allocation vector

$$w = (w_i(t_i))_{t_i\in T_i, i\in\{1,2\}}$$

that satisfies (10.22) for some positive λ and some nonnegative α can be called *virtually equitable* for Γ^β.

So Theorem 10.3 asserts that a neutral bargaining solution is an incentive-efficient mechanism that generates type-contingent expected utilities that are equal to or interim Pareto superior to a limit of virtually equitable utility allocations. That is, Theorem 10.3 suggests that, in negotiation or arbitration, players may make interpersonal equity comparisons in virtual utility terms, rather than in actual utility terms. When we first introduced virtual utility in Section 10.4, it was offered only as a convenient way of characterizing incentive-efficient mechanisms. Theorem 10.3 suggests that virtual utility scales can be used for characterizing equity as well as efficiency in bargaining problems with incomplete information.

To put it less formally, Theorem 10.3 suggests the following theory about bargaining with incomplete information, which we call the *virtual utility hypothesis*: in response to the pressure that a player feels from other players' distrust of any statements that he could make about his type, he might bargain as if he wanted to maximize some virtual utility scale that differs from his actual utility scale in a way that exaggerates the difference from the false types that jeopardize his true type. When he argues about the equity and efficiency of proposed mechanisms, his bargaining behavior may be characterized by such virtual-utility exaggeration.

Consider again Example 10.1. The neutral bargaining solution for this example is

$$Q^9(1.a) = 1/6, \quad Y^9(1.a) = 50/6, \quad Q^9(1.b) = 1, \quad Y^9(1.b) = 25,$$

which gives

$$U_1(Q^9, Y^9 | 1.a) = 1.67, \quad U_1(Q^9, Y^9 | 1.b) = 5, \quad U_2(Q^9, Y^9 | 2.0) = 4.$$

The conditions in Theorem 10.3 can be satisfied for all ε by the same λ and α that we used for this example in Section 10.4,

$$\lambda_1(1.a) = 0.3, \quad \lambda_1(1.b) = 0.7, \quad \lambda_2(2.0) = 1,$$
$$\alpha_1(1.a | 1.b) = 0.1, \quad \alpha_1(1.b | 1.a) = 0.$$

With these parameters, the only difference between virtual utility and actual utility is that $50 is the virtual value of the commodity to player 1 when his type is 1.a (instead of the actual value of $40). So the maximum sum of virtual-utility payoffs is 0 (= 50 − 50) when player 1 is type 1.a, and 10 (= 30 − 20) when player 1 is type 1.b. The virtual-equity equations (10.22) then become

$$\frac{0.3 \, w_1(1.a) - 0.1 \, w_1(1.b)}{.2} = \frac{0}{2},$$

$$\frac{(0.7 + 0.1) \, w_1(1.b)}{.8} = \frac{10}{2},$$

$$w_2(2.0) = .8 \left(\frac{10}{2} \right) + .2 \left(\frac{0}{2} \right),$$

which have the unique solution

$$w_1(1.a) = 1\tfrac{2}{3}, \quad w_1(1.b) = 5, \quad w_1(2.0) = 4.$$

In this neutral bargaining solution, player 1 sells $\frac{1}{6}$ of his supply at a price of \$50 per unit when his type is 1.a and sells all of his supply at a price of \$25 per unit when his type is 1.b. It is interesting to compare this mechanism to the first mechanism that we discussed in Section 10.3, (Q^1, Y^1), in which player 1 sells $\frac{1}{5}$ of his supply at a price of \$45 per unit when his type is 1.a and sells all of his supply at a price of \$25 per unit when his type is 1.b. The mechanism (Q^1, Y^1) is the unique incentive-efficient mechanism in which the commodity is always sold at a price per unit that is halfway between its value to the seller and its value to the buyer; so (Q^1, Y^1) might seem to be equitable and efficient in a natural intuitive sense. However, (Q^1, Y^1) is not equitable (or even individually rational) in the (λ, α) virtual bargaining problem, because it has player 1 selling some of his supply for a price of \$45 per unit when his type is 1.a and the virtual value of the commodity to him is \$50. To satisfy virtual equity in the (λ, α) virtual bargaining problem, the selling price must be \$50 per unit when 1's type is 1.a, as it is in (Q^9, Y^9).

The neutral bargaining solution (Q^9, Y^9) can be directly interpreted in terms of the logic of the random-dictatorship axiom. In Section 10.6, we argued that, if player 1 were the principal who could dictatorially choose the mechanism, he would choose (Q^4, Y^4), in which player 1 sells $\frac{1}{3}$ of his supply at a price of \$50 per unit when his type is 1.a and sells all of his supply at a price of \$30 per unit when his type is 1.b. If player 2 were the principal or dictator, she would choose the mechanism (Q^5, Y^5), in which player 1 sells nothing when his type is 1.a and sells all of his supply at a price of \$20 per unit when his type is 1.b. A .50–.50 randomization between these two dictatorial mechanisms has, for each of the two possible types of player 1, the same expected quantity sold and the same expected cash payment as (Q^9, Y^9); that is, for each t_1 in T_1,

$$Q^9(t_1) = .5Q^4(t_1) + .5Q^5(t_1) \text{ and } Y^9(t_1) = .5Y^4(t_1) + .5Y^5(t_1).$$

So the neutral bargaining solution is essentially equivalent to a random-dictator mechanism for this example.

When player 1's type is 1.a, he is in a position that can be described as surprisingly strong: "surprising" because the probability of this type is rather small; and "strong" because his willingness to trade at any given price is never greater than it would have been had his type been different. When player 1 is in such a surprisingly strong position, the terms of trade (the price per unit traded) in the neutral bargaining solution (which is based on the assumption that players 1 and 2 have equal negotiating ability) are similar to the terms of trade that would occur if player 1 were the principal (with all the negotiating ability), except that the expected quantity of trade is less. This property of neutral bargaining solutions has been observed in other examples (see Myerson, 1985b) and has been called *arrogance of strength*.

To better understand why such arrogance of strength might be reasonable to expect, it helps to think further about the question of inscrutable intertype compromise that arises in the theory of bargaining with incomplete information. Any theory that specifies an incentive-efficient mechanism for this example must implicitly make some trade-off or compromise between the two possible types of player 1, if only because there are many incentive-efficient mechanisms that all give the same expected payoff to player 2 but differ in the expected payoffs that they give to 1.a and 1.b. For example, compare (Q^9, Y^9) to

$$Q^{10}(1.a) = 4/19, \quad Y^{10}(1.a) = 45 \times 4/19,$$
$$Q^{10}(1.b) = 1, \quad Y^{10}(1.b) = 25.26,$$

which gives

$$U_1(Q^{10}, Y^{10}|1.a) = 1.05, \quad U_1(Q^{10}, Y^{10}|1.b) = 5.26,$$
$$U_2(Q^{10}, Y^{10}|2.0) = 4;$$

so it is worse than (Q^9, Y^9) for 1.a but better than (Q^9, Y^9) for 1.b. Imagine that the players are bargaining over the selection of a mediator and that they have two candidates for the job, of whom one promises to implement (Q^9, Y^9) and the other promises to implement (Q^{10}, Y^{10}). Player 2 is initially indifferent between these two candidates, but if player 2 knew which mediator was actually preferred by player 1, then player 2 would prefer the other. So it might seem that neither type of player 1 would dare to express his true preference between these two mediators, lest player 2 then insist on the other mechanism.

However, this argument disregards the fact that the strong type 1.a could benefit greatly in bargaining from revealing his type to player 2, when the possibility of other mechanisms is considered. After all, in the complete information game where player 2 knows that player 1 is type 1.a, the Nash bargaining solution would be for player 1 to sell his entire supply for $45. So, if player 1 is type 1.a, he could confidently argue for hiring the (Q^9, Y^9) mediator and add boldly, "And if you infer from my preference for (Q^9, Y^9) that my type is 1.a, then you should dispense with both of these mediators and just buy my whole supply for $45, which would be better for you and just as good for me when I am 1.a!" To maintain inscrutability then, player 1 would have to make the same argument if he were type 1.b. That is, because player 1 would actually be very eager to reveal information about his type if he were 1.a, and only 1.b really would have some incentive to conceal his type in bargaining, the inscrutable intertype compromise between the strong type 1.a and the weak type 1.b tends to get resolved in favor of the strong type.

In Example 10.2, the type 1.a has probability .9, so the strong type would not be "surprising" and there is no arrogance of strength. (To be more precise, we can say that the strong type 1.a ceases to be "surprising" when the probability of 1.a becomes high enough that player 2 prefers the mechanism in which trade occurs for sure at $40 over the mechanism in which trade occurs at $20 but only when player 1's type is 1.b.) The neutral bargaining solution for Example 10.2 is

$$Q^{11}(1.a) = 1, \quad Y^{11}(1.a) = 44, \quad Q^{11}(1.b) = 1, \quad Y^{11}(1.b) = 44,$$

(that is, all of the commodity is always sold at a price of $44), which gives

$$U_1(Q^{11}, Y^{11} | 1.a) = 4, \quad U_1(Q^{11}, Y^{11} | 1.b) = 24,$$
$$U_2(Q^{11}, Y^{11} | 2.0) = 4.$$

The virtual equity conditions of Theorem 10.3 are satisfied in Example 10.2 by giving essentially zero weight to the unlikely weak type 1.b of player 1. To be specific, the conditions of Theorem 10.3 can be satisfied for any positive ε by letting

$$\lambda_1(1.a) = 1 - 0.1\varepsilon, \quad \lambda_1(1.b) = 0.1\varepsilon, \quad \lambda_2(2.0) = 1,$$
$$\alpha_1(1.a | 1.b) = 0.1 - 0.1\varepsilon, \quad \alpha_1(1.b | 1.a) = 0.$$

With these parameters, the virtual utility payoffs differ from actual utility payoffs only in that the virtual value of the commodity for player 1 when his type is 1.a is

$$\frac{(1 - 0.1\varepsilon)40 - (0.1 - 0.1\varepsilon)20}{.9} = 42\%9 - \varepsilon\ 2\%9.$$

So the maximum possible sum of virtual utility payoffs is

$$50 - (42\%9 - \varepsilon\ 2\%9) = 7\%9 + \varepsilon\ 2\%9.$$

when player 1's type is 1.a and is $30 - 20 = 10$ when player 1's type is 1.b. So the virtual equity equations can be written:

$$\frac{(1 - 0.1\varepsilon)w_1(1.a) - (0.1 - 0.1\varepsilon)w_1(1.b)}{.9} = (7\%9 + \varepsilon\ 2\%9)/2,$$

$$\frac{(0.1\varepsilon + (0.1 - 0.1\varepsilon))w_1(1.b)}{.1} = \frac{10}{2},$$

$$w_2(2.0) = .9(7\%9 + \varepsilon\ 2\%9)/2 + .1(10/2).$$

The unique solution to these equations is

$$w_1(1.a) = \frac{4 + 0.5\varepsilon}{1 - 0.1\varepsilon}, \quad w_1(1.b) = 5, \quad w_2(2.0) = 4 + \varepsilon.$$

As ε goes to 0, these solutions converge to the limiting allocation

$$w_1(1.a) = 4, \quad w_1(1.b) = 5, \quad w_2(2.0) = 4,$$

The neutral bargaining solution is interim Pareto superior to this limiting allocation.

Using the virtual utility hypothesis, Myerson (1984b) has proposed a solution concept for multiplayer cooperative games with incomplete information that generalizes both the neutral bargaining solution (in the two-player case) and the Shapley NTU value (in the complete information case). In the definition of this generalized NTU value, fictitious transfers of virtual utility with respect to some parameters λ and α have the same role that fictitious transfers of λ-weighted utility have in the definition of the Shapley NTU value for games with complete information (see Section 9.9). Notice that, if each player has only one possible type, then the α parameters drop out of the definition of virtual utility (10.17), which then reduces to simple λ-weighted utility.

10.9 Dynamic Matching Processes with Incomplete Information

The development of the inner core in Section 9.8 can be generalized to games with incomplete information. The most convenient way to do so is in terms of a model of a game that is replicated to a dynamic matching process, for which we can prove an existence theorem that generalizes Theorem 9.6.

As in Chapter 9, let $L(N)$ be the set of all coalitions or nonempty subsets of N. For any coalition S, let $T_S = \times_{i \in S} T_i$ denote the set of all possible type profiles for the players in S, and let C_S denote the set of all jointly feasible actions that the players in S can choose if they act cooperatively. Suppose that, if a coalition S with type profile t_S in T_S cooperatively chooses action c_S in C_S, then the payoff to each player i is $u_i(c_S, t_S)$. We are assuming here that the payoffs to the players in S do not depend on the types and actions of players outside of the coalitions. (We call this an *orthogonal coalitions* assumption, like the condition discussed in Section 9.2.)

When we study a *dynamic matching process*, the set N is reinterpreted, not as the set of players, but as the set of outwardly distinguishable *classes* of players. For each i in N, the set T_i is the set of possible privately known types of class-i players. That is, we assume that each player can be described by his class i in N and his type t_i in T_i, and that his class is publicly observable and verifiable but his type is privately known and directly observable only to himself. Let us assume that each player can join only one coalition, from which he derives his utility payoff. We say that a player *exits* from the matching system when he is assigned to a coalition. The function of the matching process is to assign each player into some coalition with some combination of types that will choose some action. Between the point in time when a player enters the matching process and the point in time when he exits in a coalition, he is waiting and available to be matched.

Let us first consider a picture of the dynamic matching process at one point in time. Suppose that the waiting population is quite large, and let $f_i(t_i)$ denote the proportion of class-i type-t_i individuals in the waiting population at this point in time. Let $w_i(t_i)$ denote the expected payoff that class-i type-t_i individuals anticipate from staying in the matching process, if it operates according to some established matching plan (which we will consider in more detail later). We can write

$$w = (w_i(t_i))_{i \in N, t_i \in T_i}, \quad f = (f_i(t_i))_{i \in N, t_i \in T_i},$$

and we refer to this pair of vectors as the *waiting-population character-istics*.

Now suppose that a mediator wants to compete with this established matching plan and assign some currently waiting individuals to *blocking coalitions*. His blocking plan may be described by a vector η of the form

$$\eta = (\eta_S(c_S, t_S))_{S \in L(N), c_S \in C_S, t_S \in T_S},$$

where $\eta_S(c_S, t_S)$ denotes the number of blocking coalitions that he would form (if he successfully implemented this blocking plan) that would have type profile t_S and would choose joint action c_S, divided by the total number of waiting individuals at this point in time. Thus, under this blocking plan, the probability that a class-i type-t_i individual would be assigned to a blocking coalition with types profile t_S and that will choose joint action c_S is

$$\frac{\eta_S(c_S, t_S)}{f_i(t_i)}.$$

(As usual, t_i is understood here to be the i-component of t_S. The other components of t_S may be denoted t_{S-i}, so we can write $t_S = (t_{S-i}, t_i)$.)

We say that η is a *viable blocking plan*, relative to the waiting-population characteristics (w, f), iff η satisfies the following conditions, for every i in N, every t_i in T_i, and every r_i in T_i:

$$(10.23) \quad \sum_{S \supseteq \{i\}} \sum_{c_S \in C_S} \sum_{t_{S-i} \in T_{S-i}} \frac{\eta_S(c_S, t_S)}{f_i(t_i)} \leq 1;$$

$$(10.24) \quad \sum_{S \supseteq \{i\}} \sum_{c_S \in C_S} \sum_{t_{S-i} \in T_{S-i}} \frac{(u_i(c_S, t_S) - w_i(t_i)) \, \eta_S(c_S, t_S)}{f_i(t_i)} \geq 0;$$

$$(10.25) \quad \sum_{S \supseteq \{i\}} \sum_{c_S \in C_S} \sum_{t_{S-i} \in T_{S-i}} \frac{(u_i(c_S, t_S) - w_i(t_i)) \, \eta_S(c_S, t_S)}{f_i(t_i)}$$

$$\geq \sum_{S \supseteq \{i\}} \sum_{c_S \in C_S} \sum_{t_{S-i} \in T_{S-i}} \frac{(u_i(c_S, t_S) - w_i(t_i)) \, \eta_S(c_S, t_{S-i}, r_i))}{f_i(r_i)};$$

$$(10.26) \quad \eta_S(c_S, t_S) \geq 0, \quad \forall S \in L(N), \ \forall c_S \in C_S, \ \forall t_S \in T_S;$$

$$(10.27) \quad \sum_{S \in L(N)} \sum_{c_S \in C_S} \sum_{t_S \in T_S} \eta_S(c_S, t_S) > 0.$$

Condition (10.23) asserts that the total probability of being assigned to a blocking coalition cannot be greater than 1 for any type of individual. The inequality can be strict in (10.23), because some individuals may be left unassigned by the blocking mediator and instead may simply continue waiting for an assignment in the established matching plan. Condition (10.24), an individual-rationality constraint, asserts that no individual should expect to do worse under this blocking plan than if he continues to wait. Condition (10.25), an informational incentive constraint, asserts that no individual could increase his expected gain under the blocking plan by lying about his type. Conditions (10.26) and (10.27) assert that the relative numbers of the various kinds of coalitions must be nonnegative and not all 0. Notice that these constraints are well defined only if each $f_i(t_i)$ number is nonzero.

We can say that the waiting-population characteristics (w, f) are *strongly inhibitive* iff $f_i(t_i) > 0$ for every i and t_i, and there does not exist any viable blocking plan relative to (w, f). The following theorem extends Theorem 9.5 from the case of complete information.

THEOREM 10.4. *Waiting-population characteristics (w, f) are strongly inhibitive if and only if there exist vectors λ and α such that*

$$\lambda_i(t_i) \geq 0 \quad and \quad \alpha_i(r_i | t_i) \geq 0, \quad \forall i \in N, \ \forall r_i \in T_i, \ \forall t_i \in T_i;$$

$$\sum_{i \in S} \left((\lambda_i(t_i) + \sum_{r_i \in T_i} \alpha_i(r_i | t_i)) \, w_i(t_i) - \sum_{r_i \in T_i} \alpha_i(t_i | r_i) \, w_i(r_i) \right) / f_i(t_i)$$

$$> \sum_{i \in S} \left((\lambda_i(t_i) + \sum_{r_i \in T_i} \alpha_i(r_i | t_i)) \, u_i(c_S, t_S) \right.$$

$$\left. - \sum_{r_i \in T_i} \alpha_i(t_i | r_i) \, u_i(c_S, (t_{S-i}, r_i)) \right) / f_i(t_i),$$

$$\forall S \in L(N), \ \forall c_S \in C_S, \ \forall t_S \in T_S.$$

Proof. Notice first that (w, f) is strongly inhibitive iff there does not exist any η that satisfies (10.24)–(10.27), because a sufficiently small scalar multiple of any solution to these conditions would satisfy (10.23) as well. So consider the linear programming problem to maximize the formula in (10.27) subject to the constraints (10.24)–(10.26). The constraints in this problem always have a feasible solution (by letting η equal the zero vector), and any positive multiple of a feasible solution would also be a feasible solution. So this linear programming problem either has optimal value 0, if (w, f) is strongly inhibitive, or it has an optimal value of $+\infty$, if (w, f) is not strongly inhibitive. By the duality theorem of linear programming, this linear programming problem has a finite optimal solution iff the dual of this problem has a feasible solution. So (w,f) is strongly inhibitive iff the dual has a feasible solution. Letting λ be the vector of dual variables for the individual-rationality constraints (10.24) and letting α be the vector of dual variables for the incentive constraints (10.25), it is a straightforward exercise to show that a solution to the constraints of the dual problem exists iff there is a solution (λ,α) to the conditions in Theorem 10.4. ∎

Theorem 10.4 has a straightforward interpretation in terms of the virtual utility hypothesis. It asserts that waiting-population characteristics are strongly inhibitive iff there are virtual utility scales in which the total virtual-utility worth of every coalition that could be formed by a blocking agent would be less than the sum of virtual utility payoffs that its members could get in the established matching plan.

We say that waiting-population characteristics (w,f) are *inhibitive* iff there exists some sequence of strongly inhibitive vectors that converge to (w,f). That is, the characteristics of a waiting population are inhibitive iff, by perturbing the expected payoffs of some types and the relative numbers of individuals of the various types by arbitrarily small amounts, we could get waiting-population characteristics that would allow no viable randomized blocking plan. Thus, if a viable blocking plan is possible relative to an inhibitive waiting population, it can only be an unstable knife-edge phenomenon, with at least one type of player willing to upset the blocking plan in an essential way by refusing to participate or by lying about his type.

We must now complete the description of the dynamic matching process and the matching plans that could be established for assigning individuals into coalitions. Suppose that players of all types and all classes enter (or are "born") into the matching process at some constant rates. To complete the description of the dynamic matching process, we must specify, for each class i in N and each type t_i in T_i, a positive number $\rho_i(t_i)$ that represents the birth rate of players with type t_i in class i. So the elements of a general dynamic matching process are

$$(10.28) \quad (N, (C_S)_{S \in L(N)}, (T_i)_{i \in N}, (u_i)_{i \in N}, (\rho_i)_{i \in N}).$$

We assume here that the sets N, C_S, and T_i are all nonempty finite sets, for all i and S.

A *matching plan* may be described by a vector

$$\mu = (\mu_S(c_S,t_S))_{c_S \in C_S, t_S \in T_S, S \in L(N)},$$

where each number $\mu_S(c_S,t_S)$ represents the rate (per unit time) at which S coalitions are forming in which the type profile is t_S and the joint action is c_S. To clear the market in the long run, the rate at which players of each class and type are entering the matching system must equal the rate at which they exit the matching process in all kinds of coalitions. Thus, a matching plan must satisfy the following *balance condition*:

$$(10.29) \quad \sum_{S \supseteq \{i\}} \sum_{c_S \in C_S} \sum_{t_{S-i} \in T_{S-i}} \mu(c_S,t_S) = \rho_i(t_i), \quad \forall i \in N, \ \forall t_i \in T_i.$$

The expected payoff to a class-i type-t_i player in such a matching plan μ is

$$U_i(\mu|t_i) = \sum_{S \supseteq \{i\}} \sum_{c_S \in C_S} \sum_{t_{S-i} \in T_{S-i}} \frac{\mu(c_S,t_S) \ u_i(c_S,t_S)}{\rho_i(t_i)}$$

if he and everyone else participates honestly in the plan. On the other hand, if such a player systematically pretended that his type was r_i instead of t_i (he cannot lie about his class), then his expected payoff would be

$$U_i^*(\mu,r_i|t_i) = \sum_{S \supseteq \{i\}} \sum_{c_S \in C_S} \sum_{t_{S-i} \in T_{S-i}} \frac{\mu(c_S,(t_{S-i},r_i))u_i(c_S,t_S)}{\rho_i(r_i)} .$$

Thus, we say that a matching plan μ is *incentive compatible* iff it satisfies the informational incentive constraints

$$(10.30) \quad U_i(\mu|t_i) \geq U_i^*(\mu,r_i|t_i), \quad \forall i \in N, \ \forall t_i \in T_i, \ \forall r_i \in T_i.$$

In a stationary matching process, the expected number of type-t_i individuals who are waiting and available at any point in time is equal to the product of $\rho_i(t_i)$ times the expected waiting time for type-t_i individuals. Because different types may have different expected waiting times, the proportions $f_i(t_i)$, in the waiting population do not have to be proportional to the birth rates $\rho_i(t_i)$. Thus, we say that a matching plan μ is *competitively sustainable* iff it satisfies balance (10.29) and incentive-compatibility (10.30) and there exists some inhibitive waiting-population characteristics (w,f) such that

$$(10.31) \quad U_i(\mu|t_i) = w_i(t_i), \quad \forall i \in N, \ \forall t_i \in T_i.$$

The following existence theorem for competitively sustainable plans generalizes Theorem 9.6. The proof, using a fixed-point argument, is given by Myerson (1988) (for a somewhat more general model).

THEOREM 10.5. *At least one competitively sustainable matching plan must exist, for any dynamic matching process as in (10.28).*

Competitively sustainable plans can be interpreted as a general model of stationary market equilibria in many economic applications. For example, let us consider the dynamic matching processes that correspond to Examples 10.1 and 10.2, and let us identify the commodity with the seller's labor. Suppose that there are many sellers (class-1 players) and many buyers (class-2 players) who enter into the matching process or market during any unit interval of time. Each seller has a supply of one unit of labor to sell, and he either knows that his type is 1.a and he has high ability, or his type is 1.b and he has low ability. As before, sellers with high ability have labor that is worth \$40 per unit to themselves and \$50 per unit to a buyer, but sellers with low ability have labor that is worth \$20 per unit to themselves and \$30 per unit to a buyer. Suppose that each buyer can buy labor from at most one seller, and each seller may sell all or part of his supply to at most one buyer, who will not be able to directly observe the seller's ability level until after the transaction is over.

Let us suppose that the birth rate of buyers is higher than the birth rate of sellers,

$$\rho_2(2.0) > \rho_1(1.a) + \rho_1(1.b).$$

For this example, there is a competitively sustainable matching plan in which every low-ability seller sells all of his labor supply at a price of \$30, every high-ability seller sells $1/3$ of his labor supply at a price of \$50 per unit (for total payment \$50/3), and some excess buyers exit without trading. To formalize this result, we let $C_{\{1,2\}} = \{(q,y)\,|\,0 \le q \le 1, y \ge 0\}$ and $C_{\{1\}} = C_{\{2\}} = \{(0,0)\}$. Let the utility functions be as in (10.18)–(10.21). Then the competitively sustainable plan μ may be written

$$\mu_{\{1,2\}}((1/3,50/3),(1.a,2.0)) = \rho_1(1.a),$$

$$\mu_{\{1,2\}}((1,30),(1.b,2.0)) = \rho_1(1.b),$$

$$\mu_{\{2\}}((0,0),2.0) = \rho_2(2.0) - (\rho_1(1.a) + \rho_1(1.b)),$$

which gives

$$U_1(\mu|1.a) = 10/3, \quad U_1(\mu|1.b) = 10, \quad U_2(\mu|2.0) = 0.$$

This plan μ is essentially a translation of the seller's best safe mechanism (Q^4,Y^4) from Examples 10.1 and 10.2 into the context of the dynamic matching process. To verify that this plan μ is competitively sustainable, it suffices to recall that, in Example 10.1, the mechanism (Q^4,Y^4) is incentive-efficient and gives the same interim expected payoff allocation. Thus, for any positive ε, the conditions for strong inhibitiveness in Theorem 10.4 can be satisfied by letting

$$w_1(1.a) = 10/3, \quad w_1(1.b) = 10, \quad w_2(2.0) = \varepsilon,$$
$$f_1(1.a) = 0.1, \quad f_1(1.b) = 0.4, \quad f_2(2.0) = 0.5,$$
$$\lambda_1(1.a) = 0.15, \quad \lambda_1(1.b) = 0.35, \quad \lambda_2(2.0) = 0.5,$$
$$\alpha_1(1.a|1.b) = 0.05, \quad \alpha_1(1.b|1.a) = 0.$$

That is, μ can be sustained by waiting-population characteristics that replicate the situation in Example 10.1, by having 80% of the sellers be type 1.b and only 20% be type 1.a $(f_1(1.a)/(f_1(1.a) + f_1(1.b)) = 0.2)$. The fact that buyers get zero profits in the competitively sustainable plan is a consequence of the excess of buyers over sellers born in each generation.

Suppose that $\rho_1(1.a)/\rho_1(1.b) = 9$; so there are many more high-ability sellers than low-ability sellers born in each generation. The above demonstration that this plan μ is competitively sustainable includes this case. In particular, the sustaining waiting population with four times more low-ability sellers than high-ability sellers will be created if low-ability sellers wait and search 28 times longer than high-ability sellers in the dynamic matching process. This effect is reminiscent of *Gresham's law* in classical monetary theory, which asserts that "the bad (money) circulates more than the good."

However, with $\rho_1(1.a)/\rho_1(1.b) = 9$, there are other incentive-compatible matching plans that are better for all types than this competitively sustainable matching plan. For example, let v be the matching plan in which every seller sells all of his labor for \$47, so

$$v_{\{1,2\}}((1,47),(1.a,2.0)) = \rho_1(1.a),$$
$$v_{\{1,2\}}((1,47),(1.b,2.0)) = \rho_1(1.b),$$
$$v_{\{2\}}((0,0),2.0) = \rho_2(2.0) - (\rho_1(1.a) + \rho_1(1.b)),$$

$U_1(v|1.\text{a}) = 7,$

$U_1(v|1.\text{b}) = 27,$

$0 < U_2(v|2.0) = (3\rho_1(1.\text{a}) - 17\rho_1(1.\text{b}))/\rho_2(2.0) \le 1.$

Unfortunately, although this plan v is Pareto superior to the competitively sustainable plan μ, v is not competitively sustainable itself. The essential problem is that the high payoff that type-1.b players get in v actually makes it easier to create viable blocking plans, because it makes it easier for a blocking mediator to exclude low-ability sellers from blocking coalitions and to ask them instead to continue waiting for the high payoff that they can expect in the matching plan v. For any type distribution $\hat{f}$ and any vector $\hat{w}$ sufficiently close to the expected payoff vector $(U_1(v|1.\text{a}), U_1(v|1.\text{b}), U_2(v|2.0))$ generated by v, there is a viable blocking plan relative to $(\hat{w}, \hat{f})$ in which high-ability sellers sell 0.9 units of labor to buyers at a price of \$48.50 per unit (for a total payment of \$43.65), and all low-ability sellers stay out of the blocking coalitions and continue to wait to be matched in v. Notice that

$0.9(48.50 - 40) = 7.65 > 7,$

$0.9(48.50 - 20) = 25.65 < 27,$

$0.9(50 - 48.5) = 1.35 > 1,$

so the high-ability sellers do better in such a blocking coalition than in v, the low-ability sellers would do worse in such a blocking coalition than in v, and the buyers expect to do better in such a blocking coalition than in v.

Thus, a competitively sustainable plan can fail to be incentive-efficient in dynamic matching processes with incomplete information. In terms of Theorem 10.4, this inefficiency can occur because, when a "bad" type jeopardizes a "good" type, increasing the payoff to the bad type decreases the corresponding virtual utility for the good type and therefore makes it harder to satisfy the conditions for an inhibitive waiting population. This inefficiency result is closely related to other results in economic theory about the failures of markets with adverse selection (see the seminal papers of Spence, 1973; Rothschild and Stiglitz, 1976; and Wilson, 1977). Our general existence theorem for competitively sustainable plans relies on the possibility of having bad types wait longer than good types; so a blocking mediator at any point in time would have to recruit from a population with mostly bad-type individuals, even if the birth rate of bad types is relatively small. If we could impose the restriction that the distribution of types in the waiting population f must be proportional to the birth rates ρ, then we could guarantee that sustainable plans would be Pareto efficient among all incentive-compatible matching plans. Myerson (1988) has shown that existence of sustainable plans with this restriction can be generally guaranteed only if the equality in condition (10.31) is weakened to an inequality

$$U_i(\mu|t_i) \geq w_i(t_i),$$

but this weakened condition is hard to interpret economically.

Exercises

Exercise 10.1. For the example in Table 10.1, let μ be the mechanism such that

$$\mu(x|1.a,2.a) = 1, \quad \mu(y|1.a,2.b) = 1,$$
$$\mu(z|1.b,2.a) = 1, \quad \mu(y|1.b,2.b) = 1,$$

and all other $\mu(c|t)$ are 0. Use Theorem 10.1 to prove that this mechanism μ is incentive efficient. In your proof, let the utility weights be $\lambda_1(1.a) = \lambda_1(1.b) = \lambda_2(2.a) = \lambda_2(2.b) = 1$. (HINT: The second-to-last condition in the theorem implies that all but one of the Lagrange multipliers $\alpha_i(s_i|t_i)$ must be 0.)

Exercise 10.2. Recall the Bayesian bargaining problem from Exercise 6.3 (Chapter 6). Player 1 is the seller of a single indivisible object, and player 2 is the only potential buyer. The value of the object is $0 or $80 for player 1 and $20 or $100 for player 2. Ex ante, all four possible combinations of these values are considered to be equally likely. When the players meet to bargain, each player knows his own private value for the object and believes that the other player's private value could be either of the two possible numbers, each with probability $\frac{1}{2}$. We let the type of each player be his private value for the object, so $T_1 = \{0, 80\}$, $T_2 = \{20, 100\}$, and

$$\bar{p}_1(0) = \bar{p}_1(80) = \frac{1}{2}, \quad \bar{p}_2(100) = \bar{p}_2(20) = \frac{1}{2}.$$

The expected payoffs to the two players depend on their types $t = (t_1, t_2)$, the expected payment y from player 2 to player 1, and the probability q that the object is delivered to player 2, according to the formula

$$u_1((q,y),t) = y - t_1 q, \quad u_2((q,y),t) = t_2 q - y.$$

No-trade ($q = 0$, $y = 0$) is the disagreement outcome in this bargaining problem. We can represent a mechanism for this Bayesian bargaining problem by a pair of functions $Q{:}T_1 \times T_2 \rightarrow [0,1]$ and $Y{:}T_1 \times T_2 \rightarrow \mathbf{R}$, where $Q(t)$ is the probability that the object will be sold to player 2 and $Y(t)$ is the expected net payment from player 2 to player 1 if t is the pair of types reported by the players to a mediator.

   a. Although $Q(t)$ must be in the interval $[0,1]$, $Y(t)$ can be any real number in $\mathbf{R}$. Show that

$$\max_{(q,y)\in[0,1]\times\mathbf{R}} \sum_{i\in\{1,2\}} v_i((q,y),t,\lambda,\alpha)$$

is finite only if there is some positive constant K such that

$$\frac{\lambda_i(t_i) + \sum_{s_i \in T_i - t_i} (\alpha_i(s_i|t_i) - \alpha_i(t_i|s_i))}{\overline{p}_i(t_i)} = K, \quad \forall i \in \{1,2\}, \ \forall t_i \in T_i.$$

   b. In buyer–seller bargaining, there is usually no difficulty in preventing buyers from overstating their values or preventing sellers from understating their values. Thus, we can suppose that

$$\alpha_2(100|20) = 0 \ \text{ and } \ \alpha_1(0|80) = 0.$$

Furthermore, without loss of generality, we can let the constant K equal 1. (Changing K just requires a proportional change in all parameters.) With these assumptions, the equations from part (a) become

$$\lambda_1(0) + \alpha_1(80|0) = \lambda_1(80) - \alpha_1(80|0) = \tfrac{1}{2},$$
$$\lambda_2(100) + \alpha_2(20|100) = \lambda_2(20) - \alpha_2(20|100) = \tfrac{1}{2}.$$

With these equations, express $\sum_{i\in\{1,2\}} v_i((q,y),t,\lambda,\alpha)$ as a function of q, $\alpha_1(80|0)$, and $\alpha_2(20|100)$, for each of the four possible type profiles t in $T_1 \times T_2$.

   c. Consider a class of mechanisms (Q,Y) that depend on two parameters r and z as follows: $Q(80,20) = 0 = Y(80,20)$, $Q(0,100) = 1$, $Y(0,100) = 50$, $Q(0,20) = Q(80,100) = r$, $Y(0,20) = rz$, $Y(80,100) = r(100 - z)$. Suppose that $z \le 20$ and $r = 50/(100 - 2z)$. Show that every mechanism in this class is incentive efficient, by identifying vectors λ and α as in part (b) that satisfy the incentive-efficiency conditions in Theorem 10.1 for all these mechanisms. (HINT: If $\Sigma_{i \in \{1,2\}} v_i((q,y),t,\lambda,\alpha) = 0$ for all (q,y), then any mechanism will satisfy the condition of maximizing the sum of virtual utilities for this profile of types t.) Why are the conditions $z \leq 20$ and $r = 50/(100 - 2z)$ needed?

d. Within the class of mechanisms described in (c), show how the interim expected payoff $U_i(Q,Y|t_i)$ for each type t_i of each player i depends on the parameter z. For each player, which types prefer mechanisms with higher z? Which types prefer mechanisms with lower z? Among the mechanisms in this class, which mechanism maximizes the sum of the players' ex ante expected gains from trade? Show that this mechanism also maximizes the probability of the object being sold, among all incentive-compatible individually rational mechanisms. (Recall part (b) of Exercise 6.3.)

e. Using λ and α from part (c), find the solution w to the system of equations (10.22) in Theorem 10.3. Use this solution to show that the mechanism from part (c) with $z = 0$ and $r = \frac{1}{2}$ is a neutral bargaining solution for this Bayesian bargaining problem.

f. Suppose that, after tossing a coin, if the coin comes up Heads, then we will let player 1 make a first and final offer to sell at any price he might specify; and if the coin comes up Tails, then we will let player 2 make a first and final offer to buy at any price that she might specify. The loser of the coin toss will accept the winner's offer iff it gives the loser nonnegative gains from trade. What would be the optimal offer for each player, as a function of his or her type, if he or she won the coin toss? Compare this equilibrium of this game to the neutral bargaining solution that you found in part (e).

Bibliography · Index

Bibliography

Abreu, D. 1986. "Extremal Equilibrium of Oligopolistic Supergames." *Journal of Economic Theory* 39:191–225.

Abreu, D., P. Milgrom, and D. Pearce. 1988. "Information and Timing in Repeated Partnerships." Stanford University discussion paper.

Abreu, D., D. Pearce, and E. Stacchetti. 1986. "Optimal Cartel Equilibria with Imperfect Monitoring." *Journal of Economic Theory* 39:251–269.

Akerlof, G. 1970. "The Market for Lemons: Qualitative Uncertainty and the Market Mechanism." *Quarterly Journal of Economics* 84:488–500.

Albers, W. 1979. "Core- and Kernel-Variants Based on Imputations and Demand Profiles." In O. Moeschin and D. Pallaschke, eds., *Game Theory and Related Topics*. Amsterdam: North-Holland.

Allais, M., and O. Hagen, eds. 1979. *Expected Utility Hypothesis and the Allais Paradox*. Boston: Reidel.

Anderson, R. 1983. "Quick-Response Equilibrium." Discussion Paper No. 323, Center for Research in Management Science, University of California, Berkeley.

Anscombe, F. J., and R. J. Aumann. 1963. "A Definition of Subjective Probability." *Annals of Mathematical Statistics* 34:199–205.

Arrow, K. 1951. *Social Choice and Individual Values*. New York: Wiley.

Aumann, R. J. 1959. "Acceptable Points in General Cooperative *n*-Person Games." In H. W. Kuhn and R. D. Luce, eds., *Contributions to the Theory of Games IV*, pp. 287–324. Princeton: Princeton University Press.

Aumann, R. J. 1961. "The Core of a Cooperative Game Without Side Payments." *Transactions of the American Mathematical Society* 98:539–552.

Aumann, R. J. 1964. "Markets with a Continuum of Traders." *Econometrica* 32:39–50.

Aumann, R. J. 1967. "A Survey of Cooperative Games Without Side Payments." In M. Shubik, ed., *Essays in Mathematical Economics*, pp. 3–27. Princeton: Princeton University Press.

Aumann, R. J. 1974. "Subjectivity and Correlation in Randomized Strategies." *Journal of Mathematical Economics* 1:67–96.

Aumann, R. J. 1976. "Agreeing to Disagree." *Annals of Statistics* 4:1236–39.

Aumann, R. J. 1985. "An Axiomatization of the Non-Transferable Utility Value." *Econometrica* 53:599–612.

Aumann, R. J. 1987a. "Correlated Equilibria as an Expression of Bayesian Rationality." *Econometrica* 55:1–18.

Aumann, R. J. 1987b. "Game Theory." In J. Eatwell, M. Milgate, and P. Newman, eds., *The New Palgrave Dictionary of Economics*, pp. 460–482. London: Macmillan.

Aumann, R. J., and J. H. Dreze. 1974. "Cooperative Games with Coalition Structures." *International Journal of Game Theory* 3:217–238.

Aumann, R. J., and M. Maschler. 1964. "The Bargaining Set for Cooperative Games." In M. Dresher, L. S. Shapley, and A. W. Tucker, eds., *Advances in Game Theory*, pp. 443–447. Princeton: Princeton University Press.

Aumann, R. J., and M. Maschler. 1966. "Game Theoretic Aspects of Gradual Disarmament." Chapter 5 in *Report to the U.S. Arms Control and Disarmament Agency ST-80*. Princeton: Mathematica, Inc.

Aumann, R. J., and M. Maschler. 1968. "Repeated Games with Incomplete Information: the Zero Sum Extensive Case." Chapter 2 in *Report to the U.S. Arms Control and Disarmament Agency ST-143*. Princeton: Mathematica, Inc.

Aumann, R. J., and M. Maschler. 1972. "Some Thoughts on the Minimax Principle." *Management Science* 18(pt. 2):54–63.

Aumann, R. J., and R. B. Myerson. 1988. "Endogenous Formation of Links Between Players and of Coalitions: An Application of the Shapley Value." In A. E. Roth, ed., *The Shapley Value*, pp. 175–191. Cambridge: Cambridge University Press.

Aumann, R. J., and B. Peleg. 1960. "Von Neumann-Morgenstern Solutions to Cooperative Games Without Sidepayments." *Bulletin of the American Mathematical Society* 66:173–179.

Aumann, R. J., and L. S. Shapley. 1974. *Values of Nonatomic Games*. Princeton: Princeton University Press.

Ausubel, L. M., and R. J. Deneckere. 1989. "A Direct Mechanism Characterization of Sequential Bargaining with One-Sided Incomplete Information." *Journal of Economic Theory* 48:18–46.

Axelrod, R. 1984. *The Evolution of Cooperation*. New York: Basic Books.

Banks, J. S., and J. Sobel. 1987. "Equilibrium Selection in Signaling Games." *Econometrica* 55:647–662.

Barany, I. 1987. "Fair Distribution Protocols or How the Players Replace Fortune." CORE discussion paper 8718, Université Catholique de Louvain. To appear in *Mathematics of Operations Research*.

Bennett, E. 1983. "The Aspiration Approach to Predicting Coalition Formation and Payoff Distribution in Sidepayment Games." *International Journal of Game Theory* 12:1–28.

Benoit, J. P., and V. Krishna. 1985. "Finitely Repeated Games." *Econometrica* 53:905–922.

Benoit, J. P., and V. Krishna. 1989. "Renegotiation in Finitely Repeated Games." Harvard Business School working paper.

Ben-Porath, E., and E. Dekel. 1987. "Signaling Future Actions and the Potential for Sacrifice." Stanford University research paper.

Bergin, J. 1987. "Continuous Time Repeated Games of Complete Information." Queen's University working paper, Kingston, Ontario.

Bernheim, B. D. 1984. "Rationalizable Strategic Behavior." *Econometrica* 52:1007–28.

Bernheim, B. D., B. Peleg, and M. D. Whinston. 1987. "Coalition-Proof Nash Equilibria I: Concepts." *Journal of Economic Theory* 42:1–12.

Bernheim, B. D., and D. Ray. 1989. "Collective Dynamic Consistency in Repeated Games." *Games and Economic Behavior* 1:295–326.

Bernoulli, D. 1738. "Exposition of a New Theory of the Measurement of Risk." English translation in *Econometrica* 22(1954):23–36.

Bicchieri, C. 1989. "Self-Refuting Theories of Strategic Interaction: A Paradox of Common Knowledge." *Erkenntnis* 30:69–85.

Billera, L. J., D. C. Heath, and J. Ranaan. 1978. "Internal Telephone Billing Rates—A Novel Application of Non-Atomic Game Theory." *Operations Research* 26:956–965.

Billingsley, P. 1968. *Convergence of Probability Measures*. New York: Wiley.

Binmore, K. 1987–88. "Modeling Rational Players." *Economics and Philosophy* 3:179–214; 4:9–55.

Binmore, K., and P. Dasgupta. 1987. *The Economics of Bargaining*. Oxford: Basil Blackwell.

Binmore, K., A. Rubinstein, and A. Wolinsky. 1986. "The Nash Bargaining Solution in Economic Modeling." *Rand Journal of Economics* 17:176–188.

Blackwell, D. 1956. "An Analogue of the Minimax Theorem for Vector Payoffs." *Pacific Journal of Mathematics* 6:1–8.

Blackwell, D. 1965. "Discounted Dynamic Programming." *Annals of Mathematical Statistics* 36:226–235.

Blackwell, D., and T. S. Ferguson. 1968. "The Big Match." *Annals of Mathematical Statistics* 39:159–163.

Bondereva, O. N. 1963. "Some Applications of Linear Programming Methods to the Theory of Cooperative Games" (in Russian). *Problemy Kibernetiki* 10:119–139.

Border, K. C. 1985. *Fixed Point Theorems with Applications to Economics and Game Theory*. Cambridge: Cambridge University Press.

Borel, E. 1921. "La Théorie du Jeu et les Équations Intégrales à Noyau Symétrique." *Comptes Rendus de l'Académie des Sciences* 173:1304–1308.

Burger, E. 1963. *Introduction to the Theory of Games*. Englewood Cliffs, N.J.: Prentice-Hall.

Chatterjee, K., and W. Samuelson. 1983. "Bargaining under Incomplete Information." *Operations Research* 31:835–851.

Cho, I.-K., and D. Kreps. 1987. "Signaling Games and Stable Equilibria." *Quarterly Journal of Economics* 102:179–221.

Chvatal, V. 1983. *Linear Programming*. New York: W. H. Freeman.

Coase, R. H. 1960. "The Problem of Social Cost." *Journal of Law and Economics* 3:1–44.

Cooper, R., D. DeJong, R. Forsythe, and T. Ross. 1989. "Communication in the Battle of the Sexes Game." In *Rand Journal of Economics* 20:568–585.

Cooper, R., D. DeJong, R. Forsythe, and T. Ross. 1990. "Selection Criteria in Coordination Games: Some Experimental Results." *American Economic Review* 80:218–233.

Cramton, P., R. Gibbons, and P. Klemperer. 1987. "Dissolving a Partnership Efficiently." *Econometrica* 55:615–632.

Crawford, V. 1979. "On Compulsory-Arbitration Schemes." *Journal of Political Economy* 87:131–159.

Crawford, V. 1981. "Arbitration and Conflict Resolution in Labor-Management Bargaining." *American Economic Review: Papers and Proceedings* 71:205–210.

Crawford, V. 1985. "Efficient and Durable Decision Rules: A Reformulation." *Econometrica* 53:817–835.

Crawford, V., and J. Sobel. 1982. "Strategic Information Transmission." *Econometrica* 50:579–594.

Cross, J. 1967. "Some Theoretic Characteristics of Economic and Political Coalitions." *Journal of Conflict Resolution* 11:184–195.

Dasgupta, P., and E. Maskin. 1986. "The Existence of Equilibrium in Discontinuous Economic Games." *Review of Economic Studies* 53:1–41.

Dasgupta, P., P. Hammond, and E. Maskin. 1979. "The Implementation of Social Choice Rules: Some Results on Incentive Compatibility." *Review of Economic Studies* 46:185–216.

D'Aspremont, C., and L.-A. Gerard-Varet. 1979. "Incentives and Incomplete Information." *Journal of Public Economics* 11:25–45.

Davis, M., and M. Maschler. 1965. "The Kernel of a Cooperative Game." *Naval Research Logistics Quarterly* 12:223–259.

Debreu, G. 1959. *Theory of Value.* New York: Wiley.

Debreu, G., and H. Scarf. 1963. "A Limit Theorem on the Core of an Economy." *International Economic Review* 4:235–246.

Edgeworth, F. Y. 1881. *Mathematical Psychics.* London: Kegan Paul.

Ellsberg, D. 1961. "Risk, Ambiguity, and the Savage Axioms." *Quarterly Journal of Economics* 75:643–669.

Epstein, L. G. 1983. "Stationary Cardinal Utility and Optimal Growth under Uncertainty." *Journal of Economic Theory* 31:133–152.

Farquharson, R. 1969. *Theory of voting.* New Haven: Yale University Press.

Farrell, J. 1988. "Meaning and Credibility in Cheap-Talk Games." *Games and Economic Behavior* 5:514–531, 1993.

Farrell, J., and E. Maskin. 1989. "Renegotiation in Repeated Games." *Games and Economic Behavior* 1:327–360.

Fishburn, P. C. 1968. "Utility Theory." *Management Science* 14:335–378.

Fishburn, P. C. 1970. *Utility Theory for Decision-Making.* New York: Wiley.

Fisher, R., and W. Ury. 1981. *Getting to Yes: Negotiating Agreement Without Giving In.* Boston: Houghton Mifflin.

Forges, F. 1985. "Correlated Equilibria in a Class of Repeated Games with Incomplete Information." *International Journal of Game Theory* 14:129–150.

Forges, F. 1986. "An Approach to Communication Equilibrium." *Econometrica* 54:1375–85.

Forges, F. 1988. "Non-zero Sum Repeated Games and Information Transmission." CORE Working Paper 8825, Université Catholique de Louvain.

Forges, F. 1990. "Universal Mechanisms." *Econometrica* 58:1341–1364.

Franklin, J. 1980. *Methods of Mathematical Economics*. New York: Springer-Verlag.

Fudenberg, D., and D. M. Kreps. 1988. "A Theory of Learning, Experimentation, and Equilibrium in Games." M.I.T. and Stanford University working paper.

Fudenberg, D., D. M. Kreps, and D. K. Levine. 1988. "On the Robustness of Equilibrium Refinements." *Journal of Economic Theory* 44:354–380.

Fudenberg, D., and D. K. Levine. 1986. "Limit Games and Limit Equilibria." *Journal of Economic Theory* 38:261–279.

Fudenberg, D., D. Levine, and J. Tirole. 1985. "Infinite Horizon Models of Bargaining." In A. E. Roth, ed., *Game-Theoretic Models of Bargaining*, pp. 73–98. Cambridge: Cambridge University Press.

Fudenberg, D., and E. Maskin. 1986. "The Folk Theorem in Repeated Games with Discounting and Incomplete Information." *Econometrica* 54:533–554.

Fudenberg, D., and J. Tirole. 1984. "Preemption and Rent Equalization in the Adoption of a New Technology." *Review of Economic Studies* 52:383–401.

Fudenberg, D., and J. Tirole. 1988. "Perfect Bayesian and Sequential Equilibria." To appear in *Journal of Economic Theory*.

Geoffrion, A. M. 1971. "Duality in Nonlinear Programming: A Simplified Applications-Oriented Development." *SIAM Review* 13:1–37.

Gibbard, A. 1973. "Manipulation of Voting Schemes: A General Result." *Econometrica* 41:587–601.

Gilboa, I., E. Kalai, and E. Zemel. 1989. "On the Order of Eliminating Dominated Strategies." To appear in *Operations Research Letters*.

Glazer, J., and A. Weiss. 1990. "Pricing and Coordination: Strategically Stable Equilibria." *Games and Economic Behavior* 2:118–128.

Glicksberg, I. 1952. "A Further Generalization of the Kakutani Fixed Point Theorem with Application to Nash Equilibrium Points." *Proceedings of the American Mathematical Society* 3:170–174.

Gresik, T. A., and M. A. Satterthwaite. 1989. "The Rate at Which a Simple Market Converges to Efficiency as the Number of Traders Increases: An Asymptotic Result for Optimal Trading Mechanisms." *Journal of Economic Theory* 48:304–332.

Grossman, S., and M. Perry. 1986. "Perfect Sequential Equilibrium." *Journal of Economic Theory* 39:97–119.

Hamilton, W. D. 1964. "The Genetical Evolution of Social Behavior." *Journal of Theoretical Biology* 7:1–52.

Harris, M., and R. M. Townsend. 1981. "Resource Allocation under Asymmetric Information." *Econometrica* 49:1477–99.

Harsanyi, J. C. 1956. "Approaches to the Bargaining Problems Before and After the Theory of Games: A Critical Discussion of Zeuthen's, Hick's, and Nash's Theories." *Econometrica* 24:144–157.

Harsanyi, J. C. 1963. "A Simplified Bargaining Model for the *n*-Person Cooperative Game." *International Economic Review* 4:194–220.

Harsanyi, J. C. 1967–68. "Games with Incomplete Information Played by 'Bayesian' Players." *Management Science* 14:159–182, 320–334, 486–502.

Harsanyi, J. C. 1973. "Games with Randomly Disturbed Payoffs: A New Rationale for Mixed-Strategy Equilibrium Points." *International Journal of Game Theory* 2:1–23.

Harsanyi, J. C., and R. Selten. 1972. "A Generalized Nash Solution for Two-Person Bargaining Games with Incomplete Information." *Management Science* 18:80–106.

Harsanyi, J. C., and R. Selten. 1988. *A General Theory of Equilibrium Selection in Games.* Cambridge, Mass.: MIT Press.

Hart, S. 1985a. "Nonzero-Sum Two-Person Repeated Games with Incomplete Information." *Mathematics of Operations Research* 10:117–153.

Hart, S. 1985b. "An Axiomatization of Harsanyi's Nontransferable Utility Solution." *Econometrica* 53:1295–1313.

Hart, S., and M. Kurz. 1983. "Endogenous Formation of Coalitions." *Econometrica* 51:1047–64.

Hart, S., and A. Mas-Colell. 1989. "Potential, Value, and Consistency." *Econometrica* 57:589–614.

Herstein, I., and J. Milnor. 1953. "An Axiomatic Approach to Measurable Utility." *Econometrica* 21:291–297.

Hillas, J. 1987. "Sequential Equilibria and Stable Sets of Beliefs." I.M.S.S.S. Technical Report No. 518, Stanford University.

Holmstrom, B. 1977. "On Incentives and Control in Organizations." Ph.D. dissertation, Stanford University.

Holmstrom, B., and R. B. Myerson. 1983. "Efficient and Durable Decision Rules with Incomplete Information." *Econometrica* 51:1799–1819.

Hotelling, H. 1929. "The Stability of Competition." *Economic Journal* 39:41–57.

Howard, R. 1960. *Dynamic Programming and Markov Processes.* New York: Wiley.

Hurd, A. E., and P. A. Loeb. 1985. *An Introduction to Nonstandard Analysis.* Orlando: Academic Press.

Hurwicz, L. 1972. "On Informationally Decentralized Systems." In R. Radner and B. McGuire, eds., *Decision and Organization*, pp. 297–336. Amsterdam: North-Holland.

Kadane, J., and P. D. Larkey. 1982. "Subjective Probability and the Theory of Games." *Management Science* 28:113–120.

Kahneman, D., and A. Tversky. 1979. "Prospect Theory: An Analysis of Decision under Risk." *Econometrica* 47:263–291.

Kahneman, D., and A. Tversky. 1982. "The Psychology of Preferences." *Scientific American* 246(1):160–170.

Kahneman, D., P. Slovic, and A. Tversky, eds. 1982. *Judgment under Uncertainty: Heuristics and Biases.* Cambridge: Cambridge University Press.

Kakutani, S. 1941. "A Generalization of Brouwer's Fixed Point Theorem." *Duke Mathematical Journal* 8:457–458.

Kalai, E. 1977. "Nonsymmetric Nash Solutions and Replications of Two-Person Bargaining." *International Journal of Game Theory* 65:129–133.

Kalai, E., and D. Samet. 1984. "Persistent Equilibria." *International Journal of Game Theory* 13:129–144.

Kalai, E., and D. Samet. 1985. "Monotonic Solutions to General Cooperative Games." *Econometrica* 53:307–327.

Kalai, E., and D. Samet. 1987. "On Weighted Shapley Values." *International Journal of Game Theory* 16:205–222.

Kalai, E., and M. Smorodinsky. 1975. "Other Solutions to Nash's Bargaining Problem." *Econometrica* 45:513–518.

Kalai, E., and W. Stanford. 1985. "Conjectural Variations Strategies in Accelerated Cournot Games." *International Journal of Industrial Organization* 3:133–152.

Kandori, M. 1988. "Social Norms and Community Enforcement." University of Pennsylvania working paper.

Kaneko, M., and M. H. Wooders. 1982. "Cores of Partitioning Games." *Mathematical Social Sciences* 3:313–327.

Kohlberg, E. 1989. "Refinement of Nash Equilibrium: The Main Ideas." Harvard Business School working paper.

Kohlberg, E., and J.-F. Mertens. 1986. "On the Strategic Stability of Equilibria." *Econometrica* 54:1003–37.

Kolmogorov, A. N., and S. V. Fomin. 1970. *Introductory Real Analysis.* New York: Dover.

Knight, F. H. 1921. *Risk, Uncertainty and Profit.* Boston: Houghton Mifflin.

Kreps, D., and R. Wilson. 1982. "Sequential Equilibria." *Econometrica* 50:863–894.

Kreps, D., P. Milgrom, J. Roberts, and R. Wilson. 1982. "Rational Cooperation in the Finitely-Repeated Prisoners' Dilemma." *Journal of Economic Theory* 27:245–252.

Kuhn, H. W. 1953. "Extensive Games and the Problem of Information." In H. W. Kuhn and A. W. Tucker, eds., *Contributions to the Theory of Games* I, pp. 193–216. Princeton: Princeton University Press.

Leininger, W., P. B. Linhart, and R. Radner. 1989. "Equilibria of the Sealed-Bid Mechanism for Bargaining with Incomplete Information," *Journal of Economic Theory* 48:63–106.

Lewis, T. R., and D. E. M. Sappington. 1989. "Countervailing Incentives in Agency Problems." *Journal of Economic Theory* 49:294–313.

Lucas, W. F. 1969. "The Proof That a Game May Not Have a Solution." *Transactions of the American Mathematical Society* 137:219–229.

Lucas, W. F. 1972. "An Overview of the Mathematical Theory of Games." *Management Science* 18:3–19.

Lucas, W. F., and R. M. Thrall. 1963. "*n*-Person Games in Partition Function Form." *Naval Research Logistics Quarterly* 10:281–298.

Luce, R. D., and H. Raiffa. 1957. *Games and Decisions*. New York: Wiley.

Luenberger, D. G. 1984. *Linear and Nonlinear Programming*, 2nd Ed. Reading, Mass.: Addison-Wesley.

Machina, M. 1982. "Expected Utility without the Independence Axiom." *Econometrica* 50:277–323.

Mas-Colell, A. 1985. *The Theory of General Equilibrium: A Differentiable Approach*. Cambridge: Cambridge University Press.

Maynard Smith, J. 1982. *Evolution and the Theory of Games*. Cambridge: Cambridge University Press.

McAfee, R. P., and J. McMillan. 1987. "Auctions and Bidding." *Journal of Economic Literature* 25:699–738.

McLennan, A. 1985. "Justifiable Beliefs in a Sequential Equilibrium." *Econometrica* 53:889–904.

Mertens, J.-F. 1980. "A Note on the Characteristic Function of Supergames." *International Journal of Game Theory* 9:189–190.

Mertens, J.-F., S. Sorin, and S. Zamir. 1989. *Repeated Games*. Forthcoming.

Mertens, J.-F., and S. Zamir. 1985. "Formulation of Bayesian Analysis for Games with Incomplete Information." *International Journal of Game Theory* 14:1–29.

Milgrom, P. 1981. "Good News and Bad News: Representation Theorems and Applications." *Bell Journal of Economics* 12:380–391.

Milgrom, P. 1985. "The Economics of Competitive Bidding: A Selective Survey." In L. Hurwicz, D. Schmeidler, and H. Sonnenschein, eds., *Social Goals and Social Organization*, pp. 261–289. Cambridge: Cambridge University Press.

Milgrom, P. 1987. "Auction Theory." In T. Bewley, ed., *Advances in Economic Theory: Fifth World Congress*, pp. 1–32. Cambridge: Cambridge University Press.

Milgrom, P., D. C. North, and B. R. Weingast. 1989. "The Role of Institutions in the Revival of Trade. Part I: The Medieval Law Merchant." Stanford University discussion paper. To appear in *Economics and Politics*.

Milgrom, P., and R. J. Weber. 1982. "A Theory of Auctions and Competitive Bidding." *Econometrica* 50: 1089–1122.

Milgrom, P., and R. J. Weber. 1985. "Distributional Strategies for Games with Incomplete Information." *Mathematics of Operations Research* 10:619–632.

Miller, N. 1977. "Graph Theoretical Approaches to the Theory of Voting." *American Journal of Political Science* 21:769–803.

Miller, N. 1980. "A New Solution Set for Tournaments and Majority Voting: Further Graphical Approaches to the Theory of Voting." *American Journal of Political Science* 24:68–96.

Morgenstern, O. 1976. "The Collaboration between Oskar Morgenstern and John von Neumann on the Theory of Games." *Journal of Economic Literature* 14:805–816.

Moulin, H. 1979. "Dominance Solvable Voting Schemes." *Econometrica* 47:1337–51.

Moulin, H. 1983. *The Strategy of Social Choice.* Amsterdam: North-Holland.

Moulin, H. 1986. "Choosing from a Tournament." *Social Choice and Welfare* 3:271–291.

Moulin, H. 1988. *Axioms of Cooperative Decision Making.* Cambridge: Cambridge University Press.

Moulin, H., and J.-P. Vial. 1978. "Strategically Zero-Sum Games: The Class Whose Completely Mixed Equilibria Cannot Be Improved Upon." *International Journal of Game Theory* 7:201–221.

Myerson, R. B. 1977. "Graphs and Cooperation in Games." *Mathematics of Operations Research* 2:225–229.

Myerson, R. B. 1978a. "Refinements of the Nash Equilibrium Concept." *International Journal of Game Theory* 7:73–80.

Myerson, R. B. 1978b. "Threat Equilibria and Fair Settlements in Cooperative Games." *Mathematics of Operations Research* 3:265–274.

Myerson, R. B. 1979. "Incentive-Compatibility and the Bargaining Problem." *Econometrica* 47:61–73.

Myerson, R. B. 1980. "Conference Structures and Fair Allocation Rules." *International Journal of Game Theory* 9:169–182.

Myerson, R. B. 1981a. "Optimal Auction Design." *Mathematics of Operations Research* 6:58–73.

Myerson, R. B. 1981b. "Utilitarianism, Egalitarianism, and the Timing Effect in Social Choice Problems." *Econometrica* 49:883–897.

Myerson, R. B. 1982. "Optimal Coordination Mechanisms in Generalized Principal-Agent Problems." *Journal of Mathematical Economics* 10:67–81.

Myerson, R. B. 1983. "Mechanism Design by an Informed Principal." *Econometrica* 51:1767–97.

Myerson, R. B. 1984a. "Two-Person Bargaining Problems with Incomplete Information." *Econometrica* 52:461–487.

Myerson, R. B. 1984b. "Cooperative Games with Incomplete Information." *International Journal of Game Theory* 13:69–96.

Myerson, R. B. 1985a. "Bayesian Equilibrium and Incentive Compatibility." In L. Hurwicz, D. Schmeidler, and H. Sonnenschein, eds., *Social Goals and Social Organization,* pp. 229–259. Cambridge: Cambridge University Press.

Myerson, R. B. 1985b. "Analysis of Two Bargaining Problems with Incomplete Information." In A. E. Roth, ed., *Game-Theoretic Models of Bargaining,* pp. 115–147. Cambridge: Cambridge University Press.

Myerson, R. B. 1986a. "Acceptable and Predominant Correlated Equilibria." *International Journal of Game Theory* 15:133–154.

Myerson, R. B. 1986b. "Multistage Games with Communication." *Econometrica* 54:323–358.

Myerson, R. B. 1986c. "An Introduction to Game Theory." In S. Reiter, ed., *Studies in Mathematical Economics,* pp. 1–61. Washington, D.C.: The Mathematical Association of America.

Myerson, R. B. 1988. "Sustainable Matching Plans with Adverse Selection." Northwestern University discussion paper. To appear in *Games and Economic Behavior.*

Myerson, R. B. 1989. "Credible Negotiation Statements and Coherent Plans." *Journal of Economic Theory* 48:264–303.

Myerson, R. B., and M. A. Satterthwaite. 1983. "Efficient Mechanisms for Bilateral Trading." *Journal of Economic Theory* 29:265–281.

Myerson, R. B., G. B. Pollock, and J. M. Swinkels. 1991. "Viscous Population Equilibria." To appear in *Games and Economic Behavior.*

Nash, J. F. 1950. "The Bargaining Problem." *Econometrica* 18:155–162.

Nash, J. F. 1951. "Noncooperative Games." *Annals of Mathematics* 54:289–295.

Nash, J. F. 1953. "Two-Person Cooperative Games." *Econometrica* 21:128–140.

Okuno-Fujiwara, M., A. Postlewaite, and G. Mailath. 1991. "On Belief-Based Refinements in Signaling Games." Univ. of Penn. working paper.

Ortega-Reichert, A. 1968. "Models for Competitive Bidding under Uncertainty." Department of Operations Research Technical Report No. 8 (Ph.D. dissertation), Stanford University.

Owen, G. 1972. "A Value for Non-Transferable Utility Games." *International Journal of Game Theory* 1:95–109.

Owen, G. 1977. "Values of Games with A Priori Unions." In R. Hein and O. Moeschlin, eds., *Essays in Mathematical Economics and Game Theory,* pp. 76–88. Berlin: Springer-Verlag.

Owen, G. 1982. *Game Theory,* 2nd Ed. New York: Academic Press.

Pearce, D. G. 1984. "Rationalizable Strategic Behavior and the Problem of Perfection." *Econometrica* 52:1029–50.

Pearce, D. G. 1987. "Renegotiation-proof Equilibria: Collective Rationality and Intertemporal Cooperation." Yale University discussion paper.

Peleg, B. 1963. "Existence Theorem for the Bargaining Set $M_1^{(i)}$." *Bulletin of the American Mathematical Society* 69:109–110.

Peleg, B. 1984. *Game Theoretic Analysis of Voting in Committees.* Cambridge: Cambridge University Press.

Perles, M. A., and M. Maschler. 1981. "The Super-Additive Solution for the Nash Bargaining Game." *International Journal of Game Theory* 10:163–193.

Pollock, G. B. 1989. "Evolutionary Stability of Reciprocity in a Viscous Lattice." *Social Networks* 11:175–212.

Pratt, J. W. 1964. "Risk Aversion in the Small and in the Large." *Econometrica* 32:122–136.

Pratt, J. W., H. Raiffa, and R. Schlaiffer. 1964. "The Foundations of Decisions under Uncertainty: An Elementary Exposition." *American Statistical Association Journal* 59:353–375.

Qin, C. Z. 1989. "Strongly and Weakly Inhibitive Sets and the λε-Inner Core." Economics working paper, University of Iowa.

Radner, R. 1980. "Collusive Behavior in Oligopolies with Long but Finite Lives." *Journal of Economic Theory* 22:136–156.

Radner, R., R. Myerson, and E. Maskin. 1986. "An Example of a Repeated Partnership Game with Discounting and Uniformly Inefficient Equilibria." *Review of Economic Studies* 53:59–70.

Raiffa, H. 1968. *Decision Analysis*. Reading, Mass.: Addison-Wesley.

Raiffa, H. 1982. *The Art and Science of Negotiation*. Cambridge, Mass.: Harvard University Press.

Ramsey, F. P. 1926. "Truth and Probability." Reprinted in H. E. Kyburg, Jr. and H. E. Smokler, eds., 1964, *Studies in Subjective Probability*, pp. 62–92. New York: Wiley.

Rasmusen, E. 1989. *Games and Information: An Introduction to Game Theory*. Oxford: Basil Blackwell.

Reny, P. J. 1987. "Explicable Equilibria." University of Western Ontario working paper.

Rockafellar, R. T. 1970. *Convex Analysis*. Princeton: Princeton University Press.

Rosenthal, R. 1978. "Arbitration of Two-Party Disputes under Uncertainty." *Review of Economic Studies* 45:595–604.

Rosenthal, R. 1981. "Games of Perfect Information, Predatory Pricing, and the Chain-Store Paradox." *Journal of Economic Theory* 25:92–100.

Roth, A. E. 1979. *Axiomatic Models of Bargaining*. Berlin: Springer-Verlag.

Roth, A. E. 1980. "Values for Games Without Side-Payments: Some Difficulties with Current Concepts." *Econometrica* 48:457–465.

Roth, A. E. 1985. "Toward a Focal-Point Theory of Bargaining." In A. E. Roth, ed., *Game-Theoretic Models of Bargaining*, pp. 259–268. Cambridge: Cambridge University Press.

Roth, A. E. 1988. *The Shapley Value: Essays in Honor of Lloyd S. Shapley*. Cambridge: Cambridge University Press.

Roth, A. E., and F. Schoumaker. 1983. "Expectations and Reputations in Bargaining." *American Economic Review* 73:362–372.

Rothschild, M., and J. Stiglitz. 1976. "Equilibrium in Competitive Insurance Markets: An Essay on the Economics of Imperfect Information." *Quarterly Journal of Economics* 90:629–649.

Royden, H. 1968. *Real Analysis*. New York: Macmillan.

Rubinstein, A. 1979. "Equilibrium in Supergames with the Overtaking Criterion." *Journal of Economic Theory* 21:1–9.

Rubinstein, A. 1982. "Perfect Equilibrium in a Bargaining Model." *Econometrica* 50:97–109.

Rubinstein, A. 1987. "A Sequential Strategic Theory of Bargaining." In T. Bewley, ed., *Advances in Economic Theory: Fifth World Congress*, pp. 197–224. Cambridge: Cambridge University Press.

Rubinstein, A. 1989. "The Electronic Mail Game: Strategic Behavior under Almost Common Knowledge." *American Economic Review* 79:385–391.

Samuelson, L. 1989. "Dominated Strategies and Common Knowledge." Pennsylvania State University working paper.

Savage, L. J. 1954. *The Foundations of Statistics*. New York: Wiley.

Scarf, H. E. 1967. "The Core of an *n*-Person Game." *Econometrica* 35:50–69.

Scarf, H. E. 1973. *The Computation of Economic Equilibria.* New Haven, Conn.: Yale University Press.

Schelling, T. C. 1960. *The Strategy of Conflict.* Cambridge, Mass.: Harvard University Press.

Schmeidler, D. 1969. "The Nucleolus of a Characteristic Function Game." *SIAM Journal of Applied Mathematics* 17:1163–70.

Schmeidler, D. 1973. "Equilibrium Points of Nonatomic Games." *Journal of Statistical Physics* 7:295–300.

Selten, R. 1965. "Spieltheoretische Behandlung eines Oligopolmodells mit Nachfragetragheit." *Zeitschrift fuer die gesampte Staatswissenschaft* 121:301–324, 667–689.

Selten, R. 1975. "Reexamination of the Perfectness Concept for Equilibrium Points in Extensive Games." *International Journal of Game Theory* 4:25–55.

Selten, R. 1978. "The Chain-Store Paradox." *Theory and Decision* 9:127–159.

Sen, A. K. 1970. *Collective Choice and Social Welfare.* San Francisco: Holden-Day.

Shafer, W. 1980. "On the Existence and Interpretation of Value Allocations." *Econometrica* 48:467–477.

Shapley, L. S. 1953. "A Value for *n*-Person Games." In H. Kuhn and A. W. Tucker, eds., *Contributions to the Theory of Games* II, pp. 307–317. Princeton: Princeton University Press.

Shapley, L. S. 1967. "On Balanced Sets and Cores." *Naval Research Logistics Quarterly* 14:453–460.

Shapley, L. S. 1969. "Utility Comparison and the Theory of Games." In *La Decision*, Editions du CNRS, Paris, pp. 251–263. Reprinted in A. E. Roth, ed., 1988. *The Shapley Value*, pp. 307–319. Cambridge: Cambridge University Press.

Shapley, L. S., and M. Shubik. 1954. "A Method for Evaluating the Distribution of Power in a Committee System." *American Political Science Review* 48:787–792.

Shepsle, K., and B. Weingast. 1984. "Uncovered Sets and Sophisticated Voting Outcomes with Implications for Agenda Institutions." *American Journal of Political Science* 28:49–74.

Shubik, M. 1982. *Game Theory in the Social Sciences: Concepts and Solutions.* Cambridge, Mass.: MIT Press.

Simon, L. K. 1987. "Basic Timing Games." University of California, Berkeley, working paper.

Simon, L. K., and M. B. Stinchcombe. 1989. "Extensive Form Games in Continuous Time: Pure Strategies." *Econometrica* 57:1171–1214.

Simon, L. K., and W. R. Zame. 1990. "Discontinuous Games and Endogenous Sharing Rules." *Econometrica* 58:861–872.

Sobel, J., and I. Takahashi. 1983. "A Multistage Model of Bargaining." *Review of Economic Studies* 50:411–426.

Sorin, S. 1980. "An Introduction to Two-Person Zero Sum Repeated Games with Incomplete Information." I.M.S.S.S. Technical Report No. 312, Stanford University.

Spence, M. 1973. "Job Market Signaling." *Quarterly Journal of Economics* 87:355–374.

Stahl, I. 1972. *Bargaining Theory.* Stockholm: Stockholm School of Economics.

Sutton, J. 1986. "Noncooperative Bargaining Theory: An Introduction." *Review of Economic Studies* 53:709–724.

Tauman, Y. 1988. "The Aumann-Shapley Prices: A Survey." In A. E. Roth, ed., *The Shapley Value,* pp. 279–304. Cambridge: Cambridge University Press.

Thrall, R. M., and W. F. Lucas. 1963. "*n*-Person Games in Partition Function Form." *Naval Research Logistics Quarterly* 10:281–298.

van Damme, E. 1984. "A Relation between Perfect Equilibria in Extensive Form Games and Proper Equilibria in Normal Form Games." *International Journal of Game Theory* 13:1–13.

van Damme, E. 1987. *Stability and Perfection of Nash Equilibria.* Berlin: Springer-Verlag.

van Damme, E. 1989. "Stable Equilibria and Forward Induction." *Journal of Economic Theory* 48:476–496.

van Damme, E., R. Selten, and E. Winter. 1990. "Alternating Bid Bargaining with a Smallest Money Unit." *Games and Economic Behavior* 2:188–201.

van Huyck, J. B., R. C. Battalio, and R. O. Beil. 1990. "Tacit Coordination Games, Strategic Uncertainty, and Coordination Failure." *American Economic Review* 80:234–248.

Vickrey, W. 1961. "Counterspeculation, Auctions, and Competitive Sealed Tenders." *Journal of Finance* 16:8–37.

von Neumann, J. 1928. "Zur Theorie der Gesellschaftsspiele." *Mathematische Annalen* 100:295–320. English translation in R. D. Luce and A. W. Tucker, eds., *Contributions to the Theory of Games* IV (1959), pp. 13–42. Princeton: Princeton University Press.

von Neumann, J., and O. Morgenstern. 1944. *Theory of Games and Economic Behavior.* Princeton: Princeton University Press. Second Ed., 1947.

Wilson, C. 1977. "A Model of Insurance Markets with Incomplete Information." *Journal of Economic Theory* 16:167–207.

Wilson, R. 1971. "Computing Equilibria of *n*-Person Games." *SIAM Journal of Applied Mathematics* 21:80–87.

Wilson, R. 1987. "Game Theoretic Analyses of Trading Processes." In T. Bewley, ed., *Advances in Economic Theory: Fifth World Congress,* pp. 33–70. Cambridge: Cambridge University Press.

Wooders, M. H. 1983. "The Epsilon Core of a Large Replica Game." *Journal of Mathematical Economics* 11:277–300.

Wooders, M. H., and W. R. Zame. 1984. "Approximate Cores of Large Games." *Econometrica* 52:1327–50.

Young, H. P. 1985. "Producer Incentives in Cost Allocation." *Econometrica* 53:757–765.

Zermelo, E. 1913. "Uber eine Anwendung der Mengenlehre auf die Theorie des Schachspiels." *Proceedings Fifth International Congress of Mathematicians* 2:501–504.

Index